Springer Series in Computational Mathematics

18

W. Hackbusch

Elliptic Differential Equations

Theory and Numerical Treatment

Translated from the German
by Regine Fadiman and Patrick D. F. Ion

With 40 Figures

Springer-Verlag
Berlin Heidelberg New York
London Paris Tokyo
Hong Kong Barcelona
Budapest

Prof. Dr. Wolfgang Hackbusch
Institut für Informatik und Praktische Mathematik
Christian-Albrechts-Universität zu Kiel
Olshausenstr. 40
W-2300 Kiel 1, FRG

Translators:

Regine Fadiman
3328 Williamsburg Street
Ann Arbor
MI 48104, USA

Dr. Patrick D. F. Ion
1456 Kensington Drive
Ann Arbor
MI 48104, USA

Mathematics Subject Classification: 35Jxx, 35A40, 35Bxx, 35C15, 35Dxx, 35Q30, 35R05, 46E35, 46E39, 49R05, 49R10, 65K10, 65Nxx, 76D07

ISBN 3-540-54822-X Springer-Verlag Berlin Heidelberg New York
ISBN 0-387-54822-X Springer-Verlag New York Berlin Heidelberg

Printed in the United States of America

Typesetting: Camera ready by translators
41/3140 - 5 4 3 2 1 0 - Printed on acid-free paper

Foreword

This book has developed from lectures that the author gave for mathematics students at the Ruhr-Universität Bochum and the Christian-Albrechts-Universität Kiel. This edition is the result of the translation and correction of the German edition entitled *Theorie und Numerik elliptischer Differentialgleichungen.*

The present work is restricted to the theory of partial differential equations of elliptic type, which otherwise tends to be given a treatment which is either too superficial or too extensive. The following sketch shows what the problems are for elliptic differential equations.

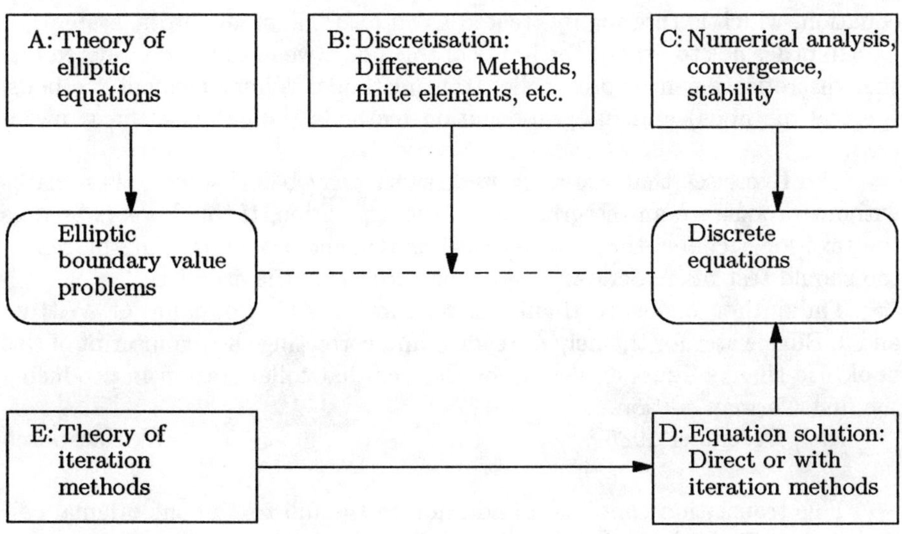

The theory of elliptic differential equations (A) is concerned with questions of existence, uniqueness, and properties of solutions. The first problem of numerical treatment is the description of the discretisation procedures (B), which give finite-dimensional equations for approximations to the solutions. The subsequent second part of the numerical treatment is numerical analysis (C) of the procedure in question. In particular it is necessary to find out if, and how fast, the approximation converges to the exact solution. The solution of the finite-dimensional equations (D, E) is in general no simple problem, since from 10^3 to 10^6 unknowns can occur. The discussion of this third area of numerical problems is skipped (one may find it, e.g., in Hackbusch [5] and [9]).

The descriptions of discretisation procedures and their analyses are closely connected with corresponding chapters of the theory of elliptic equations. In addition, it is not possible to undertake a well-founded numerical analysis without a basic knowledge of elliptic differential equations. Since the latter cannot, in general, be assumed of a reader, it seems to me necessary to present the numerical study along with the theory of elliptic equations.

The book is conceived in the first place as an introduction to the treatment of elliptic boundary-value problems. It should, however, serve to lead the reader to further literature on special topics and and applications. It is intentional that certain topics, which are often handled rather summarily, (e.g., eigenvalue problems) are treated here in greater detail.

The exposition is strictly limited to linear elliptic equations. Thus a discussion of the Navier-Stokes equations, which are important for fluid mechanics, is excluded; however, one can approach these matters via the Stokes equation, which is thoroughly treated as an example of an elliptic system.

In order not to exceed the limits of this book, we have not considered further discretisation methods (collocation methods, volume-element methods, spectral methods) and integral-equation methods (boundary-element methods).

The Exercises that are presented, which may be considered as remarks without proofs, are an integral part of the exposition. If this book is used as the text for a course they can be used as student problems. But the reader too should test his understanding of the subject on the exercises.

The author wishes to thank his collaborators G. Hofmann, G. Wittum and J. Burmeister for the help in reading and correcting the manuscript of this book. He thanks Teubner Verlag for their cordial collaboration in producing the first German edition.

Kiel, December 1985 W. Hackbusch

This translation contains, in addition to the full text of the original edition, a short Section (§3.5) on the integral-equation method. The bibliography has also been expanded.

The author wishes to thank the translators, R. Fadiman and P. D. F. Ion, for their pleasant collaboration, and Springer-Verlag for their friendly cooperation.

Kiel, March 1992 W. Hackbusch

Table of Contents

Notation

Formula Numbers.

Equations in Section $X.Y$ are numbered $(X.Y.1)$, $(X.Y.2)$, etc. The equation (3.2.1) is referred to within Section 3.2 simply as (1). In other Sections of Chapter 3 it is called (2.1).

Theorem Numbering.

All Theorems, Definitions, Lemmata etc. are numbered together. In Section 3.2 Lemma 3.2.1 is referred to as Lemma 1.

Special Symbols. The following quantities have fixed meanings:

$\mathbf{A}, \mathbf{B}, \ldots$	matrices
B, B_j	boundary differential operators (cf. (5.2.1a,b), (5.3.6))
$\mathbb{C}$	the complex numbers
$C^s(D), C^{k,1}(D)$	Hölder- and Lipschitz-continuously differentiable functions (cf. Definition 3.2.8)
$C^k(D), C^\infty(D)$	k-fold and infinitely continuously differentiable functions
$C_0^\infty(\Omega)$	infinitely differentiable functions with compact supports (cf. (6.2.3))
$d(u, V_N)$	distance of the function u from the subspace V_N (cf. Theorem 8.2.1)
$d\Gamma, d\Gamma_{\mathbf{x}}$	surface differentials in surface integrals
$\operatorname{diag}\{d_1, d_2, \ldots\}$	diagonal matrix with the diagonal elements $d_1, d_2, \ldots$
f	a function; often the right-hand side of a differential equation
$g(\cdot, \cdot)$	Green's function (cf. Section 3.2)
h	step size (cf. Sections 4.1 and 4.2)
$H^k(\Omega), H^s(\Omega), H_0^k(\Omega), H_0^s(\Omega)$	Sobolev spaces (cf. Sections 6.2.2 and 6.2.4)
$K_R(y)$	open ball about y with radius R (cf. (2.2.7), Section 6.1.1)
I	identity or inclusion (cf. Sections 6.1.2, 6.1.3)
L	a differential operator (cf. (1.2.6)) or the operator associated with a bilinear form (cf. (7.2.9′))
$\mathbf{L}$	the stiffness matrix (cf. Section 8.1)
L_h	matrix of a discrete system of equations (cf. (4.1.9a))

$L(X, Y)$	linear space of bounded operators from X to Y (cf. Section 6.1.2)
$L^\infty(\Omega)$	space of essentially bounded functions (cf. 6.1.3)
$L^2(\Omega)$	space of square-integrable functions (cf. Section 6.2.1)
$\mathbb{N}$	the natural numbers $\{1, 2, 3, \ldots\}$
$\mathbf{n} = \mathbf{n}(\mathbf{x})$	normals (cf. (2.2.3a))
$\mathbf{0}$	the zero matrix
$O(\cdot)$	Landau symbol: $f(x) = O(g(x))$ if $\lvert fx)\rvert \le \text{const}\, \lvert g(x)\rvert$
$\mathbf{P}$	cf. (8.1.6)
q_h	a grid function, right-hand side of the discrete equation (4.1.9a)
$\mathbb{R}, \mathbb{R}_+$	the real numbers, the positive real numbers
$R_h, \tilde{R}_h$	restrictions (cf. (4.5.2) and (4.5.5b))
$s(\cdot, \cdot)$	the singularity function (cf. Section 2.2)
$\mathrm{supp}(f)$	the support of the function f (cf. Lemma 6.2.2)
$u = u(\mathbf{x}) = u(x_1, \ldots, x_n)$	a function, e.g., a solution of a differential equation
u_h	a grid function (discrete solution; cf Section 4)
V_N, V_h	finite-element spaces (cf. (8.1.3) and Section 8.4.1)
$x, (x, y), (x, y, z)$	independent variables
$\mathbf{x} = (x_1, \ldots, x_n)$	a vector of independent variables
$\mathbb{Z}$	the integers
Γ	the boundary of Ω
Γ_h	the boundary points of a grid (cf. (4.2.1b), (4.8.4))
$\gamma(\cdot, \cdot)$	fundamental solution function
$\rho(\mathbf{A})$	spectral radius of the matrix $\mathbf{A}$
Ω	an open set in $\mathbb{R}^n$ or a domain (cf. Definition 2.1.1)
Ω_h	a grid (cf. (4.1.6a), (4.2.1a), (4.8.2))
Δ	the Laplace operator (cf. (2.1.1a))
Δ_h	the five-point difference operator (cf. (4.2.3))
∇	gradient (cf. (2.2.3a))
$\partial^+, \partial^-, \partial^0, \partial_x^+, \partial_y^+, \ldots$	differences (cf. (4.1.2a–c), (4.2.3))
$\partial_n^+, \partial_n^-$	differences in the normal direction (cf. (4.7.4))
$\partial/\partial n$	normal derivative (cf. (2.2.3a))
$\langle \cdot, \cdot \rangle$	scalar product (cf. (2.2.3c), (4.3.14a))
$\langle \cdot, \cdot \rangle_{X \times X'}$	duality form (cf. Section 6.1.3)
$(\cdot, \cdot)$	scalar product (cf. Section 6.1.4)
$(\cdot, \cdot)_0, (\cdot, \cdot)_{L^2(\Omega)}, \lvert \cdot \rvert_0, \lvert \cdot \rvert_{L^2(\Omega)}$	scalar product and norms on $L^2(\Omega)$
$(\cdot, \cdot)_k, (\cdot, \cdot)_s, \lvert \cdot \rvert_k, \lvert \cdot \rvert_s$	scalar products and norms on $H^k(\Omega)$, resp. $H^s(\Omega)$
$\lvert \cdot \rvert$	the Euclidean norm (cf. (2.2.2))
$\lVert \cdot \rVert_2$	the Euclidean norm and the spectral norm (cf. Section 4.3)
$\lvert \cdot \hat{\rvert}_k, \lvert \cdot \hat{\rvert}_s$	norms equivalent to $\lvert \cdot \rvert_k, \lvert \cdot \rvert_s$ (cf. (6.2.15), (6.2.16b))
$\lVert \cdot \rVert_\infty$	maximum norm (cf. (4.3.3)), row sum norm (4.3.11), or supremum norm (2.4.1), cf. also Section 6.1.1
$\mathbb{1}$	the vector $(1, 1, \ldots)$ (cf. Section 4.3)

1 Partial Differential Equations and Their Classification Into Types

1.1 Examples

An ordinary differential equation describes a function which depends on only one variable. Unfortunately, for many problems it is not possible to restrict attention to a single variable. Almost all physical quantities depend on the spatial variables x, y , and z and on time t. The time dependence might be omitted for stationary processes, and one might perhaps save one spatial dimension by special geometric assumptions, but even then there would still remain at least two independent variables. Equations that contain the first partial derivatives

$$u_{x_i} = u_{x_i}(x_1, x_2, \cdots, x_n) = \partial u(x_1, \cdots, x_n)/\partial x_i$$

where $(1 \le i \le n)$, or even higher partial derivatives $u_{x_i x_j}$, etc., are called partial differential equations.

Unlike ordinary differential equations, partial differential equations cannot be analysed all together. Rather, one distinguishes between three types of equations which have different properties and also require different numerical methods.

Before the characteristics for the types are defined, let us introduce some examples of partial differential equations.

All of the following examples will contain only two independent variables x, y. The first two examples are partial differential equations of first order, since only first partial derivatives occur.

Example 1.1.1. Find a solution $u(x, y)$ of

$$u_y(x, y) = 0. \tag{1.1.1}$$

It is obvious that $u(x, y)$ must be independent of y, i.e., the solution has the form $u(x, y) = \varphi(x)$. Thus $u(x, y) = \varphi(x)$ for some arbitrary φ is a solution of (1).

Equation (1) is a special case of

Example 1.1.2. Find a solution $u(x, y)$ of

$$cu_x - u_y = 0 \qquad (c \text{ constant}). \tag{1.1.2}$$

Let u be a solution. Introduce new coordinates $\xi = x + cy$, $\eta = y$ and define $v(\xi,\eta) := u(x(\xi,\eta), y(\xi,\eta))$ with the aid of $x(\xi,\eta) = \xi - c\eta$, $y(\xi,\eta) = \eta$. Since $v_\eta = u_x x_\eta + u_y y_\eta$ (chain rule) and $x_\eta = -c$, $y_\eta = 1$, it follows from (2) that $v_\eta(\xi,\eta) = 0$. This equation is analogous to (1), and Example 1 shows that $v(\xi,\eta) = \varphi(\xi)$. If one now replaces ξ,η by x, y one obtains

$$u(x,y) = \varphi(x + cy). \tag{1.1.3}$$

Conversely, through (3) one obviously obtains a solution of Equation (2) as long as φ is continuously differentiable.

In order to determine uniquely the solution of an ordinary differential equation $u' - f(u) = 0$ one needs an initial value $u(x_0) = u_0$. The partial differential equation (2) can be augmented by the initial-value function

$$u(x, y_0) = u_0(x) \quad \text{for} \quad x \in \mathbb{R} \tag{1.1.4}$$

on the line $y = y_0$, with y_0 a constant. The comparison of Equations (3) and (4) shows that $\varphi(x+cy_0) = u_0(x)$. Thus φ is determined by $\varphi(x) = u_0(x - cy_0)$. The unique solution of the initial value problem (2) and (4) reads

$$u(x,y) = u_0(x - c(y_0 - y)). \tag{1.1.5}$$

The following three examples involve differential equations of second order.

Example 1.1.3. (Potential equation or Laplace equation) Let Ω be an open subset of $\mathbb{R}^2$. Find a solution of

$$u_{xx} + u_{yy} = 0 \quad \text{in } \Omega. \tag{1.1.6}$$

If one identifies $(x,y) \in \mathbb{R}^2$ with the complex number $z = x + iy \in \mathbb{C}$, the solutions can be given immediately. The real and imaginary parts of any function $f(z)$ holomorphic in Ω are solutions of Equation (6). Three examples are $\operatorname{Re} z^0 = 1$, $\operatorname{Re} z^2 = x^2 - y^2$ and $\operatorname{Re}\log(z - z_0) = \log\sqrt{(x-x_0)^2 + (y-y_0)^2}$ if $z_0 \notin \Omega$. To determine the solution uniquely one needs the boundary values $u(x,y) = \varphi(x,y)$ for all (x,y) on the boundary $\Gamma = \partial\Omega$ of Ω.

Example 1.1.4. (Wave equation) All solutions of

$$u_{xx} - u_{yy} = 0 \tag{1.1.7}$$

are given by

$$u(x,y) = \varphi(x + y) + \psi(x - y) \tag{1.1.8}$$

where φ and ψ are arbitrary twice continuously differentiable functions. Suitable initial values are, for example,

$$u(x,0) = u_0(x), \quad u_y(x,0) = u_1(x), \quad (x \in \mathbb{R}) \tag{1.1.9}$$

where u_0 and u_1 are given functions. If one inserts (8) into (9), one finds $u_0 = \varphi + \psi$, $u_1 = \varphi' - \psi'$, where φ' is the derivative of φ, and infers that

$$\varphi' = (u_1 + u_0')/2, \quad \psi' = (u_0' - u_1)/2.$$

From this one can determine φ and ψ up to constants of integration. One constant can be chosen arbitrarily, for example, by $\varphi(0) = 0$, and the other is determined by $u(0,0) = u_0(0) = \varphi(0) + \psi(0)$.

Exercise 1.1.5. Prove that every solution of the wave equation (7) has the form (8). Hint: Use $\xi = x + y$ and $\eta = x - y$ as new variables.

Example 1.1.6. (Heat equation) Find the solution of

$$u_{xx} - u_y = 0 \tag{1.1.10}$$

(physical interpretation: $u =$ temperature, $y =$ time). The separation of variables $u(x,y) = v(x)w(y)$ gives that for every $c \in \mathbb{R}$

$$u(x,y) = \sin(cx) \cdot \exp(-c^2 y). \tag{1.1.11a}$$

Another solution of (10) for $y > 0$ is

$$u(x,y) = \frac{1}{\sqrt{4\pi y}} \int_{-\infty}^{\infty} u_0(\xi) \exp(-(x-\xi)^2/(4y))\, d\xi, \tag{1.1.11b}$$

where $u_0(\cdot)$ is an arbitrary continuous and bounded function. The initial condition matching Equation (10), in contrast to (9), contains only one function:

$$u(x,0) = u_0(x). \tag{1.1.12}$$

The solution (11b), which initially is defined only for $y > 0$, can be extended continuously to $y = 0$ and there satisfies the initial value requirement (12).

Exercise 1.1.7. Let u_0 be bounded in $\mathbb{R}$ and continuous at x. Then prove that the right side of Equation (11b) converges to $u_0(x)$ for $y \to 0$. Hint: First show that $u(x,y) = u_0(x) + \int_{-\infty}^{+\infty} [u_0(\xi) - u_0(x)] \exp(-(x-\xi)^2/(4y))\, d\xi / \sqrt{4\pi y}$ and then decompose the integral into subintegrals over $[x-\delta, x+\delta]$ and $(-\infty, x-\delta) \cup (x+\delta, \infty)$.

As with ordinary differential equations, equations of higher order can be described by systems of first-order equations. In the following we give some examples.

Example 1.1.8. Let the pair (u,v) be the solution of the system

$$u_x + v_y = 0, \quad v_x + u_y = 0. \tag{1.1.13}$$

If u and v are twice differentiable, the differentiation of (13) yields the equations $u_{xx} + v_{xy} = 0$, $v_{xy} + u_{yy} = 0$, which together imply that $u_{xx} - u_{yy} = 0$. Thus u is a solution of the wave equation (7). The same can be shown for v.

Example 1.1.9. (Cauchy-Riemann differential equations) If u and v satisfy the system

$$u_x + v_y = 0, \quad v_x - u_y = 0 \quad \text{in} \quad \Omega \subset \mathbb{R}^2, \tag{1.1.14}$$

then the same consideration as in Example 8 yields that both u and v satisfy the potential equation (6).

Example 1.1.10. If u and v satisfy the system

$$u_x + v_y = 0, \quad v_x + u = 0, \tag{1.1.15}$$

then v solves the heat equation (10).

A second-order system of interest in fluid mechanics can be found in

Example 1.1.11. (Stokes equations) In the system

$$u_{xx} + u_{yy} - w_x = 0, \tag{1.1.16a}$$

$$v_{xx} + v_{yy} - w_y = 0, \tag{1.1.16b}$$

$$u_x + v_y \qquad = 0 \tag{1.1.16c}$$

u and v denote the flow velocities in x- and y-directions, while w denotes the pressure.

1.2 Classification of Second-Order Equations into Types

The general linear differential equation of second order in two variables reads

$$\begin{aligned} a(x,y)u_{xx} &+ 2b(x,y)u_{xy} + c(x,y)u_{yy} \\ &+ d(x,y)u_x + e(x,y)u_y + f(x,y)u + g(x,y) = 0. \end{aligned} \tag{1.2.1}$$

Definition 1.2.1. (a) Equation (1) is said to be elliptic at (x, y) if

$$a(x,y)c(x,y) - b(x,y)^2 > 0. \tag{1.2.2a}$$

(b) Equation (1) is said to be hyperbolic at (x, y), if

$$a(x,y)c(x,y) - b(x,y)^2 < 0. \tag{1.2.2b}$$

(c) Equation (1) is said to be parabolic at (x, y) if

$$ac - b^2 = 0 \quad \text{and} \quad \operatorname{rank} \begin{bmatrix} a & b & d \\ b & c & e \end{bmatrix} = 2 \quad \text{at } (x,y). \tag{1.2.2c}$$

(d) Equation (1) is said to be elliptic (hyperbolic, parabolic) in $\Omega \subset \mathbb{R}^2$ if it is elliptic (hyperbolic, parabolic) at all $(x, y) \in \Omega$.

Occasionally the parabolic type is defined only by $ac - b^2 = 0$. But one would not want to call the equation $u_{xx}(x,y) + u_x(x,y) = 0$ parabolic, nor indeed the purely algebraic equation $u(x,y) = 0$.

Example 1.2.2. The potential equation (1.6) is elliptic, the wave equation (1.7) is of hyperbolic type, while the heat equation (1.10) is parabolic.

The definition of types can easily be generalised to the case where more than two independent variables occur, say $x_1, \cdots, x_n$. The general linear differential equation of second order in n variables $\mathbf{x} = (x_1, \cdots, x_n)$ reads

$$\sum_{i,j=1}^{n} a_{ij}(\mathbf{x}) u_{x_i x_j} + \sum_{i=1}^{n} a_i(\mathbf{x}) u_{x_i} + a(\mathbf{x}) u = f(\mathbf{x}). \tag{1.2.3}$$

Since $u_{x_i x_j} = u_{x_j x_i}$ holds for twice continuously differentiable functions, one can assume that, without loss of generality,

$$a_{ij}(\mathbf{x}) = a_{ji}(\mathbf{x}). \tag{1.2.4a}$$

Thus, the coefficients $a_{ij}(\mathbf{x})$ define a symmetric $n \times n$ matrix

$$\mathbf{A}(\mathbf{x}) = (a_{ij}(\mathbf{x}))_{i,j=1}^{n} \tag{1.2.4b}$$

which therefore has only real eigenvalues.

Definition 1.2.3. (a) Equation (3) is said to be elliptic at $\mathbf{x}$ if all n eigenvalues of the matrix $\mathbf{A}(\mathbf{x})$ have the same sign (± 1) (i.e. if $\mathbf{A}(\mathbf{x})$ is positive or negative definite).
(b) Equation (3) is said to be hyperbolic at $\mathbf{x}$ if $n-1$ eigenvalues of $\mathbf{A}(\mathbf{x})$ have the same sign (± 1) and one eigenvalue has the opposite sign.
(c) Equation (3) is said to be parabolic at $\mathbf{x}$ if one eigenvalue vanishes, the remaining $n-2$ eigenvalues have the same sign, and $\mathrm{rank}(\mathbf{A}(\mathbf{x}), a(\mathbf{x})) = n$ where $\mathbf{a}(\mathbf{x}) = (a_1(\mathbf{x}), \cdots, a_n(\mathbf{x}))^T$.
(d) Equation (3) is said to be elliptic in $\Omega \subset \mathrm{I\!R}^n$ if it is elliptic at all $\mathbf{x} \in \Omega$.

Definition 3 makes it clear that the three types mentioned by no means cover all cases. An unclassified equation occurs, for example, if $\mathbf{A}(\mathbf{x})$ has two positive and two negative eigenvalues.

In place of (3) one can also write

$$Lu = f, \tag{1.2.5}$$

where

$$L = \sum_{i,j=1}^{n} a_{i,j} \frac{\partial^2}{\partial x_i \partial x_j} + \sum_{i=1}^{n} a_i \frac{\partial}{\partial x_i} + a \tag{1.2.6}$$

is a linear differential operator of second order. The operator

$$\sum_{i,j}^{n} a_{ij}\partial^2/\partial x_i \partial x_j,$$

which contains only the highest derivatives of L, is called the principal part of L.

Remark 1.2.4. The ellipticity or hyperbolicity of Eq. (3) depends only on the principal part of the differential operator.

Exercise 1.2.5. (Invariance of the type under coordinate transformations) Let Eq. (3) be defined for $\mathbf{x} \in \Omega$. The transformation $\Phi\colon \Omega \subset \mathbb{R}^n \to \Omega' \subset \mathbb{R}^n$ is assumed to have a nonsingular Jacobian matrix $\mathbf{S} = \partial\Phi/\partial\mathbf{x} \in \mathbf{C}^1(\Omega)$ at $\mathbf{x}$. Prove that Eq. (3) does not change its type at $\mathbf{x}$ if it is written in the new coordinates $\xi = \Phi(\mathbf{x})$. Hint: the matrix $\mathbf{A} = (a_{ij})$ becomes $\mathbf{SAS}^{\mathsf{T}}$ after the transformation. Use Remark 4 and Sylvester's inertia theorem (cf. Gantmacher [1, Chapter X–§2]).

1.3 Type Classification for Systems of First Order

The examples 1.1.8–10 are special cases of the general linear system of first order in two variables.

$$\mathbf{u}_x(x,y) - \mathbf{A}(x,y)\mathbf{u}_y(x,y) + \mathbf{B}(x,y)\mathbf{u}(x,y) = \mathbf{f}(x,y). \tag{1.3.1}$$

Here, $\mathbf{u} = (u_1, \cdots, u_m)^T$ is a vector function, and $\mathbf{A}, \mathbf{B}$ are $m \times m$ matrices. In contrast to Section 1.2, $\mathbf{A}$ need not be symmetric and can have complex eigenvalues. If the eigenvalues $\lambda_1, \cdots, \lambda_m$ are real, and if there exists a decomposition $\mathbf{A} = \mathbf{S}^{-1}\mathbf{DS}$ with $\mathbf{D} = \operatorname{diag}\{\lambda_1, \cdots, \lambda_m\}$, $\mathbf{A}$ is called real-diagonalisable.

Definition 1.3.1. (a) System (1) is said to be hyperbolic at (x,y) if $\mathbf{A}(x,y)$ is real-diagonalisable. (b) System (1) is said to be elliptic at (x,y) if all the eigenvalues of $\mathbf{A}(x,y)$ are not real.

If $\mathbf{A}$ is real or possesses m distinct real eigenvalues the system is hyperbolic since these conditions are sufficient for real diagonalisability. A single real equation, in particular, is always hyperbolic.

Examples 1.1.1 and 1.1.2 are hyperbolic according to the preceding remark. System (1.13) from Example 1.1.8 has the form (1) with

$$\mathbf{A} = \begin{bmatrix} 0 & -1 \\ -1 & 0 \end{bmatrix}.$$

It is hyperbolic since $\mathbf{A}$ is real-diagonalisable:

$$\mathbf{A} = \begin{bmatrix} 1 & 1 \\ -1 & 1 \end{bmatrix}^{-1} \begin{bmatrix} -1 & 0 \\ 0 & 1 \end{bmatrix} \begin{bmatrix} 1 & 1 \\ -1 & 1 \end{bmatrix}.$$

The Cauchy-Riemann system (1.14), which is closely connected with the potential equation (1.6), is elliptic since it has the form (1) with

$$\mathbf{A} = \begin{bmatrix} 0 & -1 \\ 1 & 0 \end{bmatrix},$$

and $\mathbf{A}$ has the eigenvalues $\pm i$. The system (1.15) corresponding to the (parabolic) heat equation can be described as system (1) with

$$\mathbf{A} = \begin{bmatrix} 0 & -1 \\ 0 & 0 \end{bmatrix}.$$

The eigenvalues ($\lambda_1 = \lambda_2 = 0$) may be real but $\mathbf{A}$ is not diagonalisable. Hence, system (1.15) is neither hyperbolic nor elliptic.

A more general system than (1) is

$$\mathbf{A}_1\mathbf{u}_x + \mathbf{A}_2\mathbf{u}_y + \mathbf{B}\mathbf{u} = \mathbf{f}. \tag{1.3.2}$$

If $\mathbf{A}_1$ is invertible then multiplication by $\mathbf{A}_1^{-1}$ gives the form (1) with $\mathbf{A} = -\mathbf{A}_1^{-1}\mathbf{A}_2$. Otherwise one has to investigate the generalised eigenvalue problem $\det(\lambda\mathbf{A}_1 + \mathbf{A}_2) = 0$. However, system (2) with singular $\mathbf{A}_1$ cannot be elliptic, as can be seen from the following (cf. (4) with $\xi_1 = 1, \xi_2 = 0$).

A generalisation of (2) to n independent variables is exhibited in the system

$$\mathbf{A}_1\mathbf{u}_{x_1} + \cdots + \mathbf{A}_n\mathbf{u}_{x_n} + \mathbf{B}\mathbf{u} = \mathbf{f} \tag{1.3.3}$$

with $m \times m$ matrices $\mathbf{A}_i = \mathbf{A}_i(\mathbf{x}) = \mathbf{A}_i(x_1, \cdots, x_n)$ and $\mathbf{B} = \mathbf{B}(\mathbf{x})$. As a special case of a later definition (cf. Sect. 12.1) we obtain the

Definition 1.3.2. The system (3) is said to be e l l i p t i c at $\mathbf{x}$ if for any vector $0 \neq (\xi_1, \cdots, \xi_n) \in \mathbb{R}^n$ the following holds:

$$\det\Big(\sum_{i=1}^{n} \mathbf{A}_i(\mathbf{x})\xi_i\Big) \neq 0. \tag{1.3.4}$$

1.4 Characteristic Properties of the Different Types

The distinguishing of different types of partial differential equations would be pointless if each type did not have fundamentally different properties. When discussing the examples in Sect. 1.1 we already mentioned that the solution is uniquely determined if initial values and boundary values are prescribed.

In Example 1.1.2 the hyperbolic differential equation (1.2) is augmented by the specification (1.4) of u on the line $y = \text{const}$ (see Fig. 1a). In the case of the hyperbolic wave equation (1.7), u_y must also be prescribed (cf. (1.9)) since the equation is of second order.

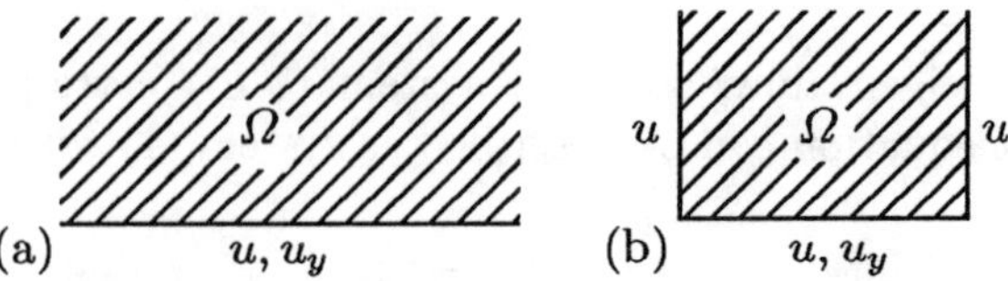

Figure 1.4.1. (a) Initial conditions and (b) initial boundary conditions for hyperbolic problems

It is also sufficient to give the values u and u_y on a finite interval $[x_1, x_2]$ if u is additionally prescribed on the lateral boundaries of the domain Ω of Fig. 1b. This prescription of initial boundary values occurs, for example, in the following physical problem. A vibrating string is described by the lateral deflection $u(x,t)$ at the point $x \in [x_1, x_2]$ at time t. The function u satisfies the wave equation (1.7) with the coordinate y corresponding to time t. At the initial instant in time, $t = t_0$, the deflection $u(x,0)$ and velocity $u_t(x,0)$ are given for $x_1 < x < x_2$. Under the assumption that the string is firmly clamped at the boundary points x_1 and x_2, one obtains the additional boundary data $u(x_1,t) = u(x_2,t) = 0$ for all t.

For parabolic equations of second order one can also formulate initial value and initial boundary value problems (cf. Fig. 2). However, as initial value only the function $u(x, y_0) = u_0(x)$ may be prescribed. An additional specification of $u_y(x, y_0)$ is not possible, since $u_y(x, y_0) = u_{xx}(x, y_0) = u_0''(x)$ is already determined by the differential equation (1.10) and by u_0.

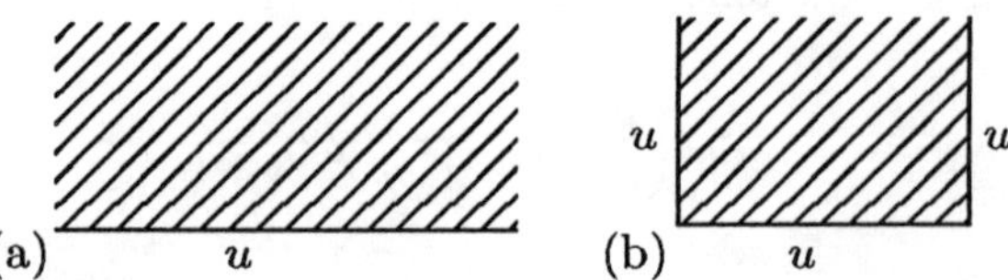

Figure 1.4.2. (a) Initial conditions and (b) initial boundary conditions for parabolic problems

The heat equation (1.10) with the initial and boundary values

$$\begin{aligned} u(x,t_0) &= u_0(x) \qquad \text{in} \quad [x_1, x_2], \\ u(x_1,t) &= \varphi_1(t), \qquad u(x_2,t) = \varphi_2(t) \qquad \text{for } t > t_0 \end{aligned} \tag{1.4.1}$$

(cf. Figure 2b) describes the temperature $u(x,t)$ of a wire whose ends at $x = x_1$ and $x = x_2$ have the temperatures $\varphi_1(t)$ and $\varphi_2(t)$. The initial temperature distribution at time t_0 is given by $u_0(x)$.

Aside from the different number of initial data functions in Figure 1 and Figure 2, there also is the following difference between hyperbolic and parabolic equations.

Remark 1.4.1. The shaded area Ω in Figure 1 corresponds to $t > t_0$, and in Figure 2 to $y > y_0$. For hyperbolic equations one can solve in the same way initial-value and initial-boundary-value problems in the domain $t \leq t_0$, whilst parabolic problems in $t < t_0$ generally do not have a solution.

If one changes the parabolic equation $u_t - u_{xx} = 0$ to $u_t + u_{xx} = 0$, the orientation is reversed: solutions exist in general only for $t \leq t_0$.

For the solution of an elliptic equation, boundary values are prescribed (cf. Example 1.1.3, Figure 3). A specification such as that in Figure 2b would not uniquely determine the solution of an elliptic problem, while the solution of a parabolic problem would be overdetermined by the boundary values of Figure 3a.

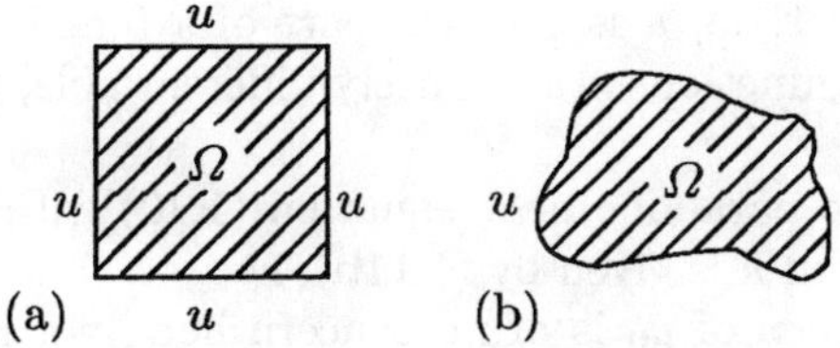

Figure 1.4.3. Boundary value conditions for an elliptic problem

An elliptic problem with specifications such as in Figure 1b in general has no solution. Let us, for example, impose the conditions $u(x,0) = u(0,y) = u(1,y) = 0$ and $u_y(x,0) = u_1(x)$ on the solution of the potential equation (1.6), where u_1 is not infinitely often differentiable. If a continuous solution u existed in $\overline{\Omega} = [0,1] \times [0,1]$ one could develop $u(x,1)$ into a sine series and the following exercise shows that u_1 would have to be infinitely often differentiable, in contradiction to the assumption.

Exercise 1.4.2. Let $\varphi \in C^0[0,1]$ have the Fourier expansion

$$\varphi(x) = \sum_{\nu=1}^{\infty} \alpha_\nu \sin(\nu\pi x).$$

Show that: (a) the solution of the potential equation (1.6) in the square $\Omega = (0,1) \times (0,1)$ with boundary values $u(0,y) = u(x,0) = u(1,y) = 0$ and $u(x,1) = \varphi(x)$ is given by

$$u(x,y) = \sum_{\nu=1}^{\infty} \frac{\alpha_\nu}{\sinh(\nu\pi)} \sin(\nu\pi x)\sinh(\nu\pi y).$$

(b) For $0 \le x \le 1$ and $0 \le y < 1$, $u(x,y)$ is differentiable infinitely often. Hint: $f(x) = \sum \beta_\nu \sin(\nu\pi x) \in C^\infty[0,1]$, if $\lim_{\nu\to\infty} \beta_\nu \nu^k = 0$ for all $k \in \mathbb{N}$.

Conversely it does not make sense to put boundary value constraints as in Fig. 3a on a hyperbolic problem. Consider as an example the wave equation (1.7) in $\overline{\Omega} = [0,1] \times [0,1/\pi]$ with the boundary values $u(x,0) = u(0,y) = u(1,y) = 0$ and $u(x,1/\pi) = \sin(\nu\pi x)$ for $\nu \in \mathbb{N}$. The solution reads $u(x,y) = \sin(\nu\pi x)\sin(\nu\pi y)/\sin\nu$. Although the boundary values, for all $\nu \in \mathbb{N}$ are bounded by 1, the solution in $\overline{\Omega}$ may become arbitrarily large since $\sup\{1/\sin\nu : \nu \in \mathbb{N}\} = \infty$. Such a boundary value problem is called "not well posed" (cf. Definition 2.4.1).

Exercise 1.4.3. Prove that the set $\{\sin\nu : \nu \in \mathbb{N}\}$ is dense in $[-1,1]$.

Another distinguishing characteristic is the regularity (smoothness) of the solution. Let u be the solution of the potential equation (1.6) in $\Omega \subset \mathbb{R}^2$. As stated in Example 1.1.3, u is the real part of a function holomorphic in Ω. Since holomorphic functions are infinitely differentiable, this property also holds for u.

In the case of the parabolic heat equation (1.10) with boundary values $u(x,0) = u_0$ the solution u is given by (1.11b). For $y > 0$, u is infinitely differentiable. The smoothness of u_0 is of no concern here, nor is the smoothness of the boundary values in the case of the potential equation.

One finds a completely different result for the hyperbolic wave equation (1.7). The solution reads $u(x,y) = \varphi(x+y) + \psi(x-y)$, where φ and ψ result directly from the initial data (1.9). Check that u is k-times differentiable if u_0 is k-times and u_1 is $(k-1)$-times differentiable.

As already mentioned in this section, in the hyperbolic and parabolic equations (1.1), (1.2), (1.7) and (1.10) the variable y often plays the role of time. Therefore one calls processes described by hyperbolic and parabolic equations nonstationary. Elliptic equations which only contain space coordinates as variables are called stationary. More clearly than in Definitions 1.2.1b,c the Definitions 1.2.3b,c distinguish the role of a single variable (time) corresponding to the eigenvalue $\lambda = 0$ in parabolic equations, and to the eigenvalue with opposite sign in hyperbolic equations.

The connection between the different types becomes more comprehensible if one relates the elliptic equations in the variables $x_1, \cdots, x_n$ to the parabolic and hyperbolic equations in the variables $x_1, \cdots, x_n, t$.

Remark 1.4.4. Let L be a differential operator (2.6) in the variables $\mathbf{x} = (x_1, \cdots, x_n)$ and let it be of elliptic type. Let L be scaled such that the matrix $\mathbf{A}(\mathbf{x})$ in (2.4b) has only negative eigenvalues. Then

$$u_t + Lu = 0 \tag{1.4.2}$$

is a parabolic equation for $u(\mathbf{x},t) = u(x_1,\cdots,x_n,t)$. In contrast

$$u_{tt} + Lu = 0 \tag{1.4.3}$$

is of hyperbolic type.

Conversely, the nonstationary problems (2) or (3) lead to the elliptic equation $Lu = 0$ if one seeks solutions of (2) or (3) that are independent of time t. One also obtains elliptic equations if one looks for solutions of (2) or (3) with the aid of a separation of variables $u(\mathbf{x},t) = \varphi(t)v(\mathbf{x})$. The results are

$$\begin{aligned} u(\mathbf{x},t) &= e^{-\lambda t}v(\mathbf{x}) && \text{in case (2),} \\ u(\mathbf{x},t) &= e^{\pm i\sqrt{\lambda}t}v(\mathbf{x}) && \text{in case (3),} \end{aligned} \tag{1.4.4}$$

where $v(\mathbf{x})$ is the solution of the elliptic eigenvalue problem

$$Lv = \lambda v. \tag{1.4.5}$$

2 The Potential Equation

2.1 Posing the Problem

The potential equation from Example 1.1.3 reads

$$\Delta u = 0 \quad \text{in } \Omega \subset \mathbb{R}^n, \tag{2.1.1a}$$

where $\Delta = \partial^2/\partial x_1^2 + \cdots + \partial^2/\partial x_n^2$ is the Laplace operator. In physics, Equation (1a) describes the potentials; for example the electric potential when Ω contains no electric charges, the magnetic potential for vanishing current density, the velocity potential, etc. Equation (1a) is also called Laplace's equation since it was described by P. S. Laplace in his five-volume work "Mécanique Céleste" (1799–1825). However, it was L. Euler who first mentioned the potential equation in 1752.

The connection between the potential equation for $n = 2$ and function theory has already been pointed out in Example 1.1.3. Not only is the Laplace operator an example of an elliptic differential operator, but it actually is the prototype (the so-called normal form). By using a transformation of variables, any elliptic differential operator of second order can be transformed so that its principal part becomes the Laplace operator (cf. Hellwig [1, II–§1.5]).

In the following Ω will always be a domain.

Definition 2.1.1. The region $\Omega \subset \mathbb{R}^n$ is called a domain if Ω is open and connected.[1]

The existence of a second derivative of u is required only in Ω, not on the boundary

$$\Gamma = \partial\Omega$$

of Ω. For a prescription of boundary values

$$u = \varphi \quad \text{on } \Gamma \tag{2.1.1b}$$

to be meaningful, one has to assume continuity of u on $\overline{\Omega} = \Omega \cup \Gamma$. These considerations lead to

1 Ω is called connected if for any $\mathbf{x}, \mathbf{y} \in \Omega$ there is a continuous curve within Ω that connects $\mathbf{x}$ and $\mathbf{y}$, that is, a $\gamma\colon s \in [0,1] \mapsto \gamma(s) \in \Omega$ continuously with $\gamma(0) = \mathbf{x}, \gamma(1) = \mathbf{y}$.

Definition 2.1.2. The function u is said to be harmonic in Ω if u belongs to $C^2(\Omega) \cap C^0(\overline{\Omega})$ and satisfies the potential equation (1a).

Here $C^0(D)$ $[C^k(D), C^\infty(D)]$ denotes the set of continuous [k-fold continuously differentiable, infinitely often differentiable] functions on D.

In general one should not expect that the solution of (1a,b) lies in $C^2(\overline{\Omega})$, as we show in the following example.

Example 2.1.3. Let $\Omega = (0,1) \times (0,1)$ (cf. Figure 1.4.3a). Let the boundary values be given by $\varphi(x,y) = x^2$ for $(x,y) \in \Gamma$. A solution of the boundary value problem exists but does not belong to $C^2(\overline{\Omega})$.

PROOF. The existence of a solution u will be discussed in Theorem 7.3.7. If $u \in C^2(\overline{\Omega})$, then it follows that $u_{xx}(x,0) = \varphi_{xx}(x,0) = 2$ for $x \in [0,1]$; in particular $u_{xx}(0,0) = 2$. From the analogous result $u_{yy}(0,0) = \varphi_{yy}(0,0) = 0$ one may conclude $\Delta u(0,0) = 2$ in contradiction to $\Delta u = 0$ in Ω. ■

In the case under discussion one can also show $u \in C^1(\overline{\Omega})$. That this statement is generally false, is shown by

Example 2.1.4. In the domain $\Omega = (-\frac{1}{2}, \frac{1}{2}) \times (-\frac{1}{2}, \frac{1}{2}) \setminus [0, \frac{1}{2}) \times [0, \frac{1}{2})$ (cf. Figure 1) introduce the polar coordinates

$$x = r\cos\varphi, \quad y = r\sin\varphi. \tag{2.1.2}$$

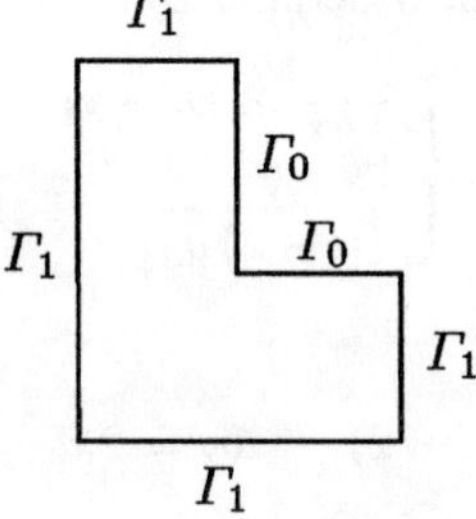

Figure 2.1.1. An L-shaped region

The function $u(r,\varphi) = r^{2/3}\sin((2\varphi - \pi)/3)$ is the solution of the potential equation (1a) and has smooth boundary values on Γ (in particular $u = 0$ holds on $\Gamma_0 \subset \Gamma$) although the first derivatives in $r = 0$ are unbounded, i.e. $u \notin C^1(\overline{\Omega})$.

This follows from the fact that, along with u_x and u_y, the radial derivative $u_r = u_x \cos\varphi + u_y \sin\varphi$ also has to be bounded. However, $u_r = O(r^{-1/3})$ holds for $r \longrightarrow 0$. In order to check that $\Delta u = 0$, use the following.

Exercise 2.1.5. Show that (a) In terms of the polar coordinates (2) in $\mathbb{R}^2$, the Laplace operator takes the form

$$\Delta = \frac{\partial^2}{\partial r^2} + \frac{1}{r}\frac{\partial}{\partial r} + \frac{1}{r^2}\frac{\partial^2}{\partial \varphi^2}. \tag{2.1.3}$$

(b) In terms of the three-dimensional polar coordinates

$$x = r\cos\varphi\sin\psi, \quad y = r\sin\varphi\sin\psi, \quad z = r\cos\psi$$

the Laplace operator is given by

$$\Delta = \frac{\partial^2}{\partial r^2} + \frac{2}{r}\frac{\partial}{\partial r} + \frac{1}{r^2}\left[\frac{1}{\sin^2\psi}\frac{\partial^2}{\partial\varphi^2} + \cot\psi\frac{\partial}{\partial\psi} + \frac{\partial^2}{\partial\psi^2}\right]. \tag{2.1.4}$$

Note. In the n-dimensional case the transformation to polar coordinates leads to

$$\Delta = \frac{\partial^2}{\partial r^2} + \frac{n-1}{r}\frac{\partial}{\partial r} + \frac{1}{r^2}B, \tag{2.1.5}$$

where the Beltrami operator B contains only derivatives with respect to the angle variables.

2.2 Singularity Function

The singularity function is defined by

$$s(\mathbf{x},\mathbf{y}) := \begin{cases} -\frac{1}{\omega_2}\log|\mathbf{x}-\mathbf{y}| & \text{for } n = 2, \\ \dfrac{|\mathbf{x}-\mathbf{y}|^{2-n}}{(n-2)\omega_n} & \text{for } n > 2. \end{cases} \tag{2.2.1a}$$

for $\mathbf{x},\mathbf{y} \in \mathbb{R}^n$, where

$$\omega_n = 2\Gamma(1/2)^n / \Gamma(n/2), \quad \omega_2 = 2\pi, \quad \omega_3 = 4\pi, \tag{2.2.1b}$$

with Γ the Gamma function, is the surface of the n-dimensional unit sphere. The Euclidean norm of $\mathbf{x}$ in $\mathbb{R}^n$ is denoted by

$$|\mathbf{x}| = \Big(\sum_{i=1}^{n} x_i^2\Big)^{1/2}. \tag{2.2.2}$$

Lemma 2.2.1. *For fixed* $\mathbf{y} \in \mathbb{R}^n$ *the potential equation in* $\mathbb{R}^n\backslash\{\mathbf{y}\}$ *is solved by* $s(\mathbf{x},\mathbf{y})$.

The proof can be carried out directly. It is simplest to introduce polar coordinates with $\mathbf{y}$ as origin and to use (1.5), since $s(\mathbf{x}, \mathbf{y})$ depends only on $r = |\mathbf{x} - \mathbf{y}|$.

For the next theorem we need to introduce the normal derivative $\partial/\partial n$. Let Ω be a domain with smooth boundary Γ. Let $\mathbf{n}(\mathbf{x}) \in \mathbb{R}^n$ denote the outer normal direction at $\mathbf{x} \in \Gamma$, i.e. $\mathbf{n}$ is a unit vector perpendicular to the tangential hyperplane at $\mathbf{x}$ and points outwards. The normal derivative of u at $\mathbf{x} \in \Gamma$ is defined as

$$\partial u(\mathbf{x})/\partial n = \langle \mathbf{n}(\mathbf{x}), \nabla u(\mathbf{x}) \rangle, \tag{2.2.3a}$$

where

$$\nabla u = (\mathbf{u}_{\mathbf{x}_1}, \cdots, \mathbf{u}_{\mathbf{x}_n})^{\mathrm{T}} \tag{2.2.3c}$$

is the gradient of $\mathbf{u}$ and

$$\langle \mathbf{x}, \mathbf{y} \rangle = \sum_{i=1}^{n} \mathbf{x}_i \mathbf{y}_i \tag{2.2.3c}$$

is the scalar product in $\mathbb{R}^n$. In the case of the sphere $K_R(\mathbf{y})$ (cf. (7)) the normal direction is radial, and $\partial u/\partial n$ becomes $\partial u/\partial r$ with respect to $r = |\mathbf{x} - \mathbf{y}|$, if one uses polar coordinates with the origin at $\mathbf{y}$. It follows from $\partial s(\mathbf{x}, \mathbf{y})/\partial r = -|\mathbf{x} - \mathbf{y}|^{1-n}/\omega_n$, that

$$\partial s(\mathbf{x}, \mathbf{y})/\partial n = -R^{1-n}/\omega_n \quad \text{for } \mathbf{x} \in \partial K_R(\mathbf{y}). \tag{2.2.4}$$

The first Green formula reads (cf. Green [1])

$$\int_\Omega u \Delta v \, dx = -\int_\Omega \langle \nabla u, \nabla v \rangle + \int_{\partial\Omega} u \frac{\partial v}{\partial n} \, d\Gamma \tag{2.2.5a}$$

and holds for $u \in C^1(\overline{\Omega})$, $v \in C^2(\overline{\Omega})$ if the domain Ω satisfies suitable conditions. Here $\int_{\partial\Omega} \cdots d\Gamma$ denotes the surface integral.

Domains for which Equation (5a) holds are called normal domains. To see sufficient conditions for this refer to Kellogg [1, Chapter IV] and Hellwig [1, I-1.2].

Functions $u, v \in C^2(\overline{\Omega})$ in a normal domain Ω satisfy the second Green formula

$$\int_\Omega u \Delta v = \int_\Omega v \Delta u + \int_{\partial\Omega} \left[u \frac{\partial v}{\partial n} - v \frac{\partial u}{\partial n} \right] d\Gamma. \tag{2.2.5b}$$

Theorem 2.2.2. *Let Ω be a normal domain, and let $u \in C^2(\overline{\Omega})$ be harmonic there. Then*

$$u(\mathbf{y}) = \int_{\partial\Gamma} \left[s(\mathbf{x}, \mathbf{y}) \frac{\partial}{\partial n} u(\mathbf{x}) - u(\mathbf{x}) \frac{\partial}{\partial n_{\mathbf{x}}} s(\mathbf{x}, \mathbf{y}) \right] d\Gamma_{\mathbf{x}} \tag{2.2.6}$$

for all $\mathbf{y} \in \Omega$. Here $\partial/\partial n_{\mathbf{x}}$ and $d\Gamma_{\mathbf{x}}$ refer to the variable $\mathbf{x}$.

PROOF. By

$$K_r(\mathbf{y}) := \{\mathbf{x} \in \mathbb{R}^n \colon |\mathbf{x} - \mathbf{y}| < r\} \tag{2.2.7}$$

we denote the ball with centre $\mathbf{y}$ and radius r. Since the singularity function $s(\cdot, \mathbf{y})$ is not differentiable on $\mathbf{x} = \mathbf{y}$, Green's formula (5b) is not directly applicable. Let

$$\Omega_\epsilon := \Omega \backslash \overline{K_\epsilon(\mathbf{y})},$$

with ϵ so small that $\overline{K_\epsilon(\mathbf{y})} \subset \Omega$. Since Ω_ϵ is again a normal domain, it follows from $\Delta u = \Delta s = 0$ (cf. Lemma 1) and (5b) with $v = s$ that

$$\int_{\partial\Omega_\epsilon} \left[u(\mathbf{x}) \frac{\partial}{\partial n_\mathbf{x}} s(\mathbf{x}, \mathbf{y}) - s(\mathbf{x}, \mathbf{y}) \frac{\partial}{\partial n} u(\mathbf{x}) \right] d\Gamma_\mathbf{x} = 0. \tag{2.2.8a}$$

We have $\partial\Omega_\epsilon = \partial\Omega \cup \partial K_\epsilon(\mathbf{y})$. However, at $\mathbf{x} \in \partial K_\epsilon(\mathbf{y})$, the normal directions of $\partial\Omega_\epsilon$ and of $\partial K_\epsilon(\mathbf{y})$ differ in their signs. The same holds for the normal derivatives, so that the integral in (8a) can be decomposed into $\int_{\partial\Omega_\epsilon} \cdots = \int_{\partial\Omega} \cdots - \int_{\partial K_\epsilon(\mathbf{y})} \cdots$. The assertion of the theorem would be proved if we could show that $\int_{\partial K_\epsilon(\mathbf{y})} \cdots \to -u(\mathbf{y})$ for $\epsilon \to 0$. The normal derivative $\partial u / \partial n$ is bounded on $\partial K_\epsilon(\mathbf{y})$, and $\int_{\partial K_\epsilon(\mathbf{y})} s(\mathbf{x}, \mathbf{y})\, d\Gamma$ converges like $O(\epsilon |\log \epsilon|)$ or $O(\epsilon)$ towards zero, as can be seen from (1) and $\int_{\partial K_\epsilon(\mathbf{y})} d\Gamma = \epsilon^{n-1}\omega_n$. Thus, we obtain

$$\int_{\partial K_\epsilon(\mathbf{y})} s(\mathbf{x}, \mathbf{y}) \frac{\partial}{\partial n} u(\mathbf{x})\, d\Gamma_\mathbf{x} \to 0 \qquad (\epsilon \to 0). \tag{2.2.8b}$$

From $\int_{\partial K_\epsilon(y)} d\Gamma = \epsilon^{n-1}\omega_n$ and (4) one infers

$$\int_{\partial K_\epsilon(\mathbf{y})} u(\mathbf{y}) \frac{\partial}{\partial n_\mathbf{x}} s(\mathbf{x}, \mathbf{y})\, d\Gamma_\mathbf{x} = -u(\mathbf{y}). \tag{2.2.8c}$$

The continuity of u in $\mathbf{y}$ yields

$$\left| \int_{\partial K_\epsilon(\mathbf{y})} (u(\mathbf{x}) - u(\mathbf{y})) \frac{\partial}{\partial n_\mathbf{x}} s(\mathbf{x}, \mathbf{y})\, d\Gamma_\mathbf{x} \right| \le \max_{\mathbf{x} \in \partial K_\epsilon(\mathbf{y})} |u(\mathbf{x}) - u(\mathbf{y})| \to 0 \text{ as } \epsilon \to 0. \tag{2.2.8d}$$

Equations 8b,c,d show that $\int_{\partial K_\epsilon(\mathbf{y})} \left[u \frac{\partial}{\partial n} s - s \frac{\partial}{\partial n} u \right] d\Gamma \to -u(y)$ $(\epsilon \to 0)$, so that (8a) proves the theorem. ■

Any function of the form

$$\gamma(\mathbf{x}, \mathbf{y}) = s(\mathbf{x}, \mathbf{y}) + \Phi(\mathbf{x}, \mathbf{y}) \tag{2.2.9}$$

is called a fundamental solution, of the potential equation, in Ω if for fixed $\mathbf{y} \in \Omega$ the function $\Phi(\cdot, \mathbf{y})$ is harmonic in Ω and belongs to $C^2(\overline{\Omega})$.

Corollary 2.2.3. *Under the conditions of Theorem 2 for every fundamental solution in Ω the following holds:*

$$u(\mathbf{y}) = \int_{\partial\Omega} \left[\gamma(\mathbf{x},\mathbf{y}) \frac{\partial}{\partial n} u(\mathbf{x}) - u(\mathbf{x}) \frac{\partial}{\partial n_{\mathbf{x}}} \gamma(\mathbf{x},\mathbf{y}) \right] d\Gamma_{\mathbf{x}}. \tag{2.2.10}$$

PROOF. (5b) implies $\int_{\partial\Omega} [\Phi\, \partial u/\partial n - u\, \partial\Phi/\partial n]\, d\Gamma = 0$. ∎

For a possible weakening of the condition $\Phi = \gamma - s \in C^2(\overline{\Omega})$ for $\Phi(\cdot,\mathbf{y}) \in C^1(\overline{\Omega}) \cap C^2(\Omega)$ refer to Hellwig [1, I–1.4].

Exercise 2.2.4. Let $\Omega = K_R(\mathbf{y})$. Define

$$\gamma(\mathbf{x},\xi) = \begin{cases} \frac{1}{(n-2)\omega_n} \left[|\mathbf{x}-\xi|^{2-n} - \left(\frac{|\xi-\mathbf{y}|}{R} |\mathbf{x}-\xi'| \right)^{2-n} \right] & \text{for } n > 2 \\ -\frac{1}{2\pi} \left[\log|\mathbf{x}-\xi| - \log\left(\frac{|\xi-\mathbf{y}|}{R} |\mathbf{x}-\xi'| \right) \right] & \text{for } n = 2 \end{cases} \tag{2.2.11a}$$

with $\mathbf{x},\xi \in \Omega$, $\xi' = \mathbf{y} + R^2 |\xi-\mathbf{y}|^{-2} (\xi - \mathbf{y})$ and show:
(a) γ is a fundamental solution in Ω,
(b) $\gamma(\mathbf{x},\xi) = \gamma(\xi,\mathbf{x})$,
(c) on the surface $\Gamma = \partial K_R(\mathbf{y})$ the following holds:

$$\frac{\partial}{\partial n_\xi} \gamma(\mathbf{x},\xi) = \frac{\partial}{\partial n_\xi} \gamma(\xi,\mathbf{x}) = -\frac{1}{R\omega_n} (R^2 - |\mathbf{x}-\mathbf{y}|^2) |\mathbf{x}-\xi|^{-n} \qquad (\xi \in \Gamma). \tag{2.2.11b}$$

2.3 The Mean Value Property and Maximum Principle

Definition 2.3.1. A function u has the mean value property in Ω if $u \in C^0(\overline{\Omega})$ and if for all $\mathbf{x} \in \Omega$ and all $R > 0$ with $K_R(\mathbf{x}) \subset \Omega$ the following equation holds

$$u(\mathbf{x}) = \frac{1}{\omega_n R^{n-1}} \int_{\partial K_R(\mathbf{x})} u(\xi) d\Gamma_\xi. \tag{2.3.1}$$

Since $\int_{\partial K_R(\mathbf{x})} d\Gamma = \omega_n R^{n-1}$ the right side in (1) is the mean value of u taken over the surface of the sphere. An equivalent characterisation results if one averages over the sphere $K_R(\mathbf{x})$.

Exercise 2.3.2. $u \in C_0(\overline{\Omega})$ has the second mean value property in Ω if $u(\mathbf{x}) = \frac{n}{R^n \omega_n} \int_{K_R(\mathbf{x})} u(\xi)\, d\xi$ for all $\mathbf{x} \in \Omega$, $R > 0$ with $K_R(\mathbf{x}) \subset \Omega$. Show that this mean value property is equivalent to the mean value property (1). Hint: $\int_{K_R(\mathbf{x})} u(\xi)\, d\xi = \int_0^R \int_{\partial K_r(\mathbf{x})} u(\xi) d\Gamma_\xi\, dr$.

Functions with the mean value property satisfy a maximum principle, as is known from the function theory for holomorphic functions:

Theorem 2.3.3. (Maximum-minimum principle) *Let Ω be a domain and let $u \in C^0(\overline{\Omega})$ be a nonconstant function which has the mean value property. Then u takes on neither a maximum nor a minimum in Ω.*

PROOF. (a) It suffices to investigate the case of a maximum since a minimum of u is a maximum of $-u$, and $-u$ also has the mean value property.
(b) For an indirect proof we assume that there exists a maximum in $\mathbf{y} \in \Omega$:

$$u(\mathbf{y}) = M \geq u(\mathbf{x}) \quad \text{for all } \mathbf{x} \in \Omega.$$

In (c) we will show $u(\mathbf{y}') = M$ for arbitrary $\mathbf{y}' \in \Omega$, i.e. $u \equiv M$ in contrast to the assumption $u \not\equiv$ const.
(c) Proof of $u(\mathbf{y}') = M$. Let $\mathbf{y}' \in \Omega$. Since Ω is connected, there exists a path connecting $\mathbf{y}$ and $\mathbf{y}'$ running through Ω, i.e. there exists a continuous $\varphi\colon [0,1] \to \Omega$ with $\varphi(0) = \mathbf{y}, \varphi(1) = \mathbf{y}'$. We set

$$I := \{s \in [0,1] \colon u(\varphi(t)) = M \text{ for all } 0 \leq t \leq s\}.$$

I contains at least 0, and is closed since u and φ are continuous. Thus there exists $s^* = \max\{s \in I\}$, and the definition of I shows that $I = [0, s^*]$. In (4) it is proved that $s^* = 1$ so that $\mathbf{y}' = \varphi(1) \in I$ and hence $u(\mathbf{y}') = M$ follows.
(d) Proof of $s^* = 1$. The opposite assumption $s^* < 1$ can be made and shown to be contradicted by proving that $u(\mathbf{x}) = M$ in a neighbourhood of $\mathbf{x}^* := \varphi(s^*)$. Since $\mathbf{x}^* \in \Omega$, there exists $R > 0$ with $K_R(\mathbf{x}^*) \subset \Omega$. Evidently, it follows that $u = M$ in $K_R(\mathbf{x})^*$ if it is shown that $u = M$ on $\partial K_r(\mathbf{x})^*$ for all $0 < r \leq R$.
(e) Proof of $u = M$ on $\partial K_r(\mathbf{x}^*)$. Equation (1) in $\mathbf{x}^*$ reads

$$M = u(\mathbf{x}^*) = \int_{\partial K_r(\mathbf{x}^*)} u(\xi)\, d\Gamma_\xi / (\omega_n r^{n-1}).$$

In general we have $u(\xi) \leq M$. If one had $u(\xi') < M$ for $\xi' \in \partial K_r(\mathbf{x}^*)$ and thus also $u < M$ in a neighbourhood of ξ', one would have on the right side a mean value smaller than M. Therefore, $u = M$ on $\partial K_r(\mathbf{x}^*)$ has been proved. ∎

Simple deductions from Theorem 3 are contained in

Corollary 2.3.4. *Let Ω be bounded. (a) A function with the mean value property takes its maximum and its minimum on $\partial\Omega$.*
(b) If two functions with the mean value property coincide on the boundary $\partial\Omega$, they are identical.

PROOF. (a) The extrema are assumed on the compact set $\overline{\Omega} = \Omega \cup \partial\Omega$. According to Theorem 3, the extremum cannot be in Ω if u is not constant on a connected component of Ω. But in this case the assertion is also obvious.
(b) If u and v with $u = v$ satisfy the mean value property on $\partial\Omega$ then the latter is also satisfied for $w := u - v$. Since $w = 0$ on $\partial\Omega$, part (a) indicates that $\max_{\overline{\Omega}} w = \min_{\overline{\Omega}} w = 0$. Thus $u = v$ in Ω. ∎

Lemma 2.3.5. *Harmonic functions have the mean value property.*

PROOF. Let u be harmonic in Ω and $\mathbf{y} \in K_R(\mathbf{y}) \subset \Omega$. We apply the representation (2.6) for $\Omega = K_R(\mathbf{y})$. The value $s(\mathbf{x}, \mathbf{y})$ is constant on $\partial K_R(\mathbf{y})$: let it be denoted by $\sigma(R)$. Because of (2.4), Equation (2.6) becomes

$$u(\mathbf{y}) = \sigma(R) \int_{\partial K_R(\mathbf{y})} \frac{\partial u}{\partial n}\, d\Gamma + \frac{1}{\omega_n R^{n-1}} \int_{\partial K_R(\mathbf{y})} u \, d\Gamma.$$

The equation agrees with (1) if the first integral vanishes. The latter follows from

Lemma 2.3.6. *Let $u \in C^2(\overline{\Omega})$ be harmonic in a normal domain Ω. Then we have*

$$\int_{\partial\Omega} \frac{\partial u}{\partial n}\, d\Gamma = 0. \tag{2.3.2}$$

PROOF. In Green's formula (2.5a) substitute 1 and u for u and v respectively. ■

Lemma 5, Theorem 3, and Corollary 4 together imply Theorems 7 and 8:

Theorem 2.3.7. (The maximum-minimum principle for harmonic functions) *Let u be harmonic in the domain Ω and nonconstant. There exists no maximum and no minimum in Ω.*

Theorem 2.3.8. (Uniqueness). *Let Ω be bounded. A function harmonic in Ω assumes its maximum and its minimum on $\partial\Omega$ and is uniquely determined by its values on $\partial\Omega$.*

The representation (1) of $u(\mathbf{y})$ by the values on $\partial K_R(\mathbf{y})$ is a special case of the following formula which will be proved at the end of this section and which provides Equation (1) for $\mathbf{x} = \mathbf{y}$.

Theorem 2.3.9. (Poisson's integral formula) *Assume we have $\varphi \in C^0(\partial K_R(\mathbf{y}))$ and $n \geq 2$. The solution of the boundary value problem*

$$\Delta u = 0 \quad \text{in } K_R(\mathbf{y}), \qquad u = \varphi \text{ on } \partial K_R(\mathbf{y}) \tag{2.3.3}$$

is given by the function

$$u(\mathbf{x}) = \frac{R^2 - |\mathbf{x} - \mathbf{y}|^2}{R\omega_n} \int_{\partial K_R(\mathbf{y})} \frac{\varphi(\xi)}{|\mathbf{x} - \xi|^{\mathbf{n}}}\, d\Gamma_\xi \quad \text{for } \mathbf{x} \in K_R(\mathbf{y}), \tag{2.3.4}$$

which belongs to $C^\infty(K_R(\mathbf{y})) \cap C^0(\overline{K_R(\mathbf{y})})$.

The mean value property only assumes $u \in C^0(\overline{\Omega})$, while harmonic functions belong to $C^2(\Omega) \cap C^0(\overline{\Omega})$. This makes the following assertion surprising:

Theorem 2.3.10. *A function is harmonic in Ω if and only if it has the mean value property there.*

PROOF. Because of Lemma 5 it remains to be shown that a function v with the mean value property is harmonic. Let $\mathbf{x} \in K_R(\mathbf{x}) \subset \Omega$ be given arbitrarily. According to Theorem 9 there exists a function u harmonic in $K_R(\mathbf{x})$ with

$$\Delta u = 0 \quad \text{in } K_R(\mathbf{x}), \qquad u = v \quad \text{on } \partial K_R(\mathbf{x}).$$

According to Lemma 5, u has the mean value property, as does v, and Corollary 4b proves that $u = v$ in $K_R(\mathbf{x})$, i.e. v is harmonic in $K_R(\mathbf{x})$. Since $K_R(\mathbf{x}) \subset \Omega$ is arbitrary, v is harmonic in Ω. ■

An important application of Theorem 10 is

Theorem 2.3.11. *(Harnack) Let $u_1, u_2, \cdots$ be a sequence of functions harmonic in Ω and converging uniformly in $\overline{\Omega}$. Then $u = \lim_{k\to\infty} u_k$ is harmonic in Ω.*

PROOF. The limit function is continuous: $u \in C^0(\overline{\Omega})$. The limit process, applied to $u_k(\mathbf{x}) = \int_{\partial K_R(\mathbf{x})} u_k(\xi)\, d\Gamma_\xi/(\omega_n R^{n-1})$ yields Equation (1) for u; i.e. u has the mean value property. According to Theorem 10, u is also harmonic in Ω. ■

Theorems 3 and 7 on the maximum-minimum principle pertain to global extrema. The proof of Theorem 3 does not yet exclude local extrema in the interior. It merely shows that u is then always constant in a circle $K_R(\mathbf{y}) \subset \Omega$. As is known from function theory one can derive from this $u(\mathbf{x}) = M$ in $\overline{\Omega}$, if u is analytic, i. e., there is a convergent power series expansion in a neighbourhood of any $\mathbf{x} \in \Omega$. Indeed, the following theorem holds whose proof can be found, for example, in Hellwig [1,III–§1.5]:

Theorem 2.3.12. *A function harmonic in Ω is analytic there.*

The proof of the Poisson formula still needs to be carried out. (a) First we must show that u in (4) is a function harmonic in $K_R(\mathbf{y})$, i. e., it satisfies $\Delta u = 0$. Since the integrand is twice continuously differentiable and the domain of integration $\Gamma = \partial K_R(\mathbf{y})$ is compact, the Laplace operator commutes with the integral sign:

$$\Delta u(\mathbf{x}) = (R\omega_n)^{-1} \int_\Gamma \varphi(\xi) \Delta[(R^2 - |\mathbf{x}-\mathbf{y}|^2) \cdot |\mathbf{x}-\xi|^{-n}]\, d\Gamma_\xi \quad \text{for } \mathbf{x} \in K_R(\mathbf{y}). \tag{2.3.5}$$

According to Exercise 2.2.4 there exists a fundamental solution $\gamma(\mathbf{x},\xi)$ such that

$$\frac{R^2-|\mathbf{x}-\mathbf{y}|^2}{|\mathbf{x}-\xi|^n R\omega_n} = -\frac{\partial}{\partial n_\xi}\gamma(\mathbf{x},\xi) = -\frac{\partial}{\partial n_\xi}\gamma(\xi,\mathbf{x}), \qquad \xi\in\Gamma, \mathbf{x}\in K_R(\mathbf{y}). \tag{2.3.6}$$

From $\Delta_\mathbf{x}\frac{\partial\gamma}{\partial n_\xi} = \frac{\partial}{\partial n_\xi}\Delta_\mathbf{x}\gamma(\mathbf{x},\xi) = 0$ and (5), one infers that $\Delta u = 0$.
(b) The expression (4) defines $u(\mathbf{x})$ at first only for $\mathbf{x}\in K_R(\mathbf{y})$. It still needs to be shown that u has a continuous extension on $\overline{K_R(\mathbf{y})} = K_R(\mathbf{y})\cup\Gamma$ (i.e., $u\in C^0(\overline{K_R(\mathbf{y})})$) and that the continuously extended values agree with the boundary values φ:

$$\lim_{\substack{\mathbf{x}\to\mathbf{z}\\ \mathbf{x}\in K_R(\mathbf{y})}} u(\mathbf{x}) = \varphi(\mathbf{z}) \qquad \text{for } \mathbf{z}\in\Gamma. \tag{2.3.7}$$

By Equation (6), putting $u\equiv 1$ in Corollary 2.2.3 gives the identity

$$\frac{R^2-|\mathbf{x}-\mathbf{y}|^2}{R\omega_n}\int_\Gamma |\mathbf{x}-\xi|^{-n}\,d\Gamma_\xi = 1 \quad \text{for all } \mathbf{x}\in K_R(\mathbf{y}). \tag{2.3.8}$$

Let $\mathbf{z}\in\Gamma$ be arbitrary. Due to Equation (8) one can then write:

$$u(\mathbf{x})-\varphi(\mathbf{z}) = \frac{R^2-|\mathbf{x}-\mathbf{y}|^2}{R\omega_n}\int_\Gamma \frac{\varphi(\xi)-\varphi(\mathbf{z})}{|\mathbf{x}-\xi|^n}\,d\Gamma_\xi. \tag{2.3.9a}$$

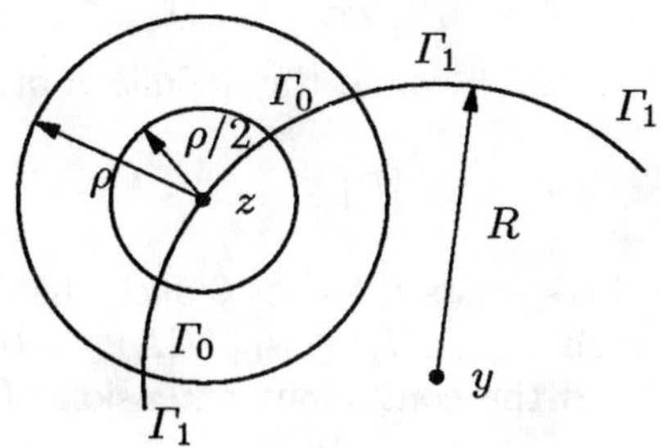

Figure 2.3.1. Construction of Γ_0, Γ_1

We define $\Gamma_0 = \Gamma\cap\overline{K_\rho(\mathbf{z})}$, $\Gamma_1 = \Gamma\backslash\Gamma_0$ (see Figure 1) and split the expression (9a) into $u(\mathbf{x})-\varphi(\mathbf{z}) = I_0+I_1$, where

$$I_i = \frac{R^2-|\mathbf{x}-\mathbf{y}|^2}{R\omega_n}\int_{\Gamma_i}\frac{\varphi(\xi)-\varphi(\mathbf{z})}{|\mathbf{x}-\xi|^n}\,d\Gamma_\xi, \quad i=0,1.$$

Since

$$\left|\int_{\Gamma_0} \frac{\varphi(\xi)-\varphi(\mathbf{z})}{|\mathbf{x}-\xi|^n}\, d\Gamma_\xi\right| \le \max_{\xi\in\Gamma_0} |\varphi(\xi)-\varphi(\mathbf{z})| \int_{\Gamma_0} \frac{d\Gamma_\xi}{|\mathbf{x}-\xi|^n}$$
$$\le \max_{\xi\in\Gamma_0} |\varphi(\xi)-\varphi(\mathbf{z})| \int_{\Gamma} \frac{d\Gamma_\xi}{|\mathbf{x}-\xi|^n},$$

it follows from Equation (8) that

$$I_0 \le \max_{\xi\in\Gamma_0} |\varphi(\xi)-\varphi(\mathbf{z})|. \tag{2.3.9b}$$

Because of the continuity of φ one can choose $\rho > 0$ such that for given $\epsilon > 0$

$$I_0 \le \epsilon/2. \tag{2.3.9c}$$

Set $C_\varphi := \max_{\xi\in\Gamma} |\varphi(\xi)|$ and choose $\mathbf{x} \in K_R(\mathbf{y})$ sufficiently close to $\mathbf{z}$ such that

$$|\mathbf{x}-\mathbf{z}| \le \delta(\epsilon) := \frac{\epsilon}{2}\left(\frac{\rho}{2}\right)^n \frac{1}{4C_\varphi R^{n-1}}$$

and $|\mathbf{x}-\mathbf{z}| \le \rho/2$. The last inequality implies

$$|\mathbf{x}-\xi| \ge \frac{\rho}{2} \quad \text{for } \xi\in\Gamma_1 \quad \text{(see Figure 1).}$$

Together with $R^2-|\mathbf{x}-\mathbf{y}|^2 = (R+|\mathbf{x}-\mathbf{y}|)(R-|\mathbf{x}-\mathbf{y}|) \le 2R(R-|\mathbf{x}-\mathbf{y}|) \le 2R|\mathbf{z}-\mathbf{x}| \le 2R\delta(\epsilon)$ one obtains

$$|I_1| = \left|\frac{R^2-|\mathbf{x}-\mathbf{y}|^2}{R\omega_n}\int_{\Gamma_1} \frac{\varphi(\xi)-\varphi(\mathbf{z})}{|\mathbf{x}-\xi|^n}\, d\Gamma_\xi\right| \le \frac{2}{\omega_n}\delta(\epsilon)\frac{2C_\varphi}{(\rho/2)^n}\int_{\Gamma_1} 1\, d\Gamma.$$

From $\int_{\Gamma_1} d\Gamma \le \int_\Gamma d\Gamma = R^{n-1}\omega_n$ and the definition of $\delta(\epsilon)$ follows

$$|I_1| \le \frac{\epsilon}{2}. \tag{2.3.9d}$$

Thus for every $\epsilon > 0$ there exists a $\delta(\epsilon) > 0$ such that $|\mathbf{x}-\mathbf{z}| \le \delta(\epsilon)$ implies the estimate $|u(\mathbf{x})-\varphi(\mathbf{z})| = |I_0+I_1| \le |I_0|+|I_1| \le \epsilon$ (cf. (9c and d)). Hence, (7) has been proved, and the continuous extension of u to $\overline{K_R(\mathbf{y})}$ leads to $u \in C^0(\overline{K_R(\mathbf{y})})$. ■

Exercise 2.3.13. Prove that the function u defined by the Poisson integral formula (4) belongs to $C^\infty(K_R(\mathbf{y}))$ and solves $\Delta u = 0$ in $K_R(\mathbf{y})$ even if φ is merely assumed to be a function integrable on $\Gamma = \partial K_R(\mathbf{y})$. For every point of continuity $\mathbf{z}\in\Gamma$ of φ we have $u(\mathbf{x}) \to \varphi(\mathbf{z})$ $(\mathbf{x}\to\mathbf{z},\ \mathbf{x}\in K_R(\mathbf{y}))$.

Exercise 2.3.14. Let Ω be bounded, and let u_1 and u_2 be harmonic in Ω with boundary values φ_1 and φ_2 on $\Gamma = \partial\Omega$. Prove that:
(a) $\varphi_1 \le \varphi_2$ on Γ implies $u_1 \le u_2$ in Ω.
(b) If, furthermore, Ω is connected and if $\varphi_1(\mathbf{x}) < \varphi_2(\mathbf{x})$ holds for at least one point $\mathbf{x}\in\Gamma$ then it follows that $u_1 < u_2$ everywhere in Ω.

2.4 Continuous Dependence on the Boundary Data

Definition 2.4.1. An abstract problem of the form

$$A(x) = y, \quad x \in \mathbf{X},\ y \in \mathbf{Y},$$

is said to be well-posed if for all $y \in \mathbf{Y}$ it has a unique solution $x \in \mathbf{X}$ and if the latter depends continuously on y.

It is important to recognise whether a mathematical problem is well-posed since otherwise essential difficulties may occur in its numerical solution. In the case of the boundary value problem (1.1a,b) $\mathbf{X} \subset C^2(\Omega) \cap C^0(\overline{\Omega})$ is the space of functions harmonic in Ω and $\mathbf{Y} = C^0(\Gamma)$ is the set of continuous boundary data on $\Gamma = \partial\Omega$.

The topologies of $\mathbf{X}$ and $\mathbf{Y}$ are given by the supremum norms:

$$||u||_\infty := \sup_{\mathbf{x}\in\Omega} |u(\mathbf{x})| \quad \text{and} \quad ||\varphi||_\infty := \sup_{\mathbf{x}\in\Gamma} |\varphi(\mathbf{x})|. \tag{2.4.1}$$

The question of the existence of a solution of (1.1a,b) will have to be postponed (see Section 7). The uniqueness, however, has been confirmed already in Theorem 2.3.8, if Ω is bounded. That the boundedness of Ω cannot be dropped without further ado is shown in

Example 2.4.2. The functions

$$u(x_1, x_2) = x_1 \quad \text{in } \Omega = (0,\infty) \times \mathbb{R} \tag{2.4.2a}$$
$$u(x_1, x_2) = \log(x_1^2 + x_2^2) \quad \text{in } \Omega = \mathbb{R}^2 \backslash K_1(0) \tag{2.4.2b}$$
$$u(x_1, x_2) = \sin(x_1)\sinh(x_2) \quad \text{in } \Omega = (0,\pi) \times (0,\infty) \tag{2.4.2c}$$

and also the trivial $u = 0$, are solutions of the boundary value problem $\Delta u = 0$ in Ω, $u = 0$ on $\Gamma = \partial\Omega$.

For bounded Ω the harmonic functions (solutions of (1.1a,b)) depend not only continuously but also Lipschitz-continuously on the boundary data:

Theorem 2.4.3. *Let Ω be bounded. If u^{I} and u^{II} are solutions of*

$$\Delta u^{\mathrm{I}} = \Delta u^{\mathrm{II}} = 0 \quad \text{in } \Omega, \qquad u^{\mathrm{I}} = \varphi^{\mathrm{I}} \text{ and } u^{\mathrm{II}} = \varphi^{\mathrm{II}} \text{ on } \gamma = \partial\Omega$$

then

$$||u^{\mathrm{I}} - u^{\mathrm{II}}||_\infty \le ||\varphi^{\mathrm{I}} - \varphi^{\mathrm{II}}||_\infty. \tag{2.4.3}$$

PROOF. $v := u^{\mathrm{I}} - u^{\mathrm{II}}$ is a solution of $\Delta v = 0$ in Ω and $v = \varphi^{\mathrm{I}} - \varphi^{\mathrm{II}}$ on Γ. According to Theorem 2.3.8, v takes its maximum and its minimum on Γ:

$$-||\varphi^{\mathrm{I}} - \varphi^{\mathrm{II}}||_\infty \le v(\mathbf{x}) \le ||\varphi^{\mathrm{I}} - \varphi^{\mathrm{II}}||_\infty \quad \text{for all } \mathbf{x} \in \overline{\Omega}.$$

The definition (1) of $\|\cdot\|_\infty$ implies (3). ■

If $\|\varphi_n - \varphi\|_\infty \to 0$ holds for a sequence of boundary values then Theorem 3 shows that the associated solutions satisfy $\|u_n - u\|_\infty \to 0$. The following theorem states that the existence of a solution of $\Delta u = 0$ in Ω, $u = \varphi$ on Γ need not be assumed.

Theorem 2.4.4. *Let Ω be bounded. Let $\varphi_n \in C^0(\Gamma)$ be a sequence of boundary data which converge uniformly to φ: let $\|\varphi_n - \varphi\|_\infty \to 0$. Let u_n be the solution of $\Delta u_n = 0$ in Ω, $u_n = \varphi_n$ on Γ. Then the functions converge uniformly in $\overline{\Omega}$ to $u \in C^2(\Omega) \cap C^0(\overline{\Omega})$, and u solves the boundary value problem $\Delta u = 0$ in Ω, $u = \varphi$ on Γ.*

PROOF. Since according to Theorem 3 $\|u_n - u_m\|_\infty \le \|\varphi_n - \varphi_m\|_\infty$, the sequence u_n is Cauchy convergent. Since $C^0(\overline{\Omega})$ is complete, u_n converges uniformly to a $u \in C^0(\overline{\Omega})$. According to Theorem 2.3.11, u is harmonic (i.e. $\Delta u = 0$). Evidently, $u = \varphi$ is also satisfied on Γ. ■

Another problem, just as important for numerical mathematics, is rarely discussed in the literature: does the solution also depend continuously on the form of the boundary Γ? Figure 1 shows domains Ω' and Ω'' which approximate Ω. A polygonal domain, as shown, for example, in Figure 1c, occurs in the method of finite elements (see Section 8.6).

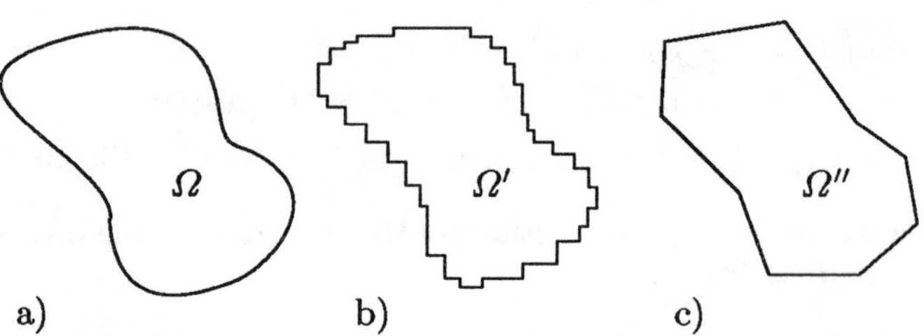

Figure 2.4.1. Approximation of Ω by Ω' and Ω''

Let Ω_n be a sequence of domains with $\Gamma_n = \partial\Omega_n$. We say that $\Gamma_n \to \Gamma$ if $\operatorname{dist}(\Gamma_n, \Gamma) \to 0$. Here, $\operatorname{dist}(\Gamma_n, \Gamma) := \sup\{\operatorname{dist}(\mathbf{x}, \Gamma) : \mathbf{x} \in \Gamma_n\}$, $\operatorname{dist}(\mathbf{x}, \Gamma) := \inf\{|\mathbf{x} - \mathbf{y}| : \mathbf{y} \in \Gamma\}$. Further, one must specify when $\varphi_n \in C^0(\Gamma_n)$ converges uniformly to $\varphi \in C^0(\Gamma)$. We make the definition :
$\varphi_n \in C^0(\Gamma_n)$ converges uniformly to $\varphi \in C^0(\Gamma)$ if, for each $\epsilon > 0$, there exist numbers $N(\epsilon)$ and $\delta(\epsilon) > 0$ such that the following implication holds:

$$n \ge N(\epsilon), \quad \mathbf{x} \in \Gamma, \quad \mathbf{y} \in \Gamma_n, \quad |\mathbf{x} - \mathbf{y}| \le \delta(\epsilon) \Rightarrow |\varphi_n(\mathbf{y}) - \varphi(\mathbf{x})| \le \epsilon. \tag{2.4.4a}$$

The sequence $u_n \in C^0(\Omega_n)$ converges uniformly to $u \in C^0(\Omega)$ if

$$\lim_{n\to\infty} \sup\{|u_n(\mathbf{x}) - u(\mathbf{x})| : \mathbf{x} \in \Omega_n \cap \Omega\} = 0. \tag{2.4.4b}$$

Remark 2.4.5. (a) Let K be a set which is compact (i.e. complete and bounded) with $\Gamma \subset K$, and $\Gamma_n \subset K$ for all n. Let $\phi_n \in C^0(K)$ converge uniformly on K to ϕ. If $\varphi_n = \phi_n$ on Γ_n and $\varphi = \phi$ on Γ then (4a) is satisfied. (b) Let $\Omega_n \subset \Omega$ hold for all n and let $\overline{u}_n$ be the following (not continuous) continuation of u_n onto $\overline{\Omega}$: $\overline{u}_n = u_n$ on $\overline{\Omega}_n$, $\overline{u}_n = u$ on $\overline{\Omega}\backslash\overline{\Omega}_n$. Then (4b) is equivalent to uniform convergence $\overline{u}_n \to u$ on $\overline{\Omega}$ in the usual sense.

Theorem 2.4.6. *Let $\Omega_n \subset \Omega$, with Ω bounded, and let $\Gamma_n \to \Gamma$. Let the functions u_n, which are harmonic in Ω_n, be solutions of*

$$\Delta u_n = 0 \quad \text{in } \Omega_n, \quad u_n = \varphi_n \quad \text{on } \Gamma_n. \tag{2.4.5a}$$

Let $\varphi_n \in C^0(\Gamma_n)$ converge uniformly in the sense of (4a) *to $\varphi \in C^0(\Gamma)$. Then the following assertions hold:*
(a) if there exists a solution $u \in C^2(\Omega) \cap C^0(\overline{\Omega})$ of

$$\Delta u = 0 \quad \text{in } \Omega, \quad u = \varphi \quad \text{on } \Gamma \tag{2.4.5b}$$

then $u_n \to u$ holds in the sense of (4b).
(b) if conversely $u_n \to u \in C^0(\overline{\Omega})$ is satisfied in the sense of (4b), *then u is the solution of* (5b).

PROOF. (a) Let the continuation $\overline{u}_n$ be defined as in Remark 5b. Since u is uniformly continuous on $\overline{\Omega}$, there exists $\delta_u(\epsilon) > 0$ for all $\epsilon > 0$ such that

$$|u(\mathbf{x}) - u(\mathbf{y})| \le \frac{\epsilon}{2}, \quad \text{if } |\mathbf{x} - \mathbf{y}| \le \delta_u(\frac{\epsilon}{2}). \tag{2.4.6a}$$

Set $\delta^*(\epsilon) := \min\left\{\delta_u\left(\frac{\epsilon}{2}\right), \delta\left(\frac{\epsilon}{2}\right)\right\}$ with δ from (4a). Because $\Gamma_n \to \Gamma$ there exists $N_\Gamma(\epsilon)$, so that $\operatorname{dist}(\Gamma_n, \Gamma) \le \delta^*(\epsilon)$ for $n \ge N_\Gamma(\epsilon)$. For $n \ge N^*(\epsilon) := \max\{N_\Gamma(\epsilon), N(\epsilon/2)\}$ (N from (4a)) we want to show that $|\overline{u}_n(\mathbf{x}) - u(\mathbf{x})| \le \epsilon$, for $\mathbf{x} \in \overline{\Omega}$. For $\mathbf{x} \in \overline{\Omega}\backslash\overline{\Omega}_n$ the estimate is trivial because $\overline{u}_n(\mathbf{x}) = u(\mathbf{x})$. For all $\mathbf{x} \in \overline{\Omega}_n \subset \Omega$, however, there holds

$$|\overline{u}_n(\mathbf{x}) - u(\mathbf{x})| = |u_n(\mathbf{x}) - u(\mathbf{x})| \le \max_{\Gamma_n} |u_n - u| \tag{2.4.6b}$$

(cf. Theorem 2.3.8), because $u_n - u$ is harmonic in Ω_n. It remains to estimate $|u_n(\mathbf{x}) - u(\mathbf{x})|$ for $\mathbf{x} \in \Gamma_n$. For $\mathbf{x} \in \Gamma_n$ with $n \ge N^*(\epsilon)$ there exists $\mathbf{y} \in \Gamma$ with $|\mathbf{x} - \mathbf{y}| \le \delta(\epsilon/2)$. Thus we obtain

$$\begin{aligned} |u_n(\mathbf{x}) - u(\mathbf{x})| &= |\varphi_n(\mathbf{x}) - u(\mathbf{x})| \le |\varphi_n(\mathbf{x}) - \varphi(\mathbf{y})| + |\varphi(\mathbf{y}) - u(\mathbf{x})| \le \\ &\le |\varphi_n(\mathbf{x}) - \varphi(\mathbf{y})| + |u(\mathbf{y}) - u(\mathbf{x})| \le \frac{\epsilon}{2} + \frac{\epsilon}{2} = \epsilon \end{aligned}$$

from (4a) and (6a). Since $\mathbf{x} \in \Gamma_n$ is arbitrary it follows that $|u_n - u| \le \epsilon$ on Γ_n, and (6b) proves the uniform convergence $\overline{u}_n \to u$ on $\overline{\Omega}$. Hence, from Remark 5b it follows that (4b) is satisfied.

(b) Let $K \subset \Omega$ be a compact set. Since $\Gamma_n \to \Gamma$ there exists a $N(K)$ such that $K \subset \Omega_n$ for $n \geq N(K)$. Thus, the sequence $\{u_n : n \geq N(K)\}$ converges uniformly in the usual sense on K to u so that one can apply Theorem 2.3.11: consequently u is harmonic in K. Since $K \subset \Omega$ may be chosen arbitrarily, it follows that $u \in C^2(\Omega)$. By assumption, we already have $u \in C^0(\overline{\Omega})$. That the boundary value $u = \varphi$ is taken on Γ is deduced from $\varphi_n \to \varphi$ and $\Gamma_n \to \Gamma$. ∎

In Theorem 4 one was able to derive the existence of a solution u of (5b) just from $\varphi_n \to \varphi$. This inference is not possible for the case of $\Omega_n \neq \Omega$ as the following example shows.

Let $\Omega_n := K_1(\mathbf{0})\backslash\overline{K_{1/n}(\mathbf{0})} \subset \Omega := K_1(\mathbf{0})\backslash\{\mathbf{0}\} \subset \mathbb{R}^2$. The boundaries are $\Gamma_n = \partial K_1(\mathbf{0}) \cup \partial K_{1/n}(\mathbf{0})$, and $\Gamma = \partial K_1(\mathbf{0}) \cup \{\mathbf{0}\}$, and satisfy $\Gamma_n \to \Gamma$. The boundary values

$$\varphi = \varphi_n = 0 \quad \text{on } \partial K_1(\mathbf{0}), \qquad \varphi_n = 1 \quad \text{on } \partial K_{1/n}(\mathbf{0}), \qquad \varphi(0,0) = 1$$

satisfy the condition $\varphi_n \to \varphi$ (cf. (4a) and Remark 5a). The solutions u_n of (5a) can be given explicitly:

$$u_n(\mathbf{x}) = \log(|\mathbf{x}|)/\log(1/n).$$

Obviously, $u_n(\mathbf{x}) \to u(\mathbf{x}) := 0$ holds pointwise, but $u = 0$ satisfies neither (4b) nor the boundary value problem (5b). Conversely, one infers from Theorem 6a the following result:

Remark 2.4.7. In $\Omega = K_1(\mathbf{0})\backslash\{\mathbf{0}\} \subset \mathbb{R}^2$ the potential equation has no solution $u \in C^2(\Omega) \cap C^0(\overline{\Omega})$ which assumes the boundary values $u(\mathbf{x}) = 0$ on $\partial K_1(\mathbf{0})$ and $u(\mathbf{x}) = 1$ in $\mathbf{x} = \mathbf{0}$.

3 The Poisson Equation

3.1 Posing the Problem

The Poisson equation reads

$$\Delta u = f \quad \text{in } \Omega \tag{3.1.1a}$$

with given $f \in C^0(\Omega)$. In the physical interpretation f is the source term [for example, the charge density in the case of an electrical potential u]. To determine the solution uniquely one needs a boundary value specification, for example, the Dirichlet condition

$$u = \varphi \quad \text{on } \Gamma. \tag{3.1.1b}$$

Definition 3.1.1. The function u is called the classical solution of the boundary value problem (1a,b) if $u \in C^2(\Omega) \cap C^0(\overline{\Omega})$ satisfies the equations (1a,b) pointwise.

Until we introduce weak solutions in Section 7, "solution" will always mean "classical solution".

The solution of the boundary value problem (1a,b) will in general no longer satisfy the mean value property and the maximum principle. But these properties still hold for the differences of two solutions u_1 and u_2 of the Poisson equation, since $\Delta(u_1 - u_2) = f - f = 0$. Thus the uniqueness of the solution of problem (1a,b) immediately follows and Theorem 2.4.3 can be brought over:

Theorem 3.1.2. *Let Ω be bounded.* (a) *The solution of* (1a,b) *is uniquely determined.* (b) *If u^{I} and u^{II} are solutions of the Poisson equation for boundary values φ^{I} and φ^{II}, then we have*

$$\|u^{\mathrm{I}} - u^{\mathrm{II}}\|_\infty \le \|\varphi^{\mathrm{I}} - \varphi^{\mathrm{II}}\|_\infty. \tag{3.1.2}$$

PROOF. (b) The proof of Theorem 2.4.3 can be repeated verbatim here.
(a) If u^{I} and u^{II} are two solutions of (1a,b) then (2) shows that $\|u^{\mathrm{I}} - u^{\mathrm{II}}\|_\infty \le \|\varphi - \varphi\|_\infty = 0$. Thus $u^{\mathrm{I}} = u^{\mathrm{II}}$. ■

Theorems 2.4.4 and 2.4.6 can be transferred likewise.

3.2 Representation of the Solution by the Green Function

Lemma 3.2.1. *Let the solution of* (1.1a,b) *belong to* $C^2(\overline{\Omega})$, *where* Ω *is a normal domain. Then* u *may be represented as*

$$u(\mathbf{x}) = -\int_\Omega \gamma(\xi,\mathbf{x}) f(\xi)\, d\xi + \int_\Gamma \left[\gamma(\xi,\mathbf{x})\frac{\partial}{\partial n}u(\xi) - u(\xi)\frac{\partial}{\partial n_\xi}\gamma(\xi,\mathbf{x})\right] d\Gamma_\xi \tag{3.2.1}$$

for every fundamental solution γ *in* (2.2.9).

The proof is the same as in Theorem 2.2.2 or in Corollary 2.2.3. The term $\int_{\Omega_\epsilon}\gamma\Delta u\, d\xi$ with $\Omega_\epsilon = \Omega\backslash\overline{K_\epsilon(\mathbf{x})}$ becomes $\int_{\Omega_\epsilon}\gamma f\, d\xi$. Since the singularity of $\gamma(\xi,\mathbf{x})$ is integrable at $\xi=\mathbf{x}$, $\int_{\Omega_\epsilon}\gamma f\, d\xi$ converges to $\int_\Omega \gamma f\, d\xi$ as $\epsilon\to 0$.

Exercise 3.2.2. (a) Let $\Omega\subset\mathbb{R}^n$ be bounded, $\mathbf{x}_0\in\Omega$, $f\in C^0(\overline{\Omega}\backslash\{\mathbf{x}_0\})$, and $|f(\mathbf{x})|\le C|\mathbf{x}-\mathbf{x}_0|^{-s}$ for $s<n$. Show that $\int_\Omega f(\mathbf{x})\, d\mathbf{x}$ exists as an improper integral.
(b) Let $\Omega\subset\mathbb{R}^n$ be bounded and let $\mathbf{x}_0(\xi)\in\Omega$ depend continuously on $\xi\in D$, with D compact. Let $f(\mathbf{x},\xi)$ be continuous in $(\mathbf{x},\xi)\in\overline{\Omega}\times D$ with $\mathbf{x}\ne\mathbf{x}_0(\xi)$ and let $|f(\mathbf{x},\xi)|\le C|\mathbf{x}-\mathbf{x}_0(\xi)|^{-s}$, $s<n$. Show that $F(\xi) := \int_\Omega f(\mathbf{x},\xi)\, d\mathbf{x}$ is continuous: $F\in C^0(D)$.

In the boundary integral in (1) one may replace $u(\xi)$ by $\varphi(\xi)$ (cf. (1.1b)). The function $\partial u/\partial n$ on Γ, however, is unknown and cannot be specified arbitrarily either, since the boundary values (1.1b) already determine the solution uniquely (cf. Theorem 3.1.2). To make $\int_\Gamma \gamma\partial u/\partial n\, d\xi$ vanish one must select the fundamental solution so that $\gamma(\xi,\mathbf{x})=0$ for $\xi\in\Gamma$, $\mathbf{x}\in\Omega$.

Definition 3.2.3. A fundamental solution g in (2.2.9) is called a Green function (of the first kind) if $g(\xi,\mathbf{x})=0$ for all $\xi\in\Gamma$, $\mathbf{x}\in\Omega$.

The existence of a Green function is closely related to the solvability of the boundary value problem for the potential equation:

Remark 3.2.4. The Green function exists if and only if for all $\mathbf{x}\in\Omega$ the boundary value problem $\Delta\Phi=0$ in Ω and $\Phi=-s(\cdot,\mathbf{x})$ on Γ has a solution $\Phi\in C^2(\overline{\Omega})$.

The above consideration results in

Theorem 3.2.5. *Let* Ω *be a normal domain. Let the boundary value problem* (1.1a,b) *have a solution* $u\in C^2(\overline{\Omega})$. *Assume the existence of a Green function of the first kind. Then one can express* u *explicitly by*

$$u(\mathbf{x}) = -\int_{\Omega} g(\xi,\mathbf{x})f(\xi)\,d(\xi) - \int_{\Gamma} \varphi(\xi)\frac{\partial}{\partial n_\xi}g(\xi,\mathbf{x})\,d\Gamma_\xi. \tag{3.2.2}$$

In the following we reverse the implication. Let the existence of the Green function be assumed. Then, does function u defined by Equation (2) represent the classical solution of the boundary value problem (1.1a,b)? Here it must be proved, in particular, that $u \in C^2(\Omega)$ and $\Delta u = f$. Firstly, it is not even clear yet whether the function $u(\mathbf{x})$ defined by Equation (2) depends continuously on $\mathbf{x}$ since the definition of a fundamental solution $\gamma(\xi,\mathbf{x})$ does not require continuity with respect to the second argument $\mathbf{x}$. Despite that, the Green function is $g(\xi,\mathbf{x})$ with respect to $\mathbf{x}$ in $C^2(\Omega\backslash\{\xi\})$, as the following result shows (cf. Leis [1, p. 67]).

Exercise 3.2.6. Let Ω be a normal domain. Let the Green function exist for Ω, and for fixed $\mathbf{y} \in \Omega$ let $g(\cdot,\mathbf{y}) \in C^2(\overline{\Omega}\backslash\{\mathbf{y}\})$ (weaker conditions are possible !). Now prove that

$$g(\mathbf{x},\mathbf{y}) = g(\mathbf{y},\mathbf{x}) \quad \text{for } \mathbf{x},\mathbf{y} \in \Omega. \tag{3.2.3}$$

Hint: Apply the Green formula (2.2.5b) with $\Omega_\epsilon = \Omega\backslash[K_\epsilon(\mathbf{x}') \cup K_\epsilon(\mathbf{x}'')]$, $\mathbf{x}',\mathbf{x}'' \in \Omega$, $u(\mathbf{x}) := g(\mathbf{x},\mathbf{x}')$, $v(\mathbf{x}) := g(\mathbf{x},\mathbf{x}'')$, and use (2.2.10).

If one tries to reverse the assertion of Theorem 5, one encounters the surprising difficulty of having to set precise conditions on the source term f. The natural requirement $f \in C^0(\overline{\Omega})$ is necessary for $u \in C^2(\overline{\Omega})$, but it is not sufficient, as the following theorem, whose proof will be appended at the end of this section, shows.

Theorem 3.2.7. *Even if the boundary Γ and the boundary values φ are sufficiently smooth and if the Green function exists, there are functions $f \in C^0(\overline{\Omega})$ to which no solutions $u \in C^2(\overline{\Omega})$ correspond.*

Theorem 7 shows that Equation (2) need not represent a classical solution for $f \in C^0(\overline{\Omega})$. However, a sufficient condition for f to do so is Hölder continuity.

Definition 3.2.8. $f \in C^0(\overline{\Omega})$ is said to be Hölder continuous in $\overline{\Omega}$ with the exponent $\lambda \in (0,1)$ if there exists a constant $C = C(f)$ such that

$$|f(\mathbf{x}) - f(\mathbf{y})| \le C|\mathbf{x}-\mathbf{y}|^\lambda \quad \text{for all } \mathbf{x},\mathbf{y} \in \overline{\Omega}. \tag{3.2.4a}$$

We write $f \in C^\lambda(\overline{\Omega})$ and define the norm $\|f\|_{C^\lambda(\overline{\Omega})}$ as the smallest constant C which satisfies (4a):

$$\|f\|_{C^\lambda(\overline{\Omega})} := \sup\left\{\frac{|f(\mathbf{x}) - f(\mathbf{y})|}{|\mathbf{x}-\mathbf{y}|^\lambda} : \mathbf{x},\mathbf{y} \in \overline{\Omega}, \mathbf{x} \neq \mathbf{y}\right\}. \tag{3.2.4b}$$

The function $f \in C^k(\overline{\Omega})$ is said to be k-fold Hölder continuously differentiable in $\overline{\Omega}$ (with the exponent λ), if $D^\nu f \in C^\lambda(\overline{\Omega})$ for all $|\nu| \le k$. Here

$$\nu = (\nu_1, \dots, \nu_n) \quad \text{with } \nu_i \in \mathbb{Z}, \quad \nu_i \ge 0, \quad |\nu| = \nu_1 + \cdots + \nu_n \tag{3.2.5a}$$

is a multi-index and

$$D^\nu = \frac{\partial^{|\nu|}}{\partial x_1^{\nu_1} \partial x_2^{\nu_2} \cdots \partial x_n^{\nu_n}} \tag{3.2.5b}$$

a $|\nu|$-fold partial derivative operator. The k-fold Hölder continuously differentiable functions form the linear space $C^{k+\lambda}(\overline{\Omega})$ with the norm

$$||f||_{C^{k+\lambda}(\overline{\Omega})} := \max\{||D^\nu f||_{C^\lambda(\overline{\Omega})} \colon |\nu| \le k\}. \tag{3.2.4c}$$

If $s = k + \lambda$ one also writes $C^s(\overline{\Omega})$ for $C^{k+\lambda}(\overline{\Omega})$. The k-fold Lipschitz continuously differentiable functions $f \in C^{k,1}(\overline{\Omega})$ are the result of the choice $\lambda = 1$ in (4a,b). For reasons of completeness let us add that

$$||f||_{C^k(\overline{\Omega})} = \max\{||D^\nu f||_\infty \colon |\nu| \le k\} \tag{3.2.4d}$$

is the norm in $C^k(\overline{\Omega})$ for integer $k \ge 0$.

Exercise 3.2.9. (a) f is said to be locally Hölder continuous in Ω if for each $\mathbf{x} \in \Omega$ there exists a neighborhood $K_\epsilon(\mathbf{x})$ such that $f \in C^\lambda(K_\epsilon(\mathbf{x}) \cap \Omega)$. Prove that if $\overline{\Omega}$ is compact then $f \in C^\lambda(\overline{\Omega})$ follows from the local Hölder continuity in $\overline{\Omega}$. Formulate and prove corresponding statements for $C^{k+\lambda}(\overline{\Omega})$ and $C^{k,1}(\overline{\Omega})$.
(b) Let $s > 0$. Show that $|\mathbf{x}|^s \in C^s(\overline{K_R(\mathbf{0})})$, if $s \notin \mathbb{N}$, otherwise $|\mathbf{x}|^s \in C^{s-1,1}(\overline{K_R(\mathbf{0})})$. Hint: $1 - t^s \le (1-t)^s$ for $0 \le t \le 1$, $s \ge 0$.

The function u from Equation (2) can be decomposed into $u_1 + u_2$ where $u_1 = -\int_\Omega gf\, d\xi$ and $u_2 = -\int_\Gamma \varphi \partial g/\partial n\, d\Gamma$; u is the solution of the boundary value problem (1.1a,b) if we are able to show that u_1 and u_2 are solutions of

$$\Delta u_1 = f \text{ in } \Omega, \quad u_1 = 0 \text{ on } \Gamma, \quad \Delta u_2 = 0 \text{ in } \Omega, \quad u_2 = \varphi \text{ on } \Gamma.$$

Theorem 3.2.10. *If the Green function exists and satisfies suitable conditions then*

$$u(\mathbf{x}) = -\int_\Gamma \varphi(\xi) \frac{\partial}{\partial n_\xi} g(\xi, \mathbf{x})\, d\Gamma_\xi \tag{3.2.6}$$

is a classical solution of $\Delta u = 0$ in Ω, with $u = \varphi$ on Γ.

The proof goes in principle just as for Theorem 2.3.9 (cf. Leis [1, p.69]).

Theorem 3.2.11. *Suppose the Green function $g(\cdot,\mathbf{x}) \in C^2(\overline{\Omega}\backslash\{\mathbf{x}\})$ for $\mathbf{x} \in \Omega$ exists, and let it be $f \in C^\lambda(\overline{\Omega})$. Then*

$$u(\mathbf{x}) = -\int_\Omega f(\xi)g(\xi,\mathbf{x})\,d\xi \tag{3.2.7}$$

is a classical solution of $\Delta u = f$ in Ω, with $u = 0$ on Γ.

PROOF. The boundary condition $u(\mathbf{x}) = 0$ for $\mathbf{x} \in \Gamma$ follows easily from $g(\mathbf{x},\xi) = 0$ and (3). The property $u \in C^1(\overline{\Omega})$ and the representation $u_{x_i}(\mathbf{x}) = -\int_\Omega f(\xi)g_{x_i}(\xi,\mathbf{x})\,d\xi$ result from the

Exercise 3.2.12. Let $\Omega \subset \mathbb{R}^n$ be bounded and $A := \{(\xi,\mathbf{x}) \in \overline{\Omega}\times\overline{\Omega} : \xi \neq \mathbf{x}\}$. For the derivatives of f with respect to $\mathbf{x}$ assume $D^\nu_{\mathbf{x}} f \in C^0(A)$ and $|D^\nu_{\mathbf{x}} f(\xi,\mathbf{x})| \le C|\mathbf{x}-\xi|^{-s}$ with $s < n$ for any $|\nu| \le k$. Prove that then $F(\mathbf{x}) := \int_\Omega f(\xi,\mathbf{x})\,d\xi \in C^k(\overline{\Omega})$ and $D^\nu F(\mathbf{x}) = \int_\Omega D^\nu_{\mathbf{x}} f(\xi,\mathbf{x})\,d\xi$, $|\nu| \le k$.

To prove $u \in C^2(\overline{\Omega})$ this step cannot be repeated since $g_{x_ix_j}(\xi,\mathbf{x}) = O(|\xi-\mathbf{x}|^{-n})$ has a singularity which is not integrable. We write the derivative in the form

$$u_{x_i}(\mathbf{x}) = -\int_\Omega [f(\xi)-f(\mathbf{x})]g_{x_i}(\xi,\mathbf{x})\,d\xi - f(\mathbf{x})\int_\Omega g_{x_i}(\xi,\mathbf{x})\,d\xi. \tag{3.2.8a}$$

Let $\partial_j F(\mathbf{x})$ be the difference quotient $(F(\mathbf{x}^\epsilon)-F(\mathbf{x}))/\epsilon$, with $x_j^\epsilon = x_j+\epsilon$ and $x_i^\epsilon = x_i$ for $i \neq j$. The product rule $\partial_j(FG)(\mathbf{x}) = G(\mathbf{x})\partial_j F(\mathbf{x}) + F(\mathbf{x}^\epsilon)\partial_j G(\mathbf{x})$ applied to Equation (8a) gives

$$\partial_j u_{x_i}(\mathbf{x}) = -\int_\Omega [f(\xi)-f(\mathbf{x}^\epsilon)]\partial_j g_{x_i}(\xi,\mathbf{x})\,d\xi - f(\mathbf{x}^\epsilon)\partial_j\frac{\partial}{\partial x_i}\int_\Omega g(\xi,\mathbf{x})\,d\xi.$$

Since $[f(\xi)-f(\mathbf{x})]g_{x_ix_j}(\xi,\mathbf{x}) = O(|\xi-\mathbf{x}|^{\lambda-n})$ is integrable, the limit $\epsilon \to 0$ results in the formula

$$u_{x_ix_j}(\mathbf{x}) = -\int_\Omega [f(\xi)-f(\mathbf{x})]g_{x_ix_j}(\xi,\mathbf{x})\,d\xi - f(\mathbf{x})\frac{\partial^2}{\partial x_i\partial x_j}\int_\Omega g(\xi,\mathbf{x})\,d\xi. \tag{3.2.8b}$$

Equation (8b) implies

$$\Delta u = \int_\Omega [f(\xi)-f(\mathbf{x})]\Delta g\,d\xi - f(\mathbf{x})\Delta\int_\Omega g\,d\xi = -f(\mathbf{x})\Delta\int_\Omega g(\xi,\mathbf{x})\,d\xi,$$

so that it remains only to show that $\Delta\int_\Omega g\,d\xi = -1$. Choose $K_R(\mathbf{z})$ so that $\mathbf{x} \in K_R(\mathbf{z}) \subset \Omega$. The Green function has the form (2.2.9): $g = s+\Phi$. The first two terms in

$$\int_\Omega g(\xi,\mathbf{x})\,d\xi = \int_{\Omega\backslash K_R(\mathbf{z})} g(\xi,\mathbf{x})\,d\xi + \int_{K_R(\mathbf{z})} \Phi(\xi,\mathbf{x})\,d\xi + \int_{K_R(\mathbf{z})} s(\xi,\mathbf{x})\,d\xi$$

are harmonic in $K_R(\mathbf{z})$, so that $\Delta\int_{K_R(\mathbf{z})} s(\xi,\mathbf{x})\,d\xi = -1$ is what has to be proved.

Let $\sigma(r)$ be defined by $s(\xi,\mathbf{x}) = \sigma(|\xi-\mathbf{x}|)$ (cf. (2.2.1)). For fixed $r>0$ set

$$v(\mathbf{x}) := \frac{1}{\omega_n r^{n-1}} \int_{\partial K_r(\mathbf{z})} s(\xi,\mathbf{z})\, d\Gamma_\xi. \tag{3.2.8c}$$

For all $\mathbf{x} \notin \partial K_R(\mathbf{z})$ (i.e., $|\mathbf{x}-\mathbf{z}| \neq r$) v is harmonic, since $s(\xi,\mathbf{x})$ is nonsingular on $\partial K_r(\mathbf{z})$ and satisfies $\Delta_{\mathbf{x}} s(\xi,\mathbf{x}) = 0$. Since $s(\cdot,\mathbf{x})$ is harmonic in $K_r(\mathbf{z})$ for $r < |\mathbf{z}-\mathbf{x}|$, the mean-value property (2.3.1) holds, which can now be written

$$v(\mathbf{x}) = s(\mathbf{z},\mathbf{x}) = \sigma(|\mathbf{z}-\mathbf{x}|) \quad \text{for} \quad |\mathbf{z}-\mathbf{x}| > r. \tag{3.2.8d}$$

Using Exercise 3.2.2b we see that $v(\mathbf{x})$ is continuous in $\mathbb{R}^n$, so that we also have

$$v(\mathbf{x}) = \sigma(r) \quad \text{for} \quad |\mathbf{z}-\mathbf{x}| = r. \tag{3.2.8e}$$

Thus v is harmonic in $K_r(\mathbf{x})$ with the constant boundary values (8e). The unique solution is therefore

$$v(\mathbf{x}) = \sigma(r) \quad \text{for} \quad |\mathbf{z}-\mathbf{x}| \le r. \tag{3.2.8f}$$

The equations (8c,d,f) yield

$$\int_{\partial K_r(\mathbf{z})} s(\xi,\mathbf{x})\, d\Gamma_\xi = \omega_n r^{n-1} \sigma(\max\{r, |\mathbf{z}-\mathbf{x}|\})$$

and then, since $0 < |\mathbf{z}-\mathbf{x}| < R$,

$$\int_{K_R(\mathbf{z})} s(\xi,\mathbf{x})\, d\xi = \int_0^R \int_{\partial K_r(\mathbf{z})} s(\xi,\mathbf{x})\, d\Gamma_\xi\, dr$$

$$\omega_n \int_0^{|\mathbf{z}-\mathbf{x}|} r^{n-1}\sigma(|\mathbf{z}-\mathbf{x}|)\, dr + \omega_n \int_{|\mathbf{z}-\mathbf{x}|}^R r^{n-1}\sigma(r)\, dr$$

$$= \omega_n \frac{|\mathbf{z}-\mathbf{x}|^n}{n}\sigma(|\mathbf{z}-\mathbf{x}|) + \omega_n \frac{r^n}{n}\sigma(r)\Big|_{|\mathbf{z}-\mathbf{x}|}^R - \omega_n \int_{|\mathbf{z}-\mathbf{x}|}^R \frac{r^n}{n}\sigma'(r)\, dr$$

$$\frac{\omega_n}{n} R^n \sigma(R) - \omega_n \int_{|\mathbf{z}-\mathbf{x}|}^R \frac{r^n}{n}\left(-\frac{r^{1-n}}{\omega_n}\right) dr = \frac{\omega_n}{n} R^n\sigma(R) + \int_{|\mathbf{z}-\mathbf{x}|}^R \frac{r}{n}\, dr$$

$$= \frac{\omega_n}{n} R^n \sigma(R) + \frac{R^2}{2n} - \frac{|\mathbf{z}-\mathbf{x}|^2}{2n}.$$

From this we see that, independently of R, n, and $\mathbf{z}$, there results

$$\Delta \int_{K_R(\mathbf{z})} s(\xi,\mathbf{x})\, d\xi = -1.$$

■

From Theorems 10 and 11 follows

Theorem 3.2.13. *Under the same assumptions as for Theorems* 10 *and* 11 *Equation* (2) *gives a representation for the classical solution of the boundary value problem* (1.1a,b).

Finally we put forward two inequalities for the Green function as exercises:

Exercise 3.2.14. In Ω, and $\Omega_1 \subset \Omega_2$, respectively, let the Green functions g, g_1 and g_2 exist. Show

(a) $\quad 0 \le g(\mathbf{x},\mathbf{y}) \le s(\mathbf{x},\mathbf{y}) \quad$ for $\mathbf{x},\mathbf{y} \in \Omega \subset \mathbb{R}^n, \quad n \ge 3$ (3.2.9)

What is the inequality for n=2?

(b) $\quad g_1(\mathbf{x},\mathbf{y}) \le g_2(\mathbf{x},\mathbf{y}) \quad$ for $\mathbf{x},\mathbf{y} \in \Omega_1 \subset \Omega_2$ (3.2.10)

Hint: Exercise 2.3.14.

Supplement: Proof of Theorem 7. If we make use of a later theorem (Theorem 6.1.13) then Theorem 7 follows from

Theorem 3.2.15. *The solution u does not depend continuously on f, if the supremum norm* (2.4.1) *is used as the norm in $C^0(\overline{\Omega})$, and that from* (4d) *is used as the norm in $C^2(\overline{\Omega})$.*

PROOF. Let $\Omega = K_1(0) \subset \mathbb{R}^2$ und $\varphi = 0$. The disk Ω is a normal region for which the Green function is known (cf. Theorem 3.3.1). By Theorem 11 there exist solutions $u^n \in C^2(\overline{\Omega})$ of $\Delta u^n = f_n$ in Ω, $u^n = 0$ on Γ for the functions

$$f_n(\mathbf{x}) = \frac{x_2^2 - x_1^2}{r^2}\rho_n(r), \; r := |\mathbf{x}|, \quad \rho_n(r) := \min\left(n \cdot r, \left|\log\frac{r}{2}\right|^{-1}\right),$$

which belong to $C^0(\overline{\Omega})$ and are uniformly bounded: $||f_n||_\infty = 1/\log 2$. By Theorem 11 we have $u^n(\mathbf{x}) = -\int_\Omega g(\xi,\mathbf{x}) f_n(\xi)\,d\xi$. Since $|f_n(\xi)| \le n|\xi|$, it follows from Exercise 12 that

$$u^n_{x_1x_1} = -\int_\Omega g_{x_1x_1} f_n\,d\xi = -\int_\Omega \Phi_{x_1x_1} f_n\,d\xi - \int_\Omega s_{x_1x_1} f_n\,d\xi \text{ at } \mathbf{x} = 0,$$

where $g = \Phi + s$. The first integral is bounded since $\Phi \in C^2(\overline{\Omega})$. The derivative of the singularity function is $s_{x_1x_1}(\xi, \mathbf{0}) = (\xi_1^2 - \xi_2^2)/|\xi|^4$. The special choice of f_n gives for $\mathbf{x} = \mathbf{0}$

$$u^n_{x_1x_1}(\mathbf{0}) = -\int_\Omega \Phi_{x_1x_1}(\xi,\mathbf{0}) f_n(\xi)\,d\xi + \int_\Omega [\xi_1^2 - \xi_2^2]^2 |\xi|^{-6} \rho_n(|\xi|)\,d\xi.$$

The surface integral $K := \int_{\partial K_r(0)} [\xi_1^2 - \xi_2^2]^2 |\xi|^{-5}\,d\xi > 0$ does not depend on $r \in (0,1]$, so that the second integral takes on the form $I_n := K\int_0^1 r^{-1}\rho_n(r)\,dr$. Since $\int_\epsilon^1 [r|\log(r/2)|]^{-1}\,dr$ diverges as $\epsilon \to 0$, we deduce $I_n \to \infty$ as $n \to \infty$. Since $||u^n||_{C^2(\overline{\Omega})} \ge |u^n_{x_1x_1}(\mathbf{0})|$, it follows that the map $f \mapsto u$ is not bounded, and thus not continuous. ■

3.3 The Green Function for the Ball

Theorem 3.3.1. *The Green function for the ball* $K_R(\mathbf{y})$ *is given by the function in* (2.2.11a). *For* $f \in C^\lambda(\overline{\Omega})$ *and* $\varphi \in C^{2+\lambda}(\Gamma)$ *with* $0 < \lambda < 1$, *the representation formula* (2.2) *defines a solution* $u \in C^{2+\lambda}(\overline{\Omega})$ *of the boundary value problem* $\Delta u = f$ *in* Ω, $u = \varphi$ *on* Γ.

The proof of the theorem follows from a result of Schauder, which is cited in Theorem 9.1.20.

In the case $n = 2$ the plane $\mathbb{R}^2$ can be identified with $\mathbb{C}$ by the correspondence $(x, y) \leftrightarrow z = x + iy$. The following considerations are based on

Exercise 3.3.2. Let the map $\varphi: z = x + iy \in \Omega \mapsto \zeta = \xi + i\eta = \Phi(z) \in \Omega'$ be holomorphic. Show

$$\Delta_z u(\Phi(z)) = |\Phi'|^2 \Delta_\zeta u(\zeta), \qquad \Phi' = \xi_x + i\eta_x \tag{3.3.1}$$

for $u \in C^2(\Omega')$.

Equation (1) shows, in particular, that a holomorphic transformation of coordinates maps harmonic functions into harmonic functions. An arbitrary simply connected region with at least two boundary points can, by the Riemann mapping theorem, be mapped by a conformal mapping $\Phi_{z_0}: z \in \Omega \mapsto \Phi_{z_0}(z) \in K_1(0)$ onto the unit disk such that $\Phi_{z_0}(z_0) = 0$ for any given $z_0 \in \Omega$. Let $g(\zeta, \zeta')$ be the Green function for $K_1(0)$. One may check that $G(z, z_0) := g(\Phi_{z_0}(z), 0)$ is again a fundamental solution. Now $z \in \partial\Omega$ implies $\Phi_{z_0}(z) \in \partial K_1(0)$, i.e., $G(z, z_0) = 0$. Thus $G(z, z_0)$ is the Green function in Ω. This proves

Theorem 3.3.3. *Let* $\Omega \subset \mathbb{R}^2$ *be simply connected with at least two boundary points. Then there exists a Green function of the first kind for* Ω.

The explicit forms of various Green functions can be found, for example, in the book by Wloka [1, Exercises 21.1–21.8]. Of numerical interest might be the fact that with conformal mapping one may remove corners which are disturbing (e.g., reentrant corners) (cf. Gladwell-Wait [1, p.70]).

Example 3.3.4. Let Ω be the L-shaped region in Example 2.1.4. Choose $\Phi = z^{2/3}: \Omega \to \Omega'$. Then Φ is conformal in Ω. The sides of the angle $\Gamma_0 \subset \partial\Omega$ (cf. Fig. 2.1.1) are mapped into a single line segment, so that Ω' has no more corners sticking in. The Poisson equation $\Delta u = f$ in Ω corresponds to the equation $\Delta v(\zeta) = \frac{9}{4}|\zeta| f(\zeta)$ in Ω'.

3.4 The Neumann Boundary Value Problem

In (1.1b) and in (2.1.1b) the boundary values $u = \varphi$ were given on Γ. These so-called Dirichlet conditions or "Boundary Conditions of the First Kind" are not the only possibility. An alternative is the Neumann Condition

$$\frac{\partial}{\partial n}u(\mathbf{x}) = \varphi(\mathbf{x}) \qquad \text{on } \Gamma. \tag{3.4.1}$$

In physics this second boundary condition, as it is also called, occurs more frequently than the Dirichlet condition. For example, if u is the velocity potential of a gas, then $\frac{\partial u}{\partial n} = 0$ means that the gas can only move tangentially at the boundary Γ. Except in some unusual cases, the boundary value problem $Lu = f$ in Ω and $\partial u/\partial n = \varphi$ on Γ has a unique solution. An exceptional case does however occur for $L = \Delta$:

Theorem 3.4.1. *Let Ω be a normal region. The Poisson equation $\Delta u = f$ with the Neumann boundary condition* (1) *is only solvable if*

$$\int_\Gamma \varphi(\xi)\, d\Gamma_\xi = \int_\Omega f(\mathbf{x})\, d\mathbf{x}. \tag{3.4.2}$$

If a solution u does exist, then $u + c$, with c any constant, is also a solution.

PROOF. (1) One may repeat the proof of Lemma 2.3.6 for $\Delta u = f$.
(2) Obviously $u + c$ satisfies the same equation. ■

Later, in Example 7.4.8, we will show that the Neumann boundary value problem for the Poisson equation has a solution if and only if (2) is satisfied, and that two solutions can differ only by a constant.

In the representation (2.1) both the values $u(\xi)$, for $\xi \in \Gamma$, and the normal derivative $\partial u/\partial n$ occur. The Green function of the first kind was chosen in such a way that $g(\xi, \mathbf{x}) = 0$ for $\xi \in \Gamma$. In the case of the second boundary conditions (1) one makes the assumption that $\partial\gamma(\xi, \mathbf{x})/\partial n_\xi = c$ (c: constant), i.e.

$$\partial\Phi(\xi, \mathbf{x})/\partial n_\xi = c - \partial s(\xi, \mathbf{x})/\partial n_\xi$$

for $\Phi = \gamma - s$. The Corollary 2.2.3 with $u \equiv 1$ and $\gamma = s$ gives

$$\int_\Gamma \partial\Phi(\xi, \mathbf{x})/\partial n_\xi \, d\Gamma_\xi = cL + 1, \qquad L := \int_\Gamma d\Gamma.$$

Since Φ must be harmonic (i.e., $f := \Delta\Phi = 0$), from Equation (2) we see that $cL + 1 = 0$ is a necessary condition for the existence of Φ. Thus the condition on the Green Function of the Second Kind for the potential equation is

$$\partial\gamma(\mathbf{x}, \xi)/\partial n_\mathbf{x} = -1/\int_\Gamma d\Gamma.$$

Thus the term $\int_\Gamma u\partial\gamma/\partial n\, d\Gamma$ in (2.1) becomes const $\cdot \int_\Gamma u\, d\Gamma$. Since u is only determined up to a constant (cf. Theorem 1), one can fix this constant with the additional condition $\int_\Gamma u\, d\Gamma = 0$. This gives the following result, if we write g for γ:

$$u(\mathbf{x}) = -\int_\Omega f(\xi)g(\xi,\mathbf{x})\, d\xi + \int_\Gamma \varphi(\xi)g(\xi,\mathbf{x})\, d\Gamma_\xi.$$

The Green function of the second kind for the ball $K_R(\mathbf{0}) \subset \mathbb{R}^3$ can be found in Leis [1, p. 79].

3.5 The Integral Equation Method

In the representation (2.1) of the Poisson solution the singularity function s can, in particular, be chosen to be γ. If in addition one imposes the given Neumann data (4.1), one obtains

$$u(x) = \int_\Gamma k(\mathbf{x},\xi)u(\xi)\, d\Gamma_\xi + g(\mathbf{x}) \text{ for } \mathbf{x} \in \Omega, \tag{3.5.1}$$

with the kernel function $k(\mathbf{x},\cdot) := -\partial s(\xi,x)/n_\xi$ and the functions

$$g(\mathbf{x}) := g_1(\mathbf{x}) + g_2(\mathbf{x}),$$

$$g_1(\mathbf{x}) := \int_\Gamma s(\xi,x)\varphi(\xi)\, d\Gamma_\xi$$

$$g_2(\mathbf{x}) := \int_\Omega s(\xi,x)f(\xi)\, d\Gamma_\xi.$$

The right-hand side in Equation (1) with the unknown boundary value $u(\xi)$, $\xi \in \Gamma$, can be used as an ansatz solution:

$$\Phi(\mathbf{x}) = \int_\Gamma k(\mathbf{x},\xi)u(\xi)\, d\Gamma_\xi + g(\mathbf{x}). \tag{3.5.2}$$

The first summand on the right of (2) is called the double-layer potential (dipole potential); g_1 is the single-layer potential, while g_2 is a volume potential.

For each $u \in C^0(\Gamma)$ the Φ in (2) is a solution of the Poisson equation (1.1a) in Ω. However, Φ is also defined for an argument $\mathbf{x}$ in the exterior domain $\mathbb{R}^n\backslash\Omega$. A closer look at the kernel function $k(\mathbf{x},\xi)$ shows that it is in fact only weakly singular for the case of smooth boundaries Γ. Thus Φ is also defined for $\mathbf{x} \in \Gamma$. The function Φ which is now defined on all $\mathbb{R}^n$ is not continuous at points of the boundary Γ. At $\mathbf{x}_0 \in \Gamma$ there exists both an interior limit $\Phi_-(\mathbf{x}_0)$ for $\mathbf{x} \to \mathbf{x}_0, \mathbf{x} \in \Omega$ and an exterior limit $\Phi_+(\mathbf{x}_0)$ for $\mathbf{x} \to \mathbf{x}_0, \mathbf{x} \in \mathbb{R}^n\backslash\overline{\Omega}$. In addition we have the third function value $\Phi(\mathbf{x}_0)$ of (2). Their connection is given by the following jump discontinuity relation (cf. Hackbusch [7, Satz 8.2.8]):

$$\Phi_+(\mathbf{x}_0) - \Phi_-(\mathbf{x}_0) = 2u(\mathbf{x}_0), \tag{3.5.3a}$$
$$\Phi_+(\mathbf{x}_0) + \Phi_-(\mathbf{x}_0) = 2\Phi(\mathbf{x}_0), \quad \text{for } x_0 \in \Gamma. \tag{3.5.3b}$$

In order that the ansatz (2) does indeed give the solution u in (1), the boundary value Φ_-, continued from the interior, must agree with the function u which is put in the integral: $\Phi_- = u$. Now one can solve Equation (3a,b) for Φ_-: $\Phi_-(\mathbf{x}_0) = \Phi(\mathbf{x}_0) - u(\mathbf{x}_0)$. The equation $\Phi_- = u$ thus leads to

$$u(\mathbf{x}_0) = \frac{1}{2}\Phi(\mathbf{x}_0) = \int_\Gamma k(\mathbf{x}_0, \xi)u(\xi)\, d\Gamma_\xi + g(\mathbf{x}_0) \text{ for } \mathbf{x}_0 \in \Gamma. \tag{3.5.4}$$

Equation (4) is called a Fredholm integral equation of the second kind for the unknown function $u \in C^0(\Gamma)$. The original Neumann boundary-value problem (1.1a), (4.1) and the integral eqation (4) are equivalent in the following sense: (a) If u is the solution of the Neumann boundary-value problem, then the boundary values, $u(\xi)$, $\xi \in \Gamma$, satisfy the integral equation (4). (b) If $u \in C^0(\Gamma)$ is a solution of the integral equation (4), then the expression (1) gives a solution of (1.1a), (4.1) in the entire domain Ω.

The transformation of a boundary-value problem into an integral equation, and the subsequent solution of the integral equation is referred to as the integral equation method. It allows, for example, a new approach to existence statements, in that one shows the solvability of (4). The integral equation (4) can also be attacked numerically. If methods similar to the finite-element method described in Section 8 are used, then the result is called the boundary-element method (BEM).

One can find references to the integral equation method in, e.g., Hackbusch [7, §§7–9] and Kress[1].

4 Difference Methods for the Poisson Equation

4.1 Introduction: The One-Dimensional Case

Before developing difference methods for the partial differential Poisson equation, let us first recall the discretisation of ordinary differential equations. The equation $a(x)u''(x) + b(x)u'(x) + c(x)u(x) = f(x)$ can be supplemented with initial conditions $u(x_1) = u_1, u'(x_1) = u'_1$ or with boundary conditions $u(x_1) = u_1, u(x_2) = u_2$. The ordinary initial value problems correspond to the hyperbolic and parabolic initial value problems, while an ordinary boundary value problem may be viewed as an elliptic boundary value problem in one variable. In particular one can view

$$-u''(x) = f(x) \quad \text{for} \quad x \in (0,1), \tag{4.1.1a}$$

$$u(0) = \varphi_0, \quad u(1) = \varphi_1 \tag{4.1.1b}$$

as the one-dimensional Poisson equation $-\Delta u = f$ in the domain $\Omega = (0,1)$ with Dirichlet conditions on the boundary $\Gamma = \{0, 1\}$.

Difference methods are characterised by the fact that derivatives are replaced by difference quotients (divided differences), in the following called, for short, "differences" . The first derivative $u'(x)$ can be approximated by several (so-called "first") differences, for example, by the forward or right difference:

$$(\partial^+ u)(x) := [u(x+h) - u(x)]/h, \tag{4.1.2a}$$

the backward or left difference

$$(\partial^- u)(x) := [u(x) - u(x-h)]/h, \tag{4.1.2b}$$

or the symmetric difference

$$(\partial^0 u)(x) := [u(x+h) - u(x-h)]/(2h), \tag{4.1.2c}$$

where $h > 0$ is called the step size. An obvious second difference for $u''(x)$ is

$$(-\partial^- \partial^+ u)(x) := [u(x+h) - 2u(x) + u(x-h)]/h^2. \tag{4.1.3}$$

One also calls ∂^+, ∂^-, ∂^0, and $\partial^-\partial^+$ index/ D7/emphdifference operators. The product $\partial^-\partial^+$ may be viewed as $\partial^- \circ \partial^+$ or as $\partial^+ \circ \partial^-$, i.e.. $(\partial^+\partial^-)u(x) = \partial^+(\partial^- u(x))$.

Lemma 4.1.1. *Let* $[x-h, x+h] \subset \overline{\Omega}$. *Then*

$$\partial^{\pm}u(x) = u'(x) + hR \text{ with } |R| \le \frac{1}{2}\|u\|_{C^2(\overline{\Omega})} \text{ if } u \in C^2(\overline{\Omega}) \quad (4.1.4a)$$

$$\partial^{0}u(x) = u'(x) + h^2R \text{ with } |R| \le \frac{1}{6}\|u\|_{C^3(\overline{\Omega})} \text{ if } u \in C^3(\overline{\Omega}) \quad (4.1.4b)$$

$$\partial^{-}\partial^{+}u(x) = u''(x) + h^2R \text{ with } |R| \le \frac{1}{12}\|u\|_{C^4(\overline{\Omega})} \text{ if } u \in C^4(\overline{\Omega}). (4.1.4c)$$

PROOF. We give the proof only for (4c). If one applies Taylor's formula

$$u(x \pm h) = u(x) \pm hu'(x) + h^2u''(x)/2 \pm h^3u'''(x)/6 + h^4R_4, \qquad (4.1.5a)$$

$$R_4 = h^{-4}\int_x^{x\pm h} u''''(\xi)(x \pm h - \xi)^3/3!d\xi = u''''(x \pm \vartheta h)/4!, \qquad (4.1.5b)$$

with $\vartheta \in (0,1)$, to Equation (3), the result is (4c) because $R = [u''''(x+\vartheta_1 h) + u''''(x-\vartheta_2 h)]/24$. ■

Ω

Ω_h **Figure** 4.1.1

$\overline{\Omega}_h$ Grid for $h = 1/8$

We replace $\Omega = (0,1)$ and $\overline{\Omega} = [0,1]$ by the grids

$$\Omega_h = \{h, 2h, \cdots, (n-1)h = 1-h\}, \qquad (4.1.6a)$$

$$\overline{\Omega}_h = \{0, h, 2h, \cdots, 1-h, 1\} \qquad (4.1.6b)$$

of step size $h = 1/n$. For $x \in \Omega_h$, $\partial^{-}\partial^{+}u(x)$ only contains the values of u at $x, x \pm h \in \overline{\Omega}_h$. Under the assumption that the solution u of Equations (1a,b) belongs to $C^4(\overline{\Omega})$, (4c) yields the equations

$$-\partial^{-}\partial^{+}u(x) = f(x) + O(h^2), \qquad x \in \Omega_h. \qquad (4.1.7)$$

If one neglects the remainder term $O(h^2)$ in Equation (7), one obtains

$$-\partial^{-}\partial^{+}u_h(x) \equiv h^{-2}[-u_h(x-h) + 2u_h(x) - u_h(x+h)] = f(x) \quad \text{for} \quad x \in \Omega_h. \qquad (4.1.8a)$$

These are $n-1$ equations in $n+1$ unknowns $\{u_n(x), x \in \overline{\Omega}_h\}$. The two missing equations are supplied by boundary conditions (1b):

$$u_h(0) = \varphi_0, \quad u_h(1) = \varphi_1. \qquad (4.1.8b)$$

u_h is a grid function defined on $\overline{\Omega}_h$. Its restriction to Ω_h yields the vector

$$u_h = (u_h(h), u_h(2h), \cdots, u_h(1-h))^{\mathsf{T}}.$$

If in (8a) one eliminates the components $u_h(0)$ and $u_h(1)$ with the aid of Equation (8b), one gets the system of equations

$$L_h u_h = q_h \tag{4.1.9a}$$

with

$$L_h = h^{-2}\begin{bmatrix} 2 & -1 & & & & \\ -1 & 2 & -1 & & & \\ & -1 & 2 & -1 & & \\ & & \ddots & \ddots & \ddots & \\ & & & -1 & 2 & -1 \\ & & & & -1 & 2 \end{bmatrix}, \tag{4.1.9b}$$

$$q_h = (f(h)+h^{-2}\varphi_0, f(2h), f(3h), \cdots, f(1-2h), f(1-h)+h^{-2}\varphi_1)^{\mathsf{T}}. \tag{4.1.9c}$$

4.2 The Five-Point Formula

First we select the unit square

$$\Omega = \{(x,y): 0 < x < 1,\ 0 < y < 1\}$$

as the fundamental domain. More general domains will be discussed in Section 4.8. In the discretisation process Ω is replaced by the grid

$$\Omega_h := \{(x,y) \in \Omega: x/h,\ y/h \in \mathbb{Z}\} \tag{4.2.1a}$$

for step size $h = \frac{1}{n}$ $(n \in \mathbb{N})$. The discrete boundary points form the set

$$\Gamma_h := \{(x,y) \in \Gamma: x/h, y/h \in \mathbb{Z}\}. \tag{4.2.1b}$$

As in (1.1b) we set

$$\overline{\Omega}_h := \Omega_h \cup \Gamma_h = \{(x,y) \in \overline{\Omega}_h: x/h,\ y/h \in \mathbb{Z}\}. \tag{4.2.1c}$$

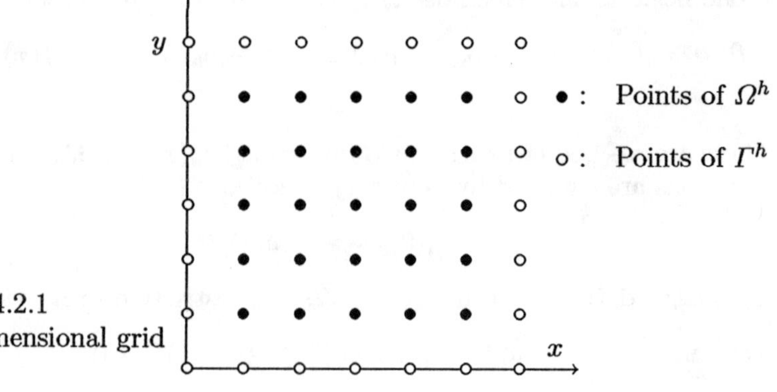

Figure 4.2.1
A two-dimensional grid

In the Poisson equation

$$-\Delta u \equiv -u_{xx} - u_{yy} = f \text{ in } \Omega, \tag{4.2.2a}$$

$$u = \varphi \text{ on } \Gamma, \tag{4.2.2b}$$

the second derivatives u_{xx} and u_{yy} can each be replaced by the respective differences (1.4c) in the x and y directions:

$$\begin{aligned}(\Delta_h u)(x,y) &:= (\partial_x^- \partial_x^+ + \partial_y^- \partial_y^+)u(x,y)\\ &= h^{-2}[u(x-h,y) + u(x+h,y) + u(x,y-h) + u(x,y+h) - 4u(x,y)].\end{aligned} \tag{4.2.3}$$

Since on the right side of (3) the function u is evaluated at five points, Δ_h is also called the Five-Point Formula. The discretisation of the boundary value problem (2a,b) using Δ_h leads to the difference equations

$$-\Delta_h u_h(\mathbf{x}) = f(\mathbf{x}) \qquad \text{for } \mathbf{x} \in \Omega_h, \tag{4.2.4a}$$

$$u_h(\mathbf{x}) = \varphi(\mathbf{x}) \qquad \text{for } \mathbf{x} \in \Gamma_h. \tag{4.2.4b}$$

Through (4a,b) one obtains one equation per grid point $\mathbf{x} = (x,y) \in \overline{\Omega}_h$, and hence one equation per component of the grid function $u_h = (u_h(\mathbf{x}))_{\mathbf{x}\in\overline{\Omega}_h}$. Except in the one-dimensional case, there exists no natural arrangement of grid points, thus one cannot immediately obtain a matrix representation as in (1.9b). The only natural indexing of u_n is that through $\mathbf{x} \in \Omega_h$ or the pair $(i,j) \in \mathbb{Z}^2$ with $\mathbf{x} = (x,y) = (ih, jh)$. Let the matrix elements be given by

$$L_{\mathbf{x}\xi} := \begin{cases} 4h^{-2} & \text{if } \mathbf{x} = \xi \in \Omega_h, \\ -h^{-2} & \text{if } \mathbf{x} \in \Omega_h, \xi \in \overline{\Omega}_h, \mathbf{x} - \xi = \binom{0}{\pm h} \text{ or } \mathbf{x} - \xi = \binom{\pm h}{0}, \\ 0 & \text{otherwise.} \end{cases} \tag{4.2.5}$$

For $\mathbf{x} = \xi$, $L_{\mathbf{xx}}$ is a diagonal element; in the second case, $L_{\mathbf{x}\xi} = -h^{-2}$, we say that $\mathbf{x}$ and ξ are neighbours. If one eliminates the components $u_h(\mathbf{x})$, $\mathbf{x} \in \Gamma_h$, with the aid of Equation (4b), then Equation (4a) assumes the following form:

$$\sum_{\xi\in\Omega_h} L_{\mathbf{x}\xi} u_h(\xi) = q_h(\mathbf{x}) \quad \text{for } \mathbf{x} \in \Omega_h, \tag{4.2.6a}$$

where

$$q_h := f_h + \varphi_h, \quad f_h(\mathbf{x}) := f(\mathbf{x}), \quad \varphi_h(\mathbf{x}) := -\sum_{\xi\in\Gamma_h} L_{\mathbf{x}\xi}\varphi(\xi). \tag{4.2.6b}$$

For the proof split the sum

$$-\Delta_h u_h(\mathbf{x}) = \sum_{\xi\in\overline{\Omega}_h} L_{\mathbf{x}\xi} u_h(\xi)$$

into $\sum_{\xi\in\Omega_h} \ldots$ and $\sum_{\xi\in\Gamma_h} \ldots$. The second partial sum is $-\varphi_h$ and is moved to the right side of the equation.

Remark 4.2.1. f_h is the restriction of f to the grid Ω_h. For all points far from the boundary we have $\varphi_h(\mathbf{x}) = 0$; here $\mathbf{x} \in \Omega_h$ is said to be far from the boundary if all neighbours $\mathbf{x} \pm (0,h)$, $\mathbf{x} \pm (h,0)$ belong to Ω_h. In the case of homogeneous boundary values $\varphi = 0$ we have $q_h = f_h$.

The system of equations (6a) can be expressed in the form (1.9a):

$$L_h u_h = q_h$$

where the matrix

$$L_h = (L_{\mathbf{x}\xi})_{\mathbf{x},\xi\in\Omega_h}$$

and the grid functions $u_h = (u_h(\mathbf{x}))_{\mathbf{x}\in\Omega_h}$ and $q_h = (q_h(\mathbf{x}))_{\mathbf{x}\in\Omega_h}$ are described by their components (cf. (5), (6b)). Strictly speaking L_h is not a matrix in the usual sense but a linear mapping since the indices $\mathbf{x} \in \Omega_h$ are not ordered.

Exercise 4.2.2. (a) An $N \times N$ matrix $\mathbf{P}$ is called a permutation matrix if $\mathbf{w} := \mathbf{P}\mathbf{v}$, for all $\mathbf{v} \in \mathbb{R}^N$, has the coefficients $w_i = v_{\pi(i)}$, $1 \le i \le N$, where π is a permutation of indices $\{1, \cdots, N\}$. Show that $\mathbf{P}$ is unitary, i.e., $\mathbf{P}^{-1} = \mathbf{P}^{\mathsf{T}}$.
(b) Let I be an index set with N elements. Let the "matrix" coefficients $A_{\alpha\beta}$, $\alpha, \beta \in I$, be given. To the arrangement $\alpha_1, \cdots, \alpha_N$ of indices corresponds the $N \times N$ matrix $\mathbf{A} = (a_{ij})_{i,j=1,\cdots,N}$ with $a_{ij} = A_{\alpha_i\alpha_j}$. Let $\hat{\mathbf{A}}$ be the matrix which belongs to a second arrangement $\hat\alpha_1, \cdots, \hat\alpha_N$ of I. Prove that there exists a permutation matrix $\mathbf{P}$ such that $\hat{\mathbf{A}} = \mathbf{P}\mathbf{A}\mathbf{P}^{\mathsf{T}}$.
(c) Let $A_{\alpha\beta} = A_{\beta\alpha}$ for all $\alpha, \beta \in I$. For each arrangement of the indices the corresponding matrix is symmetric.

A possible linear enumeration of indices $\mathbf{x} \in \Omega_h$ is lexicographical ordering

$$\begin{array}{lllll}
(h,h), & (2h,h), & (3h,h), & \cdots, & (1-h,h), \\
(h,2h), & (2h,2h), & (3h,2h), & \cdots, & (1-h,2h), \\
 & & \vdots & & \\
(h,1-h), & (2h,1-h), & (3h,1-h), & \cdots, & (1-h,1-h).
\end{array} \tag{4.2.7}$$

Generally, the point $\mathbf{x} = (x_1, \cdots, x_d)$ precedes the point $\mathbf{y} = (y_1, \cdots, y_d)$ in lexicographical order, if for a $j \in \{1, \cdots, d\}$ the conditions $x_i = y_i$ $(i > j)$ and $x_j < y_j$ hold. Each line in (7) corresponds to a so-called x-row in the grid Ω_h. A vector u_h whose $(n-1)^2$ components are enumerated in the series (7) thus separates into $n-1$ blocks (so-called x-blocks). The block decomposition of the vectors generates a block decomposition of the matrix L_h which is given in (8).

Exercise 4.2.3. (a) With the lexicographical numbering of grid points the matrix L_h has the form

$$L_h = h^{-2}\begin{bmatrix} T & -I & & & \\ -I & T & -I & & \\ & \ddots & \ddots & \ddots & \\ & & -I & T & -I \\ & & & -I & T \end{bmatrix}, T = \begin{bmatrix} 4 & -1 & & & \\ -1 & 4 & -1 & & \\ & \ddots & \ddots & \ddots & \\ & & -1 & 4 & -1 \\ & & & -1 & 4 \end{bmatrix}, \tag{4.2.8}$$

where T is an $(n-1) \times (n-1)$ matrix and L_h contains $(n-1)^2$ blocks. I is the $(n-1) \times (n-1)$ identity matrix.
(b) Let Ω be the rectangle

$$\Omega = (0,a) \times (0,b) = \{(x,y): 0 < x < a,\ 0 < y < b\}.$$

Let the step size h satisfy the conditions $a = nh$ and $b = mh$. Show that the discretisation (4a,b) in the corresponding grid Ω_h leads to a matrix which also has the form (8). But here L_h contains $(m-1)^2$ blocks of the size $(n-1) \times (n-1)$.

Another frequently used arrangement is the chequer-board ordering (or red-black ordering). To this end one chessboard pattern divides Ω_h into "red" and "black" fields:

$$\begin{aligned} \Omega_h^{\mathrm{r}} &:= \{(x,y) \in \Omega_h: (x+y)/h \text{ odd}\}, \\ \Omega_h^{\mathrm{b}} &:= \{(x,y) \in \Omega_h: (x+y)/h \text{ even}\}. \end{aligned} \tag{4.2.9}$$

First one numbers the red squares $(x,y) \in \Omega_h^{\mathrm{r}}$ lexicographically, and then those of Ω_h^{b}. The partition (9) induces a partition of vectors into 2 blocks and a partition of the matrix L_h into $2 \cdot 2 = 4$ blocks.

Exercise 4.2.4. With respect to the chequer-board ordering, the matrix L_h assumes the form

$$L_h = h^{-2}\left[\begin{array}{ccc|ccc} 4 & & & & & \\ & \ddots & & & A & \\ & & 4 & & & \\ \hline & & & 4 & & \\ & A^{\mathsf{T}} & & & \ddots & \\ & & & & & 4 \end{array}\right] \tag{4.2.10}$$

where, in general, $\mathbf{A}$ is a rectangular block matrix because for n even, Ω_h^{r} and Ω_h^{b} contain a different number of points.

The complete $(n-1)^2 \times (n-1)^2$ matrix L_h in (8) or (10) is needed neither for the theoretical investigation of the system of equations $L_h u_h = q_h$ nor for its numerical solution. All properties of L_h considered in the following are invariant with respect to re-numbering of the grid points. Even though numerical methods for the solution of $L_h u_h = q_h$ implicitly use an arrangement of grid points (with the exception of special algorithms for parallel computers), they never employ the complete $(n-1)^2 \times (n-1)^2$ matrix L_h. Every usable algorithm must take into account that L_h is sparse, i.e., it has substantially more zero than nonzero elements.

In the following we again return to indexing by $\mathbf{x} \in \Omega_h$. Nevertheless we will continue to refer to the L_h defined by (5) as a matrix.

The difference operator Δ_h is also described by the star

$$\Delta_h = h^{-2} \begin{bmatrix} & 1 & \\ 1 & -4 & 1 \\ & 1 & \end{bmatrix}. \tag{4.2.11}$$

The general definition of a difference star (with variable coefficients) reads

$$\begin{bmatrix} & & \vdots & & \\ & c_{-1,1}(x,y) & c_{0,1}(x,y) & c_{1,1}(x,y) & \\ \cdots & c_{-1,0}(x,y) & c_{0,0}(x,y) & c_{1,0}(x,y) & \cdots \\ & c_{-1,-1}(x,y) & c_{0,-1}(x,y) & c_{1,-1}(x,y) & \\ & & \vdots & & \end{bmatrix}$$

$$= \sum_{i,j=-\infty}^{\infty} c_{ij}(x,y) u_h(x+ih, y+jh), \tag{4.2.12}$$

in which the zero coefficients have not been written out.

Attention. The star (11) does not represent a submatrix of L_h! The coefficients of the star appear in each row of L_h.

Remark 4.2.5. Note that the difference operator $-\Delta_h$ cannot be equated with the matrix L_h since Δ_h does not contain information on the type or place of the boundary conditions. We shall say the matrix L_h belongs to a difference star (12) if the system of equations $L_h u_h = q_h$ results from the difference equations in $\mathbf{x} \in \Omega_h$ after elimination of the Dirichlet boundary values $u_h(\mathbf{x}) = \varphi(\mathbf{x})$, $\mathbf{x} \in \Gamma_h$. Even if the vector u_h in $L_h u_h = q_h$ contains only components $u_h(\mathbf{x})$, $\mathbf{x} \in \Omega_h$, one occasionally equates u_h with the grid function on $\overline{\Omega}_h$ which assumes the prescribed boundary values (4b) on Γ_h.

4.3 M-matrices, Matrix Norms, Positive Definite Matrices

The elements of the matrix $\mathbf{A}$ are denoted by $a_{\alpha\beta}$, $\alpha, \beta \in I$. Here $\mathbf{A}$ and the index set I assume the places of L_h and Ω_h. We write

$$\mathbf{A} \geq \mathbf{B} \quad \text{if } a_{\alpha\beta} \geq b_{\alpha\beta} \quad \text{for all } \alpha, \beta \in I,$$

and define analogously $\mathbf{A} \leq \mathbf{B}, \mathbf{A} > \mathbf{B}, \mathbf{A} < \mathbf{B}$. The zero matrix is denoted by $\mathbf{0}$.

Definition 4.3.1. $\mathbf{A}$ is called an M-matrix if

$$a_{\alpha\alpha} > 0 \quad \text{for all } \alpha \in I, \quad a_{\alpha\beta} \le 0 \quad \text{for all } \alpha \ne \beta, \tag{4.3.1a}$$

$$\mathbf{A} \text{ nonsingular and } \mathbf{A}^{-1} \ge \mathbf{0}. \tag{4.3.1b}$$

The inequalities (1a) can immediately be proved for L_h (cf. (2.5)). However we still need criteria and auxiliary results to prove (1b).

The index $\alpha \in I$ is said to be directly connected with $\beta \in I$ if $a_{\alpha\beta} \ne 0$. We say that $\alpha \in I$ is connected with $\beta \in I$ if there exists a "connection" (chain of direct connections)

$$\alpha = \alpha_0, \alpha_1, \alpha_2, \cdots, \alpha_k = \beta \quad \text{with } a_{\alpha_{i-1}\alpha_i} \ne 0 \quad (1 \le i \le k). \tag{4.3.2}$$

The index set I together with the direct connections form the graph of $\mathbf{A}$ (cf. Fig. 1). Frequently $\mathbf{A}$ has a symmetrical structure, i.e., $a_{\alpha\beta} \ne 0$ holds if and only if $a_{\beta\alpha} \ne 0$. In this case α is (directly) connected with β if and only if β is (directly) connected with α.

Definition 4.3.2. A matrix $\mathbf{A}$ is said to be irreducible if every $\alpha \in I$ is connected with every $\beta \in I$.

$$A = \begin{bmatrix} 1 & 1 & 0 \\ 0 & 1 & 1 \\ 1 & 0 & 1 \end{bmatrix}$$

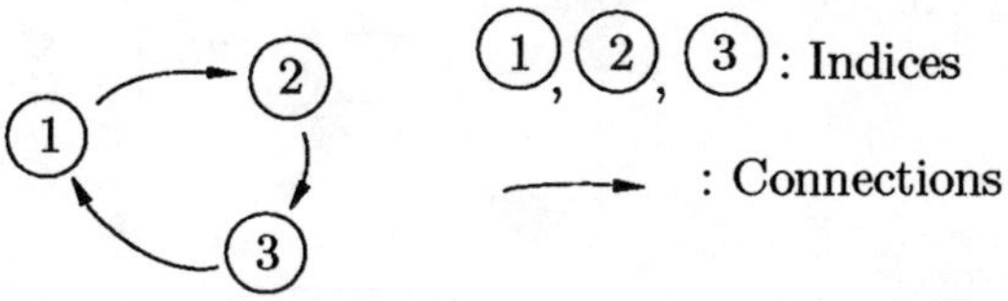

Figure 4.3.1. Graph of the irreducible matrix $\mathbf{A}$

In the case of the matrix $\mathbf{A} = L_h$, two indices $\mathbf{x}, \mathbf{y} \in \Omega_h$ are connected if and only if $\mathbf{y} = \mathbf{x}$ or if $\mathbf{y}$ is a neighbour of $\mathbf{x}$. Arbitrary $\mathbf{x}, \mathbf{y} \in \Omega_h$ can evidently be connected by a chain $\mathbf{x} = \mathbf{x}^{(0)}, \mathbf{x}^{(1)}, \cdots, \mathbf{x}^{(k)} = \mathbf{y}$ of neighbouring points. Thus L_h is irreducible.

Exercise 4.3.3. Prove that $\mathbf{A}$ is irreducible if and only if there is no ordering of the indices such that the resulting matrix has the form

$$\mathbf{A} = \begin{bmatrix} \mathbf{A}_{11} & \mathbf{A}_{12} \\ \mathbf{0} & \mathbf{A}_{22} \end{bmatrix},$$

where $\mathbf{A}_{11}$ and $\mathbf{A}_{22}$ are respectively square $n_1 \times n_1$ and $n_2 \times n_2$ matrices $(n_1 \ge 1, n_2 \ge 1)$, and $\mathbf{A}_{12}$ is an $n_1 \times n_2$ submatrix.

The important question as to whether $\mathbf{A} = L_h$ is nonsingular can be treated as a special case of the following statement.

Criterion 4.3.4. *(Gershgorin) Let $K_r(z)$ denote the open disk $\{\zeta \in \mathbb{C}: |z - \zeta| < r\}$, and let $\overline{K_r(z)} := \{\zeta \in \mathbb{C}: |z-\zeta| \le r\}$ denote the closed disk.*
(a) All eigenvalues of A lie in

$$\bigcup_{\alpha\in I} \overline{K_{r_\alpha}(a_{\alpha\alpha})} \quad \text{with } r_\alpha = \sum_{\substack{\beta\neq\alpha \\ \beta\in I}} |a_{\alpha\beta}|.$$

(b) If $\mathbf{A}$ is irreducible, the eigenvalues even lie in

$$\bigcup_{\alpha\in I} K_{r_\alpha}(a_{\alpha\alpha}) \cup \Big(\bigcap_{\alpha\in I} \partial K_{r_\alpha}(a_{\alpha\alpha})\Big).$$

PROOF. (a) Let λ be an eigenvalue of $\mathbf{A}$ and $\mathbf{u}$ a corresponding eigenvector which, without loss of generality, satisfies $||\mathbf{u}||_\infty = 1$, where

$$||\mathbf{u}||_\infty := \max\{|u_\alpha|: \alpha \in I\} \tag{4.3.3}$$

is the maximum norm. There exists (at least) one index $\gamma \in I$ with $|u_\gamma| = 1$.

Assertion 1. $|u_\gamma| = 1$ *implies*

$$|\lambda - a_{\gamma\gamma}| \le \sum_{\substack{\beta\neq\gamma \\ \beta\in I}} |a_{\gamma\beta}||u_\beta| \le \sum_{\substack{\beta\neq\gamma \\ \beta\in I}} |a_{\gamma\beta}| = r_\gamma. \tag{$*$}$$

From $(*)$ follows $\lambda \in \overline{K_{r_\gamma}(a_{\gamma\gamma})}$ and hence the statement. To prove the assertion use the equation from $\mathbf{Au} = \lambda\mathbf{u}$ associated to the index γ:

$$\lambda u_\gamma = \sum_{\beta\in I} a_{\gamma\beta}u_\beta, \quad \text{that is} \quad (\lambda - a_{\gamma\gamma})u_\gamma = \sum_{\substack{\beta\neq\gamma \\ \beta\in I}} a_{\gamma\beta}u_\beta.$$

From $|u_\gamma| = 1$ follows $|\lambda - a_{\gamma\gamma}| = |(\lambda - a_{\gamma\gamma})u_\gamma| \le |\sum_{\beta\neq\gamma} a_{\gamma\beta}u_\beta|$. By taking the modulus into the sum and by using $|u_\beta| \le ||\mathbf{u}||_\infty = 1$, $(*)$ follows.
(b) Let $\mathbf{A}$ be irreducible and let λ be an arbitrary eigenvalue of $\mathbf{A}$ with associated eigenvector $\mathbf{u}$ which in turn is again normalised by $||\mathbf{u}||_\infty = 1$. The case $\lambda \in \bigcup_{\alpha\in I} K_{r_\alpha}(a_{\alpha\alpha})$ immediately leads to the statement. Therefore let $\lambda \notin \bigcup_{\alpha\in I} K_{r_\alpha}(a_{\alpha\alpha})$ be assumed.

Assertion 2. *Let $a_{\alpha\beta} \neq 0$, i.e., γ is directly connected with β; then $|u_\gamma| = 1$ and $|\lambda - a_{\gamma\gamma}| = r_\gamma$ implies $|u_\beta| = 1$ and $|\lambda - a_{\beta\beta}| = r_\beta$.*

Part (a) proves the existence of a $\gamma \in I$ with $|u_\gamma| = 1$ and $|\lambda - a_{\gamma\gamma}| \le r_\gamma$. According to the assumption, $|\lambda - a_{\gamma\gamma}| = r_\gamma$ must hold so that assertion 2 is applicable to γ. Since $\mathbf{A}$ is irreducible, for an arbitrary $\beta \in I$ there exists a connection (2) of γ with β: $\alpha_0 = \gamma, \alpha_1, \cdots, \alpha_k = \beta$, $a_{\alpha_{i-1}\alpha_i} \neq 0$. Assertion 2 shows

$$|u_{\alpha_i}| = 1 \quad \text{and } |\lambda - a_{\alpha_i\alpha_i}| = r_{\alpha_i} \quad \text{for all } i = 0, \cdots, k;$$

in particular, $\lambda \in \partial K_{r_\beta}(a_{\beta\beta})$ for $\beta = \alpha_k$. Since β was chosen arbitrarily, it follows that $\lambda \in \bigcap_{\alpha \in I} \partial K_{r_\alpha}(a_{\alpha\alpha})$, and the statement is proved.

Proof of Assertion 2. Besides the inequality chain $(*)$ there also holds $|\lambda - a_{\gamma\gamma}| = r_\gamma$ so that all the inequalities in $(*)$ become equations. In particular

$$\sum_{\beta \neq \gamma} |a_{\gamma\beta}|\;|u_\beta| = \sum_{\beta \neq \gamma} |a_{\gamma\beta}|$$

must hold. Since $|u_\beta| \le \|\mathbf{u}\|_\infty = 1$, the identity $|a_{\gamma\beta}|\;|u_\beta| = |a_{\gamma\beta}|$ must be satisfied for each summand. Hence $a_{\gamma\beta} \neq 0$ implies $|u_\beta| = 1$. The application of Assertion 1 to β yields $|\lambda - a_{\beta\beta}| \le r_\beta$. The assumption $\lambda \notin \bigcup_{\alpha \in I} K_{r_\alpha}(a_{\alpha\alpha})$ proves $|\lambda - a_{\beta\beta}| = r_\beta$. ■

Exercise 4.3.5. Let $I_\alpha := \{\beta \in I : \text{a connection (2) exists between } \alpha \text{ and } \beta\}$. Show that the eigenvalues of $\mathbf{A}$ lie in

$$\bigcup_{\alpha \in I} \left\{ K_{r_\alpha}(a_{\alpha\alpha}) \cup \bigcap_{\beta \in I_\alpha} \partial K_{r_\beta}(a_{\beta\beta}) \right\}.$$

Definition 4.3.6. (a) $\mathbf{A}$ is said to be diagonally dominant if

$$\sum_{\beta \neq \alpha} |a_{\alpha\beta}| < |a_{\alpha\alpha}| \tag{4.3.4a}$$

for all $\alpha \in I$. (b) $\mathbf{A}$ is said to be irreducibly diagonally dominant if $\mathbf{A}$ is irreducible, the inequality (4a) holds for at least one index $\alpha \in I$ and

$$\sum_{\substack{\beta \neq \alpha \\ \beta \in I}} |a_{\alpha\beta}| \le |a_{\alpha\alpha}| \quad \text{for all } \alpha \in I. \tag{4.3.4b}$$

Note that while an irreducible and diagonally dominant matrix is irreducibly diagonally dominant, the reverse need not hold.

The matrix L_h from Section 4.2, while not diagonally dominant, is irreducibly diagonally dominant, for L_h is irreducible and satisfies (4b). At all points near the boundary—i.e., those $\mathbf{x} \in \Omega_h$, which have a boundary point $\mathbf{y} \in \Gamma_h$ as a neighbour—however, (4a) holds: $\sum_{\substack{\beta \neq \alpha \\ \beta \in I}} |a_{\alpha\beta}| \le 3h^{-2} < 4h^{-2} = a_{\alpha\alpha}$.

The spectral radius $\rho(\mathbf{A})$ of a matrix $\mathbf{A}$ is given by the eigenvalue that is largest in modulus:

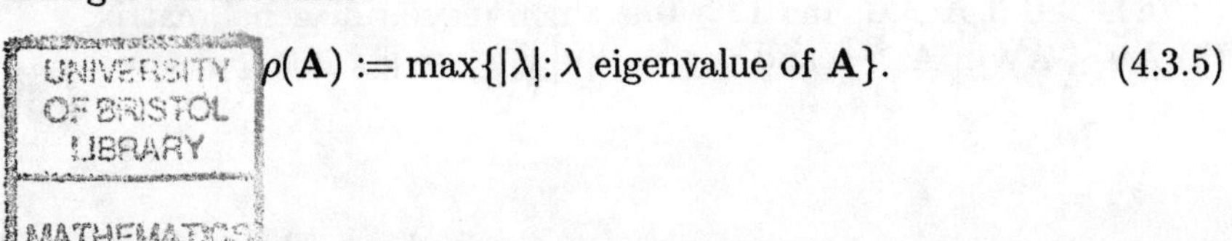

$$\rho(\mathbf{A}) := \max\{|\lambda| : \lambda \text{ eigenvalue of } \mathbf{A}\}. \tag{4.3.5}$$

In the following we split $\mathbf{A}$ into

$$\mathbf{A} = \mathbf{D} - \mathbf{B}, \quad \mathbf{D} = \text{diag}\,\{a_{\alpha\alpha}\colon \alpha \in I\}, \tag{4.3.6a}$$

where $\mathbf{D}$ is the diagonal part of $\mathbf{A}$:

$$d_{\alpha\alpha} = a_{\alpha\alpha}, \quad d_{\alpha\beta} = 0 \quad \text{for } \alpha \neq \beta. \tag{4.3.6b}$$

$\mathbf{B} := \mathbf{D} - \mathbf{A}$ is the off-diagonal part:

$$b_{\alpha\alpha} = 0, \quad b_{\alpha\beta} = -a_{\alpha\beta} \quad \text{for } \alpha \neq \beta. \tag{4.3.6c}$$

Criterion 4.3.7. *Let* (6a-c) *hold. Sufficient conditions for*

$$\rho(\mathbf{D}^{-1}\mathbf{B}) < 1 \tag{4.3.7}$$

are the diagonal dominance or the irreducible diagonal dominance of $\mathbf{A}$.

PROOF. (a) The coefficients of $\mathbf{C} := \mathbf{D}^{-1}\mathbf{B}$ read

$$c_{\alpha\beta} = -a_{\alpha\beta}/a_{\alpha\alpha} \quad (\alpha \neq \beta), \quad c_{\alpha\alpha} = 0.$$

From the diagonal dominance (4a) follows that $r_\alpha := \sum_{\beta\neq\alpha} |c_{\alpha\beta}| < 1$ for all $\alpha \in I$. By the Gershgorin Criterion 4a all eigenvalues λ of $\mathbf{C}$ lie in $\bigcup_{\alpha\in I} \overline{K_{r_\alpha}(c_{\alpha\alpha})} = \bigcup_{\alpha\in I} \overline{K_{r_\alpha}(0)}$ so that $|\lambda| \leq \max r_\alpha < 1$ and hence $\rho(\mathbf{C}) = \rho(\mathbf{D}^{-1}\mathbf{B}) < 1$ also follows.
(b) If $\mathbf{A}$ is irreducibly diagonally dominant then $r_\beta \leq 1$ for all $\beta \in I$ and $r_\alpha < 1$ for at least one α. According to Criterion 4b all eigenvalues of $\mathbf{C}$ lie in $\bigcup_{\beta\in I} K_{r_\beta}(0) \cup (\bigcap_{\beta\in I} \partial K_{r_\beta}(0))$. This set lies in $K_1(0)$ if $\bigcap_{\beta\in I} \partial K_{r_\beta}(0) \subset K_1(0)$. At first let us assume that all r_β agree: $r_\beta = r$ for all β. Since $r_\alpha < 1$ for one $\alpha \in I$, it follows that $r < 1$ and $\bigcap_\beta \partial K_{r_\beta}(0) = \partial K_r(0) \subset K_1(0)$. But if all r_β are not equal then $\bigcap_\beta \partial K_{r_\beta}(0)$ is empty. Thus in both cases $\lambda \in K_1(0)$ holds and (7) is proved. ■

Exercise 4.3.8. (a) Weaken irreducible diagonal dominance as follows: Let $\mathbf{A}$ satisfy the inequalities (4b) and for all $\beta \in I$ let the connection (2) exist for an index $\alpha \in I$ for which the strict inequality (4a) holds. Prove that even under this assumption $\rho(\mathbf{D}^{-1}\mathbf{B}) < 1$ holds.
Hint. Use Exercise 5.
(b) Show that the geometric series $\mathbf{S} = \sum_{\nu=0}^{\infty} \mathbf{C}^\nu$ converges if and only if $\rho(\mathbf{C}) < 1$. Then the following holds: $\mathbf{S} = (\mathbf{I} - \mathbf{C})^{-1}$. Hint: Represent $\mathbf{C}$ in the form $\mathbf{QRQ}^\mathsf{T}$ ($\mathbf{Q}$ a unitary, and $\mathbf{R}$ an upper triangular matrix) and show $||\mathbf{C}^\nu||_\infty \leq K[\rho(\mathbf{C})]^\nu$.
(c) Let $\mathbf{u}$ be a vector. We define $|\mathbf{u}|$ as the vector (!) with the entries $|\mathbf{u}|_\alpha := |u_\alpha|$. For two vectors one writes $\mathbf{v} \leq \mathbf{w}$ if $v_\alpha \leq w_\alpha$ $(\alpha \in I)$. Show that:
(1) $\mathbf{AB} \geq \mathbf{0}$, if $\mathbf{A} \geq \mathbf{0}$, $B \geq 0$; $\mathbf{AB} > \mathbf{0}$, if $\mathbf{A} > \mathbf{0}$, $\mathbf{B} > \mathbf{0}$;
(2) $\mathbf{AD} > \mathbf{0}$ if $\mathbf{A} > \mathbf{0}$, and $\mathbf{D} \geq \mathbf{0}$ is a nonsingular diagonal matrix;
(3) $\mathbf{Av} \leq \mathbf{Aw}$ if $\mathbf{A} \geq \mathbf{0}$ and $\mathbf{v} \leq \mathbf{w}$; $||\mathbf{v}||_\infty \leq ||\mathbf{w}||_\infty$ if $\mathbf{0} \leq \mathbf{v} \leq \mathbf{w}$;

(4) $\mathbf{Au} \le |\mathbf{Au}| \le \mathbf{A}|\mathbf{u}|$ if $\mathbf{A} \ge 0$.

The importance of inequality (7) results from

Lemma 4.3.9. *Let* $\mathbf{A}$ *satisfy* (1a). *Let* $\mathbf{D}$ *and* $\mathbf{B}$ *be defined by* (6a-c). $\mathbf{A}$ *is an* M*-matrix if and only if* $\rho(\mathbf{D}^{-1}\mathbf{B}) < 1$.

PROOF. (a) Let $\mathbf{C} := \mathbf{D}^{-1}\mathbf{B}$ satisfy $\rho(\mathbf{C}) < 1$. Then the geometric series $\mathbf{S} := \sum_{\nu=0}^{\infty} \mathbf{C}^{\nu}$ converges (cf. Exercise 8b). From $\mathbf{D}^{-1} \ge \mathbf{0}$ and $\mathbf{B} \ge \mathbf{0}$ one infers $\mathbf{C} \ge \mathbf{0}$, $\mathbf{C}^{\nu} \ge \mathbf{0}$, and $\mathbf{S} \ge \mathbf{0}$. Since $I = \mathbf{S}(\mathbf{I} - \mathbf{C}) = \mathbf{SD}^{-1}\mathbf{D} - \mathbf{B} = (\mathbf{SD})^{-1}\mathbf{A}$, $\mathbf{A}$ has the inverse $\mathbf{A}^{-1} = \mathbf{SD}^{-1}$. $\mathbf{D}^{-1} \ge \mathbf{0}$ and $\mathbf{S} \ge \mathbf{0}$ result in $\mathbf{A}^{-1} \ge \mathbf{0}$. From this (1b) also results, i.e., $\mathbf{A}$ is an M-matrix.
(b) Let $\mathbf{A}$ be an M-matrix. For an eigenvalue λ of $\mathbf{D}^{-1}\mathbf{B}$ select an eigenvector $\mathbf{u} \ne 0$. According to Exercise 8c we have

$$|\lambda|\,|\mathbf{u}| = |\lambda\mathbf{u}| = |\mathbf{D}^{-1}\mathbf{B}\mathbf{u}| \le \mathbf{D}^{-1}\mathbf{B}|\mathbf{u}|.$$

Because $\mathbf{A}^{-1}\mathbf{D} \ge \mathbf{0}$ (cf. (1a,b)) one obtains $-\mathbf{A}^{-1}\mathbf{D}\mathbf{D}^{-1}\mathbf{B}|\mathbf{u}| \le -\mathbf{A}^{-1}\mathbf{D}|\lambda|\,|\mathbf{u}|$ so that

$$\begin{aligned}|\mathbf{u}| &= \mathbf{A}^{-1}(\mathbf{D}-\mathbf{B})|\mathbf{u}| = \mathbf{A}^{-1}\mathbf{D}(\mathbf{I}-\mathbf{D}^{-1}\mathbf{B})|u| \le \mathbf{A}^{-1}\mathbf{D}|\mathbf{u}| - \mathbf{A}^{-1}\mathbf{D}|\lambda|\,|\mathbf{u}| \\ &= (1-|\lambda|)\mathbf{A}^{-1}\mathbf{D}|\mathbf{u}|\end{aligned}$$

follows. For $|\lambda| \ge 1$ we would get the inequality, $|\mathbf{u}| \le \mathbf{0}$, i.e., $\mathbf{u} = \mathbf{0}$, in contradiction to the assumption $\mathbf{u} \ne \mathbf{0}$. From this follows $|\lambda| < 1$ for every eigenvalue of $\mathbf{C} = \mathbf{D}^{-1}\mathbf{B}$, thus $\rho(\mathbf{D}^{-1}\mathbf{B}) < 1$. ∎

Criterion 7 and Lemma 9 imply

Criterion 4.3.10. *If a matrix* $\mathbf{A}$ *with the property* (1a) *is diagonally dominant or irreducibly diagonally dominant, then* $\mathbf{A}$ *is an* M*-matrix.*

Theorem 4.3.11. *An irreducible* M*-matrix* $\mathbf{A}$ *has an element-wise positive inverse:* $\mathbf{A}^{-1} > \mathbf{0}$.

PROOF. Let $\alpha, \beta \in I$ be selected arbitrarily. There exists a connection (2): $\alpha = \alpha_0, \alpha_1, \cdots, \alpha_k = \beta$. Set $\mathbf{C} := \mathbf{D}^{-1}\mathbf{B}$. Since $c_{\alpha_{i-1}\alpha_i} > 0$, it follows that

$$(\mathbf{C}^k)_{\alpha\beta} = \sum_{\gamma_1,\cdots,\gamma_{k-1}\in I} c_{\alpha\gamma_1}c_{\gamma_1\gamma_2}\cdots c_{\gamma_{k-1}\beta} \ge c_{\alpha\alpha_1}c_{\alpha_1\alpha_2}\cdots c_{\alpha_{k-1}\beta} > 0.$$

According to Lemma 9, $\rho(\mathbf{C}) < 1$ holds, so that $\mathbf{S} := \sum_{\nu=0}^{\infty}\mathbf{C}^{\nu}$ converges. Since $\mathbf{S}_{\alpha\beta} \ge (\mathbf{C}^k)_{\alpha\beta} > 0$ and $\alpha, \beta \in I$ are arbitrary, $\mathbf{S} > \mathbf{0}$ is proved. The assertion results from $\mathbf{A}^{-1} = \mathbf{SD}^{-1} > \mathbf{0}$ (cf. proof of Lemma 9). ∎

In the following we derive norm estimates for $\mathbf{A}^{-1}$.

Definition 4.3.12. Let $\mathbf{V}$ be a linear space (vector space) over the field of real numbers ($\mathbf{K} := \mathbb{R}$) or complex numbers ($\mathbf{K} := \mathbb{C}$). The functional $||\cdot||$ is called a norm in $\mathbf{V}$ if

$$||\mathbf{u}|| = 0 \quad \text{only for} \quad \mathbf{u} = \mathbf{0}, \tag{4.3.8a}$$
$$||\mathbf{u}+\mathbf{v}|| \le ||\mathbf{u}|| + ||\mathbf{v}|| \quad \text{for all } \mathbf{u}, \mathbf{v} \in \mathbf{V}, \tag{4.3.8b}$$
$$||\lambda \mathbf{u}|| = |\lambda| \; ||\mathbf{u}|| \quad \text{for all } \lambda \in \mathbf{K}, \mathbf{u} \in \mathbf{V}. \tag{4.3.8c}$$

Example. Let $\mathbf{V} = \mathbb{R}^{\#I}$ where $\#I :=$ is the number of elements of the index set I. The maximum norm defined in (3) satisfies the norm axioms (8a-c).

If one views the elements $\mathbf{u} \in \mathbf{V}$ as vectors, $||\cdot||$ is called a vector norm. But the matrices also form a linear space. In the latter case one calls $||\cdot||$ a matrix norm. A special class of matrix norms is contained in

Definition 4.3.13. Let $\mathbf{V}$ be the vector space with vector norm $||\cdot||$. Then one calls

$$|||\mathbf{A}||| := \sup\{||\mathbf{Au}||/||\mathbf{u}|| \colon \mathbf{0} \neq \mathbf{u} \in \mathbf{V}\} \tag{4.3.9}$$

the matrix norm associated with the vector norm $||\cdot||$.

Exercise 4.3.14. Let $|||\cdot|||$ be defined by (9). Show that: (a) $|||\cdot|||$ is a norm; (b) the following holds:

$$|||\mathbf{AB}||| \le |||\mathbf{A}||| \; |||\mathbf{B}|||, \tag{4.3.10a}$$
$$|||I||| = 1 \quad (I : \text{unit matrix}), \tag{4.3.10b}$$
$$||\mathbf{Au}|| \le |||\mathbf{A}||| \; ||\mathbf{u}||, \tag{4.3.10c}$$
$$|||\mathbf{A}||| \ge \rho(\mathbf{A}). \tag{4.3.10a}$$

Example. The matrix norm associated with the maximum norm $||\cdot||_\infty$ (cf. (3)) is called the row sum norm and is also denoted by $||\cdot||_\infty$. It has the explicit representation

$$||\mathbf{A}||_\infty = \max_{\alpha \in I}\{\sum_{\beta \in I} |a_{\alpha\beta}|\}. \tag{4.3.11}$$

Exercise 4.3.15. (a) Prove (11). (b) For matrices $\mathbf{0} \le \mathbf{B} \le \mathbf{C}$ there holds $||\mathbf{B}||_\infty \le ||\mathbf{C}||_\infty$.

In the next theorem we denote by $\mathbb{1}$ the vector having only ones as components:

$$\mathbb{1}_\alpha = 1 \quad \text{for all } \alpha \in I.$$

For the notation $\mathbf{v} \le \mathbf{w}$ see Exercise 8c.

Theorem 4.3.16. *Let* $\mathbf{A}$ *be an* M*-matrix and let a vector* $\mathbf{w}$ *exist with* $\mathbf{Aw} \ge \mathbb{1}$. *Then* $||\mathbf{A}^{-1}||_\infty \le ||\mathbf{w}||_\infty$.

PROOF. As in the proof of Lemma 9, let $|\mathbf{u}|$ be the vector with the components $|u_\alpha|$. For each $\mathbf{u}$ we have $|\mathbf{u}| \le ||\mathbf{u}||_\infty \mathbb{1} \le ||\mathbf{u}||_\infty \mathbf{A}\mathbf{w}$. Since $\mathbf{A}^{-1} \ge \mathbf{0}$, we obtain

$$|\mathbf{A}^{-1}\mathbf{u}| \le \mathbf{A}^{-1}|\mathbf{u}| \le ||\mathbf{u}||_\infty \mathbf{A}^{-1}\mathbf{A}\mathbf{w} = ||\mathbf{u}||_\infty \mathbf{w}$$

(cf. Exercise 8c) and $||\mathbf{A}^{-1}\mathbf{u}||_\infty / ||\mathbf{u}||_\infty \le ||\mathbf{w}||_\infty$. Definition 13 implies that $||\mathbf{A}^{-1}||_\infty \le ||\mathbf{w}||_\infty$. ∎

How to estimate with the aid of a majorising matrix is shown in

Theorem 4.3.17. *Let* $\mathbf{A}$ *and* $\mathbf{A}'$ *be* M*-matrices with* $\mathbf{A}' \ge \mathbf{A}$. *Then the following holds*

$$\mathbf{0} \le \mathbf{A}'^{-1} \le \mathbf{A}^{-1} \quad \text{and } ||\mathbf{A}'^{-1}||_\infty \le ||\mathbf{A}^{-1}||_\infty. \tag{4.3.12}$$

PROOF. $\mathbf{A}'^{-1} \le \mathbf{A}^{-1}$ follows from $\mathbf{A}^{-1} - \mathbf{A}'^{-1} = \mathbf{A}^{-1}(\mathbf{A}' - \mathbf{A})\mathbf{A}'^{-1}$ and $\mathbf{A}^{-1} \ge \mathbf{0}$, $\mathbf{A}' - \mathbf{A} \ge \mathbf{0}$, $\mathbf{A}'^{-1} \ge \mathbf{0}$. The remainder follows from Exercise 15b. ∎

Exercise 4.3.18. Prove (12) under the following weaker assumptions: $\mathbf{A}$ is an M-matrix, $\mathbf{A}'$ satisfies (1a) and $\mathbf{A}' \ge \mathbf{A}$. Hint: Repeat the considerations from the first part of the proof of Lemma 9 with the matrices $\mathbf{D}'$ and $\mathbf{B}'$ associated to $\mathbf{A}'$.

Exercise 4.3.19. Let $\mathbf{B}$ a principal submatrix of $\mathbf{A}$, that is, there exists a subset $I' \subset I$ such that $\mathbf{B}$ is given by the entries $b_{\alpha\beta} = a_{\alpha\beta}$, $\alpha, \beta \in I'$. Prove that if $\mathbf{A}$ is an M-matrix, then so is $\mathbf{B}$ and $0 \le (\mathbf{B}^{-1})_{\alpha\beta} \le (\mathbf{A}^{-1})_{\alpha\beta}$ holds for all $\alpha, \beta \in I'$. Hint: Apply Exercise 18 to the following matrix $\mathbf{A}'$: $a'_{\alpha\beta} = a_{\alpha\beta}$ $(\alpha, \beta \in I')$, $a'_{\alpha\alpha} = a_{\alpha\alpha}$ $(\alpha \in I \backslash I')$, $a'_{\alpha\beta} = 0$ otherwise.

Another well-known vector norm is the Euclidean norm

$$||\mathbf{u}||_2 := [c \sum_{\alpha \in I} |\mathbf{u}_\alpha|^2]^{1/2} \tag{4.3.13}$$

with fixed scaling constant $c > 0$ (for example, the choice $c = h^2$ in connection with the grid functions from Section 4.2 results in the fact that $c\sum_{\alpha \in I}$ represents an approximation to the integration $\int_\Omega$). The matrix norm associated to $||\cdot||_2$ is independent of the factor c. It is called the spectral norm and is also denoted by $||\cdot||_2$. The name derives from the following characterisation.

Exercise 4.3.20. Prove : (a) for symmetric matrices there holds $||\mathbf{A}||_2 = \rho(\mathbf{A})$ (cf. (5)). (b) For each real matrix holds:

$$||\mathbf{A}||_2 = [\rho(\mathbf{A}^\top \mathbf{A})]^{1/2} = [\text{maximal eigenvalue of } \mathbf{A}^\top \mathbf{A}]^{1/2}.$$

(c) For each matrix holds $||\mathbf{A}||_2^2 \le ||\mathbf{A}||_2 ||\mathbf{A}^\top||_2$. Hint: (b) and (10d).

For the proof in the exercise use the scalar product

$$\langle \mathbf{u}, \mathbf{v} \rangle := c \sum_{\alpha \in I} u_\alpha v_\alpha \tag{4.3.14a}$$

(c as in (13)) and its properties

$$\langle \mathbf{u}, \mathbf{u} \rangle = ||\mathbf{u}||_2^2, \quad \langle \mathbf{A}\mathbf{u}, \mathbf{v} \rangle = \langle \mathbf{u}, \mathbf{A}^\mathsf{T}\mathbf{v} \rangle, \quad |\langle \mathbf{u}, \mathbf{v} \rangle| \le ||\mathbf{u}||_2 ||\mathbf{v}||_2. \tag{4.3.14b}$$

Here the case $\mathbf{K} = \mathbb{R}$ is always used as basis., i.e., all matrices and vectors are real.

Definition 4.3.21. A matrix $\mathbf{A}$ is said to be positive definite if it is symmetric and

$$\langle \mathbf{A}\mathbf{u}, \mathbf{u} \rangle > 0 \quad \text{for all } \mathbf{u} \ne \mathbf{0}. \tag{4.3.15}$$

Exercise 4.3.22. Prove that (a) a symmetric matrix is positive definite if and only if all eigenvalues are positive.
(b) All principal submatrices of a positive definite matrix are positive definite (cf. Exercise 19).
(c) The diagonal elements $a_{\alpha\alpha}$ of a positive definite matrix are positive.
(d) $\mathbf{A}$ is called positive semi-definite if the inequality (15) holds with "$\ge$" instead of "$>$". A positive semidefinite matrix $\mathbf{A}$ has a unique positive semi-definite square root $\mathbf{B} = \mathbf{A}^{1/2}$, which has the property $\mathbf{B}^2 = \mathbf{A}$. If $\mathbf{A}$ is positive definite, then so is $\mathbf{A}^{1/2}$.

A corollary to Exercise 22a is

Lemma 4.3.23. *A positive definite matrix* $\mathbf{A}$ *is nonsingular and has a positive definite inverse.*

The property "$\mathbf{A}^{-1}$ is positive definite" is neither necessary nor sufficient to ensure the property "$\mathbf{A}^{-1} \ge \mathbf{0}$" of an M-matrix. In both cases, however, (irreducible) diagonal dominance is a sufficient criterion (cf. Criterion 10).

Criterion 4.3.24. *If a symmetric matrix with positive diagonal entries is diagonally dominant or irreducibly diagonally dominant then it is positive definite.*

PROOF. Since $r_\alpha < a_{\alpha\alpha}$, resp. $r_\alpha \le a_{\alpha\alpha}$, the Gershgorin circles which occur in Criterion 4 do not intersect the semi-axis $(-\infty, 0]$, so that all the eigenvalues must be positive. By Exercise 22a then $\mathbf{A}$ is positive definite. ∎

Lemma 4.3.25. *Let* $\lambda_{\min}$ *and* $\lambda_{\max}$ *be respectively the smallest and largest eigenvalues of a positive definite matrix* $\mathbf{A}$. *Then there holds*

$$||\mathbf{A}||_2 = \lambda_{\max}, \qquad ||\mathbf{A}^{-1}||_2 = 1/\lambda_{\min}. \tag{4.3.16}$$

PROOF. Exercise 20a shows that $||\mathbf{A}||_2 = \rho(\mathbf{A})$ and $||\mathbf{A}^{-1}||_2 = \rho(\mathbf{A}^{-1})$. From (5) then result $\rho(\mathbf{A}) = \lambda_{\max}$ and $\rho(\mathbf{A}^{-1}) = 1/\lambda_{\min}$, since $\lambda_{\min} > 0$. ∎

4.4 Properties of the Matrix L_h

Theorem 4.4.1. *The matrix L_h (five-point formula) defined in* (2.5) *has the following properties:*

$$L_h \text{ is an } M\text{-matrix,} \tag{4.4.1a}$$

$$L_h \text{ is positive definite,} \tag{4.4.1b}$$

$$||L_h||_\infty \le 8h^{-2}, \quad ||L_h^{-1}||_\infty \le 1/8, \tag{4.4.1c}$$

$$||L_h||_2 \le 8h^{-2}\cos^2(\pi h/2) < 8h^{-2}, \tag{4.4.1d}$$

$$||L_h^{-1}||_2 \le \frac{1}{8}h^2\sin^{-2}(\frac{\pi h}{2}) = \frac{1}{2\pi^2} + O(h^2) \le \frac{1}{16}. \tag{4.4.1e}$$

PROOF. (a) In Section 4.3 we already noticed that L_h is irreducibly diagonally dominant and satisfies the inequality (3.1a). By Criterion 4.3.10 then L_h is an M-matrix.
(b) Since L_h is symmetric and irreducibly diagonally dominant, (1b) follows from Criterion 4.3.24.
(c) That $||L_h||_\infty \le 8h^{-2}$ can be read from (2.5) and (3.11). To estimate L_h^{-1} one uses Theorem 4.3.16 with $w(x,y) = x(1-x)/2$. Then we have $L_h w \ge \mathbb{1}$ (even that $(L_h w)(x,y) = 1$ unless $y = h$ and $y = 1-h$) and $||w||_\infty \le w(1/2, y) = 1/8$.
(d) The inequalities (1d,e) result from Lemma 4.3.25 and

Lemma 4.4.2. *The $(n-1)^2$ eigenvectors of L_h are $u^{\nu\mu}$ $(1 \le \nu, \mu \le n-1)$:*

$$u^{\nu\mu}(x,y) = \sin(\nu\pi x)\sin(\mu\pi y) \qquad ((x,y) \in \Omega_h). \tag{4.4.2a}$$

The corresponding eigenvalues are

$$\lambda_{\nu\mu} = 4h^{-2}(\sin^2(\nu\pi h/2) + \sin^2(\mu\pi h/2)), \qquad 1 \le \nu, \mu \le n-1. \tag{4.4.2b}$$

PROOF. Let Ω_h^{1D} be the one-dimensional grid (1.6a) and let $u^\nu(x) := \sin(\nu\pi x)$. For each $x \in \Omega_h^{1D}$ there holds

$$\begin{aligned}\partial^-\partial^+ u^\nu(x) &= h^{-2}[\sin(\nu\pi(x-h)) + \sin(\nu\pi(x+h)) - 2\sin(\nu\pi x)] \\ &= 2h^{-2}\sin(\nu\pi x)[\cos(\nu\pi h) - 1]\end{aligned}$$

since $\sin(\nu\pi(x \pm h)) = \sin(\nu\pi x)\cos(\nu\pi h) \pm \cos(\nu\pi x)\sin(\nu\pi h)$. The identity $1 - \cos\xi = 2\sin^2(\xi/2)$ then implies

$$-\partial^-\partial^+u^\nu(x) = 4h^{-2}\sin^2(\nu\pi h/2)u^\nu(x), \qquad x \in \Omega_h^{1\mathrm{D}}. \tag{4.4.2c}$$

Let $L_h^{1\mathrm{D}}$ be the matrix (1.9b). Note that $(\partial^-\partial^+u)(h)$ also involves the boundary value $u(0)$ which $(L_h^{1\mathrm{D}}u)(h)$ does not; similarly $(\partial^-\partial^+u)(1-h)$ depends on $u(1)$. However, since $u(0) = \sin(0) = 0$ and $u(1) = \sin(\nu\pi) = 0$ we have $L_h^{1\mathrm{D}}u^\nu = -\partial^-\partial^+u^\nu$, and (2c) can be brought over:

$$L_h^{1\mathrm{D}}u^\nu = 4h^{-2}\sin^2(\nu\pi h/2)u^\nu, \qquad 1 \le \nu \le n-1. \tag{4.4.2c$'$}$$

The two-dimensional grid function $u^{\nu\mu}$ in (2a) can be written as the (tensor) product $u^\nu(x)u^\mu(y)$. Now we have that $(L_hu^{\nu\mu})(x,y)$ is equal to the sum $u^\mu(y)(L_h^{1\mathrm{D}}u^\nu)(x) + u^\nu(x)(L_h^{1\mathrm{D}}u^\mu)(y)$, so that (2b) follows from (2c′). ∎

In the sequel we want to show the analogies between the properties of the Poisson equation (2.2) and the discrete five-point formula (2.4a,b).

The analogue of the mean-value property (2.3.1) is the equation

$$u_h(x,y) = \frac{1}{4}[u_h(x-h,y)+u_h(x+h,y)+u_h(x,y-h)+u_h(x,y+h)]. \tag{4.4.3}$$

From (2.3) and (2.4a) with $f = 0$ we obtain the

Remark 4.4.3. The solution u_h of the discrete potential equation (2.4a) with $f = 0$ satisfies Equation (3) at all grid points $(x,y) \in \Omega_h$.

As in the continuous case the mean-value property (3) implies the maximum-minimum principle.

Remark 4.4.4. Let u_h be a non-constant solution of the discrete potential equation (2.4a) with $f = 0$. The extrema $\max\{u_h(\mathbf{x}) : \mathbf{x} \in \overline{\Omega}_h\}$ and $\min\{u_h(\mathbf{x}) : \mathbf{x} \in \overline{\Omega}_h\}$ are assumed not on Ω_h but on Γ_h.

PROOF. If u_h were maximal in $(x,y) \in \Omega_h$, then because of Equation (3), all neighbouring points $(x \pm h, y)$ and $(x, y \pm h)$ would have to carry the same values. Since every pair of points can be linked by a chain of neighbouring points, it follows that $u_h = \mathrm{const}$, in contradiction to the assumption. ∎

The last proof indirectly uses the fact that L_h is irreducible. The irreducibility of L_h corresponds to the assumption in Theorem 2.3.7 that Ω is a domain, i.e., connected.

The result of carrying over Theorems 2.4.3 and 3.1.2 reads as follows:

Theorem 4.4.5. *Let u_h^1 and u_h^2 be two solutions of* (2.4a): $-\Delta_h u_h^i = f$ *for different boundary values $u_h^i = \varphi^i$ $(i = 1,2)$. Then the following holds:*

$$\|u_h^1 - u_h^2\|_\infty \le \max_{\mathbf{x}\in\Gamma_h} |\varphi^1(\mathbf{x}) - \varphi^2(\mathbf{x})|, \tag{4.4.4a}$$

$$u_h^1 \le u_h^2 \quad \text{in } \Omega_h, \quad \text{if } \varphi^1 \le \varphi^2 \quad \text{on } \Gamma_h. \tag{4.4.4b}$$

PROOF. Let $w_h := u_h^2 - u_h^1$. (1) In the case that $\varphi^1 \le \varphi^2$ one has $w_h \ge 0$ on Γ_h and $-\Delta_h w_h = 0$. Remark 4 proves that $w_h = \text{const} \ge 0$ or $w_h > 0$.
(2) Let M be the right side of (4a). $-M \le w_h \le M$ on Γ_h implies the inequalities $-M \le w_h \le M$ on Ω_h. ■

The discrete analogue of the Green function $g(\mathbf{x}, \xi)$ is $h^{-2}L_h^{-1}$. Let δ_ξ be the scaled unit vector

$$\delta_\xi(\mathbf{x}) = \begin{cases} h^{-2} & \text{if } \mathbf{x} = \xi \\ 0 & \text{if } \mathbf{x} \ne \xi \end{cases} \qquad (\mathbf{x}, \xi \in \Omega_h). \tag{4.4.5a}$$

The column of the matrix $h^{-2}L_h^{-1}$ with index $\xi \in \Omega_h$ is given by

$$g_h(\cdot, \xi) := L_h^{-1}\delta_\xi \quad (\xi \in \Omega_h). \tag{4.4.5b}$$

For $\xi \in \Omega_h$ fixed, $g_h(\cdot, \xi)$ is a grid function defined on Ω_h. The domain of definition is extended to $\overline{\Omega}_h \times \overline{\Omega}_h$:

$$\overline{g}_h(\mathbf{x}, \xi) = \begin{cases} g_h(\mathbf{x}, \xi) & \text{for } \mathbf{x}, \xi \in \Omega_h \\ 0 & \text{for } \mathbf{x} \in \Gamma_h \text{ or } \xi \in \Gamma_h \end{cases} \qquad (\mathbf{x}, \xi \in \overline{\Omega}_h). \tag{4.4.5c}$$

The $g_h(\mathbf{x}, \xi)$ are entries of $h^{-2}L_h^{-1}$: $g_h(\mathbf{x}, \xi) = h^{-2}(L_h^{-1})_{\mathbf{x}\xi}$. The symmetry of L_h implies

Remark 4.4.6. $\overline{g}_h(\mathbf{x}, \xi) = \overline{g}_h(\xi, \mathbf{x})$ for all $\mathbf{x}, \xi \in \overline{\Omega}_h$ (cf. (3.2.3)).

The representation (3.2.7) is recalled by:

Remark 4.4.7. The solution u_h of the system of equations (2.4a) with boundary values $\varphi = 0$ reads

$$u_h(\mathbf{x}) = h^2 \sum_{\xi \in \Omega_h} \overline{g}_h(\xi, \mathbf{x}) f(\xi), \quad \mathbf{x} \in \overline{\Omega}_h. \tag{4.4.6}$$

Equation (6) is the component-wise representation of the equation $u_h = L_h^{-1}f_h$. The factor h^2 compensates for h^{-2} in (5a). It was introduced so that the summation $h^2\sum$ in (6) approximates the integral $\int_\Omega$. Since in Section 3.2 we consider the Poisson equation $\Delta u = f$, but here the equation $-\Delta u = f$, the right-hand sides in (3.2.7) and (6) differ in their signs.

The discrete Green function is positive also (cf. (3.2.9)):

Remark 4.4.8. $0 < g_h(\mathbf{x}, \xi) \le h^{-2}/8$ for $\mathbf{x}, \xi \in \Omega_h$.

PROOF. The upper bound follows from

$$g_h(\mathbf{x}, \xi) \le \|g_h(\cdot, \xi)\|_\infty \le \|L_h^{-1}\|_\infty \|\delta_\xi\|_\infty,$$

$||\delta_\xi||_\infty = h^{-2}$, and (1c). $g_h > 0$ can be inferred from $L_h^{-1} > 0$ (cf. Theorem 4.3.11). ■

The bound $g_h(\mathbf{x},\xi) \le h^{-2}/8$ is too pessimistic and can be improved considerably.

Remark 4.4.8′.

$$0 < g_h(\mathbf{x},\xi) \le \frac{1}{4} - \frac{\log(|\mathbf{x}-\xi|^2 + 2h^2)}{\log 16} \le \frac{|\log h|}{\log 4}. \tag{4.4.7}$$

This estimate via $O(|\log h|)$ reflects the logarithmic singularity of the singularity function $s(\mathbf{x},\xi) = -\log(|\mathbf{x}-\xi|^2)/4\pi$.

Exercise 4.4.9. For the proof of inequality (7) define

$$s_h(\mathbf{x},\xi) := 1/4 - \log(|\mathbf{x}-\xi|^2 + 2h^2)/\log 16, \quad u_h(\mathbf{x}) := s_h(\mathbf{x},\mathbf{0})$$

and carry out the following steps:
(a) $s_h(\mathbf{x},\xi) \ge 0$ for all $\mathbf{x} \in \overline{\Omega}_h$, $\xi \in \Omega_h$,
(b) $-\Delta_h u_h(\mathbf{0}) = h^{-2}$,
(c) $-\Delta_h u_h(\mathbf{x}) \ge 0$ for all $\mathbf{x} \in \mathbb{R}^2$ (longer calculation !),
(d) $-\Delta_h(s_h(\cdot,\xi) - g_h(\cdot,\xi)) \ge 0$ for fixed $\xi \in \Omega_h$,
(e) $g_h(\mathbf{x},\xi) \le s_h(\mathbf{x},\xi)$.

Let φ_h be defined as in (2.6b). The solution of the discrete potential equation

$$-\Delta_h u_h = 0 \quad \text{in } \Omega_h, \quad u_h = \varphi \quad \text{on } \Gamma_h$$

is given by $u_h := L_h^{-1}\varphi_h$ (if one continues the grid function, at first only defined on Ω_h, through φ on Γ_h; cf. Remark 4.2.5). The representation with the aid of g_h reads

$$u_h(\mathbf{x}) = h^2 \sum_{\xi\in\Omega_h} g_h(\xi,\mathbf{x})\varphi_h(\xi), \quad \mathbf{x} \in \Omega_h.$$

Since $\varphi_h(\mathbf{x})$ vanishes at all points $\mathbf{x}$ far from the boundary, it suffices to extend the sum over the points near the boundary. If one sums over the neighbouring boundary points of

$$\Gamma_h' := \{\xi \in \Gamma_h\colon \xi \text{ is not a corner point } (0,0),(1,0),(0,1), \text{ or } (1,1)\}$$

instead of over the points near the boundary, then the definition of φ_h results in the representation

$$u_h(\mathbf{x}) = -h \sum_{\xi\in\Gamma_h'} \partial_n^- g_h(\xi,\mathbf{x})\varphi(\xi), \quad \mathbf{x} \in \Omega_h, \tag{4.4.8}$$

where ∂_n^- is the backward difference with respect to the direction of the normal $\mathbf{n}$:

$$\partial_n^- \overline{g}_h(\xi,\mathbf{x}) = h^{-1}[\overline{g}_h(\xi,\mathbf{x}) - \overline{g}_h(\xi - h\mathbf{n},\mathbf{x})]$$

(note that $\overline{g}_h(\xi,\mathbf{x}) = 0$ for $\xi \in \Gamma_h' \subset \Gamma_h$). The variable $\xi - h\mathbf{n}$ ranges over all points next to the boundary.

Remark 4.4.10. Equation (8) corresponds to the representation in Theorem 3.2.10. The summation $h\sum_{\xi\in\Gamma_h'}$ approximates the integral $\int_\Gamma$.

Finally we want to take a closer look at the estimate for the solution $u_h = L_h^{-1}q_h$ through $||u_h||_\infty \le ||L_h^{-1}||_\infty ||q_h||_\infty \le ||q_h||_\infty/8$ (cf. (1c)). According to (2.6b), $q_h = f_h + \varphi_h$ contains the right-hand side $f_h(\mathbf{x}) = f(\mathbf{x})$ of the discrete Poisson equation and the boundary values $\varphi(\mathbf{x})$, $\mathbf{x} \in \Gamma_h$ hidden in φ_h. The following theorem gives a bound in which these components are separated.

Theorem 4.4.11. *According to* (2.6b), *let $q_h = f_h + \varphi_h$ be constructed from f and φ. The discrete solution $u_h = L_h^{-1}q_h$ of the Poisson boundary value problem can be bounded by*

$$||u_h||_\infty \le ||L_h^{-1}||_\infty \max_{\mathbf{x}\in\Omega_h}|f(\mathbf{x})| + \max_{\mathbf{x}\in\Gamma_h'}|\varphi(\mathbf{x})| \le \frac{1}{8}\max_{\mathbf{x}\in\Omega_h}|f(\mathbf{x})| + \max_{\mathbf{x}\in\Gamma_h'}|\varphi(\mathbf{x})|. \tag{4.4.9}$$

PROOF. Set $u_h' := L_h^{-1}f_h$ and $u_h'' := L_h^{-1}\varphi_h$. The estimate for the first summand in $u_h = u_h' + u_h''$ results in the first summand in (9). To bound u_h'' use the inequality (4a) with $u_h^1 = u_h''$, $\varphi^1 = \varphi$ and $u_h^2 = 0$, $\varphi^2 = 0$. ■

The corresponding inequality

$$||u||_\infty \le C||f||_\infty + ||\varphi||_\infty \tag{4.4.10}$$

for the solution of the boundary value problem (2.2a,b) has not been mentioned until now, but will be proved in a more general context in Section 5.1.3. The maximum norm $||\cdot||_\infty$ in (9) can be replaced by the Euclidean norms

$$||u_h||_{2,\Omega_h} := \sqrt{h^2\sum_{\mathbf{x}\in\Omega_h}|u_h(\mathbf{x})|^2}, \quad ||\varphi||_{2,\Gamma_h'} := \sqrt{h\sum_{\mathbf{x}\in\Gamma_h'}|\varphi(\mathbf{x})|^2}.$$

Here, $h^2\sum_{\Omega_h}$ and $\int_\Omega$, $h\sum_{\Gamma_h'}$ and $\int_\Gamma$ correspond to each other.

Theorem 4.4.12. *Under the assumptions of Theorem 11 there holds*

$$||u_h||_{2,\Omega_h} \le ||L_h^{-1}||_2||f||_{2,\Omega_h} + \frac{1}{\sqrt{2}}||\varphi||_{2,\Gamma_h'} \le \frac{1}{16}||f||_{2,\Omega_h} + \frac{1}{\sqrt{2}}||\varphi||_{2,\Gamma_h'}. \tag{4.4.11}$$

PROOF. (1) It suffices to consider the case of the potential equation (i.e., $f = 0$). Let the restriction of φ on Γ_h' result in the grid function Φ_h: $\Phi_h(\mathbf{x}) = \varphi(\mathbf{x})$

for $\mathbf{x} \in \Gamma_h'$. Let the mapping $\Phi_h \mapsto u_h = L_h^{-1}\varphi_h$ be given by the rectangular matrix $\mathbf{A}$: $u_h = \mathbf{A}\varphi_h$. According to Equation (8) the entries of $\mathbf{A}$ read

$$a_{\mathbf{x}\xi} = -h\partial_n^- \overline{g}_h(\xi, \mathbf{x}) = g_h(\xi - h\mathbf{n}, \mathbf{x}) = g_h(\mathbf{x}, \xi - h\mathbf{n}), \quad \mathbf{x} \in \Omega_h,\ \xi \in \Gamma_h'.$$

Since $\mathbf{A} \geq \mathbf{0}$ (cf. Remark 8), one obtains the row sum norm $||\mathbf{A}||_\infty := \max_{\mathbf{x}} \sum_\xi a_{\mathbf{x}\xi}$ as $||\mathbf{A}\varphi_h||_\infty$ for the choice of $\varphi_h(\mathbf{x}) = 1$ in all $\mathbf{x} \in \Gamma_h'$. The solution $v_h = \mathbf{A}\varphi_h$ then reads $v_h = \mathbb{1}$ (why?), so it follows that

$$||\mathbf{A}||_\infty = ||\mathbf{A}\varphi_h||_\infty = ||\mathbb{1}||_\infty = 1.$$

(2) The column sums of $\mathbf{A}$ are $s(\xi) := \sum_{\mathbf{x}\in\Omega_h} a_{\mathbf{x}\xi} = \sum_{\mathbf{x}\in\Omega_h} g_h(\mathbf{x}, \xi - h\mathbf{n})$ for $\xi \in \Gamma_n'$. The grid function $v_h := h^{-2}L_h^{-1}\mathbb{1}$ at the points $\xi - h\mathbf{n}$ near the boundary has the values

$$s(\xi) = v_h(\xi - h\mathbf{n}), \quad \xi \in \Gamma_h', \ \mathbf{n} \text{ normal direction},$$

as is implied by Remark 7. Let $\xi = (\xi_1, \xi_2) \in \Gamma_h'$ be a point of the left or right boundary (i.e., $\xi_1 = 0$ or 1). As mentioned in the proof of (1c), $L_h w_h \geq \mathbb{1}$ holds for $w_h(x, y) := x(1-x)/2$. Since $L_h^{-1} \geq \mathbf{0}$, then so is $w_h \geq L_h^{-1}\mathbb{1}$ and hence $v_h \leq h^{-2}w_h$. In particular we have the following estimate

$$s(\xi) = v_h(\xi - h\mathbf{n}) \leq h^{-2}w_h(\xi - h\mathbf{n}) = h^{-2}h(1-h)/2 \leq h^{-1}/2$$

at the point near the boundary $\xi - h\mathbf{n} = (h, \xi_2)$ or $(1-h, \xi_2)$. For a point ξ from the upper or from the lower boundary one obtains the same estimate if one uses $w_h(x, y) := y(1-y)/2$. Since the column sums $s(\xi)$ are the row sums of $\mathbf{A}^\mathsf{T}$, we have proved

$$||\mathbf{A}^\mathsf{T}||_\infty = \max\{s(\xi) : \xi \in \Gamma_h'\} \leq h^{-1}/2.$$

(3) We have $||\mathbf{A}^\mathsf{T}\mathbf{A}||_2 = \rho(\mathbf{A}^\mathsf{T}\mathbf{A}) \leq ||\mathbf{A}^\mathsf{T}\mathbf{A}||_\infty \leq ||\mathbf{A}^\mathsf{T}||_\infty ||\mathbf{A}||_\infty \leq h^{-1}/2$ (cf. Exercise 4.3.20a, (3.10d), (3.10a)) so that the solution $u_h = L_h^{-1}\varphi_h = \mathbf{A}\Phi_h$ satisfies the following estimate

$$\begin{aligned}
||u_h||_{2,\Omega_h}^2 &= h^2 \sum_{\mathbf{x}\in\Omega_h} u^2(\mathbf{x}) = h^2 \sum_{\mathbf{x}\in\Omega_h} |(\mathbf{A}\Phi_h)(\mathbf{x})|^2 \\
&= h^2 \sum_{\mathbf{x}\in\Gamma_h'} \Phi_h(\mathbf{x})(\mathbf{A}^\mathsf{T}\mathbf{A}\Phi_h)(\mathbf{x}) \\
&\leq h^2 ||\mathbf{A}^\mathsf{T}\mathbf{A}||_2 \sum_{\mathbf{x}\in\Gamma_h'} \Phi_h^2(\mathbf{x}) \\
&\leq h/2 \sum_{\mathbf{x}\in\Gamma_h'} \Phi_h^2(\mathbf{x}) = \frac{1}{2} ||\Phi_h||_{2,\Gamma_h'}^2 .
\end{aligned}$$

Since $||\varphi||_{2,\Gamma_h'} = ||\Phi_h||_{2,\Gamma_h'}$, the assertion follows. ∎

4.5 Convergence

Let U_h be the vector space of the grid functions on $\overline{\Omega}_h$. The discrete solution $u_h \in U_h$ and the continuous solution $u \in C^0(\overline{\Omega})$ cannot be compared directly because of the different domains of definition. In difference methods it is customary to compare both functions on the grid $\overline{\Omega}_h$. To this end one must map the solution u by means of a "restriction"

$$u \mapsto R_h u \in U_h \quad (u: \text{continuous solution}) \tag{4.5.1}$$

to U_h. In the following we choose R_h as restriction on $\overline{\Omega}_h$:

$$(R_h u)(\mathbf{x}) = u(\mathbf{x}) \quad \text{for all } \mathbf{x} \in \overline{\Omega}_h. \tag{4.5.2}$$

The limit $h \to 0$ is made precise as follows. Let $H \subset \mathbb{R}_+$ be a subset with accumulation point zero: $0 \in \overline{H}$. For example, the step sizes considered so far form the set $H = \{1/n : n \in \mathbb{N}\}$. For each $h \in H$ let U_h be equipped with the norm $\|\cdot\|_h$.

Definition 4.5.1. (Convergence) The discrete solutions $u_h \in U_h$ converge (with respect to the system of norms $\|\cdot\|_h, h \in H$) to u if

$$\|u_h - R_h u\|_h \to 0. \tag{4.5.3a}$$

We have convergence of order k if

$$\|u_h - R_h u\|_h = O(h^k). \tag{4.5.3b}$$

The proof of convergence is usually carried out with the aid of the concepts of "stability" and "consistency". The discretisation $\{L_h : h \in H\}$ is said to be stable with respect to $\|\cdot\|_\infty$ if

$$\sup_{h \in H} \|L_h^{-1}\|_\infty < \infty. \tag{4.5.4}$$

For the discretisation defined in Section 4.2, the stability has been proved in (4.1c) with respect to the row sum norm, and in (4.1d) with respect to the spectral norm.

The grid function f_h in $-\Delta_h u_h = f_h$ is the restriction

$$f_h = \tilde{R}_h f \tag{4.5.5a}$$

of f, where in (2.6b) $\tilde{R}_h$ was chosen as the restriction to Ω_h:

$$(\tilde{R}_h f)(\mathbf{x}) = f(\mathbf{x}) \quad \text{for all } \mathbf{x} \in \Omega_h. \tag{4.5.5b}$$

The formulation of consistency becomes simpler if instead of the matrix L_h one considers the difference operator $-\Delta_h$ on which it is based (cf. Remark 4.2.5). Let D_h be a general difference operator, and let $-D_h u_h = f_h$ discretise $-\Delta u = f$. The discretisation described by the triple $(D_h, R_h, \tilde{R}_h)$ is said to be consistent of order k with respect to $||\cdot||_\infty$ (consistent with the Laplace operator), if

$$||D_h R_h u - \tilde{R}_h \Delta u||_\infty \le K h^k ||u||_{C^{k+2}(\overline{\Omega})} \tag{4.5.6}$$

for all $u \in C^{k+2}(\overline{\Omega})$. Here K is independent of h and u.

Remark 4.5.2. Let R_h and $\tilde{R}_h$ be given by (2) and (5b). The five-point formula Δ_h is consistent of order 2: estimate (6) holds with $k=2$ and $K=1/6$.

PROOF. The expansion (1.4c) can be applied in the x and y directions and yields

$$\Delta_h R_h u(x,y) = \Delta u(x,y) + h^2(R_x + R_y) \quad \text{with } |R_x|, |R_y| \le (1/12)||u||_{C^4(\overline{\Omega})}. \tag{4.5.7}$$

■

Theorem 4.5.3. *Let the discretisation $(D_h, R_h, \tilde{R}_h)$ be consistent of order k. Let the matrix L_h associated to the difference operator D_h (cf. Remark 4.2.5) be stable with respect to $||\cdot||_\infty$. Then the method is convergent of order k if $u \in C^{k+2}(\overline{\Omega})$.*

PROOF. $w_h := u_h - R_h u$ satisfies the difference equations

$$\begin{aligned} -D_h w_h &= -D_h u_h + D_h R_h u = f_h + D_h R_h u \\ &= \tilde{R}_h f + D_h R_h u = D_h R_h u - \tilde{R}_h \Delta u \quad \text{in } \Omega_h. \end{aligned}$$

Because of the homogeneous boundary condition $w_h = 0$ on Γ_h we have that $w_h = L_h^{-1}(D_h R_h u - \tilde{R}_h \Delta u)$ (cf. Remark 4.2.1). (4) and (6) imply (3b). ■

If one substitutes the concrete values $||L_h^{-1}||_\infty \le 1/8$ and $K = 1/6$ one obtains

Corollary 4.5.4. *Let the continuous solution of the boundary value problem (2.2a,b) belong to $C^4(\overline{\Omega})$. Let u_h be the discrete solution defined in (2.4a,b). Then the convergence of u_h to u is of second order:*

$$||u_h - R_h u||_\infty \le \frac{h^2}{48}||u||_{C^4(\overline{\Omega})}. \tag{4.5.8}$$

The assumption $u \in C^4(\overline{\Omega})$ can be weakened to

Corollary 4.5.5. *Under the condition $u \in C^{3,1}(\overline{\Omega})$ we also have*

$$||u_h - R_h u||_\infty \le (h^2/48)||u||_{C^{3,1}(\overline{\Omega})}. \tag{4.5.8$'$}$$

PROOF. The remainder term R_4 in (1.5b) may also be written as

$$R_4 = h^{-4} \int_x^{x \pm h} [u'''(\xi) - u'''(x)](x \pm h - \xi)^2/2! d\xi.$$

The Lipschitz estimate $|u'''(\xi) - u'''(x)| \le |\xi - x| \, ||u||_{C^{3,1}(\overline{\Omega})}$ implies $R_4 \le ||u||_{C^{3,1}(\overline{\Omega})}/4!$ so that in (7), (6), and (8) the norm $||u||_{C^4(\overline{\Omega})}$ can be replaced by $||u||_{C^{3,1}(\overline{\Omega})}$. ∎

If, however, one further weakens $u \in C^{3,1}(\overline{\Omega})$ to $u \in C^s(\overline{\Omega})$, $2 < s < 4$, one obtains a weaker order of convergence.

Corollary 4.5.6. *Under the condition $u \in C^s(\overline{\Omega})$, $2 < s \le 4$, u_h converges of order $s - 2$:*

$$||u_h - R_h u||_\infty \le K_s h^{s-2} ||u||_{C^s(\overline{\Omega})}, \tag{4.5.8"}$$

where $K_s := 1/[2s(s-1)]$ for $2 \le s \le 3$ and $K_s := 1/[2s(s-1)(s-2)]$ for $3 \le s \le 4$.

The proof results from

Exercise 4.5.7. Show $||\Delta_h R_h u - \tilde{R}_h \Delta u||_\infty \le 8K_s h^{s-2} ||u||_{C^s(\overline{\Omega})}$, $2 < s \le 4$.

Even though the proofs of convergence are simple, the results remain unsatisfactory. As can be seen from Example 2.1.3, the continuous solution of the boundary value problem (2.2a,b) generally does not even satisfy $u \in C^2(\overline{\Omega})$, although one needs at least $u \in C^s(\overline{\Omega})$ with $s > 2$ in Corollary 6 for convergence. Stronger results can be obtained by an analysis which will be discussed in Section 9.2. That errors of the order of magnitude of $O(h^2)$ occur even under weaker conditions, is shown in

Example 4.5.8. If one solves the difference equation (2.4a, b) for $-\Delta u = 1$ in $\Omega = (0,1) \times (0,1)$ and $u = 0$ on Γ, one obtains at the centre $x = y = 1/2$ the values $u_h(\frac{1}{2}, \frac{1}{2})$, which are shown in the first column of Table 1. The exact solution $u(\frac{1}{2}, \frac{1}{2}) = 0.0736713\cdots$ results from a representation that one can find in Example 8.1.11. The quotients ϵ_h/ϵ_{2h} of the errors $\epsilon_h = u(\frac{1}{2}, \frac{1}{2}) - u_h(\frac{1}{2}, \frac{1}{2})$ approximate $1/4$. This proves $u_h(\frac{1}{2}, \frac{1}{2}) = u(\frac{1}{2}, \frac{1}{2}) + O(h^2)$, although $u \notin C^2(\overline{\Omega})$. At $(\frac{1}{2}, \frac{1}{2})$ u_h has furthermore the asymptotic expansion $u_h(\frac{1}{2}, \frac{1}{2}) = u(\frac{1}{2}, \frac{1}{2}) + h^2 e(\frac{1}{2}, \frac{1}{2}) + O(h^4)$. The error term $e(\frac{1}{2}, \frac{1}{2})$ independent of h, is eliminated by using the Richardson extrapolation

$$u_{h,2h}\left(\frac{1}{2}, \frac{1}{2}\right) := \frac{1}{3}\left[4u_h\left(\frac{1}{2}, \frac{1}{2}\right) - u_{2h}\left(\frac{1}{2}, \frac{1}{2}\right)\right].$$

The extrapolated values are already very accurate for $h = 1/16$ (cf. the last column of Table 1).

Table 4.5.1 Difference solutions for Example 8

h	$u_h(\frac{1}{2},\frac{1}{2})$	$u_{h,2h}(\frac{1}{2},\frac{1}{2})$	$\epsilon_h = u_h(\frac{1}{2},\frac{1}{2}) - u_h(\frac{1}{2},\frac{1}{2})$	Quotient	$\epsilon_{h,2h}$	Quotient
1/8	0.0727826		8.89×10^{-4}			
				0.250		
1/16	0.07344576	0.0736668	2.26×10^{-4}		4.5×10^{-6}	
				0.251		0.06
1/32	0.0736147373	0.07367106	5.66×10^{-5}		2.7×10^{-7}	
				0.249		
1/64	0.0736571855	0.07367133	1.41×10^{-5}			

4.6 Discretisations of Higher Order

The five-point formula (2.11) is of second order. Even if the solution u belongs to $C^s(\overline{\Omega})$ with $s > 4$, no better bound for $\Delta_h R_h u - \tilde{R}_h \Delta u$ than $O(h^2)$ would result. An obvious method for constructing difference methods of higher order is the following. As an ansatz for the discretisation of the one-dimensional equation $u'' = f$ choose

$$(D_h u_h)(x) = h^{-2} \sum_{\nu=-k}^{k} c_\nu u_h(x + \nu h).$$

The Taylor expansion provides

$$(D_h R_h u)(x) = \sum_{\mu=0}^{2k} a_\mu h^{\mu-2} u^\mu(x) + O(h^{2k-1}), \quad a_\mu = \Big(\sum_{\nu=-k}^{k} c_\nu \nu^\mu\Big)/\mu!.$$

The $2k+1$ equations $a_0 = a_1 = a_3 = a_4 = \cdots = a_{2k} = 0$ and $a_2 = 1$ form a linear system for the $2k+1$ unknown coefficients c_ν. For $k = 1$ one obtains the usual difference formula (1.4c); for $k = 2$ a difference of fourth order results:

$$h^2(D_h u_h)(x) = \frac{-1}{12}[u_h(x-2h)+u_h(x+2h)]+\frac{4}{3}[u_h(x-h)+u_h(x+h)]-\frac{5}{2}u_h(x).$$

If one applies this approximation for u'' to the x and y coordinates, one obtains for $-\Delta$ the difference star

$$\frac{h^{-2}}{12}\begin{bmatrix} & & 1 & & \\ & & -16 & & \\ 1 & -16 & 60 & -16 & 1 \\ & & -16 & & \\ & & 1 & & \end{bmatrix} \tag{4.6.1}$$

(cf. (2.12)). The difference scheme (1) is of fourth order, but presents difficulties at points near the boundary. To set up the difference formula at

$(h, h) \in \Omega_h$, for example, one needs the values $u_h(-h, h)$ and $u_h(h, -h)$ outside $\overline{\Omega}_h$ (cf. Figure 1). One possibility would be to use scheme (1) only at points far from the boundary and to use the five-point formula (2.11) at points near the boundary.

The above complications do not occur if one limits oneself to **compact nine-point formulae**; by this one means difference methods (2.12) which are characterised by "$c_{\alpha\beta} \neq 0$ only for $-1 \le \alpha, \beta \le 1$" (cf. Fig. 1). An ansatz with the nine free parameters $c_{\alpha\beta}, -1 \le \alpha, \beta \le 1$, leads, however, to a negative result: there is no compact nine-point formula with $(D_h u)(\mathbf{x}) = -\Delta u(\mathbf{x}) + O(h^3)$. In this sense the five-point formula is indeed optimal.

Nevertheless, nine-point procedures of fourth order can be obtained if one also selects the right side f_h of the system of equations (2.6a,b) in a suitable manner.

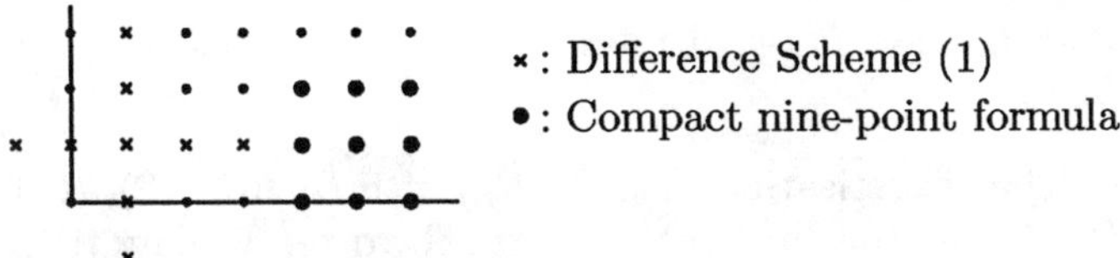

Figure 4.6.1. Scheme (1) and the compact nine-point formula

If one applies the compact nine-point scheme

$$D_h := -\frac{h^{-2}}{6}\begin{bmatrix} -1 & -4 & -1 \\ -4 & 20 & -4 \\ -1 & -4 & -1 \end{bmatrix} \tag{4.6.2}$$

to $u \in C^8(\overline{\Omega})$, the Taylor expansion results in

$$-D_h u = -\Delta u - \frac{h^2}{12}\Delta^2 u - \frac{h^4}{360}\left[\frac{\partial^4}{\partial x^4} + 4\frac{\partial^4}{\partial x^2 \partial y^2} + \frac{\partial^4}{\partial y^4}\right]\Delta u + O(h^6). \tag{4.6.3}$$

Here it is crucial that the error term can be expressed by $-\Delta u$ and hence by f.

For the special choice of the restriction $\tilde{R}_h$ via

$$f_h = \tilde{R}_h f := \frac{1}{6}\begin{bmatrix} & 1/2 & \\ 1/2 & 4 & 1/2 \\ & 1/2 & \end{bmatrix} f, \tag{4.6.4}$$

i.e.,

$$\begin{aligned} f_h(x,y) =& (\tilde{R}_h f)(x,y) \\ :=& \frac{1}{12}[f(x-h,y) + f(x+h,y) + \\ & f(x,y-h) + f(x,y+h) + 8f(x,y)], \end{aligned} \tag{4.6.4'}$$

one obtains the expansion

$$f_h(x,y) = f(x,y) + \frac{h^2}{12}\Delta f(x,y) + O(h^4), \tag{4.6.5}$$

which, because $f = -\Delta u$, agrees with (3) up to $O(h^4)$.

The matrix L_h of the system of equations which results after the elimination of the boundary values $u_h(x) = \varphi(x)$, $x \in \Gamma_h$, has the entries

$$L_{\mathbf{x}\xi} = h^{-2}\begin{cases} 20/6 & \text{if } \mathbf{x} = \xi, \\ -1/6 & \text{if } \mathbf{x} - \xi = (\pm h, \pm h) \text{ or } (\pm h, \mp h), \\ -4/6 & \text{if } \mathbf{x} - \xi = (\pm h, 0) \text{ or } (0, \pm h), \\ 0 & \text{otherwise.} \end{cases} \tag{4.6.6}$$

The right side of the system of equations $L_h u_h = q_h$ is

$$q_h := f_h + \varphi_h, \; f_h := \tilde{R}_h f \text{ according to } (4'), \; \varphi_h(x) := -\sum_{\xi \in \Gamma_h} L_{\mathbf{x}\xi}\varphi(\xi). \tag{4.6.7}$$

The discretisation $(D_h, R_h, \tilde{R}_h)$ with D_h from (2), R_h from (5.2), $\tilde{R}_h$ from $(4')$ is called the mehrstellen method (cf. Collatz [1]).

Exercise 4.6.1. Let D_h and L_h be defined respectively by (2) and (6). Prove that
(a) L_h is an M-matrix;
(b) $\|L_h^{-1}\|_\infty \le 1/8$ (stability), $\|L_h\|_\infty \le 20h^{-2}/3$;
(c) $\|D_h R_h u - \tilde{R}_h \Delta u\|_\infty \le \frac{11h^4}{180}\|u\|_{C^6(\overline{\Omega})}$, if $u \in C^6(\overline{\Omega})$ (consistency).

Theorem 4.6.2. *Let Ω_h be defined as in Section 4.2. Let u_h be the solution provided by the mehrstellen method $L_h u_h = q_h$ with L_h from (6), and q_h from (7). Let the solution u of the boundary value problem (2.2a,b) belong to $C^6(\overline{\Omega})$ or $C^{5,1}(\overline{\Omega})$. Then the following error estimates hold in the respective cases :*

$$\|u_h - R_h u\|_\infty \le \frac{11h^4}{1440}\|u\|_{C^6(\overline{\Omega})}, \quad \text{resp.} \quad \|u_h - R_h u\|_\infty \le \frac{11h^4}{1440}\|u\|_{C^{5,1}(\overline{\Omega})}. \tag{4.6.8a}$$

In the case of the potential equation, i.e. $f = 0$, we even have

$$\|u_h - R_h u\|_\infty \le K h^6 \|u\|_{C^8(\overline{\Omega})}, \quad \|u_h - R_h u\|_\infty \le K h^6 \|u\|_{C^{7,1}(\overline{\Omega})}, \tag{4.6.8b}$$

if $u \in C^8(\overline{\Omega})$, resp. $u \in C^{7,1}(\overline{\Omega})$.

PROOF. As in Theorem 4.5.3, (8a) results from Exercise (1b,c). In the case that $f = 0$ the $O(h^4)$-term also vanishes in (3). ■

4.7 The Discretisation of the Neumann Boundary Value Problem

The Dirichlet boundary condition $u(\mathbf{x}) = \varphi(\mathbf{x})$ was used directly in the difference method; a discretisation was not necessary. A different situation arises for the Neumann boundary value problem

$$-\Delta u = f \quad \text{in} \quad \Omega = (0,1) \times (0,1), \quad \partial u/\partial n = \varphi \quad \text{on} \quad \Gamma. \tag{4.7.1}$$

The normal derivative, which reads explicitly

$$\begin{aligned} \frac{\partial u}{\partial n} &= -u_y \quad \text{for } \mathbf{x} = (x,0) \in \Gamma, & \frac{\partial u}{\partial n} &= u_x \quad \text{for } \mathbf{x} = (x,1) \in \Gamma, \\ \frac{\partial u}{\partial n} &= -u_x \quad \text{for } \mathbf{x} = (0,y) \in \Gamma, & \frac{\partial u}{\partial n} &= u_y \quad \text{for } \mathbf{x} = (1,y) \in \Gamma, \end{aligned} \tag{4.7.2}$$

like the Laplace operator, must be replaced by a difference. We will investigate three different discretisations.

4.7.1 One-sided Difference for $\partial u/\partial n$

The Poisson equation leads to the $(n-1)^2 = (1/h - 1)^2$ equations

$$(-\Delta_h u_h)(\mathbf{x}) = f(\mathbf{x}) \quad \text{for all } \mathbf{x} \in \Omega_h, \tag{4.7.3}$$

which require the values of $u_h(\mathbf{x})$ for all $\mathbf{x} \in \overline{\Omega}'_h$, where

$$\overline{\Omega}'_h := \Omega_h \cup \Gamma'_h, \quad \Gamma'_h := \Gamma_h \backslash \{(0,0),(0,1),(1,0),(1,1)\}.$$

To obtain a further $4(n-1)$ equations for $\{u_h(\mathbf{x}) : \mathbf{x} \in \Gamma'_h\}$, we replace, at all $\mathbf{x} \in \Gamma'_h$, the normal derivative $\partial u/\partial n = \varphi$ by the backward difference

$$\partial_n^- u_h(\mathbf{x}) := h^{-1}[u_h(\mathbf{x}) - u_h(\mathbf{x} - h\mathbf{n})] = \varphi(\mathbf{x}) \quad \text{for } \mathbf{x} \in \Gamma'_h. \tag{4.7.4}$$

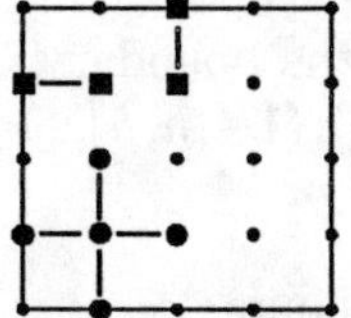

■ Discretisation of the Neumann condition
● Discretisation of the Poisson equation

Figure 4.7.1.

If one inserts the corresponding normal directions $\mathbf{n}$ for the four sides of the square one obtains

$$\left.\begin{aligned} h^{-1}[u_h(x,0)-u_h(x,h)] &= \varphi(x,0) \\ h^{-1}[u_h(x,1)-u_h(x,1-h)] &= \varphi(x,1) \end{aligned}\right\} \text{ for } x = h, 2h, \ldots, 1-h,$$

$$\left.\begin{aligned} h^{-1}[u_h(0,y)-u_h(h,y)] &= \varphi(0,y) \\ h^{-1}[u_h(1,y)-u_h(1-h,y)] &= \varphi(1,y) \end{aligned}\right\} \text{ for } y = h, 2h, \ldots, 1-h. \tag{4.7.4'}$$

Equations (3) and (4) yield $(n+1)^2 - 4$ equations for as many unknowns.

Exercise 4.7.1. After a rescaling of equations (4) to $h^{-1}\partial_n^- u_h(\mathbf{x}) = h^{-1}\varphi(\mathbf{x})$, $\mathbf{x} \in \Gamma_h'$, these equations, together with (3), form a system $L_h u_h = q_h$. Show that L_h is symmetric and satisfies (3.1a).

As in the Dirichlet problem the variables $u_h(\mathbf{x})$, $\mathbf{x} \in \Gamma_h'$, can be eliminated with the aid of (4) in (3). At the point (h,y) near the boundary, for example, Equation (3) becomes

$$h^{-2}[3u_h(h,y) - u_h(h,y-h) - u_h(h,y+h) - u_h(2h,y)] = f(h,y) + h^{-1}\varphi(0,y).$$

The star $h^{-2}\begin{bmatrix} & -1 & \\ -1 & 4 & -1 \\ & -1 & \end{bmatrix}$ thus becomes

$$h^{-2}\begin{bmatrix} & -1 & \\ 0 & 3 & -1 \\ & -1 & \end{bmatrix}, \qquad h^{-2}\begin{bmatrix} & -1 & \\ -1 & 3 & 0 \\ & -1 & \end{bmatrix},$$

$$h^{-2}\begin{bmatrix} & 0 & \\ -1 & 3 & -1 \\ & -1 & \end{bmatrix}, \qquad h^{-2}\begin{bmatrix} & -1 & \\ -1 & 3 & -1 \\ & 0 & \end{bmatrix},$$

near, respectively, the left, right, upper and lower boundaries. At the corner points one even has to replace two boundary values, so that, for example, at $(h,h) \in \Omega_h$ the star reads $h^{-2}\begin{bmatrix} & -1 & \\ 0 & 2 & -1 \\ & 0 & \end{bmatrix}$. Except for the special case $h = 1/2$, one obtains for the $(n-1)^2$ values $u_h(\mathbf{x})$, $\mathbf{x} \in \Omega_h$, the system of equations

$$L_h u_h = q_h \tag{4.7.5a}$$

with $$L_{\mathbf{x}\mathbf{x}} = h^{-2}\begin{cases} 4 & \text{if } \mathbf{x} \in \Omega_h \text{ is far from the boundary,} \\ 2 & \text{if } \mathbf{x} = (h,h), (h,1-h), (1-h,h), (1-h,1-h), \\ 3 & \text{otherwise} \end{cases} \tag{4.7.5b}$$

$$L_{\mathbf{x}\xi} = \begin{cases} -h^{-2} & \text{if } \mathbf{x}, \xi \in \overline{\Omega}_h \text{ are neighbours,} \\ 0 & \text{otherwise for } \mathbf{x} \neq \xi \end{cases}$$

and $$q_h = f_h + \varphi_h, \quad f_h(\mathbf{x}) = (\tilde{R}_h f)(\mathbf{x}) := f(\mathbf{x}), \quad \varphi_h(\mathbf{x}) = -h\textstyle\sum_{\xi \in \Gamma_h'} L_{\mathbf{x}\xi}\varphi(\xi). \tag{4.7.5c}$$

Remark 4.7.2. (a) L_h is symmetric and satisfies the sign condition (3.1a). (b) With lexicographical arrangement of the grid points of Ω_h, L_h has the form

$$L_h = h^{-2}\begin{bmatrix} T-I & -I & & & \\ -I & T & -I & & \\ & \ddots & \ddots & \ddots & \\ & & -I & T & -I \\ & & & -I & T-I \end{bmatrix},$$

$$T = \begin{bmatrix} 3 & -1 & & & \\ -1 & 4 & -1 & & \\ & \ddots & \ddots & \ddots & \\ & & -1 & 4 & -1 \\ & & & -1 & 3 \end{bmatrix}.$$

The matrix L_h is singular because the system $L_h u_h = q_h$, like the continuous boundary value problem (1), is, in general, not solvable. The analogue of Theorem 3.4.1 reads as follows.

Theorem 4.7.3. *The system of equations* (5a) *is solvable if and only if*

$$-h^2 \sum_{\mathbf{x}\in\Omega_h} f(\mathbf{x}) = h \sum_{\mathbf{x}\in\Gamma_h'} \varphi(\mathbf{x}). \tag{4.7.6}$$

Any two solutions of (5a) *can only differ by a constant:* $u_h^1 - u_h^2 = c\mathbb{1}$, $c \in \mathbb{R}$.

PROOF. Evidently, $L_h\mathbb{1} = 0$ holds, i.e., $\mathbb{1} \in \mathrm{kernel}(L_h)$. Furthermore, Theorem 4 will then imply $\dim(\mathrm{kernel}\,(L_h)) = 1$. This proves

$$\mathrm{kernel}(L_h) = \{c\mathbb{1} : c \in \mathbb{R}\} \tag{4.7.7}$$

and thus the second part of the assertion. (5a) is solvable if and only if the scalar product $\langle v, q_h\rangle$ vanishes for all $v \in \mathrm{kernel}\,(L_h^\mathsf{T}) = \mathrm{range}\,(L_h)^\perp$. Because of $L_h^\mathsf{T} = L_h$ and (7)

$$\langle \mathbb{1}, q_h\rangle = 0 \quad \text{or} \quad \sum_{\mathbf{x}\in\Omega_h} q_h(\mathbf{x}) = 0 \tag{4.7.6$'$}$$

is sufficient and necessary. According to Definition (5c), (6) and (6′) agree. ∎

Let condition (6) be satisfied. System (5a) can be solved as follows. Select an arbitrary $\mathbf{x}_0 \in \Omega_h$ and normalise the solution u_h (determined except for one constant) by

$$u_h(\mathbf{x}_0) = 0. \tag{4.7.8}$$

Let $\hat{u}_h$ be the vector u_h without the component $u_h(\mathbf{x}_0)$. Let $\hat{L}_h$ be the submatrix of L_h in which the row and column with index $\mathbf{x}_0$ have been left out. Let $\hat{q}_h$ be constructed likewise. Then

$$\hat{L}_h\hat{u} = \hat{q}_h \tag{4.7.9}$$

is a system with $(n-1)^2 - 1$ equations and unknowns.

Theorem 4.7.4. *The system of equations* (9) *is solvable; in particular, $\hat{L}_h$ is a symmetric M-matrix. Under condition* (6), $\hat{u}_h = \hat{L}_h^{-1}\hat{q}_h$, *supplemented by* (8), *yields the solution u_h of system* (5a).

PROOF. (a) As a principal submatrix of L_h, $\hat{L}_h$ is symmetric. In $\Omega_h\backslash\{\mathbf{x}_0\}$ any two grid points can be connected by a chain of neighbouring points so that $\hat{L}_h$ is irreducible. For all $\mathbf{x} \in \Omega_h\backslash\{\mathbf{x}_0\}$ there holds (3.4b); at neighbouring points of $\mathbf{x}_0$ we even have (3.4a), so that $\hat{L}_h$ is irreducibly diagonally dominant. According to Criterion 4.3.10, $\hat{L}_h$ is an M-matrix, thus nonsingular.
(b) If u_h is the solution of (5a), one can assume (8) without loss of generality, so that u_h restricted to $\Omega_h\backslash\{\mathbf{x}_0\}$ also solves Equation (9) and has to agree with the unique solution $\hat{u}_h$. ■

As a corollary of Theorem 4 one obtains that $\mathrm{rank}(L_h) \geq \mathrm{rank}(\hat{L}_h) = (n-1)^2 - 1$, i.e., $\dim(\mathrm{kernel}\ (L_h)) = 1$ and hence (7) holds.

Another possibility for the solution of Equation (5a) is to pass to an extended system of equations

$$\overline{L}_h\overline{u}_h = \overline{q}_h \tag{4.7.10a}$$

with $$\overline{L}_h = \begin{bmatrix} L_h & \mathbb{1} \\ \mathbb{1}^\mathsf{T} & 0 \end{bmatrix}, \qquad \overline{u}_h = \begin{bmatrix} u_h \\ \lambda \end{bmatrix}, \qquad \overline{q}_h = \begin{bmatrix} q_h \\ \sigma \end{bmatrix}, \tag{4.7.10b}$$

where σ can be prescribed arbitrarily.

Theorem 4.7.5. *Equation* (10a) *is always solvable. If for the last component of the solution $\overline{u}_h$ we have $\lambda = 0$, then condition* (6) *is satisfied and u_h represents the solution of system* (5a) *which is normalised by $\mathbb{1}^\mathsf{T}u_h = \sum_{\mathbf{x}\in\Omega_h} u_h(\mathbf{x}) = \sigma$. However, if $\lambda \neq 0$ holds one can interpret u_h as solution of $L_hu_h = \tilde{q}_h$ where $\tilde{q}_h = q_h - \lambda\mathbb{1}$ belongs to $\tilde{f}(\mathbf{x}) := f(\mathbf{x}) - \lambda$ and $\tilde{f}$ and φ satisfy condition* (6).

PROOF. $\mathbb{1}$ is linearly independent of the columns of L_h so that $\mathrm{rank}\ [L_h, \mathbb{1}] = \mathrm{rank}\ (L_h) + 1 = (n-1)^2$. Likewise, $(\mathbb{1}^\mathsf{T}, 0)$ is linearly independent of the rows of $[L_h, \mathbb{1}]$ so that $\mathrm{rank}\ (\tilde{L}_h) = (n-1)^2 + 1$, i.e., $\tilde{L}_h$ is nonsingular. The other statements can be read from (10b). ■

Attention. One should either use Equation (10a,b) or Equation (9), after first replacing q_h by $\tilde{q}_h := q_h - (\mathbb{1}^\mathsf{T}q_h/\mathbb{1}^\mathsf{T}\mathbb{1})\mathbb{1}$. As a justification of this recommendation note that the condition for solvability of the continuous problem is $\int_\Omega f dx + \int_\Gamma \varphi d\Gamma = 0$ and that this does not at all imply the discrete solvability condition (6). For smooth functions f and φ Equation (6) can be shown to hold up to a remainder of order $O(h)$. Thus it is generally unavoidable to replace f_h and q_h by $f_h - \lambda\mathbb{1}$ and $q_h - \lambda\mathbb{1}$ ($\lambda = \mathbb{1}^\mathsf{T} q_h / \mathbb{1}^\mathsf{T}\mathbb{1}$). In the case of Equation (10a,b) this correction is carried out implicitly. If, however, (9) is used without any correction, the resulting solution can be interpreted as a solution of $L_h u_h = \tilde{q}_h$ with $\tilde{q}_h(\mathbf{x}) = q_h(\mathbf{x})$ for $\mathbf{x} \neq \mathbf{x}_0$ and $\tilde{q}_h(\mathbf{x}_0) := -\sum_{\mathbf{x}\neq\mathbf{x}_0} q_h(\mathbf{x})$. That is, here too, an implicit correction of q_h is carried out, with the difference that the correction is not distributed over all components as before, but is concentrated on $q_h(\mathbf{x_0})$. If Equation (6) is satisfied up to order $O(h)$, then $q_h(\mathbf{x_0})$ and $\tilde{q}_h(\mathbf{x_0})$ differ by $O(h^{-1})$. Therefore the solution $\hat{u}_h$ of Equation (9) contains a singularity at the point $\mathbf{x_0}$.

Theorem 4.7.6. (Convergence) *Let $u \in C^{3,1}(\overline{\Omega})$ be the solution of the Neumann problem* (1). *Let $\overline{u}_h = \begin{pmatrix} u_h \\ \lambda \end{pmatrix}$ be the solution of Equation* (10a). *Then there exists a $c \in \mathbb{R}$ and constants C, C' independent of u and h such that*

$$|\lambda| \leq C' h \|u\|_{C^{0,1}(\overline{\Omega})}, \quad \|u_h - R_h u - c\mathbb{1}\|_\infty \leq C[h\|u\|_{C^{1,1}(\overline{\Omega})} + h^2\|u\|_{C^{3,1}(\overline{\Omega})}]. \tag{4.7.11}$$

PROOF. (a) We have

$$\lambda = \mathbb{1}^\mathsf{T} q_h / \mathbb{1}^\mathsf{T}\mathbb{1} = (h^2 \sum_{\Omega_h} f(\mathbf{x}) + h \sum_{\Gamma_h'} \varphi(\mathbf{x})) h^{-2} (n-1)^{-2}.$$

Because $\int_\Omega f\,d\mathbf{x} + \int_\Gamma \varphi\,d\Gamma = 0$, the first factor consists only of the quadrature error $O(h\|u\|_{C^{0,1}(\overline{\Omega})})$.

(b) According to Theorem 5, u_h is the solution of $L_h u_h = \tilde{q}_h := q_h - \lambda\mathbb{1}$. This corresponds to the difference equations

$$-\Delta_h u_h = \tilde{f}_h := f_h - \lambda\mathbb{1} \quad \text{in } \Omega_h, \qquad \partial_n^- u_h = \varphi \quad \text{on } \Gamma_h'.$$

The difference $w_h := u_h - R_h u$ satisfies

$$\begin{aligned} -\Delta_h w_h &= -\Delta_h u_h + \Delta_h R_h u = \Delta_h R_h u + f_h - \lambda\mathbb{1} \\ &= \Delta_h R_h u - \tilde{R}_h \Delta u - \lambda\mathbb{1} =: c_h \quad \text{in } \Omega_h, \end{aligned} \tag{4.7.12a}$$

$$\partial_n^- w_h = \partial_n^- u_h - \partial_n^- R_h u = \varphi - \partial_n^- R_h u = \frac{\partial u}{\partial n} - \partial_n^- R_h u =: \psi \quad \text{on } \Gamma_h'. \tag{4.7.12b}$$

The errors of consistency are

$$\begin{aligned} \|c_h\|_\infty &\leq \frac{1}{6} h^2 \|u\|_{C^{3,1}(\overline{\Omega})} && \text{(cf. Remark 4.5.2)}, \\ |\psi(\mathbf{x})| &\leq \frac{1}{2} h \|u\|_{C^{1,1}(\overline{\Omega})} + |\lambda| && \text{(cf. Lemma 4.1.1)}. \end{aligned}$$

Since the solution w_h exists, condition (6) is satisfied with c_h and ψ:

$$h^2 \sum_{\Omega_h} c_h(\mathbf{x}) + h \sum_{\Gamma_h'} \psi(\mathbf{x}) = 0.$$

Then $\tilde{w}_h := w_h - c\mathbb{1}$ with $c = \mathbb{1}^\mathsf{T} w_h / \mathbb{1}^\mathsf{T}\mathbb{1}$ is the solution of (12a,b) normalised by $\mathbb{1}^\mathsf{T} w_h = 0$. The application of the following Theorem 4.7.7 to Equation (12a,b) provides the inequality (11). ∎

Theorem 4.7.7. (Stability) *Let condition* (6) *be satisfied. Let the solution* u_h *of* (3), (4) *be normalised by* $\mathbb{1}^\mathsf{T} u_h = 0$. *Then there exist constants* C_1, C_2 *independent of* u *and* h *such that*

$$||u_h||_\infty \le C_1 \max_{\mathbf{x}\in\Omega_h} |f(\mathbf{x})| + C_2 \max_{\mathbf{x}\in\Gamma_h'} |\varphi(\mathbf{x})|. \tag{4.7.13}$$

The proof of this theorem, which corresponds to Theorem 4.4.11, will be supplied in Sect. 4.7.4.

4.7.2 Symmetric Difference for $\partial u/\partial n$

As can be seen from Theorem 6, the one-sided difference ∂_n^- causes the error term $O(h)$. It seems obvious to replace ∂_n^- by a symmetric difference. To this end the five-point discretisation is set up at all points $\mathbf{x} \in \overline{\Omega}_h = \Omega_h \cup \Gamma_h$ (cf. (2.1c)):

$$-\Delta_h u_h = f_h := \tilde{R}_h f \quad \text{in } \overline{\Omega}_h, \tag{4.7.14a}$$

where $\tilde{R}_h$ is the restriction to $\overline{\Omega}_h$. The symmetric difference ∂_n^0 is defined by

$$(\partial_n^0 u_h)(\mathbf{x}) := (2h)^{-1}[u_h(\mathbf{x} + h\mathbf{n}) - u_h(\mathbf{x} - h\mathbf{n})] = \varphi(\mathbf{x}) \text{ in } \mathbf{x} \in \Gamma_h. \tag{4.7.14b}$$

Here, we assign two normal directions to the corner points, so that for each of the corner points two equations of the form (14b) can be set up. In the corner $\mathbf{x} = (0,0)$ one has, for example, the normals

$$\mathbf{n} = \begin{pmatrix} 0 \\ -1 \end{pmatrix}, \qquad \mathbf{n} = \begin{pmatrix} -1 \\ 0 \end{pmatrix}.$$

The corresponding equations (14b) also contain different (!) values $\varphi(0+,0) = \lim_{x\to 0} \varphi(x,0)$ and $\varphi(0,0+) = \lim_{y\to 0} \varphi(0,y)$. For $\mathbf{x} \in \Gamma_h$ the difference formula (14a) needs the values in points $\mathbf{x} + h\mathbf{n}$ outside of $\overline{\Omega}_h$. These can be eliminated with the aid of (14b) so that a system of equations $L_h u_h = q_h$ remains for the $(n+1)^2$ components $u_h(\mathbf{x})$, $\mathbf{x} \in \overline{\Omega}_h$.

Exercise 4.7.8. (a) If the grid points of $\overline{\Omega}_h$ are arranged in lexicographical order, L_h has the form

$$L_h = h^{-2} \begin{bmatrix} T & -2I & & & \\ -I & T & -I & & \\ & \ddots & \ddots & \ddots & \\ & & -I & T & -I \\ & & & -2I & T \end{bmatrix}, \quad T = \begin{bmatrix} 4 & -2 & & & \\ -1 & 4 & -1 & & \\ & \ddots & \ddots & \ddots & \\ & & -1 & 4 & -1 \\ & & & -2 & 4 \end{bmatrix}.$$

(b) L_h is not symmetric, but $D_h L_h$ with $D_h = \operatorname{diag}\{d(x)d(y)\}$, $d(0) = d(1) = 1/2$, $d(\cdot) = 1$ otherwise, is symmetric.

The analogue of Theorem 3 reads:

Theorem 4.7.9. *Equation* (14a,b) *is solvable if and only if* $\mathbb{1}^\mathsf{T} D_h q_h = 0$ *for* D_h *in Exercise* 8b. *Any two solutions may differ by only a constant. The formulation of* $\mathbb{1}^\mathsf{T} D_h q_h = 0$ *with the aid of* f *and* φ *reads:*

$$-h^2 \sum_{(x,y)\in\overline{\Omega}_h} d(x)d(y)f(x,y) = 2h \sum_{(x,y)\in\Gamma_h} d(x)d(y)\varphi(x,y), \tag{4.7.15}$$

where the summand for the corner points occurs twice in the second sum, and both the different limits for φ *are taken into account.*

Remark 4.7.10. The sums in (15) represent summed trapezoidal formulae. Thus, from $\int_\Omega f dx + \int_\Gamma \varphi d\Gamma = 0$ follows Equation (15), except for a remainder $O(h^2 \|u\|_{C^{1,1}(\overline{\Omega})})$.

Theorems 4 and 5 can be transferred without difficulty. The convergence theorem, Theorem 6, becomes

Theorem 4.7.11. *Let* $u \in C^{3,1}(\overline{\Omega})$ *be a solution of* (1). *Let* $\overline{u}_h = \begin{pmatrix} u_h \\ \lambda \end{pmatrix}$ *be the solution of* $\overline{L}_h \overline{u}_h = \begin{pmatrix} D_h q_h \\ 0 \end{pmatrix}$ *with* $\overline{L}_h = \begin{bmatrix} D_h L_h & \mathbb{1} \\ \mathbb{1}^\mathsf{T} & 0 \end{bmatrix}$. *We have convergence of second order:*

$$|\lambda| \le C' h^2 \|u\|_{C^{1,1}(\overline{\Omega})}, \qquad \|u_h - R_h u - c\mathbb{1}\|_\infty \le C h^2 \|u\|_{C^{3,1}(\overline{\Omega})}. \tag{4.7.16}$$

The proof is essentially the same as for Theorem 6. An additional technical difficulty is the fact that the consistency error $\Delta_h R_h u - \tilde{R}_h \Delta u$ must also be determined in $\mathbf{x} \in \Gamma_h$ although u is only defined in $\overline{\Omega}$. Instead of treating the difference equation and the boundary discretisation separately as in (12a,b) one should directly analyse the equations $L_h u_h = q_h$ from which the values $u_h(\mathbf{x} + h\mathbf{n})$ outside of $\overline{\Omega}$ have already been eliminated.

4.7.3 Symmetric Difference for $\partial u/\partial n$ on an Offset Grid

If we offset the above grid by $h/2$ in the x and y directions, we obtain the grid

$$\Omega_h := \left\{(x,y) \in \Omega \colon \frac{x}{h} - \frac{1}{2}, \frac{y}{h} - \frac{1}{2} \in \mathbb{Z}\right\}$$

in Figure 2. The points near the boundary of Ω_h are at a distance $h/2$ from Γ. We set

$$\Gamma_h := \left\{(x,y) \in \Gamma: \frac{x}{h} - \frac{1}{2} \in \mathbb{Z} \text{ or } \frac{y}{h} - \frac{1}{2} \in \mathbb{Z}\right\}$$

(cf. Figure 2). To each point near the boundary $\mathbf{x} - h\mathbf{n}/2$ ($\mathbf{x} \in \Gamma_h$) corresponds an outlying neighbour $\mathbf{x} + h\mathbf{n}/2$.

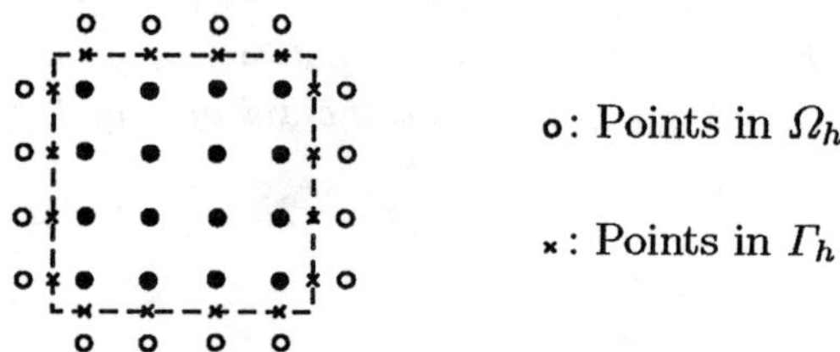

Figure 4.7.2. Offset grid

The discretisation of the Poisson equation (1) is

$$-\Delta_h u_h(\mathbf{x}) = f(\mathbf{x}) \quad \text{in } \mathbf{x} \in \Omega_h, \tag{4.7.17a}$$

$$h^{-1}[u_h(\mathbf{x} + h\mathbf{n}/2) - u_h(\mathbf{x} - h\mathbf{n}/2)] = \varphi(\mathbf{x}) \quad \text{in } \mathbf{x} \in \Gamma_h. \tag{4.7.17b}$$

The difference (17b) is symmetric with respect to the boundary point $\mathbf{x}$ and nevertheless agrees with the backward difference ∂_n^- at the grid point $\mathbf{x}+h\mathbf{n}/2$.

Remark 4.7.12. After eliminating the values $u_h(\mathbf{x} + h\mathbf{n}/2)$, $\mathbf{x} \in \Gamma_h$, we obtain a system of equations $L_h u_h = q_h$, where Remark 2 is also valid for this matrix L_h. In contrast to Section 4.7.1, L_h is of size $n^2 \times n^2$.

The Theorems 3, 4, 5, and 7 hold analogously. Theorem 6 holds with the inequality (16) instead of (11).

4.7.4 Proof of the Stability Theorem 7

In the case of Dirichlet boundary values the stability statement of Theorem 4.4.11 follows immediately from the maximum principle and the bound for L_h^{-1}. The corresponding statement of Theorem 7 for Neumann boundary values, however, cannot be proved that easily. In the literature one can only find weaker estimates which on the right-hand side of Equation (13) contain an additional factor $|\log h|$. However, this factor is not to be avoided if one uses Equation (9), $\hat{L}_h \hat{u}_h = \hat{q}_h$, without condition (6) being satisfied.

Let the discrete Green function $g_h(\mathbf{x}, \xi)$, $\mathbf{x} \in \Omega_h \cup \Gamma_h$, $\xi \in \Omega_h$ be defined by

$$-\Delta_h g_h(\mathbf{x}, \xi) = \delta(\mathbf{x}, \xi) \tag{4.7.18a}$$

$$:= \begin{Bmatrix} h^{-2} & \text{for } \mathbf{x} = \xi \\ 0 & \text{otherwise} \end{Bmatrix} - \begin{Bmatrix} h^{-1}/(4-4h) & \text{if } \mathbf{x} \text{ is near the boundary} \\ 0 & \text{if } \mathbf{x} \text{ is far from the boundary} \end{Bmatrix},$$

$$\partial_n^- g_h(\mathbf{x},\xi) = 0 \quad \text{for} \quad \mathbf{x} \in \Gamma_h' \tag{4.7.18b}$$

where Δ_h and ∂_n^- act on $\mathbf{x}$. g_h exists since $\sum_{\mathbf{x}\in\Omega_h} \delta(\mathbf{x},\xi) = 0$ proves the solvability condition (6′).

Lemma 4.7.13. *For arbitrary q_h with $\mathbb{1}^\mathsf{T} q_h = 0$ (cf. (6′)),*

$$u_h(\mathbf{x}) := h^2 \sum_{\xi\in\Omega_h} q_h(\xi) g_h(\mathbf{x},\xi) \tag{4.7.19}$$

represents a solution of $L_h u_h = q_h$.

PROOF. At points far from the boundary $\mathbf{x} \in \Omega_h$, $(L_h u_h)(\mathbf{x}) = -\Delta_h u_h(\mathbf{x}) = q_h(\mathbf{x})$. At points near the boundary $\mathbf{x} \in \Omega_h$, from $\partial_n^- u_h = 0$ and (6′), one has the identity

$$(L_h u_h)(\mathbf{x}) = (-\Delta_h u_h)(\mathbf{x}) = q_h(\mathbf{x}) - \frac{h^{-1}}{4-4h} \sum_{\xi\in\Omega_h} q_h(\xi) = q_h(\mathbf{x}). \qquad ■$$

Equation (18a,b) determine g_h up to a constant.

Theorem 4.7.14. *The Green function $g_h(\mathbf{x},\xi)$ can be so selected that*

$$|g_h(\mathbf{x},\xi)| \le C[1 + |\log(|\mathbf{x}-\xi| + h)|] \quad \text{for} \ \mathbf{x},\xi \in \Omega_h. \tag{4.7.20}$$

This inequality corresponds to the bound (4.7) in the Dirichlet case. Before Theorem 14 is proved, we want to show that Theorem 7 follows from it. Analogously to $\int_\Omega [1 + |\log|\mathbf{x}-\xi||]\,d\mathbf{x} \le K_1$ and $\int_\Gamma [1 + |\log|\mathbf{x}-\xi||]\,d\Gamma_\mathbf{x} \le K_2$ we obtain

$$h^2 \sum_{\mathbf{x}\in\Omega_h} [1 + |\log(|\mathbf{x}-\xi| + h)|] \le K_1',\ h \sum_{\mathbf{x}\in\Gamma_h'} [1 + |\log(|\mathbf{x}-\xi| + h)|] \le K_2'.$$

From (19), (20), and $q_h = f_h + \varphi_h$ (cf. (5c)) thus follows the estimate

$$|u_h(\mathbf{x})| \le K_1 \|f_h\|_\infty + hk_2 \|\varphi_h\|_\infty \le K_1 \max_{\mathbf{x}\in\Omega_h} |f(\mathbf{x})| + 2K_2 \max_{\mathbf{x}\in\Gamma_h'} |\varphi(\mathbf{x})|.$$

If $\tilde{u}_h := u_h - \mathbb{1}^\mathsf{T} u_h / \mathbb{1}^\mathsf{T}\mathbb{1}$ is the solution of $L_h u_h = q_h$ normalised by $\mathbb{1}^\mathsf{T} \tilde{u}_h = 0$, one can see that

$$\|\tilde{u}_h\|_\infty \le 2 \inf\{\|u_h - c\mathbb{1}\|_\infty : c \in \mathbb{R}\} \le 2\|\tilde{u}_h\|_\infty,$$

so that the inequality (11) and Theorem 7 are proved. ■

It remains to prove Theorem 14.

Lemma 4.7.15. *Let* $\mathbf{e}$ *be one of the unit vectors* $(1,0)$ *or* $(0,1)$. *Let* $\xi \in \Omega_h$ *be such that* $\xi + h\mathbf{e} \in \Omega_h$ *also. For all these* $\mathbf{e}$ *and* ξ *let there exist a function* $G_h(\mathbf{x}) = G_h(\mathbf{x};\xi,\mathbf{e})$ *with the properties*

$$-\Delta_h G_h(\mathbf{x}) = h^{-2}\left\{\begin{array}{rl} 1 & \text{if } \mathbf{x}=\xi \\ -1 & \text{if } \mathbf{x}=\xi+h\mathbf{e} \\ 0 & \text{otherwise} \end{array}\right\} \text{ in } \Omega_h, \tag{4.7.21a}$$

$$\partial_n^- G_h = 0 \quad \text{on } \Gamma_h', \tag{4.7.21b}$$

$$|G_h(\mathbf{x})| \le hC'/(|\mathbf{x}-\xi|+h), \tag{4.7.21c}$$

where C' *does not depend on* $\mathbf{e}$ *and* ξ. *Then Theorem* 14 *holds, and thus also Theorem* 7.

PROOF. For $\xi, \xi' \in \Omega_h$ let $g_h(\mathbf{x},\xi,\xi')$ be defined as a solution of

$$\partial_n^- g_h(\cdot,\xi,\xi') = 0 \text{ on } \Gamma_h', \quad -\Delta_h g_h(\mathbf{x},\xi,\xi') = h^{-2}\left\{\begin{array}{rl} 1 & \text{if } \mathbf{x}=\xi \\ -1 & \text{if } \mathbf{x}=\xi' \\ 0 & \text{otherwise} \end{array}\right\} \text{ in } \Omega_h.$$

For $\xi' = \xi + h\mathbf{e}$, $g_h(\cdot,\xi,\xi')$ agrees with G_h in Lemma 15. For arbitrary ξ, ξ', one finds a connection $\xi = \xi^0, \xi^1, \cdots, \xi^l = \xi'$ with $\xi^{k+1} - \xi^k = \pm h\mathbf{e}^k$, $\mathbf{e}^k = (1,0)$ or (0,1). Since $g_h(\cdot,\xi,\xi') = \sum_{k=1}^{l} g_h(\cdot,\xi^{k-1},\xi^k)$, (21c) implies the estimate

$$|g_h(\mathbf{x},\xi,\xi')| \le h\sum_{k=1}^{l} C'/(|\mathbf{x}-\xi^{k-1}|+h).$$

If one considers first the case $x_1 = \xi_1 = \xi_1'$, $x_2 \le \xi_2 < \xi_2'$ and one applies $\sum_{k=k_1}^{k_2} 1/k \le \text{const} \cdot (1 + \log(k_2 h) - \log(k_1 h))$, one obtains

$$|g_h(\mathbf{x},\xi,\xi')| \le C''[1 + |\log(|\mathbf{x}-\xi|+h)| + |\log(|\mathbf{x}-\xi'|+h)|].$$

In the general case, this bound also comes out, but one must choose the connection ξ^k such that $|\mathbf{x}-\xi^k| \ge \min\{|\mathbf{x}-\xi|, |\mathbf{x}-\xi'|\}$. Since we know that $h\sum_{\xi'\in\Gamma_h'} |\log(|\mathbf{x}-\xi'|+h)| \le \text{const}$ for all $\mathbf{x} \in \Omega_h$,

$$g_h(\mathbf{x},\xi) := \frac{h}{4-4h}\sum_{\xi'\in\Gamma_h'} g_h(\mathbf{x},\xi,\xi')$$

satisfies the inequality (20). Because $\sum_{\Gamma_h'} 1 = (4-4h)/h$ the equations (18a,b) are also satisfied, i.e., $g_h(\mathbf{x},\xi)$ is the Green function. ■

We shall construct the function G_h needed in Lemma 15 explicitly. The basic building block is the discrete singularity function $s_h(\mathbf{x},\xi)$ on the grid in $\mathbb{R}^2$

$$Q_h := \{(x,y) \in \mathbb{R}^2 : x/h, y/h \in \mathbb{Z}\}.$$

Lemma 4.7.16. *The singularity function defined by* $s_h(\mathbf{x},\xi) := \sigma_h(\mathbf{x}-\xi)$ *and*

$$\sigma_h(\mathbf{x}) := \frac{1}{16\pi^2}\int_{-\pi}^{\pi}\int_{-\pi}^{\pi}\frac{e^{i(x_1\eta_1+x_2\eta_2)/h}-1}{\sin^2(\eta_1/2)+\sin^2(\eta_2/2)}\,d\eta_1\,d\eta_2$$
$$= -\frac{1}{8\pi^2}\int_{-\pi}^{\pi}\int_{-\pi}^{\pi}\frac{\sin^2((x_1\eta_1+x_2\eta_2)/(2h))}{\sin^2(\eta_1/2)+\sin^2(\eta_2/2)}\,d\eta_1\,d\eta_2$$

for all $\mathbf{x},\xi\in Q_h$ *has the property* $-\Delta_h s_h(\mathbf{x},\xi)=h^{-2}$ *for* $\mathbf{x}=\xi$ *and* $-\Delta_h s_h = 0$ *otherwise.*

PROOF. Let $e(\mathbf{x},\eta) := \exp(i(x_1\eta_1+x_2\eta_2)/h)$. Note that

$$-\Delta_h(e(\mathbf{x},\eta)-1) = 4h^{-2}e(\mathbf{x},\eta)[\sin^2(\eta_1/2)+\sin^2(\eta_2/2)]$$

and

$$\int_{-\pi}^{\pi}\int_{-\pi}^{\pi} e(\mathbf{x},\eta)\,d\eta = \begin{Bmatrix} 4\pi^2 & \text{for } \mathbf{x}=\mathbf{0} \\ 0 & \text{for } \mathbf{x}\neq\mathbf{0}\end{Bmatrix}. \qquad \blacksquare$$

For multi-indices $\alpha=(\alpha_1,\alpha_2)\in\mathbb{Z}^2$ with $\alpha_1\ge 0$, $\alpha_2\ge 0$ one defines the partial difference operators of order $|\alpha|=\alpha_1+\alpha_2$ by

$$\partial^\alpha = (\partial_x^+)^{\alpha_1}(\partial_y^+)^{\alpha_2}. \tag{4.7.22}$$

Starting with the representation in Lemma 16, Thomée [1] proves

Lemma 4.7.17. $|(\partial^\alpha\sigma_h)(\mathbf{x})| \le C(|\mathbf{x}|+h)^{-|\alpha|}$ *for all* $\mathbf{x}\in Q_h$.

For the construction of the function G_h in Lemma 15 we select, without loss of generality, $\mathbf{e}=(1,0)$ and keep $\xi\in\Omega_h$ with $\xi+h\mathbf{e}\in\Omega_h$. The function

$$G_h'(\mathbf{x}) := h\partial_x^-\sigma_h(\mathbf{x}-\xi) = \sigma_h(\mathbf{x}-\xi)-\sigma_h(\mathbf{x}-\xi-h\mathbf{e}) \tag{4.7.23}$$

satisfies (21a) but not the boundary condition (21b). Let the symmetrisation operators S_x and S_y be defined by

$$S_x u_h(x,y) := \frac{1}{2}[u_h(x,y)+u_h(h-x,y)],\ S_y u_h(x,y) := \frac{1}{2}[u_h(x,y)+u_h(x,h-y)]$$

for $(x,y)\in Q_h$. The function

$$G_h'' := 4S_xS_yG_h' \tag{4.7.24}$$

is symmetric with respect to the axes $y=h/2$ and $x=h/2$. Thus we have (21b) on the left and lower boundary:

$$\partial_n^- G_h''(0,y) = \partial_n^- G_h''(x,0) = 0 \tag{4.7.25}$$

Furthermore, G_h'' satisfies condition (21a) just as G_h' does. For each $\beta\in\mathbb{Z}^2$ we define the operator P_β by

$$(P_\beta u_h)(x,y) = \frac{1}{4}[u_h(x+\beta_1 L, y+\beta_2 L) + u_h(x-\beta_1 L, y+\beta_2 L)$$
$$+u_h(x+\beta_1 L, y-\beta_2 L) + u_h(x-\beta_1 L, y-\beta_2 L)],$$

where $L = 2-2h$, and set

$$G_h^\beta(\mathbf{x}) := (P_\beta G_h'')(\mathbf{x}) - (P_\beta G_h'')(\mathbf{0}). \tag{4.7.26}$$

Lemma 4.7.18. *For $\beta = (\beta_1, \beta_2)$ with $\|\beta\|_\infty \geq 2$ there holds*

$$|\partial^\alpha G_h^\beta(\mathbf{x})| \leq hK/|\beta|^3 \quad \text{for all } |\alpha| \leq 2, \ \mathbf{x} \in \Omega_h. \tag{4.7.27}$$

Here, K is independent of the choice of the $\xi, \xi + h\mathbf{e} \in \Omega_h$.

PROOF. According to the definition we have

$$G_h^\beta(0,0) = 0. \tag{4.7.28a}$$

The operator P_β preserves the symmetry, i.e., $u_h = S_x u_h$ implies $P_\beta u_h = S_x P_\beta u_h$. Thus we have that $G_h^\beta = S_x G_h^\beta = S_y G_h^\beta$ and consequently

$$\partial_x^+ G_h^\beta(0,0) = \partial_y^+ G_h^\beta(0,0) = 0. \tag{4.7.28b}$$

$G_h^\beta(\mathbf{x})$ is the linear combination of $h\partial_x^- \sigma_h(\mathbf{x}-\tilde{\xi})$ for different $\tilde{\xi}$ with $|\mathbf{x}-\tilde{\xi}|+h \geq K'|\beta|$, if $\mathbf{x} \in \Omega_h$. According to Lemma 17,

$$|(\partial^\alpha G_h^\beta)(\mathbf{x})| \leq hCK''|\beta|^{-3} \quad \text{for all } |\alpha| = 2, \mathbf{x} \in \Omega_h, \tag{4.7.28c}$$

so that (27) follows for $|\alpha| = 2$. Let $|\alpha| = 1$. $\partial^\alpha G_h(\mathbf{x})$ can be written in the form

$$\partial^\alpha G_h^\beta(\mathbf{0}) + \sum_{k=1}^{l} [\partial^\alpha G_h^\beta(\mathbf{x}^k) - \partial^\alpha G_h^\beta(\mathbf{x}^{k-1})]$$

with $\mathbf{x}^0 = \mathbf{x}$, $\mathbf{x}^l = \mathbf{x}$, $\|\mathbf{x}^k - \mathbf{x}^{k-1}\|_\infty = h$. Each summand has the form $\pm h\partial^\gamma G_h^\beta$ with $|\gamma| = 2$ so that (28b,c) lead to the estimate (28d)

$$|(\partial^\alpha G_h^\beta)(\mathbf{x})| \leq 2hCK''|\beta|^{-3} \quad \text{for } |\alpha| = 1, \ \mathbf{x} \in \Omega_h. \tag{4.7.28d}$$

Likewise one infers from (28a) and (28d) the inequality (27) for $|\alpha| = 0$, which proves Lemma 18. ■

Since $\sum_{\beta \in \mathbb{Z}^2 \setminus \{0\}} |\beta|^{-3} < \infty$,

$$G_h(\mathbf{x}) := \sum_{\beta \in \mathbb{Z}^2} G_h^\beta(\mathbf{x}) \tag{4.7.29}$$

exists. Since $-\Delta_h G_h^\beta(\mathbf{x}) = 0$ in Ω_h for all $\beta \neq (0,0)$, G_h also satisfies Equation (21a). As already mentioned in the proof of (28b), $G_h = S_x G_h = S_y G_h$

holds so that G_h also satisfies Equation (25): $\partial_n^- G_h = 0$ at the left and lower boundary. The proof of $\partial_n^- G_h = 0$ at the lower boundaries requires

Lemma 4.7.19. *G_h is L-periodic: $G_h(x,y) = G_h(x+L,y) = G_h(x,y+L)$.*

PROOF. Let $G_h''(x+\beta_1 L, y+\beta_2 L) - G_h''(\beta_1 L, \beta_2 L)$ be abbreviated to $\gamma_\beta(x,y)$. Definition (29) says that

$$G_h = \lim_{k\to\infty} \sum_{|\beta_1|\le k} \sum_{|\beta_2|\le k} \gamma_\beta. \tag{4.7.30a}$$

Now γ_β can be written as a sum over the differences $h\partial_h^+ G_h''(x+\beta_1 L, \beta_2 L+\mu h)$ from $\mu h = 0$ to $\mu h = x$, and over $h\partial_h^+ G_h''(\beta_1 L + \nu h, \beta_2 L)$ from $\nu h = 0$ to $\nu h = y$. For $|\beta_1| \ge 2$ the distance between the arguments of ξ and $\xi + he$ is always $\le (|\beta_1|-1)L$. Each summand is thus, by Lemma 17, bounded by $O(h^2/(|\beta_1|-1)^2) = O(h^2/\beta_1^2)$. As a sum of such terms then γ_β can be estimated by

$$|\gamma_\beta| \le O(h/\beta_1^2) \quad \text{for} \quad |\beta_1| \ge 2. \tag{4.7.30b}$$

We want now to show that the following equation (30c) also holds:

$$G_h = \lim_{k\to\infty} \sum_{|\beta_1-1|\le k} \sum_{|\beta_2|\le k} \gamma_\beta = \lim_{k\to\infty} \sum_{1-k\le\beta_1\le 1+k} \sum_{|\beta_2|\le k} \gamma_\beta. \tag{4.7.30c}$$

The sums (30a) and (30c) differ by $D_h := \sum_{|\beta_2|\le k}[\gamma_{(1+k,\beta_2)} - \gamma_{(-k,\beta_2)}]$. The values $1+k$ and $-k$ that can be given to β_1 satisfy $|\beta_1| \le 2k$. The absolute value of the sum D_h is then bounded, from (30b), by $\sum_{|\beta_2|\le k} O(h/k^2) = (2k+1)O(h/k^2) = O(h/k)$, so that the limits of the double sums in (30a) and (30c) are the same. Changing the variable β_1 to $\beta_1 - 1$ then transforms (30c) into

$$G_h(x,y) = \lim_{k\to\infty} \sum_{-k\le\beta_1,\beta_2\le k} [G_h''(x+\beta_1 L + L, y+\beta_2 L) - G_h''(\beta_1 L + L, \beta_2 L)] \tag{4.7.30d}$$

The first part of the identity

$$\begin{aligned}
&\sum_{\beta_1,\beta_2=-k}^{k} [G_h''(\beta_1 L, \beta_2 L) - G_h''(\beta_1 L + L, \beta_2 L)] \\
&= \sum_{\beta_2=-k}^{k} [G_h''(-kL, \beta_2 L) - G_h''(kL+L, \beta_2 L)] \\
&= \sum_{\beta_2=-k}^{k} [G_h''(h+kL, \beta_2 L) - G_h''(kL+L, \beta_2 L)]
\end{aligned} \tag{4.7.30e}$$

is elementary; the second results from the symmetry $G_h'' = S_x G_h''$. As above, the summands of the last sum can be bounded by $O(h/k^2)$ so that (30e) vanishes for $k \to \infty$. Together with (30d) one obtains

$$G_h(x,y) = \lim_{k\to\infty} \sum_{-k\le\beta_1,\beta_2\le k} [G_h''(x+\beta_1 L + L, y+\beta_2 L) - G_h''(\beta_1 L, \beta_2 L)]$$

$$= \lim_{k\to\infty} \sum_{-k\le\beta_1,\beta_2\le k} \gamma_\beta(x+L,y) = G_h(x+L,y).$$

The proof of $G_h(x,y) = G_h(x,y+L)$ is analogous. ■

Since $L = 2-2h$, the symmetry $G_h = S_x G_h$ and the periodicity yield

$$G_h(1,y) = G_h(h-1,y) = G_h(h-1+L,y) = G_h(1-h,y),$$

i.e., $\partial_n^- G_h(1,y) = 0$ on the upper boundary $(1,y) \in \Gamma_h'$. Likewise we show that $\partial_n^- G_h(x,1) = 0$ on the right boundary. Thus (21b) is also proved.
It remains to show (21c): $|G_h(\mathbf{x})| \le hC/(|\mathbf{x}-\xi| + h)$ in Ω_h. The G_h^β for $||\beta||_\infty \le 1$ are linear combinations of $h\partial_x^- \sigma_h(\mathbf{x}-\tilde{\xi})$ for $\tilde{\xi} = \xi$ and other $\tilde{\xi} \notin \Omega_h$, which are generated by S_x, S_y, and P_β. For all ξ there holds $|\mathbf{x}-\tilde{\xi}| \ge |\mathbf{x}-\xi|$, so that from Lemma 17 follows

$$|G_h^\beta(\mathbf{x})| \le hC'/(|\mathbf{x}-\xi| + h) \quad \text{for } ||\beta||_\infty \le 1, \quad \mathbf{x} \in \Omega_h.$$

On the other hand, Lemma 18 shows that

$$|\sum_{||\beta||_\infty \ge 2} G_h^\beta(\mathbf{x})| \le h \sum_{||\beta||_\infty \ge 2} K/|\beta|^3 = hK' \le hK''/(|\mathbf{x}-\xi| + h).$$

The assumptions of Lemma 15 are hence satisfied: G_h with (21a-c) exists. Thus theorems 14 and 7 are proved. ■

4.8 Discretisation in an Arbitrary Domain

4.8.1 Shortley-Weller Approximation

Let the boundary value problem

$$-\Delta u = f \quad \text{in } \Omega, \qquad u = \varphi \quad \text{on } \Gamma \tag{4.8.1}$$

be given for an arbitrary domain Ω. If one places a square grid of step size h over Ω one obtains

$$\Omega_h := \{(x,y) \in \Omega \colon x/h \in \mathbb{Z} \quad \text{and } y/h \in \mathbb{Z}\} \tag{4.8.2}$$

as the set of grid points. By contrast, the set Γ_h of boundary points must be defined differently from the case of the square. The left neighbour point of $(x,y) \in \Omega_h$ reads $(x-h,y)$. If the connecting segment $\{(x-\vartheta h, y) \colon \vartheta \in (0,1]\}$ does not lie completely in Ω, there exists a boundary point

$$(x-sh,y) \in \Gamma \text{ with } (x,y) \in \Omega_h, s \in (0,1], (x-\vartheta h, y) \in \Omega \text{ for all } \vartheta \in [0,s), \tag{4.8.3a}$$

which now, instead of $(x-h, y)$, is called the left neighbour point of (x, y) (cf. Figure 1). Likewise the right, lower, and upper neighbour points can be boundary points of the following form:

$$(x+sh, y) \in \Gamma \text{ with } (x,y) \in \Omega_h, s \in (0,1], (x+\vartheta h, y) \in \Omega \text{ for all } \vartheta \in [0,s), \tag{4.8.3b}$$

$$(x, y-sh) \in \Gamma \text{ with } (x,y) \in \Omega_h, s \in (0,1], (x, y-\vartheta h) \in \Omega \text{ for all } \vartheta \in [0,s), \tag{4.8.3c}$$

$$(x, y+sh) \in \Gamma \text{ with } (x,y) \in \Omega_h, s \in (0,1], (x, y+\vartheta h) \in \Omega \text{ for all } \vartheta \in [0,s). \tag{4.8.3d}$$

We set

$$\Gamma_h := \{\text{boundary points which satisfy (3a), (3b), (3c), or (3d)}\}. \tag{4.8.4}$$

A grid point $(x, y) \in \Omega_h$ possessing a neighbour from Γ_h, is said to be near the boundary. All other points of Ω_h are said to be far from the boundary. As can be seen in Figure 1, a point may be near the boundary although $(x \pm h, y)$ and $(x, y \pm h)$ belong to Ω_h (namely if not all connecting segments lie completely in Ω).

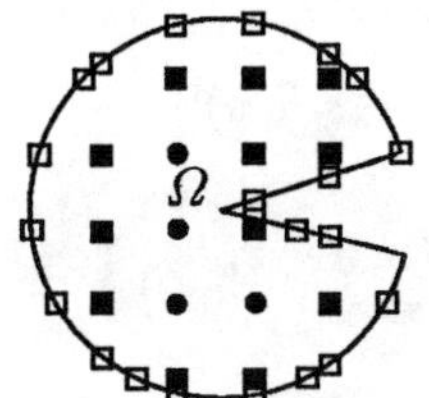

Figure 4.8.1. Ω_h and Γ_h

Exercise 4.8.1. For a convex domain Ω show that

$$\Gamma_h = \{(x, y) \in \Gamma \colon x/h \in \mathbb{Z} \text{ or } y/h \in \mathbb{Z}\}.$$

If one wishes to approximate the second derivative $u''(x)$ with the aid of the values of u at $x' < x < x''$, one can use Newton's divided differences:

$$u''(x) = 2\left[\frac{u(x'') - u(x)}{x'' - x} - \frac{u(x) - u(x')}{x - x'}\right] / (x'' - x') + \text{Rem}. \tag{4.8.5}$$

Exercise 4.8.2. Show that (a) the Taylor expansion implies

$$\begin{aligned} |\text{Rem}| &\le \frac{1}{3}\frac{(x''-x)^2 + (x-x')^2}{x''-x'} ||u||_{C^3([x',x''])} \\ &\le \frac{\max\{x''-x, x-x'\}}{3} ||u||_{C^3([x',x''])} \end{aligned} \tag{4.8.6}$$

for the remainder in Equation (5) if $u \in C^3([x', x''])$.

(b) In Equation (6) one can replace the norm of $C^3([x',x''])$ by that of $C^{2,1}([x',x''])$.
(c) If $x''=x+h$ and $x'=x-h$ the difference in (5) agrees with the usual second difference $\partial^-\partial^+u(x)$.

To set up the difference equation for $-\Delta u=f$ at $(x,y)\in\Omega_h$ we use the four neighbouring points

$$(x-s_lh,y),\ (x+s_rh,y),\ (x,y-s_uh),\ (x,y+s_oh)\in\Omega_h\cup\Gamma_h,\ 0<s_*\le 1,$$

as defined above. The subscripts to the s are suggest respectively left, right, under and over. For points far from the boundary $s_*=1$ holds; for points near the boundary (x,y) at least one neighbour lies on Γ_h and the corresponding distance s_*h may be smaller than h. Equation (5) with $x'=x-s_lh$ and $x''=x+s_rh$ provides an approximation for u_{xx}. Analogously one can replace u_{yy} by a divided difference. One obtains the difference scheme of Shortley and Weller [1]:

$$\begin{aligned}-D_hu(x,y):=h^{-2}\Big\{&\Big(\frac{2}{s_ls_r}+\frac{2}{s_os_u}\Big)u(x,y)-\frac{2}{s_r(s_r+s_l)}u(x+s_rh,y)\\&-\frac{2}{s_l(s_r+s_l)}u(x-s_lh,y)-\frac{2}{s_o(s_o+s_u)}u(x,y+s_oh)\\&-\frac{2}{s_u(s_o+s_u)}u(x,y-s_uh)\Big\}.\end{aligned}\tag{4.8.7}$$

Remark 4.8.3. If $s_l=s_r=s_u=s_o=1$, D_h agrees with Δ_h.

The discrete boundary value problem assumes the form

$$-D_hu_h=f_h:=\tilde{R}_hf\quad\text{on }\Omega_h\tag{4.8.8a}$$
$$\text{with }(\tilde{R}_hf)(\mathbf{x})=f(\mathbf{x})\quad\text{for }\mathbf{x}\in\Omega_h,$$
$$u_h=\varphi\quad\text{on }\Gamma_h.\tag{4.8.8b}$$

The five coefficients on the right side of Equation (7) define the matrix elements $L_{\mathbf{x}\xi}$ for $\xi=\mathbf{x}$ and for the four neighbours ξ of $\mathbf{x}$. Otherwise we set $L_{\mathbf{x}\xi}=0$. The right side of the system of equations

$$L_hu_h=q_h,\tag{4.8.9a}$$

in which u_h is interpreted as a grid function on Ω_h (not $\Omega_h\cup\Gamma_h$) is given by

$$q_h=f_h+\varphi_h,\quad \varphi_h(\mathbf{x}):=-\sum_{\xi\in\Gamma_h}L_{\mathbf{x}\xi}\varphi(\xi).\tag{4.8.9b}$$

Again $\varphi_h(\mathbf{x})=0$ holds for points $\mathbf{x}\in\Omega_h$ far from the boundary.

Theorem 4.8.4. *Let Ω be bounded and contained in a strip $(x_0, x_0+d)\times\mathbb{R}$ or $\mathbb{R}\times(y_0, y_0+d)$ of width d. For the matrix L_h belonging to the Shortley-Weller discretisation the following holds:*
(a) *L_h is generally not symmetric;*
(b) *L_h is M-matrix with*

$$||L_h^{-1}||_\infty \leq d^2/8. \tag{4.8.10}$$

PROOF. (a) Let $\mathbf{x} = (x,y) \in \Omega_h$ be near the boundary, but its neighbour $\mathbf{x}' = (x+h,y) \in \Omega_h$ be far from the boundary. Then it holds that $L_{\mathbf{x}\mathbf{x}'} = -2h^{-2}/(1+s_l) \neq -h^{-2} = L_{\mathbf{x}'\mathbf{x}}$, if $s_l < 1$. Other than in Exercise 4.7.1, in general no scaling can be found so that D_hL_h (D_h diagonal) becomes symmetric.
(b) L_h need not necessarily be irreducible and hence irreducibly diagonally dominant. But the weaker condition in Exercise 4.3.8 is satisfied and proves the M-matrix property.
(c) For the proof of (10) we use Theorem 4.3.16. If the domain Ω lies in the strip $(x_0, x_0+d)\times\mathbb{R}$ we select $w_h(x,y) := R_hw$, $w := (x-x_0)(x_0+d-x)/2$. The remainder in Equation (5) contains only third derivatives that vanish for w. Thus D_hw_h agrees with $\Delta w = -1$:

$$-D_hw_h = \mathbb{1} \quad \text{in } \Omega_h, \quad w_h \geq 0 \quad \text{on } \Gamma_h.$$

The corresponding system of equations reads $L_hw_h = q_h := f_h + \varphi_h$ with $f_h = \mathbb{1}$ and $q_h \geq 0$. Thus we have $L_hw_h \geq \mathbb{1}$ and Theorem 4.3.16 proves $||L_h^{-1}||_\infty \leq ||w_h||_\infty \leq (d/2)^2/2 = d^2/8$. ∎

Exercise 4.8.5. Prove the analogue of the derivative (4.4.9):

$$||u_h||_\infty \leq ||L_h^{-1}||_\infty||f_h||_\infty + \max_{\xi\in\Gamma_h}|\varphi(\xi)| \leq \frac{d^2}{8}\max_{\mathbf{x}\in\Omega_h}|f(\mathbf{x})| + \max_{\xi\in\Gamma_h}|\varphi(\xi)|.$$

With (10) stability has been proved. The order of consistency however is only 1, for at points near the boundary

$$c_h := D_hR_hu - \tilde{R}_h\Delta u$$

is of the order of magnitude $O(h^1)$. Here, R_h is the restriction to $\Omega_h \cup \Gamma_h$. $\tilde{R}_h$ has been defined in (8a). We want to show that nevertheless there is convergence of order 2. The difference $w_h := u_h - R_hu$ between the discrete solution $u_h = L_h^{-1}q_h$ and the solution $u \in C^{3,1}(\overline{\Omega})$ of (1) satisfies

$$\begin{aligned} -D_hw_h &= -D_hu_h + D_hR_hu = \tilde{R}_hf + D_hR_hu \\ &= D_hR_hu - \tilde{R}_h\Delta u = c_h \qquad \text{in } \Omega_h, \end{aligned} \tag{4.8.11a}$$

$$w_h = 0 \quad \text{on } \Gamma_h, \tag{4.8.11b}$$

so that $w_h = L_h^{-1} c_h$ follows. c_h can be written as $c_h^x + c_h^y$, where c_h^x (resp. c_h^y) is the error of discretisation of the x difference (resp. y difference). In turn, c_h^x is split into $c_h^x = c_h^{x,1} + c_h^{x,2}$:

$$c_h^{x,2}(x,y) := \left\{ \begin{matrix} c_h^x(x,y) & \text{if } s_l = s_r = 1 \\ 0 & \text{otherwise} \end{matrix} \right\}, \qquad c_h^{x,1} := c_h^x - c_h^{x,2}.$$

Analogously one defines $c_h^{y,1}$ and $c_h^{y,2}$ and sets

$$c_h^1 := c_h^{x,1} + c_h^{y,1}, \quad c_h^2 := c_h^{x,2} + c_h^{y,2}, \quad w_h^i := L_h^{-1} c_h^i.$$

The errors c_h^2 are described by (5.7):

$$\|w_h^2\|_\infty \le \|L_h^{-1}\|_\infty \|c_h^2\|_\infty, \quad \|c_h^2\|_\infty \le \frac{1}{6} h^2 \|u\|_{c^{3,1}(\overline{\Omega})}. \tag{4.8.12a}$$

With $K := \frac{1}{3} h^3 \|u\|_{C^{2,1}(\overline{\Omega})}$ define

$$v_h = K\mathbb{1} \quad \text{in } \Omega_h, \quad v_h = 0 \quad \text{on } \Gamma_h, \quad \tilde{c}_h := L_h v_h.$$

For $\mathbf{x} \in \Omega_h$ far from the boundary $\tilde{c}_h(\mathbf{x}) = 0$; for $\mathbf{x} \in \Omega_h$ near the boundary, however, we have

$$\tilde{c}_h(\mathbf{x}) = K \sum_{\xi \in \Omega_h} L_{\mathbf{x}\xi} = -K \sum_{\xi \in \Gamma_h} L_{\mathbf{x}\xi} \quad (\mathbf{x} \in \Omega_h \text{ near boundary}).$$

Consider, for example, the case $\mathbf{x} = (x,y) \in \Omega_h$ near the boundary with $\xi = (\mathbf{x} - s_l h, y) \in \Gamma_h$. According to Exercise (2a,b) the x difference has the error

$$|c_h^{x,1}(\mathbf{x})| \le \frac{h}{3} \frac{s_r^2 + s_l^2}{s_r + s_l} \|u\|_{C^{2,1}(\overline{\Omega})} = h^{-2} \frac{s_r^2 + s_l^2}{s_r + s_l} K \le K \frac{2h^{-2}}{s_l(s_r + s_l)} = -K L_{\mathbf{x}\xi}.$$

The analogous estimate for $c_h^{y,1}(\mathbf{x})$ gives $|c_h^1(\mathbf{x})| \le \tilde{c}_h(\mathbf{x})$. Because $c_h^1(\mathbf{x}) = \tilde{c}_h(\mathbf{x}) = 0$ at the $\mathbf{x} \in \Omega_h$ far from the boundary one has $-\tilde{c}_h \le c_h^1 \le \tilde{c}_h$. Since $L_h^{-1} \ge \mathbf{0}$ it follows that $-v_h \le w_h^1 \le v_h$, i.e.,

$$\|w_h^1\|_\infty \le K = \frac{1}{3} h^3 \|u\|_{C^{2,1}(\overline{\Omega})}. \tag{4.8.12b}$$

Equations (12a,b) together with $w_h^1 + w_h^2 = w_h = u_h - R_h u$ prove

Theorem 4.8.6. *Let Ω satisfy the assumption of Theorem 4. The convergence for the Shortley-Weller method is second order:*

$$\begin{aligned} \|u_h - R_h u\|_\infty &\le \frac{1}{3} h^3 \|u\|_{C^{2,1}(\overline{\Omega})} + \|L_h^{-1}\|_\infty \frac{1}{6} h^2 \|u\|_{C^{3,1}(\overline{\Omega})} \\ &\le \left(\frac{1}{3} h^3 + \frac{d^2}{48} h^2\right) \|u\|_{C^{3,1}(\overline{\Omega})}, \end{aligned} \tag{4.8.13}$$

if $u \in C^{3,1}(\overline{\Omega})$.

Exercise 4.8.7. Show that if one uses the Shortley-Weller discretisation for all grid points near the boundary, but the mehrstellen method from Section 4.6 for all points far from the boundary, we obtain a method with convergence of third order: $||u_h - R_h u||_\infty = O(h^3)$.

4.8.2 Interpolation at Points near the Boundary

Instead of discretising the Poisson equation at grid points $\mathbf{x} \in \Omega_h$ near the boundary, one could also try to determine $u_h(\mathbf{x})$ by interpolation from the neighbouring points. If, for example, $\mathbf{x} = (x, y) \in \Omega_h$ is near the boundary, $(x - s_l h, y) \in \Gamma_h$ and $(x + s_r h, y) \in \Omega_h \cup \Gamma_h$, then linear interpolation yields

$$u_h(x,y) = [s_l u_h(x + s_r h, y) + s_r u_h(\mathbf{x} - s_l h, y)]/(s_r + s_l).$$

Thus, at the point $\mathbf{x}$ we set up the equation

$$(s_l + s_r)u_h(x,y) - s_l u_h(x + s_r h, y) - s_r u_h(\mathbf{x} - s_l h, y) = 0. \tag{4.8.14a}$$

Since $\xi = (x - s_l h, y)$ should be a boundary point, one can replace $u_h(\xi)$ by $\varphi(\xi)$. If, however, $(x, y + s_o h)$ or $(x, y - s_u h)$ is a boundary point, we choose interpolation in the y direction:

$$(s_u + s_o)u_h(x,y) - s_u u_h(x, y + s_o h) - s_o u_h(x, y - s_u h) = 0. \tag{4.8.14b}$$

At any $\mathbf{x} \in \Omega_h$ far from the boundary the five-point formula (2.3) is used:

$$-(\Delta_h u_h)(\mathbf{x}) = f_h(\mathbf{x}) = \tilde{R}_h f(\mathbf{x}) = f(\mathbf{x}) \quad (\mathbf{x} \text{ far from boundary}). \tag{4.8.14c}$$

Theorem 4.8.8. *Let Ω satisfy the assumptions of Theorem 4. Let the discretisation be given by* (14a-c) *with the choice between* (14a) *and* (14b) *being made in such a way that always (at least) one boundary point is used for the interpolation. Let the system of equations resulting after the elimination of the boundary values $u_h(\xi) = \varphi(\xi)$, $\xi \in \Gamma_h$ be $L_h u_h = q_h$. L_h is an (in general unsymmetric) M-matrix which satisfies the estimate* (10). *The discrete solutions u_h converge with the order 2, if $u \in C^{3,1}(\overline{\Omega})$:*

$$||u_h - R_h u||_\infty \le h^2 ||u||_{C^{1,1}(\overline{\Omega})} + ||L_h^{-1}||_\infty \frac{1}{6} h^2 ||u||_{C^{3,1}(\overline{\Omega})}. \tag{4.8.15}$$

PROOF. The M-matrix property and (10) are proved as in Theorem 4.2. Let $\mathbf{x} \in \Omega_h$ be a point near the boundary at which (14a) is used. The error of interpolation is

$$c_h^1(x,y) := (s_l + s_r)R_h u(x,y) - s_l R_h u(x + s_r h, y) - s_r R_h u(x - s_l h, y),$$

$$|c_h^1(x,y)| \le \frac{1}{2} s_r s_l (s_l + s_r) h^2 ||u||_{C^{1,1}(\overline{\Omega})}.$$

With this c_h^1 and $K := h^2||u||_{C^{1,1}(\bar{\Omega})}$, one can essentially just repeat the proof of Theorem 6. ■

By rescaling and adding the equations (14a,b), one obtains

$$h^{-2}\left\{\left(\frac{s_l+s_r}{s_l s_r}+\frac{s_o+s_u}{s_o s_u}\right)u_h(x,y)-\frac{1}{s_l}u_h(x-s_l h,y)\right.$$
$$-\frac{1}{s_r}u_h(x+s_r h,y)-\frac{1}{s_o}u_h(x,y+s_o h)$$
$$\left.-\frac{1}{s_u}u_h(x,y-s_u h)\right\}=0. \tag{4.8.16}$$

Using this device one can obtain a symmetric matrix L_h, even at arbitrary Ω.

Exercise 4.8.9. Show that the discretisation (14c), (16) leads to a symmetric M-matrix. The estimate of convergence reads

$$||u_h - R_h u||_\infty \le 2h^2||u||_{C^{1,1}(\bar{\Omega})} + \frac{1}{6}d^2h^2||u||_{C^{3,1}(\bar{\Omega})}.$$

In (14a,b) linear interpolation was chosen because the values at the neighbouring points were sufficient for it. Constant interpolation by

$$u(x,y) = u(x-s_l h,y) = \varphi(x-s_l h,y), \quad \text{if } (x,y)\in\Omega_h, (x-s_l h,y)\in\Gamma_h,$$

is less desirable since it only provides first-order convergence : $||u_h - R_h u||_\infty = O(h)$. By contrast, interpolation of higher order is very applicable indeed (cf. Pereyra–Proskurowski–Widlund [1]). However, it is described by an equation which also contains points at an distance of $\ge 2h$. Higher boundary approximations are in particular then necessary if one wants to apply extrapolation methods (cf. Marchuk-Shaidurov [1, p. 162 ff.]).

5 General Boundary Value Problems

5.1 Dirichlet Boundary Value Problems for Linear Differential Equations

5.1.1 Posing the Problem

In Section 1.2 we have already formulated the general linear differential equation of second order:

$$Lu = f \quad \text{in } \Omega \tag{5.1.1a}$$

with

$$L = \sum_{i,j=1}^{n} a_{ij}(\mathbf{x}) \frac{\partial^2}{\partial x_i \partial x_j} + \sum_{i=1}^{n} a_i(\mathbf{x}) \frac{\partial}{\partial x_i} + a(\mathbf{x}). \tag{5.1.1b}$$

We mentioned that, without loss of generality, one can assume

$$a_{ij}(\mathbf{x}) = a_{ji}(\mathbf{x}) \tag{5.1.1c}$$

so that the matrix

$$\mathbf{A}(\mathbf{x}) := (a_{ij}(\mathbf{x}))_{i,j=1,\cdots,n} \tag{5.1.1d}$$

is symmetric. A differential operator apparently more general than (1b) is

$$L = \sum_{i,j=1}^{n} \left[a_{ij}^{\mathrm{I}} \frac{\partial^2}{\partial x_i \partial x_j} + \frac{\partial}{\partial x_j} a_{ij}^{\mathrm{II}} \frac{\partial}{\partial x_i} + \frac{\partial^2}{\partial x_i \partial x_j} a_{ij}^{\mathrm{III}} \right] + \sum_{i=1}^{n} \left[a_i^{\mathrm{I}} \frac{\partial}{\partial x_i} + \frac{\partial}{\partial x_i} a_i^{\mathrm{II}} \right] + a. \tag{5.1.2}$$

But since, for example, $\frac{\partial}{\partial x_j}(a_{ij}^{\mathrm{II}} \frac{\partial}{\partial x_i} u) = a_{ij}^{\mathrm{II}} u_{x_i x_j} + (\partial a_{ij}^{\mathrm{II}} / \partial x_j) u_{x_i}$, the operator (2) can be described in the form (1b), provided the coefficients are sufficiently often differentiable. According to Definition 1.2.3, Equation (1) is elliptic in Ω if all eigenvalues of $\mathbf{A}(\mathbf{x})$ have the same sign. One can assume without loss of generality that all eigenvalues are positive so that $\mathbf{A}(\mathbf{x})$ is positive definite (cf. Exercise 4.3.22a).

Thus, L is elliptic in Ω if

$$\sum_{i,j=1}^{n} a_{ij}(\mathbf{x}) \xi_i \xi_j > 0 \quad \text{for all } \mathbf{x} \in \Omega, \quad 0 \neq \xi \in \mathbb{R}^n. \tag{5.1.3a}$$

For any $\mathbf{x} \in \Omega$ there exists $c(\mathbf{x}) := \min\{a_{ij}(\mathbf{x})\xi_i\xi_j : |\xi| = 1\}$ and it must be positive ($c(\mathbf{x})$ is the smallest eigenvalue of $\mathbf{A}(\mathbf{x})$!). Hence one can also write (3a) in the form (3a′):

$$\sum_{i,j=1}^{n} a_{ij}(\mathbf{x})\xi_i\xi_j \geq c(\mathbf{x})|\xi|^2, \qquad c(\mathbf{x}) > 0 \text{ for all } \mathbf{x} \in \Omega,\ \xi \in \mathbb{R}^n. \qquad (5.1.3a')$$

Definition 5.1.1. The equation (1a), or the operator L, is defined to be uniformly elliptic in Ω if

$$\inf\{c(\mathbf{x}) : \mathbf{x} \in \Omega\} > 0 \quad (c(\mathbf{x}) \text{ from (3a')}). \qquad (5.1.3b)$$

On $\Gamma = \partial\Omega$ we impose the following Dirichlet boundary value condition:

$$u = \varphi \quad \text{on } \Gamma. \qquad (5.1.4)$$

5.1.2 Maximum Principle

In general, the maximum principle does not hold for the equation $Lu = f$, nor is the solution of the boundary value problem (1a), (4) uniquely determined.

Example 5.1.2. Let $\Omega = (0,\pi) \times (0,\pi)$, $\varphi = 0$, $f = 0$, $Lu = \Delta u + 2u$. Then both $u = 0$ and $u(x,y) = \sin(x)\sin(y)$ are solutions of the boundary value problem. The second solution assumes its maximum at the interior point $(\pi/2, \pi/2) \in \Omega$.

In the above example the coefficient $a(\mathbf{x}) = 2$ (cf. (1b)) has the wrong sign. As soon as $a \leq 0$, we have

Theorem 5.1.3. (Maximum-minimum principle) *Let $a(\mathbf{x}) \leq 0$ in the domain Ω. Assume the coefficients of the elliptic operator* (1b) *are continuous in Ω. Let $u \in C^2(\Omega)$ satisfy $Lu = f$ and be nonconstant. Then we have:*
(a) *if $f \leq 0$ in Ω, no negative minimum of u exists in Ω;*
(b) *if $f \geq 0$ in Ω, no positive maximum exists in Ω.*

PROOF. The proof is based on Hopf's lemma, which can be studied, for example, in Hellwig [1,III–1.1]. Here we give a shorter proof, which requires, however, the stronger condition $a(\mathbf{x}) < 0$ instead of $a(\mathbf{x}) \leq 0$. We assume that u has a minimum at $x^* \in \Omega$ although $f \leq 0$. Consequently we have $u_{x_i}(\mathbf{x}^*) = 0$, i. e.,

$$f(\mathbf{x}^*) = (Lu)(\mathbf{x}^*) = \sum_{i,j} a_{ij}(\mathbf{x}^*)u_{x_ix_j}(\mathbf{x}^*) + a(\mathbf{x}^*)u(\mathbf{x}^*), \qquad (5.1.5a)$$

and the Hesse matrix $\mathbf{B} := (u_{x_i x_j}(\mathbf{x})^*))_{i,j=1,\cdots,n}$ must be positive semidefinite (here, a matrix $\mathbf{B}$ is said to be positive semidefinite if it is symmetric and satisfies $\langle \xi, \mathbf{B}\xi \rangle \geq 0$ for all $\xi \in \mathbb{R}^n$). From Exercise 4 follows

$$\mathrm{tr}(\mathbf{A}(\mathbf{x}^*)\mathbf{B}) = \sum_{i,j=1}^{n} a_{ij}(\mathbf{x}^*) u_{x_i x_j}(\mathbf{x}^*) \geq 0. \tag{5.1.5b}$$

If the minimum $u(\mathbf{x}^*)$ were negative, one would obtain from (5a,b) the result $f(\mathbf{x}^*) \geq a(\mathbf{x}^*)u(\mathbf{x}^*) > 0$ in contradiction to the assumption $f \leq 0$. Part (b) is proved analogously. ■

Exercise 5.1.4. The trace is defined by $\mathrm{tr}(\mathbf{A}) := \sum_{i=1}^{n} a_{ii}$. Prove that:
(a) $a_{ii} \geq 0$ and $\mathrm{tr}(\mathbf{A}) \geq 0$ if $\mathbf{A}$ is positive semidefinite;
(b) $\mathrm{tr}(\mathbf{AB}) = \mathrm{tr}(\mathbf{BA}) = \sum_{i,j=1}^{n} a_{ij} b_{ji}$;
(c) $\mathrm{tr}(\mathbf{AB}) \geq 0$, if $\mathbf{A}$ and $\mathbf{B}$ are positive semidefinite.
Hint for (c): $\mathbf{B}^{1/2}\mathbf{A}\mathbf{B}^{1/2}$ is positive semidefinite; Exercise 4.3.22d.

As in the case of the potential equation, from Theorem 3 follows the

Corollary 5.1.5. *Let Ω be bounded (not necessarily a domain); furthermore, assume that the conditions of Theorem 3 hold. If $f \leq 0$ in Ω [$f \geq 0$ in Ω] and if there exists a negative minimum [positive maximum] of u in $\overline{\Omega}$ it must lie on the boundary $\partial\Omega$.*

Remark 5.1.6. The continuity of the coefficients a_{ij}, a_i, and a of L in Theorem 3 and Corollary 5 can be replaced by the assumption $a < 0$ in Ω or by the stipulation: In every compact set $K \subset \Omega$ let a_{ij}, a_i and a be bounded and let L be uniformly elliptic.

5.1.3 Uniqueness of the Solution and Continuous Dependence

Lemma 5.1.7. *Let Ω be bounded, let the coefficients of L be continuous and let (3a) be valid and $a \leq 0$ in Ω. Let $u_1, u_2 \in C^2(\Omega) \cap C^0(\overline{\Omega})$ be solutions of the boundary value problems*

$$Lu_i = f_i \quad \text{in } \Omega, \qquad u_i = \varphi_i \quad \text{on } \Gamma \quad (i = 1, 2). \tag{5.1.6}$$

If $f_1 \geq f_2$ in Ω and $\varphi_1 \leq \varphi_2$ on Γ, then also $u_1 \leq u_2$ in $\overline{\Omega}$.

PROOF. $v := u_2 - u_1$ satisfies the equations $Lv = f_2 - f_1 \leq 0$ in Ω and $v = \varphi_2 - \varphi_1 \geq 0$ on Γ. From Corollary 5 and the fact that $v \geq 0$ on Γ one infers that no negative minimum of v exists. Thus $v \geq 0$, i. e., $u_2 \geq u_1$. ■

Theorem 5.1.8. (Uniqueness) *Under the conditions of Lemma 7 the solution of the boundary value problem $Lu = f$ in Ω and $u = \varphi$ on Γ is uniquely determined.*

PROOF. Let u_1, u_2 be two solutions. Lemma 7 with $f_1 = f_2 = f$, $\varphi_1 = \varphi_2 = \varphi$ shows that $u_1 \le u_2$ as well as $u_2 \le u_1$. Thus, $u_1 = u_2$. ■

The next theorem states that the solution depends Lipschitz-continuously on f and φ.

Theorem 5.1.9. *Let L be uniformly elliptic in Ω. Under the conditions of Lemma 7 the following holds:*

$$||u_1 - u_2||_\infty \le ||\varphi_1 - \varphi_2||_\infty + M||f_1 - f_2||_\infty \tag{5.1.7}$$

for solutions u_1 and u_2 of (6). In this inequality the number M depends only on $K := \sup\{|a_{ij}(\mathbf{x})|,\ |a_i(\mathbf{x})|,\ |a(\mathbf{x})| : \mathbf{x} \in \Omega\}$, on the ellipticity constant defined by $m := \inf\{c(\mathbf{x}) : \mathbf{x} \in \Omega\} > 0$ (cf. (3b)) and on the diameter of the domain Ω.

PROOF. Let $\Omega \subset K_R(\mathbf{z})$. For all $\mathbf{x} \in \Omega$ we have $z_1 - R \le x_1 \le z_1 + R$. We select $\alpha \ge 0$ so that

$$m\alpha^2 - K(\alpha + 1) \ge 1$$

and we define

$$w(\mathbf{x}) := ||\varphi_1 - \varphi_2||_\infty + (e^{2R\alpha} - e^{\alpha(x_1 - z_1 + R)})||f_1 - f_2||_\infty.$$

We compare w with the solution $v := u_1 - u_2$ of $Lv = f_1 - f_2$ in Ω, $v = \varphi_1 - \varphi_2$ on Γ. From the choice of $K_R(z)$ we have

$$w(\mathbf{x}) \ge v(\mathbf{x}) \quad \text{for } \mathbf{x} \in \Gamma.$$

Furthermore, the selection made of α ensures:

$$\begin{aligned}
&(Lw)(\mathbf{x}) = \\
&\quad \underbrace{a(\mathbf{x})||\varphi_1 - \varphi_2||_\infty}_{\le 0} + \\
&\quad \{\underbrace{ae^{2R\alpha}}_{\le 0} - \underbrace{e^{\alpha(x_1 - z_1 + R)}}_{\ge 1}[\underbrace{a_{11}(\mathbf{x})}_{\ge m}\alpha^2 + \underbrace{a_1(\mathbf{x})}_{\ge -K}\alpha + \underbrace{a(\mathbf{x})}_{\ge -K}]\}||f_1 - f_2||_\infty \\
&\quad \le -\{m\alpha^2 - K(\alpha + 1)\}||f_1 - f_2||_\infty \\
&\quad \le -||f_1 - f_2||_\infty \le f_1(\mathbf{x}) - f_2(\mathbf{x}) = (Lv)(\mathbf{x}).
\end{aligned}$$

If one uses Lemma 7 with $u_1 = v$, $u_2 = w$, the result is $v \le w$ in Ω, i .e.,

$$u_1(\mathbf{x}) - u_2(\mathbf{x}) = v(\mathbf{x}) \le w(\mathbf{x}) \le ||w||_\infty \le ||\varphi_1 - \varphi_2||_\infty + M||f_1 - f_2||_\infty,$$

where $M := e^{2R\alpha}$. Analogously one proves $-w \le v$ so that (7) follows. ■

Exercise 5.1.10. (a) Let Ω be bounded; let the coefficients of L be continuous in $\overline{\Omega}$. Now show that L is uniformly elliptic in Ω if and only if L is elliptic in $\overline{\Omega}$.
(b) Theorem 9 holds for the special case $f_1 = f_2$ without the assumption of uniform ellipticity.
(c) Let a strip in $\mathbb{R}^n$ be described by

$$S = \{\mathbf{x} \in \mathbb{R}^n : 0 \le \langle \eta, \mathbf{x} - \mathbf{x}^* \rangle \le \delta\},$$

where $\mathbf{x}^* \in \mathbb{R}^n$ is a boundary point of s, $\eta \in \mathbb{R}^n$ with $|\eta| = 1$ is a unit vector, and δ is the strip width. Show that inequality (7) holds with $M := e^{\delta\alpha}$, if $\Omega \subset S$ and α are as in the proof of Theorem 9.
(d) Under the conditions of Theorem 9 show that

$$\|u\|_\infty \le \|\varphi\|_\infty + M\|f\|_\infty. \tag{5.1.8}$$

Exercise 5.1.11. Let the coordinate transformation $\Phi: \mathbf{x} \in \overline{\Omega} \mapsto \xi \in \overline{\Omega'}$ and the inverse $\Phi^{-1}: \overline{\Omega'} \to \overline{\Omega}$ be continuously differentiable. To the operator L (in the x coordinates) let L' correspond in the ξ coordinates. Show that if L satisfies the conditions of Lemma 7, or of Theorem 9, then so does L'.

The right side f and the boundary values φ are not the only parameters on which the solution u depends. We next investigate how the solution depends on the coefficients a_{ij}, a_i, and a of the differential operator L.

Theorem 5.1.12. *Let the coefficients of L^{I} and L^{II} be $a^{\mathrm{I}}_{ij}, a^{\mathrm{I}}_i, a^{\mathrm{I}}$, resp. $a^{\mathrm{II}}_{ij}, a^{\mathrm{II}}_i, a^{\mathrm{II}}$. Let u^{I} and u^{II} be solutions of*

$$L^{\mathrm{I}}u^{\mathrm{I}} = L^{\mathrm{II}}u^{\mathrm{II}} = f \quad \text{in } \Omega, \quad u^{\mathrm{I}} = u^{\mathrm{II}} = \varphi \quad \text{on } \Gamma.$$

Let L^{I} satisfy the conditions of Theorem 9 and let u^{II} belong to $C^2(\overline{\Omega})$. With M from (7) we then have

$$\begin{aligned}\|u^{\mathrm{I}} - u^{\mathrm{II}}\|_\infty \le M\Big\{&\Big(\sum_{i,j=1}^n \|a^{\mathrm{I}}_{ij} - a^{\mathrm{II}}_{ij}\|_\infty\Big)\|u^{\mathrm{II}}\|_{C^2(\overline{\Omega})} \\ &+ \Big(\sum_{i=1}^n \|a^{\mathrm{I}}_i - a^{\mathrm{II}}_i\|_\infty\Big)\|u^{\mathrm{II}}\|_{C^1(\overline{\Omega})} \\ &+ \|a^{\mathrm{I}} - a^{\mathrm{II}}\|_\infty \|u^{\mathrm{II}}\|_\infty\Big\}.\end{aligned} \tag{5.1.9}$$

If $a^{\mathrm{I}}_{ij} = a^{\mathrm{II}}_{ij}$, the condition $u^{\mathrm{II}} \in C^1(\overline{\Omega}) \cap C^2(\Omega)$ is sufficient; if also $a^{\mathrm{I}}_i = a^{\mathrm{II}}_i$ then just $u^{\mathrm{II}} \in C^0(\overline{\Omega}) \cap C^2(\Omega)$ will do.

PROOF. Set $f' := L^{\mathrm{I}}u^{\mathrm{II}}$. Then $\|f'-f\|_\infty = \|(L^{\mathrm{I}}-L^{\mathrm{II}})u^{\mathrm{II}}\|_\infty$ can be bounded by the right side of (9) without the factor M. Theorem 9 applied to $L^{\mathrm{I}}u^{\mathrm{II}} = f'$ and $L^{\mathrm{I}}u^{\mathrm{I}} = f$ implies $\|u^{\mathrm{I}} - u^{\mathrm{II}}\|_\infty \le M\|f' - f\|_\infty$. ■

5.1.4 Difference Methods for the General Differential Equation of Second Order

For notational reasons we limit ourselves to the two-dimensional case $n = 2$. General domains $\Omega \in \mathbb{R}^2$ require special discretisations at the boundary as explained in Section 4.8. Here we only want to discuss the difference formulae at points far from the boundary. Therefore it will suffice to base our comments on the unit square

$$\Omega = (0,1) \times (0,1).$$

Let L be given by (1b). If one wishes to obtain a difference method of consistency order 2, the following choice suggests itself, which will later be modified for reasons of stability.

$$\begin{aligned} &a_{11}(x,y)\partial_x^+\partial_x^- + 2a_{12}(x,y)\partial_x^0\partial_y^0 + a_{22}(x,y)\partial_y^+\partial_y^- \\ &\quad + a_1(x,y)\partial_x^0 + a_2(x,y)\partial_y^0 + a(x,y) \qquad (5.1.10)\\ &= h^{-2}\begin{bmatrix} -a_{12}(x,y)/2 & a_{22}(x,y) & a_{12}(x,y)/2 \\ a_{11}(x,y) & -2[a_{11}(x,y)+a_{22}(x,y)] & a_{11}(x,y) \\ a_{12}(x,y)/2 & a_{22}(x,y) & -a_{12}(x,y)/2 \end{bmatrix} \\ &\quad + \frac{1}{2}h^{-1}\begin{bmatrix} 0 & a_2(x,y) & 0 \\ -a_1(x,y) & 0 & a_1(x,y) \\ 0 & -a_2(x,y) & 0 \end{bmatrix} + \begin{bmatrix} 0 & 0 & 0 \\ 0 & a(x,y) & 0 \\ 0 & 0 & 0 \end{bmatrix}. \end{aligned}$$

Remark 5.1.13. The difference method (10) is a nine-point scheme of consistency order 2.

Let L_h be the matrix of the system of difference equations in Ω_h associated to (10) (cf. Remark 4.2.5):

$$L_h u_h = q_h. \qquad (5.1.11)$$

The solvability of Equation (11), for arbitrary q_h, is equivalent to uniqueness. In the continuous case, uniqueness was essentially given by the condition $a \le 0$ and the ellipticity condition $a_{11}a_{22} > a_{12}^2$ (cf. Theorem 8). These conditions are in general not sufficient to guarantee the solvability of Equation (11). We thus replace (10) with another discretisation.

Theorem 5.1.14. *For the coefficients of L let* (3a,b) *hold, and assume that*

$$|a_{12}(x,y)| \le \min\{a_{11}(x,y), a_{22}(x,y)\} \qquad (5.1.12)$$

and $a(x,y) \le 0$ in Ω. Then there exists for L a seven-point difference method of consistency order 1 such that the associated matrix L_h is an M-matrix except for the sign. In particular, the resulting Equation (11) *is solvable.*

Note that Condition (12) follows from (3a) only if $a_{11} = a_{22} > 0$.

PROOF. At the grid point $(x,y) \in \Omega_h$ (cf. (4.2.1a)) we abbreviate the coefficients $a_{ij}(x,y)$, $a_i(x,y)$, and $a(x,y)$ to a_{ij}, a_i, and a. The principal part $\sum a_{ij}\partial^2/\partial x_i \partial x_j$ ($x_1 = x, x_2 = y$) is discretised by the following difference stars:

$$\frac{\partial^2}{\partial x_1^2} : h^{-2}\begin{bmatrix} & 0 & \\ 1 & -2 & 1 \\ & 0 & \end{bmatrix}, \qquad \frac{\partial^2}{\partial x_2^2} : h^{-2}\begin{bmatrix} & 1 & \\ 0 & -2 & 0 \\ & 1 & \end{bmatrix},$$

$$\frac{\partial^2}{\partial x_1 \partial x_2} : \begin{cases} \frac{1}{2}h^{-2}\begin{bmatrix} 0 & -1 & 1 \\ -1 & 2 & -1 \\ 1 & -1 & 0 \end{bmatrix}, & \text{if } a_{12} \ge 0, \\ \frac{1}{2}h^{-2}\begin{bmatrix} -1 & 1 & 0 \\ 1 & -2 & 1 \\ 0 & 1 & -1 \end{bmatrix}, & \text{if } a_{12} < 0. \end{cases} \tag{5.1.13}$$

For $a_{12} \ge 0$, we thus obtain

$$\sum_{i,j=1}^{2} a_{ij}\frac{\partial^2}{\partial x_i \partial x_j} : h^{-2}\begin{bmatrix} 0 & a_{22}-a_{12} & a_{12} \\ a_{11}-a_{12} & 2(a_{12}-a_{11}-a_{22}) & a_{11}-a_{12} \\ a_{12} & a_{22}-a_{12} & 0 \end{bmatrix}.$$

If one introduces $a_{12}^+ := \max\{a_{12}, 0\}$ and $a_{12}^- := \min\{a_{12}, 0\}$ then both for $a_{12} \ge 0$ and for $a_{12} < 0$ we get the seven-point star

$$h^{-2}\begin{bmatrix} -a_{12}^- & a_{22}-|a_{12}| & a_{12}^+ \\ a_{11}-|a_{12}| & 2(|a_{12}|-a_{11}-a_{22}) & a_{11}-|a_{12}| \\ a_{12}^+ & a_{22}-|a_{12}| & -a_{12}^- \end{bmatrix}. \tag{5.1.13'}$$

Exercise 5.1.15. Let L_h be the matrix belonging to the difference formula (13′) (note that the coefficients in (13′) depend on the spatial coordinates). Assume (3) and (12) hold. Prove that $-L_h$ satisfies the conditions of Exercise 4.3.8a.

The first derivative terms $a_i\partial/\partial x_i$ are replaced by the forward and backward differences, respectively, $a_i\partial_{x_i}^{\pm}$ if $a_i \gtrless 0$:

$$\sum_{i=1}^{2} a_i\frac{\partial}{\partial x_i} + a : h^{-1}\begin{bmatrix} & a_2^+ & \\ -a_1^- & -|a_1|-|a_2| & a_1^+ \\ & -a_2^- & \end{bmatrix} + \begin{bmatrix} & 0 & \\ 0 & a & 0 \\ & 0 & \end{bmatrix}, \tag{5.1.14}$$

where $a_i^{\pm}$ is defined in the same way as $a_{12}^{\pm}$. Because $a \le 0$, the diagonal element is negative, the others nonnegative. If one adds these terms to (13′), the resulting matrix $-L_h$ satisfies the conditions of Exercise 4.3.8a. Therefore $-L_h$ is an M-matrix.

It is easy to see that the difference formula (13′) is of second order of consistency; (14), however, contains the one-sided differences so that the total discretisation is only of first order. ∎

The condition (12) can be avoided if one allows larger difference stars, which also contain the values $u(x_1 \pm \nu h,\ x_2 \pm \mu h)$ for fixed $\nu, \mu \in \mathbb{Z}$ (but in general $|\nu|, |\mu| > 1$) (cf. Bramble-Hubbard [1]). Layton–Morley [1] point out that with weaker conditions than (12) one may still obtain a matrix L_h, which, though not an M-matrix, does have a positive inverse.

To obtain a method of consistency order 2, one must discretise $\sum a_i \partial/\partial x_i$ as in (10). The following corollary shows that $-L_h$ is also an M-matrix when ha_i is sufficiently small.

Corollary 5.1.16. *In addition to the assumptions of Theorem* 14 *let the following hold:*

$$a_{ii} > |a_{12}| + \frac{h}{2}|a_i| \quad (i = 1, 2). \tag{5.1.15}$$

Then the discretisation of $\sum a_i \partial/\partial x_i$ *from* (10) *together with* (13′) *leads to a seven-point difference method of second order of consistency such that* $-L_h$ *is an* M*-matrix.*

PROOF. $-L_h$ satisfies (4.3.1a) and is irreducibly diagonally-dominant. ∎

Exercise 5.1.17. The condition $a_{ii} \ge |a_{12}| + h|a_i|/2$ instead of (15) is not sufficient. Construct a counterexample with $a_{11} = a_{22} = 1$, $a_{12} = 0$, $h = 1/3$, $a_i(x, y)$ variable, and $|a_i| = 6$ so that L_h is singular.

Considerably weaker conditions for the nonsingularity of L_h than in Theorem 14 and Corollary 16 are needed in Section 9.2 (cf. Exercise 9.2.6, Corollary 11.3.5).

In general, L_h is not a symmetric matrix. Symmetry of L_h is to be expected only if L is also symmetric: $L = L'$. Here the *formally adjoint differential operator* L' which is associated to L in (2), is defined by

$$L' = \sum_{i,j} \left[a_{ij}^{\mathrm{III}} \frac{\partial^2}{\partial x_i \partial x_j} + \frac{\partial}{\partial x_i} a_{ij}^{\mathrm{II}} \frac{\partial}{\partial x_j} + \frac{\partial^2}{\partial x_i \partial x_j} a_{ij}^{\mathrm{I}} \right] - \sum_{i=1}^{n} \left[a_i^{\mathrm{II}} \frac{\partial}{\partial x_i} + \frac{\partial}{\partial x_i} a_i^{\mathrm{I}} \right] + a. \tag{5.1.16}$$

It is easy to see that a *symmetric operator* can always be written in the form

$$L = \sum_{i,j=1}^{n} \frac{\partial}{\partial x_i} a_{ij}(x) \frac{\partial}{\partial x_j} + a(x), \quad a_{ij} = a_{ji}. \tag{5.1.17}$$

A difference method for this is given for the case $n = 2$ and $a_{12} = 0$ by the five-point star

$$h^{-2}\begin{bmatrix} 0 & a_{22}(x,y+\frac{h}{2}) & 0 \\ a_{11}(x-\frac{h}{2},y) & \begin{matrix}[-a_{11}(x-\frac{h}{2},y)-a_{11}(x+\frac{h}{2},y)\\ -a_{22}(x,y-\frac{h}{2})-a_{22}(x,y+\frac{h}{2})]\end{matrix} & a_{11}(x+\frac{h}{2},y) \\ 0 & a_{22}(x,y-\frac{h}{2}) & 0 \end{bmatrix}$$

$$+\begin{bmatrix} 0 & 0 & 0 \\ 0 & a(x,y) & 0 \\ 0 & 0 & 0 \end{bmatrix}. \tag{5.1.18}$$

Theorem 5.1.18. *Let $a_{11}, a_{22} \in C^{2,1}(\overline{\Omega})$. The difference method (18) is consistent of order 2. The associated matrix L_h is symmetric. If $a_{ii} > 0$ (ellipticity) and $a \le 0$, then $-L_h$ is a positive definite M-matrix.*

PROOF. (a) For the proof of consistency, expand

$$v(x+h/2) := a_{11}(x+h/2,y)[u(x+h,y)-u(x,y)] \quad (y \text{ fixed})$$

around $x+h/2$, and then expand $v(x+h/2)-v(x-h/2)$ around x.
(b) The symmetry results from $L_{\mathbf{x}\xi} = L_{\xi\mathbf{x}}$, for $\mathbf{x},\xi \in \Omega_h$ (cf. Exercise 4.2.2c).
(c) L_h is irreducibly diagonally dominant so that Criteria 4.3.10 and 4.3.24 are applicable. ∎

The mixed term $\frac{\partial}{\partial x}a_{12}(x,y)\frac{\partial}{\partial y} + \frac{\partial}{\partial y}a_{12}(x,y)\frac{\partial}{\partial x}$ can be discretised by

$$\frac{1}{2}h^{-2}\begin{bmatrix} -a_{12}(\mathbf{x}_A)-a_{12}(\mathbf{x}_B) & a_{12}(\mathbf{x}_C)+a_{12}(\mathbf{x}_B) & 0 \\ 2a_{12}(\mathbf{x}_A) & -2[a_{12}(\mathbf{x}_A)+a_{12}(\mathbf{x}_C)] & 2a_{12}(\mathbf{x}_C) \\ 0 & a_{12}(\mathbf{x}_A)+a_{12}(\mathbf{x}_D) & -a_{12}(\mathbf{x}_C)-a_{12}(\mathbf{x}_D) \end{bmatrix} \tag{5.1.19}$$

with $\mathbf{x}_A = (x-\frac{h}{2},y)$, $\mathbf{x}_B = (x-\frac{h}{2},y+h)$, $\mathbf{x}_C = (x+\frac{h}{2},y)$, and $\mathbf{x}_D = (x+\frac{h}{2},y-h)$.

Another way of writing (19) is:

$$\frac{1}{2}\left[\partial_y^+ b(\cdot-\frac{h}{2},\cdot)\partial_x^- + \partial_x^+ b(\cdot-\frac{h}{2},\cdot)\partial_y^- + \partial_y^- b(\cdot+\frac{h}{2},\cdot)\partial_x^+ + \partial_x^- b(\cdot+\frac{h}{2},\cdot)\partial_y^+\right], \tag{5.1.19'}$$

where $b := a_{12}$.

Exercise 5.1.19. (a) For the coefficients of L in (17) let the following hold: $a_{11} > 0$, $a_{22} > 0$, $a_{12} = a_{21} \le 0$, $a_{11} + a_{12} > 0$, and also $2a_{22}(x,y+h/2) + a_{12}(x-h/2,y+h) + a_{12}(x+h/2,y) > 0$, and $a \le 0$. Show that the difference scheme which is described by (18) and (19) has consistency order 2 and that the associated matrix $-L_h$ is a symmetric, irreducibly diagonally-dominant and positive definite M-matrix.
(b) What is the suitable discretisation for the case $a_{12} \ge 0$?

Exercise 5.1.20. The difference formula from Theorem 14 for the operator L reads $a_{11}\partial_x^+\partial_x^- + a_{12}(\partial_x^+\partial_y^+ + \partial_x^-\partial_y^-) + a_{22}\partial_y^+\partial_y^- + a_1\partial_x^+ + a_2\partial_y^+ + a$ when

$a_{12} \geq 0$, $a_1 \geq 0$, $a_2 \geq 0$. Let the associated matrix be L_h. Then prove that the transposed matrix L_h^{T} describes a difference method for the adjoint operator L' and also possesses consistency order 1.

In general it is possible to show for regular difference methods that L_h^{T} is a discretisation of L'. The role of regularity is demonstrated in

Example 5.1.21. Let $Lu := u'' + au'$ in $\Omega = (-1, 1)$ with $a(x) \leq 0$ for $x \leq 0$ and $a(x) \geq 0$ for $x > 0$. According to (14) au' is discretised for $x \leq 0$ by $a(x)\partial^+ u(x)$ and for $x > 0$ by $a(x)\partial^- u(x)$. Let the associated matrix be $L_h = L_{h,2} + L_{h,1}$, where $L_{h,2}$ and $L_{h,1}$ correspond to the terms u'' and au' respectively. According to the above, $L_{h,1}^{\mathsf{T}} v_h$ should be a discretisation of $-(av)'$. But the differences $L_{h,1}^{\mathsf{T}} v_h$ at $x = 0$ and $x = h$ are

$$\begin{aligned} h^{-1}[a(-h)v_h(-h) - a(0)v_h(0) - a(h)v_h(h)] &= -(av)' - h^{-1}a(0)v_h(0) + O(h), \\ h^{-1}[a(0)v_h(0) + a(h)v_h(h) - a(2h)v_h(2h)] &= -(av)' + h^{-1}a(0)v_h(0) + O(h); \end{aligned}$$

thus they are not consistent. Nevertheless, L_h^{T} is a possible discretisation of $L'v = v'' - (av)'$, for it can be shown that the error $v_h - R_h v$ has order of magnitude $O(h)$.

To prove stability one has to show $\|L_h^{-1}\|_\infty \leq \text{const}$. Obviously it is sufficient to prove this inequality for sufficiently small h. In the proof of Theorem 9 we used the fact that

$$-Lw \geq 1 \quad \text{in } \Omega, \qquad w \geq 0 \quad \text{on } \Gamma$$

for $w(\mathbf{x}) := \exp(2R\alpha) - \exp(\alpha(x_1 - z_1 + R))$. Let $D_h u_h(\mathbf{x})$ be the difference equations from which $L_h u_h$ results after elimination of the boundary values. We set $w_h := 2R_h w$; i. e., $w_h(\mathbf{x}) = 2w(\mathbf{x})$ for $\mathbf{x} \in \overline{\Omega}_h$. The following holds:

$$-D_h w_h = -2(D_h R_h - \tilde{R}_h L)w - 2\tilde{R}_h Lw \geq 2 - 2(D_h R_h - \tilde{R}_h L)w.$$

Each consistent difference method satisfies $\|(D_h R_h - \tilde{R}_h L)w\|_\infty \to 0$. For sufficiently small h_0 we thus have

$$-D_h w_h(\mathbf{x}) \geq 1 \quad (\mathbf{x} \in \Omega_h, \ h \leq h_0).$$

This inequality agrees for $\mathbf{x} \in \Omega_h$ far from the boundary with $-L_h w_h(\mathbf{x}) \geq 1$. For $\mathbf{x} \in \Omega_h$ close to the boundary, $-D_h w_h(\mathbf{x})$ also contains the sum $-\sum_{\xi \in \Gamma_h} L_{\mathbf{x}\xi} w_h(\xi)$, which is not contained in $-L_h w_h(\mathbf{x})$. For the discretisations from Theorem 14, Corollary 16, Theorem 18, and Exercise 19, however, $L_{\mathbf{x}\xi} \geq 0$ $(\mathbf{x} \in \Omega_h,\ \xi \in \Gamma_h)$ holds, so that because $w_h \geq 0$ on Γ_h,

$$-L_h w_h \geq -D_h w_h \geq \mathbb{1} \quad (h \leq h_0)$$

holds for all grid points. Theorem 4.3.16 yields

Theorem 5.1.22. *The discretisations from Theorem* 14, *Corollary* 16, *Theorem* 18, *and Exercise* 19 *are stable under the conditions posed there, i.e.,* $\|L_h^{-1}\|_\infty \leq \text{const}$ *for all* $h \in H = \{1/n{:}\, n \in \mathbb{N}\}$. *According to Theorem* 4.5.3 *the methods converge. The order of convergence agrees with the corresponding order of consistency.*

5.1.5 Green's Function

The idea of representing the solution by the Green function can be repeated for the general differential equation (1a,b). The Green function (of the first kind) $g(\xi, \mathbf{x})$ is singular at $\xi = \mathbf{x}$ and satisfies

$$L_{\mathbf{x}} g(\xi, \mathbf{x}) = 0, \quad L'_\xi g(\xi, \mathbf{x}) = 0 \quad \text{for } \mathbf{x}, \xi \in \Omega,\ \mathbf{x} \neq \xi;$$

$$g(\xi, \mathbf{x}) = 0 \quad \text{for } \mathbf{x} \in \Gamma \text{ or } \xi \in \Gamma.$$

Here, L' is the adjoint differential operator (16). If $L \neq L'$, then g is no longer symmetric: $g(\mathbf{x}, \xi) \neq g(\xi, \mathbf{x})$. Under suitable conditions the solution of (1a–c), (4) can be represented as

$$u(\mathbf{x}) = -\int_\Omega g(\xi, \mathbf{x}) f(\xi)\, d\xi - \int_\Gamma \varphi(\xi) B_\xi g(\xi, \mathbf{x})\, d\Gamma_\xi$$

where $B = B_\xi = \sum_{i,j=1}^n n_j \frac{\partial}{\partial \xi_i} a_{ij}$ is a boundary differential operator (n_j are the components of the normal vector $\mathbf{n} = \mathbf{n}(\xi)$, $\xi \in \Gamma$). Only when the principal part of L agrees with Δ is B the normal derivative.

In the discrete case the inverse L_h^{-1} again corresponds to the Green function $g(\cdot, \cdot)$.

5.2 General Boundary Conditions

5.2.1 Formulating the Boundary Value Problem

Let the differential equation be given by (1.1a,b). The Dirichlet boundary condition (1.4) can be written in the form

$$Bu = \varphi \quad \text{on } \Gamma \tag{5.2.1a}$$

where B is the identity (to be precise: the trace on Γ). In more general settings B can be an operator — a so-called boundary differential operator — of order 1:

$$B = \sum_{i=1}^n b_i(\mathbf{x}) \partial/\partial x_i + b_0(\mathbf{x}), \quad \mathbf{x} \in \Gamma. \tag{5.2.1b}$$

If one introduces the vector $\mathbf{b}(\mathbf{x}) = (b_1(\mathbf{x}), \cdots, b_n(\mathbf{x}))^\mathsf{T}$, Bu can be written in the form

$$Bu = \langle \mathbf{b}(\mathbf{x}), \nabla u(\mathbf{x})\rangle + b_0(\mathbf{x})u(\mathbf{x}) \quad \text{or } B = \mathbf{b}^\mathsf{T}\nabla + b_0. \tag{5.2.1b'}$$

Example 5.2.1. (a) From $\mathbf{b} = \mathbf{0}$, $b_0(x) \neq 0$ there results what is known as the Dirichlet condition $u = \varphi/b_0$ on Γ, also known as the boundary condition of the first kind.
(b) The choice $\mathbf{b} = \mathbf{n}$, $b_0 = 0$ characterises the Neumann condition, also called the boundary condition of the second kind.
(c) Equation (1a) with $\langle \mathbf{b}, \mathbf{n}\rangle \neq 0$, $b_0 \neq 0$, is known as the mixed, or the boundary condition of the third kind. Occasionally one just means by a mixed boundary condition: $B = \sigma\mathbf{n}^\mathsf{T}\nabla + b_0 = \sigma\partial/\partial n + b_0$ with $\sigma(\mathbf{x}) \neq 0$.

Remark 5.2.2. The case $\langle \mathbf{b}, \mathbf{n}\rangle = 0$ is excluded in general. For $\langle \mathbf{b}, \mathbf{n}\rangle = 0$, $\mathbf{b}^\mathsf{T}\nabla$ is a tangential derivative. The boundary condition $Bu = \varphi$ is then very similar to a Dirichlet condition. The condition "$u = \varphi$ on Γ" implies "$Bu = \tilde{\varphi} := B\varphi$ on Γ" (Why is $B\varphi$ defined?).

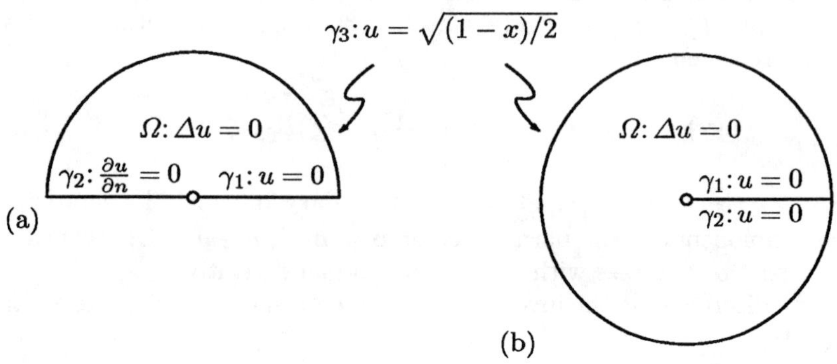

Figure 5.2.1.
(a) Boundary value problem with changing boundary condition type
(b) Dirichlet problem in a disk with a cut

The normal derivative $B = \partial/\partial n$ (i.e. $\mathbf{b} = \mathbf{n}$) is important in connection with $L = -\Delta$. For the general operator L in (1.1b) the so-called conormal derivative B with

$$\mathbf{b} = \mathbf{A}\mathbf{n} \quad (\mathbf{A} = (a_{ij})_{i,j=1,\cdots,n},\ a_{ij} \text{ as in (1.1b)})$$

is of greater importance, as we will see in Section 7.4.

Statements about existence and uniqueness of the solution always depend on L and B. We have seen already that for $L = -\Delta$, $B = I$ (Dirichlet condition) uniqueness is guaranteed (cf. Theorem 3.1.2), while the problem associated to $L = -\Delta$, $B = \mathbf{n}^\mathsf{T}\nabla = \partial/\partial n$ is, in general, not solvable (cf. Theorem 3.4.1).

The coefficients of B depend on position. Of course, $B(\mathbf{x}) = 0$, i.e., $\mathbf{b}(\mathbf{x}) = \mathbf{0}$ and $b_0(\mathbf{x}) = 0$, must not occur for any $\mathbf{x} \in \Gamma$. But it is possible that $\mathbf{b}(\mathbf{x}) = \mathbf{0}$ (and $b_0 \neq 0$) in $\gamma \subset \Gamma$ and $\mathbf{b}(\mathbf{x}) \neq \mathbf{0}$ in $\Gamma\backslash\gamma$. Then there is a Dirichlet condition $u = \varphi/b_0$ on the piece γ and a boundary condition of first order on the remaining boundary piece $\Gamma\backslash\gamma$. At the points of contact between γ and $\Gamma\backslash\gamma$ the solution generally is not smooth (it has singularities in the derivatives).

Example 5.2.3. Let Ω be the upper semicircle around $x = y = 0$ with radius 1. Let the differential equation and boundary conditions be given as in Figure 1a. The boundary condition changes its order at $x = y = 0$. The solution in polar coordinates reads: $u = r^{1/2}\sin(\varphi/2)$ (cf. (2.1.3)). Check that $u_x = O(r^{-1/2})$ and $u_y = O(r^{-1/2})$.

The same singularity as in Example 3 occurs in the problem described in Figure 1b; the solution is also $r^{1/2}\sin(\varphi/2)$. This Dirichlet problem and Example 3 are closely connected with each other.

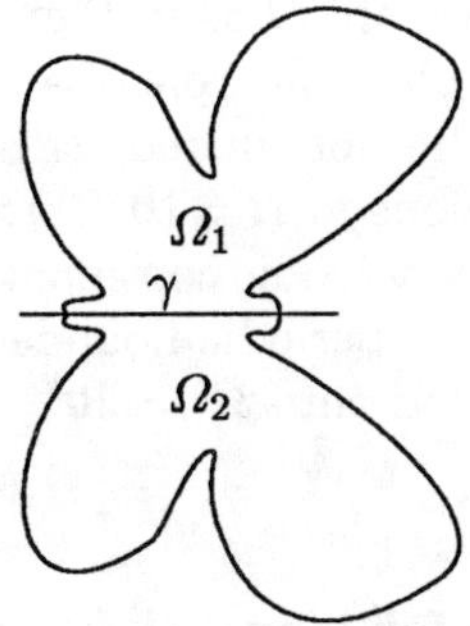

Figure 5.2.2. A domain symmetric with respect to γ

Example 5.2.4. Let $\Omega = \Omega_1 \cup \Omega_2 \cup \gamma$ be as in Figure 2: let the reflection of Ω_1 in γ result in Ω_2. If one seeks a solution of $Lu = f$ in Ω, $Bu = \varphi$ on $\partial\Omega$, and if together with u the function $\overline{u}$ reflected in γ is also a solution, one expects $u = \overline{u}$. This solution then satisfies $Lu = f$ in Ω_1, $Bu = \varphi$ on $\partial\Omega_1\backslash\gamma$, $\partial u/\partial n = 0$ on γ.

While the boundary condition $Bu = \varphi$ on $\partial\Omega$ may be of physical origin, Example 4 shows that a Neumann condition may also have a geometric basis.

Another geometrically justified boundary condition is the following. Let Ω be given as in Figure 3: γ_1 and γ_2 are parts of $\Gamma = \partial\Omega$ with

$$\gamma_i = \{(x_i, y): y_1 \le y \le y_2\}, \quad i = 1, 2.$$

Then, in addition to $Bu = \varphi$ on $\Gamma \backslash (\gamma_1 \cup \gamma_2)$, we can require the periodic boundary condition

$$u(x_1, y) = u(x_2, y), \quad u_x(x_1, y) = u_x(x_2, y) \quad \text{for } y_1 \le y \le y_2, \tag{5.2.2}$$

on γ_1 and γ_2. The solution is periodically continuable in the x-direction (with period $x_2 - x_1$). The origin of periodic boundary conditions is discussed in Example 5.

Example 5.2.5. (a) Let Ω' be an annulus which is described by the polar coordinates $r \in (r_1, r_2)$, $\varphi \in [0, 2\pi)$. Transformation of the differential equation to polar coordinates gives as the image domain the rectangle $\Omega = (r_1, r_2) \times (0, 2\pi)$. The original boundary conditions on Ω' become boundary conditions at the upper and lower boundaries while Equation (2) describes the periodicity of the angular variable $x \in (0, 2\pi)$.
(b) Instead of on $\Omega' \subset \mathbb{R}^2$, one can also define a boundary value problem on a part of the 2-dimensional surface of a 3-dimensional body. If Ω' lies, for example, on the surface of the cylinder $\{\xi \in \mathbb{R}^3 : \xi_1^2 + \xi_2^2 \le r^2, \ \xi_3 \in \mathbb{R}\}$, the unfolding of Ω' results in a domain Ω as in Figure 3. Here too, Equation (2) is justified by the fact that x plays the role of an angle variable.
(c) If one seeks solutions in the unbounded strip $\Omega' = \mathbb{R} \times (y_1, y_2)$ one can instead look for periodic solutions in $\Omega = (0, 2\pi) \times (y_1, y_2)$ with the boundary condition (2), since these (after periodic continuation) are also solutions of the original problem. Solutions with periodic boundary conditions in the x and y directions can even be continued onto $\Omega' = \mathbb{R}^2$.

5.2.2 Difference Methods for General Boundary Conditions

The boundary conditions posing the least difficulties are the periodic boundary conditions. Define Ω_h as the set of grid points in Ω and on γ_1 (but not on γ_2; cf. Figure 3).

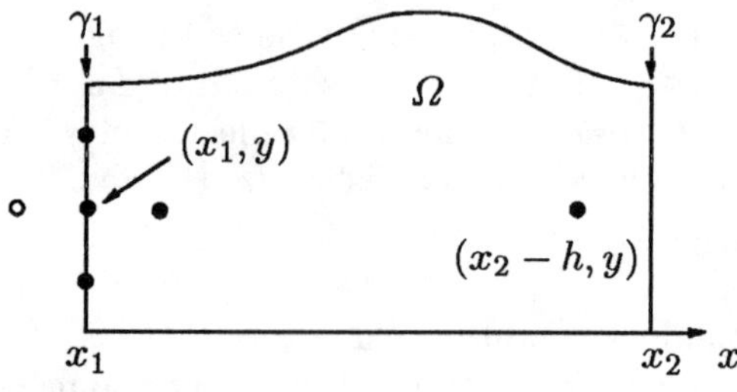

Figure 5.2.3. Discretisation for periodic boundary conditions

The difference equation at a grid point $(x_1, y) \in \gamma_1$ has a left neighbour $(x_1 - h, y)$ outside $\overline{\Omega}$. Replace $u_h(x_1 - h, y)$ in the difference equation by $u_h(x_2 - h, y)$. By doing so we have transferred the periodicity onto the difference solution without explicitly discretising Equation (2). Of course, the step size must be chosen so that $x_2 - x_1$ is a multiple of h.

Exercise 5.2.6. For periodic boundary conditions the structure of the matrix L_h changes. Let $L = -\Delta$ in the square $\Omega = (0,1) \times (0,1)$. Assume Dirichlet conditions for the upper and lower boundaries, and for the lateral boundaries the periodicity condition (2). In analogy to (4.2.8) exhibit the form of the matrix L_h for a lexicographical arrangement of the grid points.

In Section 4.7 we described the discretisation of the boundary condition $B = \mathbf{n}^\mathsf{T}\nabla = \partial/\partial n$ for the case that Ω is a rectangle and Γ coincides with the grid. In the same situation one can discretise the general boundary condition (1a,b) as follows.

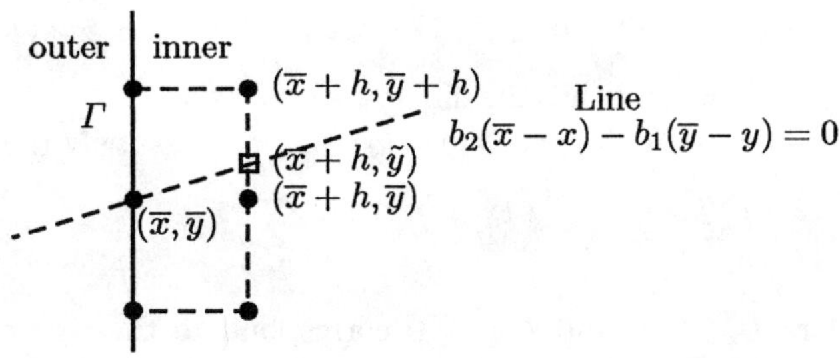

Figure 5.2.4. Discretisation of $Bu = \varphi$

Let $(\overline{x}, \overline{y}) \in \Gamma_h = \overline{\Omega}_h \cap \Gamma$ be a grid point on Γ. The point of intersection $(\overline{x}+h, \tilde{y})$ drawn in Figure 4 is given by $\tilde{y} = \overline{y} + hb_2(\overline{x}, \overline{y})/b_1(\overline{x}, \overline{y})$. Note that $b_1 \neq 0$, since $\mathbf{b}^\mathsf{T}\nabla$ is not to be a tangential derivative. For $u \in C^2(\overline{\Omega})$ the directional derivative $\mathbf{b}^\mathsf{T}\nabla u = \langle \mathbf{b}, \nabla u \rangle$ in $(\overline{x}, \overline{y})$ is approximated by

$$[u(\overline{x}, \overline{y}) - u(\overline{x}+h, \tilde{y})]/\sqrt{h^2 + (\overline{y} - \tilde{y})^2} = (\mathbf{b}^\mathsf{T}\nabla u)(\overline{x}, \overline{y}) + O(h). \quad (5.2.3a)$$

On the other hand, one obtains $u(\overline{x}+h, \tilde{y})$ under the same conditions by interpolation up to order $O(h^2)$:

$$u(\overline{x}+h, \tilde{y}) = \frac{\tilde{y} - \overline{y}}{h} u(\overline{x}+h, \overline{y}+h) + \frac{\overline{y} + h - \tilde{y}}{h} u(\overline{x}+h, \overline{y}) + O(h^2). \quad (5.2.3b)$$

The combination of (3a) and (3b) leads to the difference formula

$$(B_h u_h)(\overline{x},\overline{y}) := \frac{\left\{u_h(\overline{x},\overline{y}) - \frac{\overline{y}+h-\tilde{y}}{h}u_h(\overline{x}+h,\overline{y}) - \frac{\tilde{y}-\overline{y}}{h}u_h(\overline{x}+h,\overline{y}+h)\right\}}{\sqrt{h^2+(\tilde{y}-\overline{y})^2}} + b_0(\overline{x},\overline{y})u_h(\overline{x},\overline{y}) = \varphi(\overline{x},\overline{y}) \tag{5.2.3c}$$

with $\tilde{y}-\overline{y} = hb_2(\overline{x},\overline{y})/b_1(\overline{x},\overline{y})$.

Lemma 5.2.7. *Let R_h be the restriction to the grid points $\overline{\Omega}_h$. For a function $u \in C^2(\overline{\Omega})$ one has*

$$B_h(u_h - R_h u)(\overline{x},\overline{y}) = \varphi(\overline{x},\overline{y}) - B_h R_h u(\overline{x},\overline{y}) = O(h\|u\|_{C^2(\overline{\Omega})}),$$

i.e., the discretisation of the boundary condition has consistency order 1.

Equation (3c) has the form

$$(B_h u_h)(\overline{x},\overline{y}) = c_0 u_h(\overline{x},\overline{y}) - c_1 u_h(\overline{x}+h,\overline{y}) - c_2 u_h(\overline{x}+h,\overline{y}+h). \tag{5.2.3c'}$$

We assume that the difference equations $\sum_{\xi\in\overline{\Omega}_h} L_{x\xi}u_h(\xi) = f_h(\mathbf{x})$ (before the elimination of the values $u_h(\xi)$, $\xi \in \Gamma_h$) satisfy the inequalities

$$L_{\mathbf{x}\xi} \le 0 \quad (\mathbf{x} \ne \xi), \quad L_{\mathbf{xx}} = -\sum_{\xi\ne\mathbf{x}} L_{\mathbf{x}\xi} \quad (\mathbf{x} \in \Omega_h, \xi \in \overline{\Omega}_h). \tag{5.2.4a}$$

Here $L_{\mathbf{xx}} > 0$ and $L_{\mathbf{x}\xi} \le 0$ correspond to the sign condition (4.3.1a). The second inequality in (4a) agrees with (4.3.4b). The corresponding inequalities for (3c′) read:

$$c_1 \ge 0, \quad c_2 \ge 0, \quad c_0 \ge c_1 + c_2. \tag{5.2.4b}$$

Let the difference equations in $\mathbf{x} \in \Omega_h$ and the boundary equations (3c′) given for $(\overline{x},\overline{y}) \in \Gamma_h$ be combined into $A_h u_h = g_h$. (4a,b) implies $a_{\mathbf{x}\xi} \le 0$, $a_{\mathbf{xx}} \ge -\sum_{\xi\ne\mathbf{x}} a_{\mathbf{x}\xi}(\mathbf{x},\xi \in \overline{\Omega}_h)$.

Therefore, A_h is an M-matrix if A_h is irreducible and (4.3.4a) holds for at least one $\mathbf{x} \in \overline{\Omega}_h$.

The above considerations explain why the boundary discretisation should satisfy the conditions (4b). (4b) holds for the case $b_0 = 0$ if and only if $b_2/b_1 \in [0,1]$. If $b_2/b_1 \in [-1,0]$, (4b) can be satisfied if one interpolates between $\overline{y}$ and $\overline{y}-h$ (instead of $\overline{y}$ and $\overline{y}+h$). If, however, $|b_2| > |b_1|$, i.e., if the tangential component is greater, one could interpolate between $(\overline{x}+h,\overline{y})$ and $(\overline{x}+h,\overline{y}\pm kh)$, where $|b_2/b_1| \le k$. For the case $b_2/b_1 \ge 1$, however, it is more practical to interpolate at the point $(\tilde{x},\overline{y}+h)$ between $(\overline{x},\overline{y}+h)$ and $(\overline{x}+h,\overline{y}+h)$. Generally one should choose as interpolation point the point of intersection of the line with the dotted straight line in Figure 4.

In the preceding discussion we started with the case $\Omega = (0,1) \times (0,1)$. If Ω is a general region, two discretisation techniques offer themselves (cf. Sections 4.8.1 and 4.8.2).

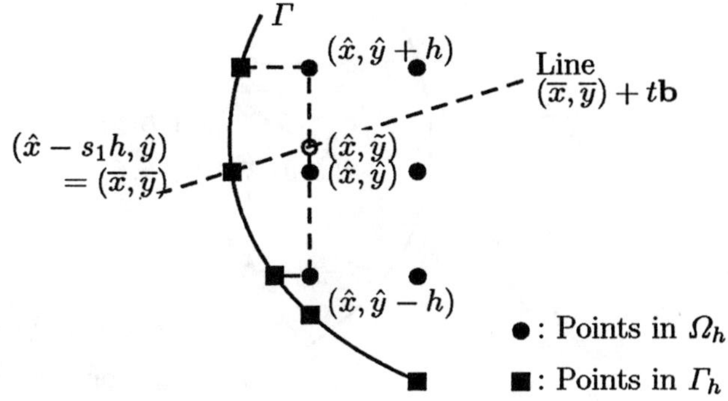

Figure 5.2.5. First boundary discretisation

First option for discretisation:

Let Ω_h and Γ_h be chosen as in Section 4.8.1. At near-boundary points the differential equation is approximated by a difference scheme which in the case of $L = -\Delta$ corresponds to the Shortley-Weller method. For this one needs the values of u_h at the boundary points $\xi \in \Gamma_h$. As in Figure 5 let $(\overline{x}, \overline{y}) = (\hat{x} - s_l h, \hat{y}) \in \Gamma_h$ be a boundary point. Again we can set up an equation analogous to (3a–c). In Figure 5 $(\hat{x}, \tilde{y})$ has the same position as $(\overline{x} + h, \tilde{y})$ in Figure 4. In general one has to use the point of intersection of the line which is also presented by $(\overline{x}, \overline{y}) + t\mathbf{b}$ $(t \in \mathbb{R})$, and the dotted straight line in Figure 5.

Second option for discretisation:

Let $\overline{\Omega}_h$ be the grid Ω_h used above. Now let Ω_h consist of all points of $\overline{\Omega}_h$ that are far from the boundary. For all $(x, y) \in \Omega_h$ difference equations (with equidistant step size) are declared which involve $\{u_h(\xi) : \xi \in \overline{\Omega}_h\}$. For each point $(\hat{x}, \hat{y}) \in \Gamma_h := \overline{\Omega}_h \backslash \Omega_h$ a boundary discretisation must be found (cf. Figure 6). Equation (3a) can be set up with $(\hat{x}, \hat{y})$ and $(\hat{x} + h, \tilde{y})$ instead of $(\overline{x}, \overline{y})$ and $(\overline{x} + h, \tilde{y})$. The arguments of the coefficients b_1 and b_2 are $(\overline{x}, \overline{y})$. This point results implicitly from

$$b_2(\overline{x}, \overline{y})(\overline{x} - \hat{x}) = b_1(\overline{x}, \overline{y})(\overline{y} - \hat{y}), \quad (\overline{x}, \overline{y}) \in \Gamma. \tag{5.2.5}$$

Remark 5.2.8. (a) At the grid points adjacent to the boundary, the first discretisation requires a difference scheme for $Lu = f$ with nonequidistant stepsizes. The second discretisation requires an approximation of the nonlinear problem (5). From a programming viewpoint both procedures are undesirable because of the case distinctions. (b) If the vector $\mathbf{b}$ from B approaches the tangential direction, both methods fail since the straight line no longer intersects the dotted straight line from Figures 5 and 6. Here, the second discretisation fails earlier than the first one.

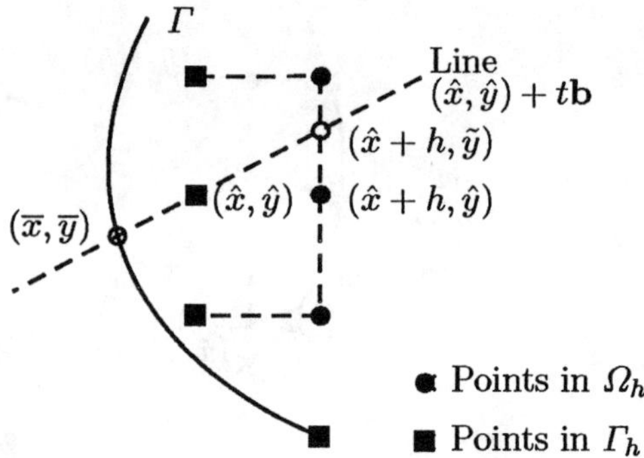

Figure 5.2.6. Second boundary discretisation

Another option for avoiding the difficulties described above consists in using variational difference equations, at least near the boundary (cf. Remark 8.6.1).

Occasionally it is also possible to simplify the boundary conditions by using coordinate transformations. Let $Bu = \varphi$ be prescribed on $\gamma \subset \Gamma$. If one finds a transformation $(x,y) \leftrightarrow (\xi,\eta)$ such that the equations $\xi = 0$ and $b_2 x_\xi = b_1 y_\xi$ are satisfied on γ, one obtains for the transformed problem $Bu \equiv \sigma \partial u/\partial n + b_0 u = \varphi$ on a vertical boundary piece ($\partial/\partial n = -\partial/\partial\xi$). The discretisation may be carried out as described in Section 4.7.

The same reasoning as in the preceding sections proves stability and convergence:

Remark 5.2.9. Let the difference method $D_h u_h = f_h$ satisfy the conditions of Theorem 5.1.22 so that $-D_h \tilde{w}_h \geq \mathbf{1}$ holds for $\tilde{w}_h := K\mathbf{1} + w_h$ ($K > 0$ is constant; w_h is as in the proof of Theorem 5.1.22). Let the boundary discretisation (3c′) satisfy

$$c_0 - c_1 - c_2 = b_0(\mathbf{x}) \geq \epsilon > 0 \text{ for all } \mathbf{x} \in \Gamma_h.$$

For sufficiently large K, one then obtains $(B_h \tilde{w}_h)(\mathbf{x}) \geq 1$ for $x \in \Gamma_h$. Hence it follows that $A_h \tilde{w}_h \geq \mathbf{1}$, where A_h is the M-matrix defined immediately after Equation (4b). Since $\|A_h^{-1}\|_\infty \leq$ const (cf. Theorem 4.3.16), stability has been proved. The order of convergence is the minimum of 1 (order of consistency of B_h ; cf. Lemma 7) and the order of consistency of $D_h u_h = f_h$.

In general it is not possible to approximate $\mathbf{b}^\mathsf{T}\nabla$ by symmetric differences. To construct a discretisation of $Bu = \varphi$ with consistency order 2 despite this fact, set:

$$B_h u_h(\overline{x},\overline{y}) := c_0 u_h(\overline{x},\overline{y}) - \sum_{i=1}^{3} c_i u_h(x_i,y_i) = g(\overline{x},\overline{y}), \quad (\overline{x},\overline{y}) \in \Gamma_h,$$

where (x_i, y_i) are three points in Ω_h. If, in order to remove the first two terms in the Taylor expansion, one uses the differential equation $Lu = f$ in $(\overline{x}, \overline{y})$ and the tangential derivative of $Bu = \varphi$, one obtains a discretisation of order 2. For the special case $L = \Delta$, $B = \partial/\partial n + b_0$, Bramble-Hubbard [2] proved that the three points (x_i, y_i) may be chosen such that the inequalities $c_i \geq 0$, $c_0 \geq c_1 + c_2 + c_3$ hold and guarantee stability. However, in general, the (x_i, y_i) do not lie in the direct neighbourhood of $(\overline{x}, \overline{y}) \in \Gamma_h$. Nonetheless, the construction of the discretisation appears too complicated to be recommended for practical purposes.

5.3 Boundary Problems of Higher Order

5.3.1 The Biharmonic Differential Equation

In elastomechanics the free vibration of rods leads to (ordinary) differential equations of second order if longitudinal vibrations (= compression waves) or torsional vibrations are involved. By contrast, transversal vibrations (= bending waves) result in an equation of fourth order. Correspondingly, the bending vibration of a plate leads to a partial differential equation of fourth order. This is the biharmonic equation (plate equation)

$$\Delta^2 u = f \quad \text{in } \Omega \tag{5.3.1}$$

where $\Delta^2 = \partial^4/\partial x^4 + 2\partial^4/\partial x^2 \partial y^2 + \partial^4/\partial y^4$. Here u describes the deflection of the plate perpendicular to the surface. If the plate is firmly clamped at the edge, one obtains the boundary conditions

$$u = \varphi_1 \quad \text{and} \quad \partial u/\partial n = \varphi_2 \quad \text{on } \Gamma \tag{5.3.2}$$

with $\varphi_1 = \varphi_2 = 0$. A biharmonic problem (1), (2) also results from the transformation of the Stokes equations in $\Omega \subset \mathbb{R}^2$ (cf. Remark 12.1.5).

The differential equation (1) may be combined with other boundary values than (2). An example is:

$$u = \varphi_1 \quad \text{and} \quad \Delta u = \varphi_2 \quad \text{on } \Gamma \tag{5.3.3}$$

(simply supported plate). Other examples can be found in (7.4.12b,e).

Exercise 5.3.1. Show that if one solves the Poisson equations $\Delta v = f$ in Ω, $v = \varphi_2$ on Γ and, subsequently, $\Delta u = v$ in Ω and $u = \varphi_1$ on Γ, then u is the solution of the boundary value problem (1), (3). Why can Problem (1), (2) not be handled likewise?

Remark 5.3.2. The solutions of $\Delta^2 u = 0$ do not satisfy a maximum-minimum principle (counterexample: $u = x^2 + y^2$ in $\Omega = K_R(\mathbf{0})$).

5.3.2 General Linear Differential Equations of Order $2m$

The partial derivative D^α ($\alpha \in \mathbf{Z}^n$: multi-index) of order $|\alpha| = \alpha_1 + \cdots + \alpha_n$ is defined in (3.2.5). A differential operator of order $2m$ has the form

$$L = \sum_{|\alpha| \le 2m} a_\alpha(\mathbf{x}) D^\alpha \quad (\mathbf{x} \in \Omega) \tag{5.3.4a}$$

and defines the differential equation of order $2m$:

$$Lu = f \quad \text{in } \Omega. \tag{5.3.4b}$$

Ellipticity has been explained thus far only for equations of second order (cf. Definition 1.2.3).

Definition 5.3.3. The differential operator L (with real-valued coefficients a_α) is said to be elliptic (of order $2m$) at $\mathbf{x} \in \Omega$ if

$$\sum_{|\alpha| = 2m} a_\alpha(\mathbf{x}) \xi^\alpha \ne 0 \quad \text{for all } 0 \ne \xi \in \mathbb{R}^n. \tag{5.3.5a}$$

Here, ξ^α is an abbreviation for the polynomial $\xi_1^{\alpha_1} \xi_2^{\alpha_2} \cdot \ldots \cdot \xi_n^{\alpha_n}$ of degree $|\alpha|$. We set $P(\mathbf{x}, \xi) := \sum_{|\alpha|=2m} a_\alpha(\mathbf{x}) \xi^\alpha$. Evidently (5a) is equivalent to $P(\mathbf{x}, \xi) \ne 0$ for all $\xi \in \mathbb{R}^n$ with $|\xi| = 1$. For reasons of continuity either $P(\mathbf{x}, \xi) > 0$ or $P(\mathbf{x}, \xi) < 0$ must hold. Without loss of generality, we may assume that $P(\mathbf{x}, \xi) > 0$; otherwise we scale with the factor -1 (changing from $Lu(\mathbf{x}) = f(\mathbf{x})$ to $-Lu(\mathbf{x}) = -f(\mathbf{x})$). Since the set $\{\xi \in \mathbb{R}^n\colon |\xi| = 1\}$ is compact it follows that $c(\mathbf{x}) := \min\{P(\mathbf{x}, \xi)\colon |\xi| = 1\} > 0$ and this justifies the formulation of (5a) as

$$\sum_{|\alpha| = 2m} a_\alpha(\mathbf{x}) \xi^\alpha \ge c(\mathbf{x}) |\xi|^{2m} \quad \text{for all } \xi \in \mathbb{R}^n \text{ with } c(\mathbf{x}) > 0. \tag{5.3.5b}$$

Definition 5.3.4. The differential operator L is said to be uniformly elliptic in Ω if $\inf\{c(\mathbf{x})\colon \mathbf{x} \in \Omega\} > 0$ for $c(\mathbf{x})$ from (5b).

Exercise 5.3.5. (a) Translate L from (1.1b) into the notation of (4a). What are the coefficients a_α for $L = \Delta^2$?
(b) Prove that the biharmonic operator Δ^2 is uniformly elliptic.
(c) Let a_α be real-valued. Why are there no elliptic operators L of odd order?
(d) If the coefficients a_α are sufficiently smooth one can write L from (4a) in the form $L = \sum_{|\alpha| \le m} \sum_{|\beta| \le m} (-1)^{|\beta|} D^\beta a_{\alpha\beta}(\mathbf{x}) D^\alpha$ (cf. (1.1b) and (1.2)).

For $m = 1$ (equation of order 2) we have used one boundary condition; for the biharmonic equation ($m = 2$) two boundary conditions occur. In general one needs m boundary conditions

$$B_j u := \sum_{|\alpha| \le m_j} b_{j\alpha} D^\alpha u = \varphi_j \quad (j = 1, 2, \cdots, m) \quad \text{on } \Gamma \tag{5.3.6}$$

with boundary differential operators B_j of order $0 \le m_j < 2m$.

Remark 5.3.6. The boundary operators B_j cannot be chosen arbitrarily, but must be independent of each other (they must form a so-called "normal system"; cf. Lions-Magenes [1, p. 113] and Wloka [1, Definition 14.1]). In particular the orders m_j must differ pairwise.

We have a Dirichlet boundary condition if $B_j = \partial^{j-1}/\partial n^{j-1} = (\partial/\partial n)^{j-1}$ for $j = 1, \cdots, m$ (cf. (2)).

The representation of the solution using the Green function will not be discussed at this point. However, it is remarkable that the Green function (and in particular the singularity function) for L is continuous whenever $2m > n$. For $L = \Delta^2$ in $\mathbb{R}^2$, for example, the singularity function is $\frac{1}{8\pi}|\mathbf{x} - \mathbf{y}|^2 \log |\mathbf{x} - \mathbf{y}|$ (cf. Wloka [1, Exercise 21.9]).

5.3.3 Discretisation of the Biharmonic Differential Equation

The simplest difference formula for $L = \Delta^2$ is the 13-point difference method

$$D_h = \Delta_h^2 = h^{-4} \begin{bmatrix} & & 1 & & \\ & 2 & -8 & 2 & \\ 1 & -8 & 20 & -8 & 1 \\ & 2 & -8 & 2 & \\ & & 1 & & \end{bmatrix}, \tag{5.3.7}$$

which can be represented as the square of the five-point star Δ_h from (4.2.3). Let the grid $\Omega_h \subset \Omega$ and the boundaries $\Gamma_h \subset \Gamma$, $\gamma_h \subset \mathbb{R}^2 \backslash \overline{\Omega}$ be defined as in Figure 1. The difference equation

$$D_h u_h = f_h \quad \text{in } \Omega_h \tag{5.3.8a}$$

requires the values of u_h in $\overline{\Omega}_h := \Omega_h \cup \Gamma_h \cup \gamma_h$. These are determined by the boundary conditions (cf. (2))

$$u_h(\mathbf{x}) = \varphi_1(\mathbf{x}) \qquad \text{for } \mathbf{x} \in \Gamma, \tag{5.3.8b}$$

$$\partial_n^0 u_h(\mathbf{x}) := [u_h(\mathbf{x} + h\mathbf{n}) - u_h(\mathbf{x} - h\mathbf{n})]/(2h) = \varphi_2(\mathbf{x}) \quad \text{for } \mathbf{x} \in \Gamma_h, \mathbf{x} + h\mathbf{n} \in \gamma_h. \tag{5.3.8c}$$

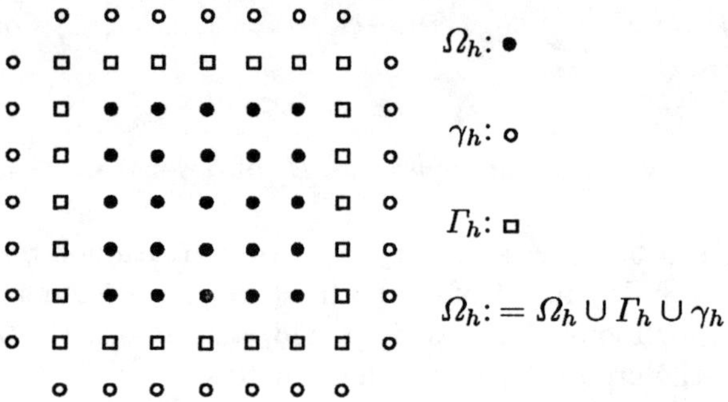

Ω_h: •

γ_h: ∘

Γ_h: □

$\bar{\Omega}_h := \Omega_h \cup \Gamma_h \cup \gamma_h$

Figure 5.3.1. Grid for the biharmonic equation in $\Omega = (0,1) \times (0,1)$

As in Equation (4.7.14b) we assign to the corner points two normal directions each. Correspondingly, to each corner point belong two equations (8c). Note that two ($m = 2$) boundary layers occur, Γ_h and γ_h, and that two boundary conditions are given.

Remark 5.3.7. If with the aid of (8b,c) one eliminates in Equation (8a) the unknowns $\{u_h(\xi): \xi \in \Gamma_h \cup \gamma_h\}$, one obtains a system of equations $L_h u_h = q_h$ for $\{u_h(\xi): \xi \in \Omega_h\}$. L_h is not an M-matrix since the sign condition (4.3.1a) is violated. But L_h^{-1} also is in general not $\geq O$ (cf. (4.3.1b)).

Thus the methods of the proof in Section 4 cannot be used here. Instead we will prove the analogue of Theorem 4.4.12.

Theorem 5.3.8. *Let u_h be the discrete solution of the biharmonic equation in the square $\Omega = (0,1) \times (0,1)$ defined by (8a–c). The matrix L_h described in Remark 7 is symmetric and positive definite. It satisfies*

$$\|L_h\|_2 \leq 64h^{-4}, \quad \|L_h^{-1}\|_2 \leq 1/256. \tag{5.3.9a}$$

If $\varphi_1 = 0$ on Γ_h (cf. (8b)), then u_h can be estimated by

$$\|u_h\|_{\Omega_h} \leq \frac{1}{256}\|f_h\|_{\Omega_h} + \frac{h^{-1/2}}{16\sqrt{2}}|\varphi_2|_{\Gamma_h}, \tag{5.3.9b}$$

where the norms are defined as follows:

$$\|u_h\|_{\Omega_h} := [h^2 \sum_{\mathbf{x}\in\Omega_h} |u_h(\mathbf{x})|^2]^{1/2}, \quad |\varphi_2|_{\Gamma_h} := [h \sum_{\mathbf{x}\in\Gamma_h} |\varphi_2(\mathbf{x})|^2]^{1/2}. \tag{5.3.9c}$$

Let the sum $\sum_{\mathbf{x}\in\Gamma_h}$ in $|\varphi_2|_{\Gamma_h}$ contain the corner points twice, where, for example, the value $\varphi_2(\mathbf{x})$ at $\mathbf{x} = (0,0)$ should be interpreted once as $\varphi_2(0,0+)$, and once as $\varphi_2(0,0+)$ (cf. comment on (8c)).

PROOF. We limit ourselves to the most important part, the inequality (9b). In the infinite grid $Q_h := \{(x,y) \in \mathbb{R}^2 : x/h, y/h \in \mathbb{Z}\}$ we set

$$\overline{u}_h(\mathbf{x}) := u_h(\mathbf{x}) \quad \text{for } \mathbf{x} \in \Omega_h, \quad \overline{u}_h(\mathbf{x}) := 0 \quad \text{for } \mathbf{x} \in Q_h \backslash \Omega_h.$$

After partial summation one obtains

$$\sum_{\mathbf{x}\in\Omega_h} u_h(\mathbf{x}) f_h(\mathbf{x}) = \sum_{\mathbf{x}\in Q_h} \overline{u}_h(\mathbf{x}) \Delta_h^2 u_h(\mathbf{x}) = \sum_{\mathbf{x}\in Q_h} (\Delta_h \overline{u}_h)(\mathbf{x})(\Delta_h u_h)(\mathbf{x})$$

(First consider the identities $\sum v(x)\partial^+\partial^- u(x) = -\sum(\partial^- v)(x)(\partial^- u)(x) = \sum(\partial^+\partial^- v)(x)u(x)$ in the one-dimensional case.) From the definition of $\overline{u}_h$ it follows that $\Delta\overline{u}_h = 0$ on $Q_h \backslash (\Omega_h \cup \Gamma_h)$. On Γ_h one has $u_h = \overline{u}_h = 0$, and

$$\overline{u}_h(\mathbf{x} - h\mathbf{n}) = u_h(\mathbf{x} - h\mathbf{n}), \quad \overline{u}_h(\mathbf{x} + h\mathbf{n}) = 0,$$
$$u_h(\mathbf{x} + h\mathbf{n}) = u_h(\mathbf{x} - h(\mathbf{n}) + 2h\varphi_2(\mathbf{x})$$

for $\mathbf{x} \in \Gamma_h$. If one sets $\partial_n^- u_h(\mathbf{x}) := [u_h(\mathbf{x}) - u_h(\mathbf{x} - h\mathbf{n})]/h = -u_h(\mathbf{x} - h\mathbf{n})/h$ at $\mathbf{x} \in \Gamma_h$, then

$$\Delta_h \overline{u}_h(\mathbf{x}) = -h^{-1}\partial_n^- u_h(\mathbf{x}), \quad \Delta_h u_h(\mathbf{x}) = -2h^{-1}\partial_n^- u_h(\mathbf{x}) + 2h^{-1}\varphi_2(\mathbf{x}),$$

for $\mathbf{x} \in \Gamma_h$, implies that

$$(u_h, L_h u_h)_{\Omega_h} = (u_h, f_h)_{\Omega_h} = ||\Delta_h u_h||_{\Omega_h}^2 + 2h^{-1}|\partial_n^- u_h|_{\Gamma_h}^2 - 2h^{-1}(\partial_n^- u_h, \varphi_2)_{\Gamma_h},$$

where $(\cdot,\cdot)_{\Omega_h}$ and $(\cdot,\cdot)_{\Gamma_h}$ are the scalar products belonging to the norms (9c). According to

$$ab \le a^2 + b^2/4 \quad (a, b \in \mathbb{R}) \tag{5.3.10}$$

one estimates

$$(\partial_n^- u_h, \varphi_2)_{\Gamma_h} \le |\partial_n^- u_h|_{\Gamma_h} |\varphi_2|_{\Gamma_h} \le |\partial_n^- u_h|_{\Gamma_h}^2 + |\varphi_2|_{\Gamma_h}^2/4.$$

Thus one obtains

$$||\Delta_h u_h||_{\Omega_h}^2 + 2h^{-1}|\partial_n^- u_h|_{\Gamma_h}^2 \le (u_h, f_h)_{\Gamma_h} + 2h^{-1}(\partial_n^- u_h, \varphi_2)_{\Gamma_h} \tag{5.3.11}$$
$$\le ||u_h||_{\Omega_h}||f_h||_{\Omega_h} + 2h^{-1}|\partial_n^- u_h|_{\Gamma_h}^2 + h^{-1}|\varphi_2|_{\Gamma_h}^2/2.$$

Theorem 4.4.12 shows that $||u_h||_{\Omega_h} \le ||\Delta_h u_h||_{\Omega_h}/16$. Thus for $\varphi_2 = 0$ (11) implies $||u_h||_{\Omega_h}^2 \le 16^{-2}||\Delta_h u_h||_{\Omega_h}^2 \le 16^{-2}||u_h||_{\Omega_h}||f_h||_{\Omega_h}$, so that

$$||u_h||_{\Omega_h} \le 16^{-2}||f_h||_{\Omega_h}. \tag{5.3.12a}$$

For $f_h = 0$, (11) implies the inequality $||\Delta_h u_h||_{\Omega_h}^2 \le (2h)^{-1}|\varphi_2|_{\Gamma_h}^2$, therefore

$$||u_h||_{\Omega_h} \le \frac{1}{16}||\Delta_h u_h||_{\Omega_h} \le \frac{1}{16}(2h)^{-1/2}|\varphi_2|_{\Gamma_h}. \tag{5.3.12b}$$

In the general case we write u_h as $u_h^{\mathrm{I}} + u_h^{\mathrm{II}}$, where $\varphi_2^{\mathrm{I}} = 0$, $\varphi_2^{\mathrm{II}} = 0$, $f_h^{\mathrm{I}} = f_h$ and $f_h^{\mathrm{II}} = 0$. $||u_h||_{\Omega_h} \le ||u_h^{\mathrm{I}}||_{\Omega_h} + ||u_h^{\mathrm{II}}||_{\Omega_h}$ together with (12a,b) implies (9b). ∎

Theorem 5.3.9. (Convergence) *Let the solution of the biharmonic boundary value problem* (1), (2) *in* $\Omega = (0,1) \times (0,1)$ *satisfy* $u \in C^{5,1}(\overline{\Omega})$. *Then the discrete solution of equations* (8a–c) *is convergent (at least) of order* $\frac{3}{2}$:

$$||u_h - R_h u||_{\Omega_h} \le C_1 h^2 ||u||_{C^{5,1}(\overline{\Omega})} + C_2 h^{3/2} ||u||_{C^{2,1}(\overline{\Omega})}, \tag{5.3.13}$$

where R_h *is the restriction to* $\Omega_h \cup \Gamma_h$.

PROOF. The Taylor expansion shows that

$$\Delta u_h R_h u = \Delta u + \frac{h^2}{4!}(u_{xxxx} + u_{yyyy}) + h^4 R, \quad |R| \le \frac{1}{360}||u||_{C^{5,1}(\overline{\Omega})} \text{in } \Omega_h. \tag{5.3.14}$$

Outside Ω_h one can specify $R_h u$ arbitrarily since these values do not appear in (13). In γ_h we set

$$R_h u(\mathbf{x} + h\mathbf{n}) :=$$
$$u(\mathbf{x} - h\mathbf{n}) + 2h\varphi_2(\mathbf{x}) + \frac{h^3}{3} u_{nnn} - \frac{h^4}{4!} u_{nnnn} + \frac{2h^5}{5!}(\frac{\partial}{\partial \mathbf{n}})^5 u(\mathbf{x}), \quad \mathbf{x} \in \Gamma_h.$$

The choice is made so that (14) also holds for $\mathbf{x} \in \Gamma_h$. Applying Δ_h to $\Delta_h R_h u$ yields

$$\Delta_h^2 R_h u = D_h R_h u = \tilde{R}_h \Delta^2 u + O(h^2 ||u||_{C^{5,1}(\overline{\Omega})} \quad \text{in } \Omega_h$$

($\tilde{R}_h$ is the restriction to Ω_h). The difference $w_h := u_h - R_h u$ (defined in $\overline{\Omega}_h$) satisfies

$$\begin{aligned} D_h w_h &= D_h u_h - D_h R_h u \\ &= \tilde{R}_h (f - \Delta^2 u) + O(h^2 ||u||_{C^{5,1}(\overline{\Omega})}) =: g_h, \end{aligned} \tag{5.3.15a}$$

$$\begin{aligned} w_h(\mathbf{x}) &= u_h(\mathbf{x}) - (R_h u)(\mathbf{x}) \\ &= \varphi_1(\mathbf{x}) - u(\mathbf{x}) = 0 \text{ for } \mathbf{x} \in \Gamma_h, \end{aligned} \tag{5.3.15b}$$

$$\begin{aligned} \partial_n^0 w_h(\mathbf{x}) &= \partial_n^0 u_h(\mathbf{x}) - \partial_n^0 R_h u(\mathbf{x}) \\ &= \varphi_2(\mathbf{x}) - [\varphi_2(\mathbf{x}) + O(h^2 ||u||_{C^{2,1}(\overline{\Omega})})] \\ &=: \psi_h \text{ in } \mathbf{x} \in \Gamma_h. \end{aligned} \tag{5.3.15c}$$

If one sets $g_h = O(h^2 ||u||_{C^{5,1}(\overline{\Omega})})$ and $\psi_h = O(h^2 ||u||_{C^{2,1}(\overline{\Omega})})$ instead of f_h and φ_2 in Theorem 8 (note that $\varphi_1 = 0$!) then (9b) implies assertion (13). ■

By using more complicated methods one can replace the factor $h^{-1/2}$ in (13) by $C(\epsilon)h^{-\epsilon}$ ($\epsilon > 0$ arbitrary) so that the order of convergence is $O(h^{2-\epsilon})$, i.e., almost 2.

Remark 5.3.10. Inequality (9a) shows $\text{cond}_2(L_h) := ||L_h||_2 ||L_h^{-1}||_2 = O(h^{-4})$. In general a difference method for a differential equation of order $2m$ leads to the condition

$$\text{cond}(L_h) = O(h^{-2m}). \tag{5.3.16}$$

This indicates greater sensitivity to roundoff errors at higher orders $2m$ (cf. §4.4 in Stoer[1] and Bulirsch–Stoer[1]).

Difference methods for boundary value problems of fourth order with variable coefficients and general domains Ω are discussed in an paper by Zlámal [1]. There too convergence of order $O(h^{3/2})$ is shown. By contrast, $O(h^2)$-convergence was proved by Bramble [1] for the 13-point star (7) in a general domain Ω if the boundary conditions are suitably discretised.

6 Tools from Functional Analysis

To enable readers without previous knowledge of functional analysis to follow the next chapters we summarise here all definitions and results that will be needed later on.

6.1 Banach Spaces and Hilbert Spaces

6.1.1 Normed Spaces

Let X be a linear space (alternative term: vector space) over $\mathbf{K}$ where $\mathbf{K} = \mathbb{R}$ or $\mathbf{K} = \mathbb{C}$. In the following the normal case $\mathbf{K} = \mathbb{R}$ is always intended. $\mathbf{K} = \mathbb{C}$ occurs only in connection with Fourier transforms.

The notion of a norm $||\cdot||: X \to [0,\infty)$ is explained in Definition 4.3.12. The linear space X equipped with a norm is called a normed space and is denoted by the pair $(X, ||\cdot||)$. Whenever it is clear which norm belongs to X, this norm is called $||\cdot||_X$ and one writes X instead of $(X, ||\cdot||_X)$.

Example 6.1.1. (a) The Euclidean norm $|\cdot|$ from (4.3.13) and the maximum norm (4.3.3) are norms on $\mathbb{R}^n$.
(b) The continuous functions form the (infinite-dimensional) space $C^0(\overline{\Omega})$. If Ω is bounded, all $u \in C^0(\overline{\Omega})$ are bounded so that the supremum norm (2.4.1) is defined and satisfies the norm axioms. If Ω is unbounded, the bounded, continuous functions form a proper subset $(C^0(\overline{\Omega}), ||\cdot||_{C^0(\overline{\Omega})})$ of $C^0(\overline{\Omega})$. Instead of $||\cdot||_{C^0(\overline{\Omega})}$, we use the traditional notation $||\cdot||_\infty$.
(c) The Hölder-continuous functions introduced in Definition 3.2.8 form the normed space $(C^s(\overline{\Omega}), ||\cdot||_{C^s(\overline{\Omega})})$.

The norm defines a topology on X: $A \subset X$ is open if for all $x \in A$ there exists an $\epsilon > 0$ so that the "ball" $K_\epsilon(x) = \{y \in X: ||x-y|| < \epsilon\}$ is contained in A. We write

$$x_n \to x \quad \text{or } x = \lim_{n\to\infty} x_n, \quad \text{if } ||x_n - x|| \to 0.$$

Example 6.1.2. Let $f_n, f \in C^0(\overline{\Omega})$. The limit process $f_n \to f$ (with respect to $||\cdot||_\infty$) denotes the uniform convergence known from analysis.

Exercise 6.1.3. The norm $||\cdot||: X \to [0,\infty)$ is continuous; in particular the reversed triangle inequality holds

$$|\,||x|| - ||y||\,| \le ||x-y|| \quad \text{for } x, y \in X. \tag{6.1.1}$$

As Example 1a shows, several norms can be defined on X. Two norms $||\cdot||$ and $|||\cdot|||$ on X are said to be equivalent if there exists a number $0 < C < \infty$ such that

$$\frac{1}{C}||x|| \le |||x||| \le C||x|| \quad \text{for all } x \in X. \tag{6.1.2}$$

Exercise 6.1.4. Equivalent norms lead to the same topology on X.

6.1.2 Operators

Let X and Y be normed spaces with the norms $||\cdot||_X$ resp. $||\cdot||_Y$. A linear mapping $T: X \to Y$ is called an operator. If the operator norm

$$||T||_{Y\leftarrow X} := \sup\{||Tx||_Y / ||x||_X : 0 \ne x \in X\} \tag{6.1.3}$$

is finite, T is said to be bounded (cf. (4.3.9)).

Exercise 6.1.5. (a) T is bounded if and only if it is continuous.
(b) $||\cdot||_{Y\leftarrow X}$ is a norm.

With the addition $(T_1+T_2)x = T_1x + T_2x$ and the multiplication by scalars, the bounded operators form a linear space which is denoted by $L(X,Y)$. $(L(X,Y), ||\cdot||_{Y\leftarrow X})$ is a normed space. With the additional multiplication $(T_1T_2)x = T_1(T_2x)$, $L(X,X)$ even forms an algebra with unit I, since $||I||_{X\leftarrow X} = 1$. I always denotes the identity: $Ix = x$.

Exercise 6.1.6. Prove that

$$||Tx||_Y \le ||T||_{Y\leftarrow X}||x||_X \quad \text{for all } x \in X,\ T \in L(X,Y). \tag{6.1.4a}$$

If $T_1 \in L(Y,Z)$ and $T_2 \in L(X,Y)$, then $T_1T_2 \in L(X,Z)$ and

$$||T_1T_2||_{Z\leftarrow X} \le ||T_1||_{Z\leftarrow Y}||T_2||_{Y\leftarrow X}. \tag{6.1.4b}$$

We write $T_n \to T$ for $T, T_n \in L(X,Y)$ if $||T - T_n||_{Y\leftarrow X} \to 0$ (convergence in the operator norm).

6.1.3 Banach Spaces

A sequence $\{x_n \in X: n \geq 1\}$ is said to be Cauchy convergent, or is called a Cauchy sequence, if $\sup\{||x_n - x_m||_X: n, m \geq k\} \to 0$ for $k \to \infty$. A space X is said to be complete if any Cauchy-convergent sequence converges to an $x \in X$. A Banach space is a complete normed space.

Example 6.1.7. (a) $\mathbb{R}^n$ is complete, but $\mathbb{Q}^n$ is not ($\mathbb{Q}$: rational numbers).
(b) Let $D \subset \mathbb{R}^n$. $(C^0(D), ||\cdot||_\infty)$ is a Banach space.
(c) $(C^k(D), ||\cdot||_{C^k(D)})$ for $k \in \mathbb{N}$ and the Hölder spaces $(C^s(D), ||\cdot||_{C^s(D)})$, $s > 0$ as well as $(C^{k,1}(D), ||\cdot||_{C^{k,1}(D)})$ are complete, thus Banach spaces.

PROOF of (b). If $\{u_n\}$ is Cauchy convergent then there exists a limit $u^*(x) := \lim_{n\to\infty} u_n(x)$ for all $x \in D$. Since u_n converges uniformly to u^*, u^* must be continuous, i.e., $u^* \in C^0(D)$. ■

Denote by $L^\infty(D)$ the set of functions that are bounded and locally integrable on D. Here we do not distinguish between functions which agree almost everywhere. In this case the supremum norm is defined by

$$||u||_{L^\infty(D)} := \inf\{\sup\{|u(x)|: x \in D\backslash A\}: A \text{ set of measure zero}\}$$

$(L^\infty(D), ||\cdot||_{L^\infty(D)})$ is a Banach space.

Exercise 6.1.8. Let X be a normed space and Y a Banach space. Show that $L(X,Y)$ is a Banach space.

Exercise 6.1.9. Let X be a Banach space and $Z \subset X$ a closed subspace. The quotient space X/Z has as elements the classes $\tilde{x} = \{x + z: z \in Z\}$. Show that X/Z with the norm $||\tilde{x}|| = \inf\{||x+z||_X: z \in Z\}$ is a Banach space.

A set A is said to be dense in $(X, ||\cdot||_X)$ if $A \subset X$ and $\overline{A} = X$, i.e., every $x \in X$ is the limit of a sequence $a_n \in A$. If $(X, ||\cdot||_X)$ is a normed, but not complete, space, one calls $(\tilde{X}, ||\cdot||_{\tilde{X}})$ the completion of X, if X is dense in $\tilde{X}$ and $||x||_{\tilde{X}} = ||x||_X$ for all $x \in X$. The completion is uniquely determined up to isomorphism, and can be constructed in the same way as one constructs $\mathbb{R}$ as the completion of $\mathbb{Q}$.

Lemma 6.1.10. *Let $(X, ||\cdot||_X)$ be a normed linear space and subspace of a Banach space $(Y, ||\cdot||_Y)$ with $||x||_Y \leq C||x||_X$ for all $x \in X \subset Y$. Let each Cauchy sequence $\{x_n\}$ in X with $||x_n||_Y \to 0$ also satisfy $||x_n||_X \to 0$. Then there exists a completion $\tilde{X}$ of X in Y: $X \subset \tilde{X} \subset Y$.*

Frequently it is not easy to describe the image Tx of an operator $T \in L(X,Y)$ for all elements x of the Banach space X. The following theorem permits a considerable simplification: It suffices to investigate T on a dense set $X_0 \subset X$.

Theorem 6.1.11. *Let X_0 be a dense subspace (or just a dense subset) of the normed space X. Let Y be a Banach space.*
(a) An operator $T_0 \in L(X_0, Y)$ defined on the subspace X_0 with $||T_0||_{Y \leftarrow X_0} = \sup\{||T_0 x||_Y / ||x||_X : 0 \neq x \in X_0\}$ has a unique continuation $T \in L(X, Y)$, i.e., $Tx = T_0 x$ for all $x \in X_0$.
(b) For $x_n \to x$ $(x_n \in X_0, x \in X)$ holds $Tx = \lim_{n\to\infty} T_0 x_n$.
(c) $||T||_{Y \leftarrow X} = ||T_0||_{T \leftarrow X_0}$.

PROOF. For $x \in X_0$, T is defined by $Tx = T_0 x$, while for $x \in X \backslash X_0$ there exists a sequence $x_n \to x, x_n \in X_0$ and $Tx = \lim_{n\to\infty} Tx_n = \lim_{n\to\infty} T_0 x_n$. It remains to show that $\lim T_0 x_n$ exists and is independent of the choice of the sequence $x_n \in X_0$. Since $||T_0 x_n - T_0 x_m||_Y \leq ||T_0||_{Y \leftarrow X} ||x_n - x_m||_X$, the sequence $T_0 x_n$ is Cauchy convergent. Because of the completeness of Y there exists $y \in Y$ with $T_0 x_n \to y$. Similarly, for a second sequence $x_n' \in X_0$ with $x_n' \to x$, there exists for the same reason a $y' \in Y$ such that $T_0 x_n' \to y'$. Since $y' - y = \lim(T_0 x_n' - T_0 x_n)$ and $||T_0 x_n' - T_0 x_n||_Y \leq ||T_0(x_n' - x_n)||_Y \leq ||T_0||_{Y \leftarrow X_0} ||x_n' - x_n||_X \to 0$, Tx is well defined by (b). Part (c) follows from $||Tx||_Y / ||x||_X = \lim ||T_0 x_n||_Y \backslash ||x_n||_X$ for $x_n \to x, x_n \in X_0$. ■

Exercise 6.1.12. Let X_0 be dense in $(X, ||\cdot||)$. Let $|||\cdot|||$ be a second norm on X_0, that is equivalent with $||\cdot||$ on X_0. Show that the completion of X_0 with respect to $|||\cdot|||$ results in $(X, |||\cdot|||)$ and $||\cdot||$ and $|||\cdot|||$ are also equivalent on X.

A corollary of the "open mapping theorem" (cf. Yosida [1, §II.5]) is the following, important result:

Theorem 6.1.13. *Let X, Y be Banach spaces. Let $T \in L(X, Y)$ be injective and surjective. Then $T^{-1} \in L(Y, X)$ also holds.*

Exercise 6.1.14. Let X, Y be Banach spaces, and $T \in L(X, Y)$ be injective. Show that:
(a) $Y_0 := \text{range}(T) := \{Tx : x \in X\}$ with $||\cdot||_Y$ as norm is a Banach subspace of Y.
(b) T^{-1} exists on Y_0 and is bounded: $T^{-1} \in L(Y_0, X)$.

Let X and Y be Banach spaces with $X \subset Y$. Obviously the inclusion denoted by $I : x \in X \mapsto x \in Y$ is a linear mapping. If it is bounded, i.e.,

$$I \in L(X, Y) \quad \text{that is, } ||x||_Y \leq C||x||_X \quad \text{for all } x \in X, \tag{6.1.5}$$

then X is said to be continuously embedded in Y. If furthermore X is dense in $(Y, ||\cdot||_Y)$, then X is said to be densely and continuously embedded in Y.

Exercise 6.1.15. Let $X \subset Y \subset Z$ be Banach spaces. Show that if X is [densely and] continuously embedded in Y and Y is [densely and] continuously embedded in Z, then X is [densely and] continuously embedded in Z.

6.1.4 Hilbert Spaces

A mapping $(\cdot,\cdot): X \times X \to \mathbf{K}$ ($\mathbf{K} = \mathbb{R}$ or $\mathbf{K} = \mathbb{C}$) is called a scalar product on X if

$$(x,x) > 0 \quad \text{for all } 0 \neq x \in X, \tag{6.1.6a}$$

$$(\lambda x + y, z) = \lambda(x,z) + (y,z) \quad \text{for all } \lambda \in \mathbf{K}, x, y, z \in X, \tag{6.1.6b}$$

$$(x,y) = \overline{(y,x)} \quad \text{for all } x, y \in X, \tag{6.1.6c}$$

where $\overline{\lambda}$ denotes the complex conjugate of λ.

Exercise 6.1.16. Show that (a) $\|x\| := (x,x)^{1/2}$ is a norm on X. For the proof use
(b) the Schwarz inequality:

$$|(x,y)| \leq \|x\| \, \|y\| \quad \text{for all } x, y \in X. \tag{6.1.7}$$

Hint: consider $\|x - \lambda y\| \geq 0$ for $\lambda = \|x\|^2/(x,y)$, if $(x,y) \neq 0$.
(c) $(\cdot,\cdot): X \times X \to \mathbf{K}$ is continuous.

A Banach space X is called a Hilbert space if there exists a scalar product $(\cdot,\cdot)_X$ on X such that $\|x\|_X = (x,x)_X^{1/2}$ for all $x \in X$.
$x \in X$ and $y \in X$ are said to be orthogonal ($x \perp y$) if $(x,y)_X = 0$. If $A \subset X$ is a subset of the Hilbert space X then the orthogonal space $A^\perp := \{x \in X: (x,a)_X = 0 \text{ for all } a \in A\}$ defines a closed subspace of X.

Lemma 6.1.17. *Let U be a closed subspace of a Hilbert space X. Then X can be decomposed into the direct sum $X = U \oplus U^\perp$, i.e., any $x \in X$ has a unique decomposition $x = u + v$ with $u \in U$, $v \in U^\perp$. Furthermore, $\|x\|_X^2 = \|u\|_X^2 + \|v\|_X^2$.*

Exercise 6.1.18. Let A be a subset of the Hilbert space X. Prove the equivalence of the following statements: (a) A is dense in X; (b) $A^\perp = \{0\}$; (c) $x \in X$ and $(a,x)_X = 0$ for all $a \in A$ imply $x = 0$; (d) for every $0 \neq x \in X$ there exists an $a \in A$ with $(a,x)_X \neq 0$.

6.2 Sobolev Spaces

In the following Ω is always an open subset of $\mathbb{R}^n$.

6.2.1 $L^2(\Omega)$

$L^2(\Omega)$ consists of all Lebesgue-measurable functions whose squares on Ω are Lebesgue-integrable. Two functions $u, v \in L^2(\Omega)$ are considered to be equal ($u = v$) if $u(x) = v(x)$ for almost all $x \in \Omega$, i.e., for all $x \in \Omega\backslash A$, where the exceptional set A has Lebesgue measure $\mu(A) = 0$.

Theorem 6.2.1. *$L^2(\Omega)$ forms a Hilbert space with the scalar product*

$$(u, v)_0 := (u, v)_{L^2(\Omega)} := \int_\Omega u(\mathbf{x})\overline{v(\mathbf{x})}\, d\mathbf{x} \tag{6.2.1}$$

and the norm

$$|u|_0 := ||u||_{L^2(\Omega)} := \sqrt{\int_\Omega |u(x)|^2\, dx}. \tag{6.2.2}$$

Lemma 6.2.2. *The spaces $C^\infty(\Omega) \cap L^2(\Omega)$ and $C_0^\infty(\Omega)$ are dense in $L^2(\Omega)$. Here*

$$C_0^\infty(\Omega) := \{u \in C^\infty(\Omega)\colon \operatorname{supp}(u) \ \textit{compact},\ \operatorname{supp}(u) \subset\subset \Omega\}. \tag{6.2.3}$$

$\operatorname{supp}(u) := \overline{\{x \in \Omega\colon u(x) \neq 0\}}$ *denotes the* s u p p o r t *of u. The double inclusion $\Omega' \subset\subset \Omega$ indicates that $\overline{\Omega'}$ lies in the interior of Ω (i.e., $\Omega' \subset \Omega$ and $\overline{\Omega}' \cap \partial\Omega = \emptyset$).*

Let D^α be the partial derivative operator (3.2.5b). In the following we need so-called weak derivatives $D^\alpha u$ which are defined in a nonclassical way.

Definition 6.2.3. $u \in L^2(\Omega)$ has a (weak) derivative $v := D^\alpha u \in L^2(\Omega)$ if for the latter $v \in L^2(\Omega)$ holds:

$$(w, v)_0 = (-1)^{|\alpha|}(D^\alpha w, u)_0 \quad \text{for all } w \in C_0^\infty(\Omega). \tag{6.2.4}$$

Exercise 6.2.4. Show the following: (a) Let u have a weak derivative $D^\alpha u \in L^2(\Omega)$. If the classical derivative $D^\alpha u$ exists in $\Omega' \subset \Omega$, it coincides there (almost everywhere) with the weak one.
(b) If u has the weak derivative $v_\alpha = D^\alpha u \in L^2(\Omega)$, and if v_α has the weak derivative $v_{\alpha+\beta} = D^\beta v_\alpha \in L^2(\Omega)$ then $v_{\alpha+\beta}$ is also the weak $D^{\alpha+\beta}$-derivative of u; i.e., $D^{\alpha+\beta}u = D^\beta(D^\alpha u)$.
(c) Let $\Omega \subset \mathbb{R}^n$ be bounded and $\mathbf{0} \in \Omega$. $u(x) := |x|^\sigma$ ($\sigma \in \mathbb{R}$) has weak first derivatives in $L^2(\Omega)$ if $\sigma = 0$ or $2\sigma + n > 2$.

(d) For $u_\nu \in C^\infty(\Omega)$ let $u_\nu \to u \in L^2(\Omega)$ and $D^\alpha u_\nu \to v \in L^2(\Omega)$ in the $L^2(\Omega)$ norm. Then $v = D^\alpha u$.

For further applications we consider

Example 6.2.5. Let $\overline{\Omega} = \bigcup_{i=1}^N \overline{\Omega}_i$, where the bounded subdomains Ω_i are disjoint and have piecewise smooth boundaries. Let $k \in \mathbb{N}$. A function $u \in C^{k-1}(\overline{\Omega})$ with restrictions $u|_{\Omega_i} \in C^k(\overline{\Omega}_i)$ $(1 \le i \le N)$ has a (weak) kth derivative $v_\alpha = D^\alpha u \in L^2(\Omega)$, $|\alpha| \le k$, which coincides with the classical one in $\bigcup \Omega_i$.

PROOF. The $(k-1)$st derivatives exist as classical derivatives so that the assertion need only be shown for $k=1$. Let $D^\alpha = \partial/\partial x_j$, $w \in C_0^\infty(\Omega)$. The assertion is obtained via integration by parts:

$$\begin{aligned}-(D^\alpha w, u)_0 &= -(w_{x_j}, u)_0 = -\int_\Omega w_{x_j} u \, d\mathbf{x} \\ &= -\sum_i \int_{\Omega_i} w_{x_j} u \, d\mathbf{x} = \sum_i \left[\int_{\Omega_i} w u_{x_j} \, d\mathbf{x} - \int_{\partial\Omega_i} w u n_j^{(i)} \, d\Gamma\right] \\ &= \sum_i \int_{\Omega_i} w u_{x_j} \, d\mathbf{x} = \sum_i \int_{\Omega_i} w v_\alpha \, d\mathbf{x} = \int_\Omega w v_\alpha \, d\mathbf{x} = (w, v_\alpha)_0,\end{aligned}$$

since every $\mathbf{x} \in \partial\Omega_i \backslash \partial\Omega$ also belongs to $\partial\Omega_k \backslash \partial\Omega$ for some $k \ne i$ with opposite normal direction $\mathbf{n}^{(k)} = -\mathbf{n}^{(i)}$. ∎

The inequality $|(u,v)_0| \le |u|_0 |v|_0$ (cf. (1.7)) reads explicitly

$$|\int_\Omega u(\mathbf{x}) v(\mathbf{x}) \, d\mathbf{x}| \le [\int_\Omega |u(\mathbf{x})|^2 \, d\mathbf{x} \int_\Omega |v(\mathbf{x})|^2 \, d\mathbf{x}]^{1/2}. \tag{6.2.5a}$$

For $v = 1$ the result is

$$|\int_\Omega u(\mathbf{x}) \, d\mathbf{x}| \le \mu(\Omega)^{1/2} [\int_\Omega |u(\mathbf{x})|^2 \, d\mathbf{x}]^{1/2} \quad (\mu(\Omega)\text{: measure of } \Omega). \tag{6.2.5b}$$

For $a \in L^\infty(\Omega), u, v \in L^2(\Omega)$ one now has

$$|\int_\Omega a(\mathbf{x}) u(\mathbf{x}) v(\mathbf{x}) \, d\mathbf{x}| \le \|a\|_{L^\infty(\Omega)} \|u\|_{L^2(\Omega)} \|v\|_{L^2(\Omega)}. \tag{6.2.5c}$$

6.2.2 $H^k(\Omega)$ and $H_0^k(\Omega)$

Let $k \in \mathbb{N} \cup \{0\}$. Let $H^k(\Omega) \subset L^2(\Omega)$ be the set of all functions having weak derivatives $D^\alpha u \in L^2(\Omega)$ for $|\alpha| \le k$:

$$H^k(\Omega) := \{u \in L^2(\Omega) \colon D^\alpha u \in L^2(\Omega) \quad \text{for } |\alpha| \le k\}.$$

The Sobolev space denoted here by $H^k(\Omega)$ is denoted by $W_2^k(\Omega)$ in some other places. Usually, there $H^k(\Omega)$ is a space which only coincides for sufficiently smoothly bounded Ω with $W_2^k(\Omega)$.

Theorem 6.2.6. *$H^k(\Omega)$ forms a Hilbert space with the scalar product*

$$(u,v)_k := (u,v)_{H^k(\Omega)} := \sum_{|\alpha|\le k} (D^\alpha u, D^\alpha v)_{L^2(\Omega)} \tag{6.2.6}$$

and the (Sobolev) norm

$$|u|_k := \|u\|_{H^k(\Omega)} := \sqrt{\sum_{|\alpha|\le k} \|D^\alpha u\|^2_{L^2(\Omega)}}. \tag{6.2.7}$$

Lemma 6.2.7. *$C^\infty(\Omega)\cap H^k(\Omega)$ lies densely in $H^k(\Omega)$.*

Lemma 7, whose proof can be found, for example, in Wloka [1, Theorem 3.5] permits a second definition of the Sobolev space $H^k(\Omega)$:

Remark 6.2.8. Let $X_0 := \{u \in C^\infty(\Omega): |u|_k < \infty\}$. The completion of X_0 in $L^2(\Omega)$ with respect to norm (7) results in $H^k(\Omega)$.

Definition 6.2.9. The completion of $C_0^\infty(\Omega)$ in $L^2(\Omega)$ with respect to the norm (7) is denoted by $H_0^k(\Omega)$.

Theorem 6.2.10. *The Hilbert space $H_0^k(\Omega)$ is a subspace of $H^k(\Omega)$ with the same scalar product* (6) *and same norm* (7). *$C_0^\infty(\Omega)$ is dense in $H_0^k(\Omega)$. For $k=0$ there holds*

$$H_0^0(\Omega) = H^0(\Omega) = L^2(\Omega). \tag{6.2.8}$$

PROOF. (a) $C_0^\infty(\Omega) \subset X_0$ (cf. Remark 8) implies $H_0^k(\Omega) \subset H^k(\Omega) \subset L^2(\Omega)$. (b) According to the definition $C_0^\infty(\Omega)$ is dense in $H_0^k(\Omega)$. (c) For $k=0$ the norms (2) and (7) coincide. Lemma 2 proves $H_0^0(\Omega) = L^2(\Omega)$. Because of $H_0^k(\Omega) \subset H^k(\Omega) \subset L^2(\Omega)$ (k arbitrary) (8) follows. ■

Lemma 6.2.11. *For Ω bounded, $\|\cdot\|_{H^k(\Omega)}$ and*

$$|u|_{k,0} := \Big[\sum_{|\alpha|=k} \|D^\alpha u\|^2_{L^2(\Omega)}\Big]^{1/2} \tag{6.2.9}$$

are equivalent norms in $H_0^k(\Omega)$.

PROOF. (a) Evidently, $|\cdot|_{k,0} \le \|\cdot\|_{H^k(\Omega)}$.

(b) Let $u \in C_0^\infty(\Omega)$. For an α with $|\alpha| = k-1$ set $v := D^\alpha u \in C_0^\infty(\Omega)$. There exists an R with $\Omega \subset K_R(\mathbf{0})$. For each $\mathbf{x} \in \Omega$, $x_1 \in [-R, R]$; thus

$$|v(x)|^2 = \left| \int_{-R}^{x_1} v_{x_1}(\xi, x_2, \ldots, x_n)\, d\xi \right|^2 \le (x_1 + R) \int_{-R}^{x_1} |v_{x_1}(\xi, \ldots)|^2\, d\xi$$
$$\le 2R \int_{-R}^{+R} |v_{x_1}(\xi, x_2, \ldots, x_n)|^2\, d\xi$$

(cf. (5b)). Integration over $\mathbf{x} \in \Omega$ yields $|v|_0^2 \le 4R^2 |v_{x_1}|_0^2 \le 4R^2 |u|_{k,0}^2$, since v_{x_1} is k-fold derivative. Summation over all α with $|\alpha| = k-1$ yields $|u|_{k-1,0}^2 \le C_{k-1}|u|_{k,0}^2$. Now $|u|_{j-1,0}^2 \le C_{j-1}|u|_{j,0}^2$ follows likewise for all $1 \le j \le k$, and thus $|u|_{j,0}^2 \le C_j C_{j+1} \cdot \ \ldots \ \cdot C_{k-1} |u|_{k,0}^2$. Since $|u|_k^2 = \sum_{j=0}^k |u|_{j,0}^2 \le C|u|_{k,0}^2$, for all $u \in C_0^\infty(\Omega)$, the statement follows from Exercise 6.1.12. ∎

Exercise 6.2.12. Show that (a) for bounded Ω and $k \ge 1$, $H^k(\Omega)$ and $H_0^k(\Omega)$ are different. Hint: Consider the constant function $u(\mathbf{x}) = 1$ and use Lemma 11.
(b) Lemma 11 holds even if Ω is bounded in one direction, i.e., if Ω lies on a strip $\{\mathbf{x} \in \mathbb{R}^n : |x_k| < R\}$ $(k \in \{1, \cdots, n\})$.

Theorem 6.2.13. *Let $m \ge 1$. There exist constants $C = C(m)$ and $\eta(\epsilon) = \eta(\epsilon, m)$ such that*

$$|u|_k \le C|u|_m^{k/m} |u|_0^{(m-k)/m} \quad \textit{for all } 0 \le k \le m, u \in H_0^m(\Omega), \tag{6.2.10a}$$
$$|u|_k \le \epsilon |u|_m + \eta(\epsilon)|u|_0 \quad \textit{for all } \epsilon > 0, 0 \le k < m, u \in H_0^m(\Omega). \tag{6.2.10b}$$

PROOF. (a) Partial integration for $|\alpha| = 1$ shows that

$$|D^\alpha u|_0^2 = (D^\alpha u, D^\alpha u)_0 = -(D^{2\alpha} u, u)_0 \le |D^{2\alpha} u|_0 |u|_0 \le |u|_2 |u|_0.$$

Since also $|u|_0^2 \le |u|_2 |u|_0$, it follows that $|u|_1^2 \le (n+1)|u|_2|u|_0$. If one replaces u by $D^\beta u$ with $|\beta| = l$, one obtains in the same way

$$|u|_l^2 \le \tilde{C} |u|_{l+1} |u|_{l-1} \quad (1 \le l < m;\ u \in H_0^m(\Omega)). \tag{6.2.10c}$$

(b) Let $u \ne 0$ be fixed. Set $v_l := \log |u|_l$, $w_l := [l v_m + (m-l) v_0]/m$ and $z_l := v_l - w_l$. (10c) leads to $2z_l - z_{l-1} - z_{l+1} \le \tilde{c} := \log \tilde{C}$, where $z_0 = z_m = 0$. For $\mathbf{z} = (z_1, \cdots, z_{m-1})^\mathsf{T}$ one obtains $\mathbf{A}\mathbf{z} \le \tilde{c}\mathbb{1}$. Here $\mathbf{A}$ is the M-matrix (4.1.9b) for $h = 1$. $\mathbf{A}^{-1} \ge \mathbf{0}$ proves that $\mathbf{z} \le \mathbf{c} := \tilde{c} A^{-1} \mathbb{1}$. $v_l = z_l + w_l \le c_l + w_l$ $(1 \le l < m)$ implies $|u|_l = \exp(v_l) \le \exp(c_l + w_l)$, i.e., (10a) with $C := \exp(\tilde{c}\|\mathbf{A}^{-1}\mathbb{1}\|_\infty)$.
(c) Elementary calculation shows that for each $\epsilon > 0$, $0 \le \Theta \le \Theta_0 < 1$ there exists an $\eta(\epsilon, \Theta_0)$ such that

$$a^\Theta b^{1-\Theta} \le \epsilon a + \eta(\epsilon) b \quad \text{for all } a, b \ge 0. \tag{6.2.10d}$$

Formula (10b) is a corollary of (10a,d). ∎

By similar means, together with (5.3.10), one proves

$$(D^\alpha u, D^\beta u)_0 \le \epsilon |u|_m^2 + (4\epsilon)^{-1} |u|_k^2$$
$$(\epsilon > 0,\ |\alpha| \le m,\ ||\beta|| \le k \le m,\ u \in H^m(\Omega)), \tag{6.2.10e}$$
$$(D^\alpha u, D^\beta u)_0 \le \epsilon |u|_m^2 + \eta(\epsilon) |u|_0^2$$
$$(\epsilon > 0,\ |\alpha| \le m,\ |\beta| < m,\ u \in H_0^m(\Omega)). \tag{6.2.10f}$$

Remark 6.2.14. The set $\{u \in C^\infty(\Omega)\colon \mathrm{supp}(u) \text{ compact}, |u|_k < \infty\}$ is dense in $H^k(\Omega)$.

PROOF. Let $u \in H^k(\Omega)$, $\epsilon > 0$. According to Lemma 7 there exists a function $u_\epsilon \in C^\infty(\Omega)$ with $|u - u_\epsilon|_k \le \epsilon/2$. There exists $a \in C^\infty(\mathbb{R}^n)$ with $a(\mathbf{x}) = 1$ for $|\mathbf{x}| \le 1$, $a(\mathbf{x}) = 0$ for $|\mathbf{x}| \ge 2$. For sufficiently large R, one also has $|u_\epsilon(\mathbf{x}) - a(\mathbf{x}/R) u_\epsilon(\mathbf{x})|_k \le \epsilon/2$. Thus there exists $v(\mathbf{x}) = a(\mathbf{x}/R) u_\epsilon(\mathbf{x}) \in C_0^\infty(\Omega)$ with $|u - v|_k \le \epsilon$. ■

Since "supp(u) compact" already implies "supp(u) $\subset\subset \mathbb{R}^n$" we obtain

Corollary 6.2.15. $H_0^k(\mathbb{R}^n) = H^k(\mathbb{R}^n)$ *for all* $k \ge 0$.

The Leibniz rule for derivatives of products proves

Theorem 6.2.16. $||au||_{H^k(\Omega)} \le C_k ||a||_{C^k(\overline{\Omega})} ||u||_{H^k(\Omega)}$ *for all* $a \in C^k(\overline{\Omega})$, $u \in H^k(\Omega)$.

Theorem 16 together with the substitution rule for volume integrals shows

Theorem 6.2.17. (Transformation theorem). *Let* $T\colon \Omega \to \Omega'$ *be a one-to-one mapping onto* Ω' *with* $T \in C^{\max(k,1)}(\overline{\Omega})$ *and* $|\det dT/d\mathbf{x}| \ge \delta > 0$ *in* Ω. *We write* $v = u \circ T$ *for* $v(\mathbf{x}) = u(T(\mathbf{x}))$. *Then* $u \in H^k(\Omega')$ $[u \in H_0^k(\Omega')]$ *also implies* $u \circ T \in H^k(\Omega)$ $[\in H_0^k(\Omega)]$ *and*

$$||u \circ T||_{H^k(\Omega)} \le C_k ||T||_{C^k(\overline{\Omega})} ||u||_{H^k(\Omega')} \Big/ \sqrt{\delta}\,. \tag{6.2.11}$$

6.2.3 Fourier Transformation and $H^k(\mathbb{R}^n)$

For $u \in C_0^\infty(\mathbb{R}^n)$ one defines the Fourier-transformed function $\hat{u}$ by

$$\hat{u}(\xi) := (\mathfrak{F}u)(\xi) := (2\pi)^{-n/2} \int_{\mathbb{R}^n} e^{-i\langle \xi, \mathbf{x} \rangle} u(\mathbf{x})\, d\mathbf{x}. \tag{6.2.12}$$

Note that $\hat{u}$ is described by a proper integral since the support of u is bounded.

Lemma 6.2.18. *Let $u \in C_0^\infty(\mathbb{R}^n)$. For $R \to \infty$*

$$I_R(u;\mathbf{y}) := (2\pi)^{-n} \int_{|\xi|_\infty \le R} \Big[\int_{\mathbb{R}^n} e^{-i\langle \xi, \mathbf{x}-\mathbf{y}\rangle} u(\mathbf{x})\, d\mathbf{x}\Big]\, d\xi$$

converges uniformly to $u(\mathbf{y})$ on $\operatorname{supp}(u)$.

PROOF. It suffices to discuss the case $n = 1$ (by Fubini's theorem). Integration with respect to ξ results in

$$I_R(u;y) = (1/\pi) \int_{\mathbb{R}} (x-y)^{-1} \sin(R(x-y)) u(x)\, dx.$$

Then $I_R(1;y) = (1/\pi) \int_{\mathbb{R}} t^{-1} \sin(t)\, dt = 1$ for all $R > 0$. Since $u \in C^\infty(\mathbb{R}^n)$, then also $w(x,y) := [u(x)-u(y)]/(x-y) \in C^\infty(\mathbb{R}^{2n})$. The estimates $w(x,y) = O(1/|x|)$ and $w_x(x,y) = O(1/x^2)$ hold uniformly for $y \in \operatorname{supp}(u)$. Partial integration yields

$$\begin{aligned} I_R(u(\cdot) - u(y); y) &= I_R((\cdot - y)w(\cdot, y); y) = (1/\pi) \int_{\mathbb{R}} \sin(R(x-y)) w(x,y)\, dx \\ &= -(1/\pi) \int_{\mathbb{R}} \cos(R(x-y)) w_x(x,y)\, dx/R = O(1/R). \end{aligned}$$

The statement follows from

$$I_R(u;y) = u(y) I_R(1;y) + I_R(u(\cdot) - u(y); y) = u(y) + O(1/R). \qquad \blacksquare$$

Lemma 6.2.19. *$\hat{u} \in L^2(\mathbb{R}^n)$ and $|\hat{u}|_0 = |u|_0$ for all $u \in C_0^\infty(\mathbb{R}^n)$.*

PROOF. Lemma 18 shows

$$\begin{aligned} &\int_{|\xi|_\infty \le R} |\hat{u}(\xi)|^2\, d\xi \\ &= (2\pi)^{-n} \int_{|\xi|_\infty \le R} \Big[\int_{\mathbb{R}^n} e^{-i\langle \xi, x\rangle} u(x)\, dx\Big] \overline{\Big[\int_{\mathbb{R}^n} e^{-i\langle \xi, y\rangle} u(y)\, dy\Big]}\, d\xi, \\ &= \int_{\mathbb{R}^n} I_R(u;y) \overline{u(y)}\, dy \to \int_{\mathbb{R}^n} |u(y)|^2\, dy. \end{aligned} \qquad \blacksquare$$

Lemma 6.2.20. *The inverse Fourier transformation $\mathcal{F}^{-1}\hat{u} = u$ is defined for $u \in C_0^\infty(\mathbb{R}^n)$ by (13):*

$$\begin{aligned} (\mathcal{F}^{-1}\hat{u})(\mathbf{x}) &:= (2\pi)^{-n/2} \int_{\mathbb{R}^n} e^{i\langle \xi, \mathbf{x}\rangle} \hat{u}(\xi) d\xi \\ &:= \lim_{R\to\infty} (2\pi)^{-n/2} \int_{|\xi|_\infty \le R} e^{i\langle \xi, \mathbf{x}\rangle} \hat{u}(\xi) d\xi. \end{aligned} \tag{6.2.13}$$

PROOF. We have from Lemma 18

$$(2\pi)^{-n/2}\int_{|\xi|_\infty\le R} e^{i\langle\xi,\mathbf{x}\rangle}\hat u(\xi)d\xi = (2\pi)^{-n}\int_{|\xi|_\infty\le R}\int_{\mathbb{R}^n} e^{i\langle\xi,\mathbf{x}-\mathbf{y}\rangle}u(\mathbf{y})\,d\mathbf{y}\,d\xi$$
$$= u(\mathbf{x}) + O(1/R). \qquad \blacksquare$$

Theorem 6.2.21. *$\mathfrak{F},\mathfrak{F}^{-1}\in L(L^2(\mathbb{R}^n),L^2(\mathbb{R}^n))$ with $||\mathfrak{F}||_{L^2(\mathbb{R}^n)\leftarrow L^2(\mathbb{R}^n)} = ||\mathfrak{F}^{-1}||_{L^2(\mathbb{R}^n)\leftarrow L^2(\mathbb{R}^n)} = 1$, i.e., $\mathfrak{F}$ is an isometric mapping of $L^2(\mathbb{R}^n)$ onto itself. The scalar product satisfies $(u,v)_0 = (\hat u,\hat v)_0$ for all $u,v\in L^2(\mathbb{R}^n)$.*

PROOF. Since $C_0^\infty(\mathbb{R}^n)$ is dense in $L^2(\mathbb{R}^n)$ (cf. Lemma 2), $\mathfrak{F}$ can be continued to $\mathfrak{F}\colon L^2(\mathbb{R}^n)\to L^2(\mathbb{R}^n)$ (cf. Theorem 6.1.11). The norm estimate follows from Lemma 19. The roles of $\mathfrak{F}$ and $\mathfrak{F}^{-1}$ are interchangeable (cf. (12) and (13)); thus $\mathfrak{F}^{-1}\in L(L^2(\mathbb{R}^n),L^2(\mathbb{R}^n))$ also holds.

The second statement results from

$$(u,v)_0 = \frac{1}{2}(|u+v|_0^2 - |u|_0^2 - |v|_0^2) = \frac{1}{2}(|\hat u+\hat v|_0^2 - |\hat u|_0^2 - |\hat v|_0^2) = (\hat u,\hat v)_0. \qquad \blacksquare$$

Exercise 6.2.22. Prove that: (a) With $\xi^\alpha = \xi_1^{\alpha_1}\cdots\xi_n^{\alpha_n}$, there holds

$$\mathfrak{F}(D^\alpha u)(\xi) = i^{|\alpha|}\xi^\alpha\hat u(\xi) \quad \text{for } u\in C_0^\infty(\mathbb{R}^n). \tag{6.2.14}$$

(b) There exists $C=C(k)$ such that $\frac{1}{C}(1+|\xi|^2)^k \le \sum_{|\alpha|\le k}|\xi^\alpha|^2 \le C(1+|\xi|^2)^k$ for all $\xi\in\mathbb{R}^n$.

Lemma 6.2.23. (a) *$|u|_k = |\sqrt{\sum_{|\alpha|\le k}|\xi^\alpha|^2}\hat u(\xi)|_0$ holds for all $u\in H^k(\mathbb{R}^n)$.*
(b) *A norm on $H^k(\mathbb{R}^n)$ equivalent to $|\cdot|_k$ is*

$$|u|\hat{_k} := |(1+|\xi|^2)^{k/2}\hat u(\xi)|_0. \tag{6.2.15}$$

PROOF. (a) It suffices to show the statement for $u\in C_0^\infty(\mathbb{R}^n)$ (cf. Corollary 15, Theorem 6.1.11):

$$|u|_k^2 = \sum_{|\alpha|\le k}|D^\alpha u|_0^2 = \sum_{|\alpha|\le k}|\mathfrak{F}D^\alpha u|_0^2 = \sum_{|\alpha|\le k}|\xi^\alpha\hat u(\xi)|_0^2 = \left|\sqrt{\sum_{|\alpha|\le k}|\xi^\alpha|^2}\hat u(\xi)\right|_0^2.$$

(b) The statement follows from Exercise 22b. $\blacksquare$

Lemma 6.2.24. *Let $\partial_{h,j}$ be the difference operator $\partial_{h,j}u(\mathbf{x}) := [u(\mathbf{x}+\frac{h}{2}\mathbf{e}_j) - u(\mathbf{x}-\frac{h}{2}\mathbf{e}_j)]/h$, where $\mathbf{e}_j$ is jth unit vector. If $u\in H^k(\mathbb{R}^n)$ and $|\partial_{h,j}u|_k\le C$ for all $h>0$, $1\le j\le n$, then $u\in H^{k+1}(\mathbb{R}^n)$ holds. Conversely, $|\partial_{h,j}u|_k\le|u|_{k+1}$ holds for all $u\in H^{k+1}(\mathbb{R}^n)$, $h>0$.*

PROOF. (a) From $\mathcal{F}(u(\cdot+\delta \mathbf{e}_j))(\xi) = e^{-i\delta\xi_j}\hat{u}(\xi)$ follows $\mathcal{F}(\partial_{h,j}u)(\xi) = \frac{2i}{h}\sin(\xi_j h/2)\hat{u}(\xi)$. Hence, since $4h^{-2}\sin^2(\xi_j h/2) \geq \xi_j^2$ for $h \leq 1/|\xi|$, follows $|\mathcal{F}(\partial_{h,j}u)(\xi)|^2 = \xi_j^2|\hat{u}(\xi)|^2$ for $|\xi| \leq 1/h$. Summation over $j = 1,\dots,n$ and integration over ξ then gives

$$\begin{aligned}
(|u\hat{|}_{k+1})^2 &= \int_{\mathbb{R}^n} (1+|\xi|^2)^{k+1}|\hat{u}(\xi)|^2\,d\xi \\
&\leq \int_{|\xi|\geq 1/h} \dots + (|u\hat{|}_k)^2 + \int_{|\xi|\leq 1/h} (1+|\xi|^2)^k|\xi|^2|\hat{u}|^2\,d\xi \\
&\leq \int_{|\xi|\geq 1/h} \dots + (|u\hat{|}_k)^2 + \sum_{j=1}^{n}(|\partial_{h,j}u\hat{|}_k)^2 \\
&\leq \int_{|\xi|\geq 1/h} \dots + C_k|u|_k^2 + nC^2.
\end{aligned}$$

The integral over $|\xi| \geq 1/h$ vanishes for $h \to 0$ so that the statement follows.
(b) For the converse use $(\partial_{h,j}u)(\mathbf{x}) = \int_{-1/2}^{1/2} u_{x_j}(\mathbf{x}+th\mathbf{e}_j)\,dt$. ■

6.2.4 $H^s(\Omega)$ for Real $s \geq 0$

Let $s \geq 0$. For $\Omega = \mathbb{R}^n$ one can define the following scalar product (16a) and the Sobolev norm (16b) for all $u \in C_0^\infty(\mathbb{R}^n)$:

$$(u,v)\hat{}_s := \int_{\mathbb{R}^n} (1+|\xi|^2)^s\hat{u}(\xi)\overline{\hat{v}(\xi)}\,d\xi, \tag{6.2.16a}$$

$$|u\hat{|}_s := \|(1+|\xi|^2)^{s/2}\hat{u}(\xi)\|_{L^2(\mathbb{R}^n)}. \tag{6.2.16b}$$

The completion in $L^2(\mathbb{R}^n)$ also defines the Sobolev space $H^s(\mathbb{R}^n)$ for non-integer order s. On the basis of Lemma 23 and Exercise 6.1.12 the newly defined $H^s(\mathbb{R}^n)$ for $s \in \mathbb{N}\cup\{0\}$ agree with the Sobolev spaces used until this point. Let $\Omega \subset \mathbb{R}^n$. The number $s \geq 0$ can be decomposed as $s = k+\lambda$ with $k \in \mathbb{N}\cup\{0\}$ and $0<\lambda<1$. We define

$$\begin{aligned}
(u,v)_s := \sum_{|\alpha|\leq k} \Bigg[\int_\Omega D^\alpha u(\mathbf{x})D^\alpha v(\mathbf{x})\,d\mathbf{x} \\
+ \int\int_{\Omega\times\Omega} \frac{[D^\alpha u(\mathbf{x}) - D^\alpha u(\mathbf{y})][D^\alpha v(\mathbf{x}) - D^\alpha v(\mathbf{y})]}{|\mathbf{x}-\mathbf{y}|^{n+2\lambda}}\,d\mathbf{x}\,d\mathbf{y}\Bigg],
\end{aligned} \tag{6.2.17a}$$

$$|u|_s := \|u\|_{H^s(\Omega)} := \sqrt{(u,u)_s} \quad (s = k+\lambda,\ 0<\delta<1). \tag{6.2.17b}$$

The norm $|\cdot|_s$ is called the Sobolev-Slobodeckiĭ norm. One can define the Hilbert spaces $H^s(\Omega)$ and $H_0^s(\Omega)$ in the same way as in the case $s = k \in \mathbb{N}$. The properties of these spaces are summarised in the following theorem (cf. Adams [1], Wloka [1]):

Theorem 6.2.25. *Let $s \geq 0$.* (a) *For $\Omega = \mathbb{R}^n$ the norms* (16b) *and* (17b) *are equivalent, i.e., both norms define the same space $H^s(\mathbb{R}^n) = H_0^s(\mathbb{R}^n)$.*
(b) $\{u \in C^\infty(\Omega)$: supp(u) compact , $|u|_s < \infty\}$ *is dense in $H^s(\Omega)$.*
(c) *$C_0^\infty(\Omega)$ is dense in $H_0^s(\Omega)$.*
(d) *$aD^\alpha(bu) \in H^{s-|\alpha|}(\Omega)$, if $|\alpha| \leq s$, $u \in H^s(\Omega)$, $a \in C^{t-|\alpha|}(\overline{\Omega})$, $b \in C^t(\overline{\Omega})$, where $t = s \in \mathbb{N} \cup \{0\}$ or $t > s$.*
(e) *$H^s(\Omega) \subset H^t(\Omega)$, $H_0^s(\Omega) \subset H_0^t(\Omega)$ for $s \geq t$.*
(f) *In* (10a,b) *k and m can be real.*
(g) *In* (11) *replace $||T||_{C^k(\overline{\Omega})}$ by $||T||_{C^t(\overline{\Omega})}$, with $t > k$ if $k \notin \mathbb{N}$.*

Exercise 6.2.26. Check, using the norm (16b), as well as (17b), that the characteristic function $u(x) = 1$ in $[-1,1]$, $u(x) = 0$ for $|x| > 1$ belongs to $H^s(\mathbb{R})$ if and only if $0 \leq s < 1/2$.

6.2.5 Trace and Extension Theorems

The nature of boundary value problems requires that one can form boundary values $u|_\Gamma$ (restriction of u to $\Gamma = \partial\Omega$, or trace of u on $\Gamma = \partial\Omega$) in a meaningful way. As can be seen easily, a Hölder-continuous function $u \in C^s(\overline{\Omega})$ has a restriction $u|_\Gamma \in C^s(\Gamma)$ if only Γ is sufficiently smooth. But, from $u \in H^s(\Gamma)$ does not necessarily follow $u|_\Gamma \in H^s(\Gamma)$. Since the equality $u = v$ on $H^s(\Omega)$ only means that $u(\mathbf{x}) = v(\mathbf{x})$ almost everywhere in Ω, and Γ is a set of measure zero, $u(\mathbf{x}) \neq v(\mathbf{x})$ may hold everywhere on Γ. Also, the boundary value $u(\mathbf{x})$ ($\mathbf{x} \in \Gamma$) cannot be defined by a continuous extension either since, for example, $u \in H^1(\Omega)$ need not be continuous (cf. Exercise 4c).

The inverse problem for the definition of $u|_\Gamma$ is extension: does there exist, for a given boundary value φ on Γ, a function $u \in H^s(\Omega)$ such that $\varphi = u|_\Gamma$? If the answer is negative there exists no solution $u \in H^s(\Omega)$ to the Dirichlet boundary value problem.

First we study these problems on the half-space

$$\Omega = \mathbb{R}_+^n := \{(x_1, \cdots, x_n) \in \mathbb{R}^n : x_n > 0\} \quad \text{with } \Gamma = \partial\Omega = \mathbb{R}^{n-1} \times \{0\}. \tag{6.2.18}$$

Functions $u \in H^s(\mathbb{R}_+^n)$ will first be continued to $\overline{u} \in H^s(\mathbb{R}^n)$, and then $\overline{u}$ restricted to Γ.

Theorem 6.2.27. *Let $s \geq 0$. There exists an extension operator $\Phi_s \in L(H^s(\mathbb{R}_+^n), H^s(\mathbb{R}^n))$ such that for all $u \in H^s(\mathbb{R}_+^n)$ the extension $\overline{u} = \Phi_s u$ and u coincide on $\mathbb{R}_+^n$.*

PROOF. For $s = 0$, set $\overline{u} = u$ on $\mathbb{R}_+^n$ and $\overline{u} = 0$ otherwise. Since $||\overline{u}||_{L^2(\mathbb{R}^n)} = ||u||_{L^2(\mathbb{R}_+^n)}$, the mapping defined by $\Phi_0 u := \overline{u}$ is bounded: $||\Phi_0||_{L^2(\mathbb{R}^n) \leftarrow L^2(\mathbb{R}_+^n)} = 1$. For $s \leq 1$ define $\Phi_s u = \overline{u}$ by

$$\overline{u} = u \quad \text{on } \mathbb{R}_+^n, \quad \overline{u}(x_1, \cdots, x_{n-1}, -x_n) = u(x_1, \cdots, x_{n-1}, x_n) \quad \text{for } x_n > 0$$

(i.e., by reflection on Γ). For $s = 1$ one obtains, for example, $|\overline{u}|_1 = \sqrt{2}|u|_1$, i.e., $\Phi_s \in L(H^s(\mathbb{R}^n_+), H^s(\mathbb{R}^n))$. For larger s one uses higher interpolation formulae for $\overline{u}(\cdots, -x_n)$ (cf. Exercise 9.1.13, cf. Wloka [1, p. 101]). ■

In the following the restriction $u|_\Gamma$ is written in the form γu. At first, γ is defined only on $C_0^\infty(\mathbb{R}^n)$:

$$\gamma: C_0^\infty(\mathbb{R}^n) \to C_0^\infty(\Gamma) \subset L^2(\mathbb{R}^{n-1}), \quad (\gamma u)(\mathbf{x}) := u(\mathbf{x}) \quad \text{for all } (\mathbf{x}) \in \Gamma. \tag{6.2.19}$$

We write $\mathbf{x} = (\mathbf{x}', x_n)$ with $\mathbf{x}' = (x_1, \cdots, x_{n-1}) \in \mathbb{R}^{n-1}$. Γ is identified with $\mathbb{R}^{n-1}$: $H^s(\Gamma) = H^s(\mathbb{R}^{n-1})$.

Theorem 6.2.28. *Let $s > 1/2$. Then γ from (19) can be continued to $\gamma \in L(H^s(\mathbb{R}^n), H^{s-1/2}(\mathbb{R}^{n-1}))$. Thus we have in particular $|\gamma u|_{s-1/2} \le C_s|u|_s$ for $u \in H^s(\mathbb{R}^n)$. In the case $n = 1$, i.e., $\gamma u = u(0)$, we have $|\gamma u| \le C_s|u|_s$.*

PROOF. It suffices to show $|\gamma u|\hat{}_{s-1/2} \le C_s'|u|\hat{}_s$ for $u \in C_0^\infty(\mathbb{R}^n)$ (cf. Theorem 6.1.11 and Theorem 25a). Let the Fourier transforms of $u \in C_0^\infty(\mathbb{R}^n)$ and $w := \gamma u \in C_0^\infty(\mathbb{R}^{n-1})$ be $\hat{u} = \mathcal{F}_n u$ and $\hat{w} = \mathcal{F}_{n-1} w$ ($\mathcal{F}_k$: k-dimensional Fourier transform). $\mathcal{F}_n$ can be written as product $\mathcal{F}_1 \circ \mathcal{F}_{n-1}$ where $\mathcal{F}_{n-1}$ acts on $\mathbf{x}' \in \mathbb{R}^{n-1}$ and $\mathcal{F}_1$ on x_n. Therefore $\hat{W}(\cdot, x_n) := \mathcal{F}_{n-1}u(\cdot, x_n)$ has the properties $\hat{u}(\xi', \cdot) = \mathcal{F}_1 \hat{W}(\xi', \cdot)$ and $\hat{w} = \hat{W}(\cdot, 0)$. According to Lemma 20, $\hat{w}(\xi') = \hat{W}(\xi', 0) = [\mathcal{F}_1^{-1}\hat{u}(\xi', \cdot)]\big|_{x_n=0}$ has the representation

$$\hat{w}(\xi') = (2\pi)^{-1/2} \int_{\mathbb{R}} \hat{u}(\xi', \xi_n)\, d\xi_n \quad \text{for all } \xi' \in \mathbb{R}^{n-1}. \tag{6.2.20}$$

Since $u \in H^s(\mathbb{R}^n)$, $\hat{U}(\xi', \xi_n) := (1 + |\xi'|^2 + \xi_n^2)^{s/2}\hat{u}(\xi', \xi_n)$ lies in $L^2(\mathbb{R}^n)$ (cf. Lemma 23b). Inequality (5a) yields

$$2\pi|\hat{w}(\xi')|^2 = |\int_{\mathbb{R}} \hat{u}(\xi', \xi_n)\, d\xi_n|^2 \le \int_{\mathbf{R}} (1 + |\xi'|^2 + \xi_n^2)^{-s}\, d\xi_n \int_{\mathbb{R}} |\hat{U}(\xi', \xi_n)|^2\, d\xi_n.$$

The first factor may be computed as $K_s(1 + |\xi'|^2)^{1/2-s}$ with $K_s = \int_{\mathbb{R}}(1 + x^2)^{-s}\, dx < \infty$, since $s > 1/2$. The second is $\tilde{U}(\xi') := ||\hat{U}(\xi', \cdot)||_{L^2(\mathbb{R})} \in L^2(\mathbf{R}^{n-1})$ because $||\tilde{U}||_{L^2(\mathbb{R}^{n-1})} = ||\hat{U}||_{L^2(\mathbb{R}^n)}$ (Fubini's theorem). Together we have:

$$(1 + |\xi'|^2)^{s-1/2}|\hat{w}(\xi')|^2 \le (K_s/(2\pi))\tilde{U}(\xi')^2 \quad \text{for } \xi' \in \mathbb{R}^{n-1}.$$

Integration over $\xi' \in \mathbb{R}^{n-1}$ results in

$$\begin{aligned}\int_{\mathbb{R}^{n-1}} (1 + |\xi'|^2)^{s-1/2}|\hat{w}(\xi')|^2\, d\xi' &\le \frac{K_s}{2\pi}|\tilde{U}|_0^2 = \frac{K_s}{2\pi}|\hat{U}|_0^2 \\ &= \frac{K_s}{2\pi}\int_{\mathbb{R}^n} (1 + |\xi|^2)^{s/2}|\hat{u}(\xi)|^2\, d\xi.\end{aligned}$$

Thus $|w|\hat{}_{s-1/2} \le C'_s|u|\hat{}_s$ is proved with $C'_s = (K_s/(2\pi))^{1/2}$ (cf. (16b)). In the case $n = 1$, $\hat{w}(\xi')$ already represents $\gamma u = u(0)$, and the integration over ξ' is not required. ■

Theorem 28 describes the restriction $u(\cdot, 0) = \gamma u$ to $x_n = 0$. Evidently, similarly we have $|u(\cdot, x_n)|_{s-1/2} \le C_s|u|_s$ for any other $x_n \in \mathbb{R}$, with the same constant C_s. The mapping $x_n \mapsto u(\cdot, x_n)$ is continuous [resp. Hölder-continuous] in the following sense.

Theorem 6.2.29. *For $s > 1/2$ the following statements hold:*

$$\lim_{y_n \to x_n} ||u(\cdot, x_n) - u(\cdot, y_n)||_{H^{s-1/2}(\mathbb{R}^{n-1})} = 0 \quad \text{for all } x_n \in \mathbb{R}, u \in H^s(\mathbb{R}^n), \tag{6.2.21a}$$

$$||u(\cdot, x_n) - u(\cdot, y_n)||_{H^{s-1/2-\lambda}(\mathbb{R}^{n-1})} \le K_{s,\lambda}|x_n - y_n|^\lambda ||u||_{H^s(\mathbb{R}^n)} \tag{6.2.21b}$$

for all $u \in H^s(\mathbb{R}^n)$, $0 \le \lambda < 1$, $\lambda \le s - 1/2$ (and for $\lambda = 1$ if $s > 3/2$).

PROOF. (a) Let $u_\nu \in C_0^\infty(\mathbb{R}^n)$ be a sequence with $u_\nu \to u \in H^s(\mathbb{R}^n)$ and set $\varphi_\nu(x) := ||u_\nu(\cdot, x)||_{H^{s-1/2}(\mathbb{R}^{n-1})}$. The function φ_ν is continuous in $\mathbb{R}$ and converges uniformly to $||u(\cdot, x)||_{H^{s-1/2}(\mathbb{R}^{n-1})}$ since $|u_\nu(\cdot, x) - u(\cdot, x)|_{s-1/2} \le C_s|u_\nu - u|_s$ for all $x \in \mathbb{R}$. Thus (21a) follows.
(b) $u_\epsilon(\cdot, x_n) := u(\cdot, x_n + \epsilon) - u(\cdot, x_n)$ has the Fourier transform $\hat{u}_\epsilon(\xi) = \hat{u}_\epsilon(\xi', \xi_n) = [\exp(i\xi_n\epsilon) - 1]\hat{u}(\xi)$ so that $|\hat{u}_\epsilon(\xi)|^2 = 4\sin^2(\xi_n\epsilon/2)|\hat{u}(\xi)|^2$. As in the proof for Theorem 28 set $\hat{W}(\cdot, x_n) := \mathfrak{F}_{n-1}u_\epsilon(\cdot, x_n)$, $\hat{w} = \hat{W}(\cdot, 0)$. The first integral in the estimate of $2\pi|\hat{w}(\xi')|^2$ now reads

$$\int_{\mathbb{R}} (1 + |\xi'|^2 + \xi_n^2)^{-s} \sin^2(\xi_n\epsilon/2)\, d\xi_n = (1 + |\xi'|^2)^{1/2-s} \int_{\mathbb{R}} (1 + t^2)^{-s} \sin^2(\eta t)\, dt$$

with $\eta = \frac{\epsilon}{2}(1 + |\xi'|^2)^{1/2}$. The decomposition of the last integral into subintegrals over $|t| \le 1/\eta$ and $|t| \ge 1/\eta$ shows $\int_{\mathbb{R}}(1 + t^2)^{-s}\sin^2(\eta t)\, dt \le C_{s,\lambda}\eta^{2\lambda}$. The remainder of the argument follows the same lines as in the proof of Theorem 28. ■

Up to this point we have obtained $H^s(\mathbb{R}^n)$ by completion of $C_0^\infty(\mathbb{R}^n)$ in $L^2(\mathbb{R}^n)$. The next theorem shows that for sufficiently large s one can also complete in $C^0(\mathbb{R}^n) \cap L^2(\mathbb{R}^n)$ so that $H^s(\mathbb{R}^n)$ contains only classical functions (i.e. continuous, Hölder-continuous, [Hölder-] continuously differentiable functions).

Theorem 6.2.30. (Sobolev's lemma) $H^s(\mathbb{R}^n) \subset C^k(\mathbb{R}^n)$ *holds for* $k \in \mathbb{N} \cup \{0\}$, $s > k + n/2$ *and* $H^s(\mathbb{R}^n) \subset C^t(\mathbb{R}^n)$ *for* $0 < t \notin \mathbb{N}$, $s \ge t + n/2$.

PROOF. (a) Let $s \ge t + n/2$, $0 < t < 1$, $u \in H^s(\mathbb{R}^n)$. For given $\mathbf{x}, \mathbf{y} \in \mathbb{R}^n$ we want to show $|u(\mathbf{x}) - u(\mathbf{y})| \le C|\mathbf{x} - \mathbf{y}|^t$ with C independent of $\mathbf{x}, \mathbf{y}$. The coordinates of the $\mathbb{R}^n$ can be rotated so that $\mathbf{x} = (x_1, 0, \cdots, 0)$, $\mathbf{y} = (y_1, 0, \cdots, 0)$. An $(n-1)$-fold application of Theorem 28 to $u(\cdot)$, $u(\cdot, 0)$, $u(\cdot, 0, 0)$

, etc., results in $u(\cdot, 0, 0, \cdots, 0) \in H^{s-(n-1)/2}(\mathbb{R})$. Theorem 29 provides the desired estimate with $C = K_{s-(n-1)/2,t}|u(\cdot, 0, \cdots, 0)|_{s-(n-1)/2} \le C'|u|_s$; thus $u \in C^t(\mathbb{R}^n)$ and $\|\cdot\|_{C^t(\mathbb{R}^n)} \le C\|\cdot\|_{H^s(\mathbb{R}^n)}$. Then $u \in C^0(\mathbb{R}^n)$ follows from $C^t(\mathbb{R}^n) \subset C^0(\mathbb{R}^n)$.
(b) Let $s \ge t + n/2$, $1 < t < 2$, $u \in H^s(\mathbb{R}^n)$. Part (a) is applicable to $D^\alpha u \in H^{s-1}(\mathbb{R}^n)$ (cf. Theorem 25d,e) with $|\alpha| \le 1$: $D^\alpha u \in C^{t-1}(\mathbb{R}^n)$ for $|\alpha| \le 1$. Thus $u \in C^t(\mathbb{R}^n)$, etc. ■

We return to the statements of Theorems 27 and 28. For all $u \in C_0^\infty(\mathbb{R}^n_+)$ the restriction $\gamma u = u(\cdot, 0)$ agrees with $\gamma \Phi_s u$. Completion in $H^s(\mathbb{R}^n_+)$ yields

$$\|\gamma\|_{H^{s-1/2}(\mathbb{R}^{n-1}) \leftarrow H^s(\mathbb{R}^n_+)} \le \|\gamma\|_{H^{s-1/2}(\mathbb{R}^{n-1}) \leftarrow H^s(\mathbb{R}^n)} \|\Phi_s\|_{H^s(\mathbb{R}^n) \leftarrow H^s(\mathbb{R}^n_+)}.$$

This proves the following corollary:

Corollary 6.2.31. *Let $s > 1/2$. For the restriction $\gamma u := u(\cdot, 0)$ we have $\gamma \in L(H^s(\mathbb{R}^n_+), H^{s-1/2}(\mathbb{R}^{n-1}))$.*

With the restriction to $x_n = 0$ one evidently loses half an order of differentiability. Conversely one gains half an order if one continues $w \in H^{s-1/2}(\mathbb{R}^{n-1})$ suitably in $\mathbb{R}^n$.

Theorem 6.2.32. *Let $s > 1/2$, $w \in H^{s-1/2}(\mathbb{R}^{n-1})$. There exists a function $u \in H^s(\mathbb{R}^n)$ $[u \in H^s(\mathbb{R}^n_+)]$ such that $|u|_s \le C_s |w|_{s-1/2}$ and $\gamma u = w$, i.e., $w = u(\cdot, 0)$.*

PROOF. Let $\hat{u} = \mathcal{F}_n u$ and $\hat{w} = \mathcal{F}_{n-1} w$ be the Fourier transforms. $\gamma u = w$ is equivalent to (20). For

$$\hat{u}(\xi) = \hat{u}(\xi', \xi_n) := \hat{w}(\xi')(1 + |\xi'|^2)^{s-1/2}(1 + |\xi'|^2 + \xi_n^2)^{-s}/K_s,$$
$$K_s := \int_{\mathbb{R}} (1 + t^2)^{-s} dt,$$

one checks that (20) and $|u|_s^{\wedge} = K_s^{-1/2} |w|_{s-1/2}^{\wedge}$ hold. Restriction of $u \in H^s(\mathbb{R}^n)$ to $\mathbb{R}^n_+$ proves the parenthetical addition. ■

If one replaces $\mathbb{R}^n_+$ with a general domain $\Omega \subset \mathbb{R}^n$, then $\mathbb{R}^{n-1} \cong \mathbb{R}^{n-1} \times \{0\} = \partial \mathbb{R}^n_+$ becomes $\Gamma = \partial\Omega$, and the necessity arises of defining the Sobolev space $H^s(\Gamma)$. We begin with the

Definition 6.2.33. Let $0 < t \in \mathbb{R} \cup \{\infty\}$ [resp. $k \in \mathbb{N} \cup \{0\}$]. We write $\Omega \in C^t$ $[\Omega \in C^{k,1}]$, if for every $\mathbf{x} \in \Gamma := \partial\Omega$ there exists a neighbourhood $U \subset \mathbb{R}^n$ such that there exists a bijective mapping $\varphi: U \to K_1(\mathbf{0}) = \{\xi \in \mathbb{R}^n : |\xi| < 1\}$ with

$$\varphi \in C^t(\overline{U}), \quad \varphi^{-1} \in C^t(\overline{K_1(\mathbf{0})})$$
$$[\varphi \in C^{k,1}(\overline{U}), \quad \varphi^{-1} \in C^{k,1}(\overline{K_1(\mathbf{0})})], \tag{6.2.22a}$$
$$\varphi(U \cap \Gamma) = \{\xi \in K_1(\mathbf{0}) \colon \xi_n = 0\}, \tag{6.2.22b}$$
$$\varphi(U \cap \Omega) = \{\xi \in K_1(\mathbf{0}) \colon \xi_n > 0\}, \tag{6.2.22c}$$
$$\varphi(U \cap (\mathbb{R}^n \backslash \Omega)) = \{\xi \in K_1(\mathbf{0}) \colon \xi_n < 0\}. \tag{6.2.22d}$$

Here $K_1(\mathbf{0})$ is a circle (ball) if $|\cdot|$ is the Euclidean norm. For the maximum norm $|\cdot|_\infty$, $K_1(\mathbf{0})$ is a square (cube). Likewise, $K_1(\mathbf{0})$ can be replaced by another circle $K_R(\mathbf{z})$ or any rectangle $(x_1', x_1'') \times \cdots \times (x_n', x_n'')$.

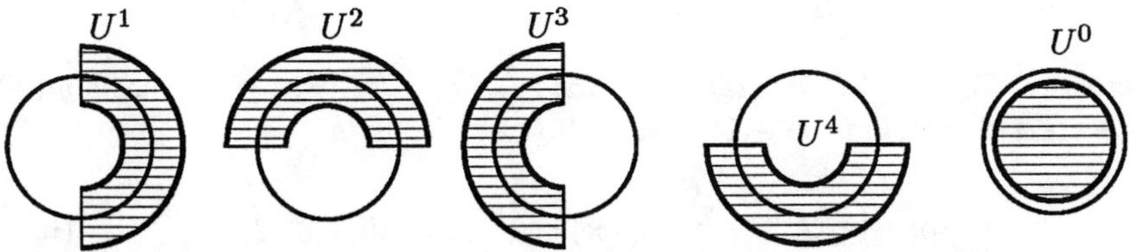

Figure 6.2.1. Covering neighbourhoods of Γ and Ω

Example 6.2.34. Let Ω be the circle $K_1(\mathbf{0}) \subset \mathbb{R}^2$. A neighbourhood of $\mathbf{x}^* = (1,0)$ is U^1 from Fig. 1. The mapping $\mathbf{x} \in U^1 \mapsto \varphi(\mathbf{x}) := \xi \in (-1,1) \times (-1,1)$ with $x_1 = (1 - \xi_2/2)\cos(\pi\xi_1/2)$, $x_2 = (1 - \xi_2/2)\sin(\pi\xi_1/2)$ is bijective and satisfies (22a-d) with $t = \infty$. The same holds for any $\mathbf{x} \in \Gamma$. Thus $\Omega \in C^\infty$.

Exercise 6.2.35. (a) The rectangle $\Omega = (x_1', x_1'') \times (x_2', x_2'')$ and the L-domain from Example 2.1.4 are domains $\Omega \in C^{0,1}$.
(b) The cut circle in Figure 5.2.1b does not belong to $C^{0,1}$.

Lemma 6.2.36. *Let $\Omega \in C^t$ $[\Omega \in C^{k,1}]$ be a bounded domain. Then there exists $N \in \mathbb{N}$, $U^i (0 \le i \le N)$, U_i, α_i $(1 \le i \le N)$ with*

$$U^i \text{ open, bounded } (0 \le i \le N), \quad \bigcup_{i=0}^{N} U^i \supset \overline{\Omega}, \quad U^0 \subset\subset \Omega, \tag{6.2.23a}$$

$$U_i := U^i \cap \Gamma \quad (1 \le i \le N), \quad \bigcup_{i=1}^{N} U_i = \Gamma \tag{6.2.23b}$$

$$\alpha_i \colon U_i \to \alpha_i(U_i) \subset \mathbb{R}^{n-1} \text{ bijective for } i = 1, \cdots, N, \tag{6.2.23c}$$
$$\alpha_i \circ \alpha_j^{-1} \in C^t(\overline{\alpha_j(U_i \cap U_j)}) \; [\text{resp. } \alpha_i \circ \alpha_j^{-1} \in C^{k,1}(\overline{\alpha_j(U_i \cap U_j)})]. \tag{6.2.23d}$$

On U^i $(1 \le i \le N)$ are defined mappings φ_i with the properties (22a-d).

PROOF. For every $\mathbf{x} \in \Gamma$ there exist $U = U(\mathbf{x})$ and $\varphi = \varphi_{\mathbf{x}}$ according to Definition 33. Let $\alpha_{\mathbf{x}}$ be the restriction of $\varphi_{\mathbf{x}}$ to $U(\mathbf{x}) \cap \Gamma$. Let V^i ($i \in \mathbb{N}$) be the open sets $\{\mathbf{x} \in \Omega: \text{dist}\,(\mathbf{x}, \Gamma) > 1/i\} \subset\subset \Omega$. Let $\bigcup_{\mathbf{x}\in\Gamma} U(\mathbf{x}) \cup \bigcup_i V^i$ be an open covering of the compact set $\overline{\Omega}$. Therefore there exists a finite covering through $U^i := U(\mathbf{x}_i)$ $(1 \le i \le N)$ and at most one V^j which is denoted by U^0. If one sets $U_i := U^i \cap \Gamma$, $\alpha_i = \alpha_{\mathbf{x}_i}$, $\varphi_i = \varphi_{\mathbf{x}_i}$, the statements follow from (22a-d). ■

A set of pairs $\{(U_i, \alpha_i): 1 \le i \le N\}$ fulfilling the conditions (23b-d) is called a C^t- [resp. $C^{k,1}$]- coordinate system for Γ.

In Example 34 one has $N = 4$. The map α_1^{-1} is given by $\alpha_1^{-1}(\xi_1) = (\cos(\pi\xi_1/2),\ \sin(\pi\xi_1/2)) \in U^1$; and likewise $\alpha_4^{-1}(\xi_1) = \cos((3+\xi_1)\pi/2)$, $\sin((3+\xi_1)\pi/2) \in U^4$, where in each case $-1 < \xi_i < 1$. On $\alpha_1(U^1 \cap U^4)$ one obtains $\alpha_4(\alpha_1^{-1}(\xi_1)) = \xi_1 + 1$.

Lemma 6.2.37. (*Partition of unity*) *Let* $\{U^i: 0 \le i \le N\}$ *satisfy* (23a). *There exist functions* $\sigma_i \in C_0^\infty(\mathbb{R}^n)$, $0 \le i \le N$, *with*

$$\operatorname{supp}(\sigma_i) \subset U^i, \quad \sum_{i=0}^{N} \sigma_i^2(\mathbf{x}) = 1 \quad \textit{for all } \mathbf{x} \in \overline{\Omega}. \tag{6.2.24}$$

The general construction of the σ_i can be found, for example, in Wloka [1, §1.2]. In the special case of Figure 1 one may proceed as follows. Let $\sigma(t) := 0$ for $|t| \ge 1$ and $\sigma(t) := \exp(1/(t^2-1))$ for $t \in (-1,1)$. Then $\sigma \in C_0^\infty(\mathbb{R})$ and $\operatorname{supp}(\sigma) = [-1,1]$. In U^i from Figure 1 one defines, for example,

$$\psi_0(\mathbf{x}) := \sigma(9|\mathbf{x}|^2/4), \quad \psi_1(\mathbf{x}) := \sigma(2r-2)\sigma(2\varphi/\pi) \quad \text{for } \mathbf{x} = r\begin{pmatrix}\cos\varphi\\ \sin\varphi\end{pmatrix}, \text{ etc.}$$

These $\sigma_i(\mathbf{x}) := \psi_i(\mathbf{x})/\sqrt{\sum \psi_i^2(\mathbf{x})}$ satisfy (24).

A function u on Γ can be written in the form $\sum \sigma_i^2 u$. Each summand $\sigma_i^2 u$ is parametrisable over $\alpha_i(U_i) \subset \mathbb{R}^{n-1}$: $(\sigma_i^2 u) \circ \alpha_i^{-1}: \alpha_i(U_i) \subset\subset \mathbb{R}^{n-1} \to \mathbb{R}$. This in turn makes possible

Definition 6.2.38. Let $\Omega \in C^t$ [$\in C^{k,1}$]. Assume (U_i, α_i) and σ_i satisfy (23b-d) and (24). Let $s \le t \in \mathbb{N}$ [$s \le k+1$] or $s < t \notin \mathbb{N}$, $t > 1$. The Sobolev space $H^s(\Gamma)$ is defined as the set of all functions $u: \Gamma \to \mathbb{R}$ such that $(\sigma_i u) \circ \alpha_i^{-1} \in H_0^s(\mathbb{R}^{n-1})$ $(1 \le i \le N)$.

Theorem 6.2.39. (a) *$H^s(\Gamma)$ is a Hilbert space with the scalar product*

$$(u,v)_s := (u,v)_{H^s(\Gamma)} := \sum_{i=1}^{N} ((u\sigma_i) \circ \alpha_i^{-1}, (v\sigma_i) \circ \alpha_i^{-1})_{H^s(\mathbb{R}^{n-1})}.$$

(b) *If* $\{(\tilde{U}_i, \tilde{\alpha}_i): 1 \le i \le N\}$ *is another* C^t- [$C^{k,1}$-] *coordinate system of* Γ *and* $\{\tilde{\sigma}_i\}$ *another partition of unity, then the space* $\tilde{H}^s(\Gamma)$ *defined by this is equal to* $H^s(\Gamma)$ *as a set . The norms of* $H^s(\Gamma)$ *and* $\tilde{H}^s(\Gamma)$ *are equivalent.*

PROOF. For (b): Transformation Theorem 17 [resp. 25g] ; for $\Omega \in C^{0,1}$, cf. Wloka [1, Lemma 4.5]. ■

The trace and extension theorems (Corollary 31 and Theorem 32) can be extended to any domain with a sufficiently smooth boundary. γ now denotes the restriction to $\Gamma = \partial\Omega$: $\gamma u = u\,|_\Gamma$.

Theorem 6.2.40. *Let $\Omega \in C^t$ with $1/2 < s \le t \in \mathbb{N}$ or $1/2 < s < t$ [resp. $\Omega \in C^{k,1}$, $1/2 < s = k+1 \in \mathbb{N}$].*
(a) The restriction γu of $u \in H^s(\Omega)$ belongs to $H^{s-1/2}(\Gamma)$: $\gamma \in L(H^s(\Omega), H^{s-1/2}(\Gamma))$.
(b) For each $w \in H^{s-1/2}(\Gamma)$ there exists an extension $u \in H^s(\Omega)$ with $w = \gamma u$, $|u|_s \le C_s |w|_{s-1/2}$.
(c) For each $w \in H^s(\Omega)$ there exists a continuation $Ew \in H^s(\mathbb{R}^n)$ with $E \in L(H^s(\Omega), H^s(\mathbb{R}^n))$.

PROOF. The proofs follow the same pattern. Let U^i, U_i and φ_i, α_i be as in Lemma 36. The summand $u_i = \sigma_i^2 u$ from $u = \sum \sigma_i^2 u$ [resp. $w_i = \sigma_i^2 w$] has its support in U^i [resp. U_i] and can be mapped via φ_i (resp. α_i) onto $\mathbb{R}^n_+$ [resp. $\mathbb{R}^{n-1} \cong \mathbb{R}^{n-1} \times \{0\} = \partial\mathbb{R}^n_+$]. There Corollary 31 and Theorem 32 hold. The restriction to $\mathbb{R}^{n-1}$ [resp. continuation to $\mathbb{R}^n_+$ or $\mathbb{R}^n$] can be mapped back again. The first statement of the theorem is proved in detail as follows.
Let $u_i := \sigma_i^2 u$ and $\tilde{u}_i := u_i \circ \varphi_i^{-1}$. On the grounds of Theorems 17, 25g the function $\tilde{u}_i$ belongs to $H^s(\mathbb{R}^n_+)$. Thus the restriction $\gamma_+ \tilde{u}_i := \tilde{u}_i(\cdot, 0)$ lies in $H^{s-1/2}(\mathbb{R}^{n-1})$ (cf. Corollary 31) and has $\alpha_i(U_i)$ as support. Set $w_i := (\gamma_+ \tilde{u}_i) \circ \alpha_i$ on U_i, $w_i := 0$ on $\Gamma \backslash U_i$. According to Definition 38, $w := \sum w_i$ belongs to $H^{s-1/2}(\Gamma)$. Since α_i represents the restriction of φ_i on U_i one finds for all $u \in C^t(\overline{\Omega})$ ($C^t(\overline{\Omega})$ is dense in $H^s(\Omega)$):

$$w(\mathbf{x}) = \sum_i w_i(\mathbf{x}) = \sum_i (\gamma_+ \tilde{u}_i(\alpha_i(\mathbf{x})) = \sum_i \tilde{u}_i(\varphi_i(\mathbf{x})) = \sum_i (\sigma_i^2 u)(\mathbf{x}) = u(\mathbf{x}),$$

for $\mathbf{x} \in \Gamma$. Since all the partial mappings are bounded, one finds that

$$|\gamma u|_{s-1/2} = |w|_{s-1/2} \le C_s |u|_s. \quad ■$$

Remark 6.2.41. (a) Under the conditions of Theorem 40 and the additional condition $s > |\alpha| + 1/2$ there exists a restriction $\gamma D^\alpha u \in H^{s-|\alpha|-1/2}(\Gamma)$ of the derivative of $u \in H^s(\Omega)$.
(b) For each $u \in H_0^s(\Omega)$ with $s < l + 1/2$ one has $\gamma D^\alpha u = 0$ if $|\alpha| \le l$.

Theorem 6.2.42. *For $\Omega \in C^{0,1}$ holds $H_0^1(\Omega) = \{u \in H^1(\Omega): u|_\Gamma = 0\}$.*

PROOF. (a) $u \in H_0^1(\Omega)$ is the limit of $u_\nu \in C_0^\infty(\Omega)$, all of which satisfy $u_\nu\big|_\Gamma = 0$. Since $u_\nu\big|_\Gamma \to u\big|_\Gamma$ with respect to $H^{1/2}(\Gamma)$, it follows that $u \in H^1(\Omega)$ and $u\big|_\Gamma = 0$.

(b) Conversely, let $u \in H^1(\Omega)$ and $u|_{\Gamma} = 0$. The proof that $u \in H_0^1(\Omega)$ can be divided up as follows:
(ba) By using the partition of unity over $\{U^i: 0 \le i \le N\}$ (cf. Lemma 37), the statement reduces to the case $\Omega = \mathbb{R}^n_+$.
(bb) Without loss of generality one can assume $n = 1$: $\Omega = \mathbb{R}_+$.
(bc) According to Remark 14, for each $\epsilon > 0$ there exists a $\tilde{u}_\epsilon \in C^\infty(\mathbb{R}_+)$ with finite support such that $|u - \tilde{u}_\epsilon|_1 \le \eta/2$. For each $\eta > 0$ there exists $\varphi_\eta \in C^\infty(\mathbb{R}_+)$ with

$$\varphi_\eta(x) = 0 \quad \text{for } x \le \eta/2, \quad \varphi_\eta(x) = 1 \quad \text{for } x \ge \eta,$$

$$|\varphi'_\eta(x)| \le 3/\eta, \qquad 0 \le \varphi_\eta(x) \le 1.$$

The function $u_\epsilon := \tilde{u}_\epsilon - (1 - \varphi_1)\tilde{u}_\epsilon(0)$ satisfies $u_\epsilon \in C^\infty(\mathbb{R}_+)$, $u_\epsilon(0) = 0$, $\mathrm{supp}(u_\epsilon)$ finite, and $|u - u_\epsilon|_1 \le C\epsilon$. Thus $X := \{v \in C^\infty(\mathbb{R}_+): v(0) = 0, \mathrm{supp}(v) \text{ finite}\}$ is dense in $\{u \in H^1(\mathbb{R}_+): u(0) = 0\}$.
(bd) The statement would be proved if $C_0^\infty(\mathbb{R}_+)$ were also dense in X with respect to $|\cdot|_1$. Let $v \in X$. Evidently, $v_\eta := \varphi_\eta v \in C_0^\infty(\mathbb{R}_+)$ holds for all $\eta > 0$. Set $\Theta(\eta) := ||v'||_{L^2(0,\eta)}$ and note that $\Theta(\eta) \to 0$ for $\eta \to 0$. Since $v(x) = v_\eta(x)$ for $x \ge \eta$, it remains to estimate the variables $||v - v_\eta||_{L^2(0,\eta)}$ and $||v' - v'_\eta||_{L^2(0,\eta)}$. Because of $v'_\eta - v' = \varphi'_\eta v + (\varphi_\eta - 1)v'$, one obtains

$$\begin{aligned} ||v'_\eta - v'||_{L^2(0,\eta)} &\le ||\varphi'_n||_{L^\infty(0,\eta)}||v||_{L^2(0,\eta)} + ||\varphi_\eta - 1||_{L^\infty(0,\eta)}||v'||_{L^2(0,\eta)} \\ &\le (3/\eta)||v||_{L^2(0,\eta)} + \Theta(\eta). \end{aligned}$$

Since $v(x) = \int_0^x v'(\xi)\,d\xi$, (5b) implies the estimate $|v(x)| \le \sqrt{\eta}\Theta(\eta)$ for all $0 \le x \le \eta$, and hence $||v||_{L^2(0,\eta)} \le \eta\Theta(\eta)$. Then the statement follows from $|v - v_\eta|_1^2 \le ||v - v_\eta||^2_{L^2(0,\eta)} + ||v' - v'_\eta||^2_{L^2(0,\eta)} \le C\Theta^2(\eta) \to 0$. ■

When $\Omega \in C^1$, the normal direction $\mathbf{n}$ exists at all boundary points. Analogously to Theorem 42 one proves

Corollary 6.2.43. *For $\Omega \in C^1$ and $k \in \mathbb{N}$ holds*

$$\begin{aligned} H_0^k(\Omega) &= \{u \in H^k(\Omega): \partial^l u/\partial n^l|_\Gamma = 0 \quad \text{for } 0 \le l \le k-1\} \\ &= \{u \in H^k(\Omega): D^\alpha u|_\Gamma = 0 \quad \text{for } 0 \le |\alpha| \le k-1\}. \end{aligned}$$

6.3 Dual Spaces

6.3.1 Dual Space of a Normed Space

Let X be a normed, linear space over $\mathbb{R}$. As a dual space, X' denotes the space of all bounded, linear mappings of X onto $\mathbb{R}$:

$$X' := L(X, \mathbb{R}).$$

According to Exercise 6.1.8, X' is a Banach space with the norm (dual norm)

$$\|x'\|_{X'} := \|x'\|_{\mathbb{R}\leftarrow X} = \sup\{|x'(x)|/\|x\|_X : 0 \neq x \in X\}. \tag{6.3.1}$$

The elements $x' \in X'$ are called linear functionals on X. Instead of $x'(x)$ (application of x' to x) one also writes $\langle x, x'\rangle_{X\times X'}$, or $\langle x', x\rangle_{X'\times X}$, and calls $\langle\cdot,\cdot\rangle_{X\times X'}$ the dual form on $X \times X'$:

$$\langle x, x'\rangle_{X\times X'} = \langle x', x\rangle_{X'\times X} = x'(x).$$

Lemma 6.3.1. *Let the Banach space X be embedded densely and continuously in the Banach space Y. Then Y' is continuously embedded in X'.*

PROOF. (a) Let $y' \in Y'$. Because $X \subset Y$, y' is defined on X.
(b) Since X is a dense subspace of Y, according to Theorem 6.1.11 and (1.5), for each $y' \in Y'$ holds:

$$\|y'\|_{Y'} = \sup_{x\in X} |y'(x)|/\|x\|_Y \geq \frac{1}{C}\sup_x |y'(x)|/\|x\|_X = \frac{1}{C}\|y'\|_{X'},$$

i.e., Y' is embedded continuously in X'. ■

To the transposed matrix in the finite-dimensional case corresponds the dual mapping (or the dual operator).

Lemma 6.3.2. *Let X and Y be normed and let $T \in L(X,Y)$. For each $y' \in Y'$*

$$\langle Tx, y'\rangle_{Y\times Y'} = \langle x, x'\rangle_{X\times X'} \quad \text{for all } x \in X \tag{6.3.2}$$

defines a unique $x' \in X'$. The linear mapping $y' \mapsto x'$ defines the dual operator $T': Y' \to X'$ with $T'y' = x'$. The following holds: $T' \in L(Y', X')$ and

$$\|T'\|_{X'\leftarrow Y'} = \|T\|_{Y'\leftarrow X}. \tag{6.3.3}$$

PROOF. If one writes (2) in the form $y'(Tx) = x'(x)$, one can see that $x' = y' \circ T$. (3) follows from the definitions of the norms:

$$\|T'\|_{X'\leftarrow Y'} = \sup_{y'\neq 0} \|T'y'\|_{X'}/\|y'\|_{Y'} = \sup_{x\neq 0, y'\neq 0} |\langle x, T'y'\rangle_{X\times X'}/[\|x\|_X\|y'\|_{Y'}]$$

$$= \sup_{x\neq 0, y'\neq 0} |\langle Tx, y'\rangle_{Y\times Y'}|/[\|x\|_X\|y'\|_Y] = \sup_{x\neq 0}\|Tx\|_Y/\|x\|_X = \|T\|_{Y\leftarrow X}.$$

■

Example 6.3.3. Let $\Omega = (0,1)$, $X = (C^0(\overline{\Omega}), \|\cdot\|_\infty)$, and $\mathbf{x} \in \Omega$. The mapping $\delta_{\mathbf{x}}: u \in C^0(\overline{\Omega}) \mapsto u(\mathbf{x}) \in \mathbb{R}$ is a functional: $\delta_{\mathbf{x}} \in C^0(\overline{\Omega})'$ (the so-called delta function). The Laplace operator Δ belongs to $L(C^2(\overline{\Omega}), C^0(\overline{\Omega}))$.

The dual mapping $\Delta' \in L(C^0(\overline{\Omega})', C^2(\overline{\Omega})')$ is applicable to $\delta_{\mathbf{x}}$: $\Delta'\delta_{\mathbf{x}}$ is characterised by $(\Delta'\delta_{\mathbf{x}})u = \Delta u(\mathbf{x})$ for all $u \in C^2(\overline{\Omega})$.

Exercise 6.3.4. Let $S \in L(X,Y)$, $T \in L(Y,Z)$. Show that $(TS)' = S'T'$.

Exercise 6.3.5. Show that if $T \in L(X,Y)$ is surjective, then T' is injective.

6.3.2 Adjoint Operators

Let X be a Hilbert space (over $\mathbb{R}$). Every $y \in X$ defined by

$$f_y(x) := (x,y)_X$$

is a linear functional $f_y \in X'$ with $\|f_y\|_{X'} = \|y\|_X$. The converse also holds (cf. Yosida [1, §III.6]).

Theorem 6.3.6. (Riesz representation theorem) *Let X be a Hilbert space and $f \in X'$. There exists a unique $y_f \in X$ such that*

$$f(x) = (x, y_f)_X \quad \textit{for all } x \in X \quad \textit{and } \|f\|_{X'} = \|y_f\|_X.$$

Corollary 6.3.7. *Let X be a Hilbert space.*
(a) *There exists a one-to-one correspondence* (the Riesz isomorphism) $J_X \in L(X,X')$ *with* $J_X y = f_y$, $J_X^{-1} f = y_f$ *that preserves the norm, i.e.,* $\|J_X\|_{X' \leftarrow X} = \|J_X^{-1}\|_{X \leftarrow X'} = 1$.
(b) *X' is a Hilbert space with the scalar product* $(x',y')_{X'} = (J_X^{-1}x', J_X^{-1}y')_X$. *The dual norm $\|x'\|_{X'}$ from* (1) *agrees with the norm induced by* $(x',x')_X^{1/2}$.
(c) *One always identifies X with X'' because $x(x') := x'(x)$. From this follows* $J_{X'} = J_X^{-1}$, $J_X = J_X'$, $T'' = T$ *for* $T \in L(X,Y)$ *if* $Y = Y''$ *is also a Hilbert space.*
(d) *One can identify X and X':* $X = X'$, $J_X = I$.

Let X, Y be Hilbert spaces and $T \in L(X,Y)$. The mapping defined by $T^* := J_X^{-1} T' J_Y \in L(Y,X)$ is called the operator adjoint to T and satisfies

$$(Tx,y)_y = (x, T^*y)_X \text{ for all } x \in X,\ y \in Y; \quad \|T\|_{Y \leftarrow X} = \|T^*\|_{X \leftarrow Y}. \tag{6.3.4}$$

The adjoint and the dual operator only coincide (i.e., $T^* = T'$) if X' is identified with X and Y' with Y.
$T \in L(X,X)$ is said to be selfadjoint (or symmetric) if $T = T^*$.
$T \in L(X,X)$ is called a projection if $T^2 = T$. It is an orthogonal projection if furthermore T is selfadjoint.

Remark 6.3.8. Let X_0 be a closed subspace of the Hilbert space X. An orthogonal projection is given by $Tx := y \in X_0$ with

$$\|x-y\|_X = \inf\{\|x-\xi\|_X : \xi \in X_0\}. \qquad (6.3.5).$$

If conversely $T \in L(X,X)$ is an orthogonal projection with the range $X_0 := \{Tx : x \in X\}$ one has (5) for $y = Tx$. An orthogonal projection always has the norm $\|T\|_{T\leftarrow X} \le 1$.

PROOF. (a) x can be decomposed uniquely into $x = y + z$ ($y \in X_0$, $z \in X_0^{\perp}$) (cf. Lemma 6.1.17). y is the unique solution of (5). $x \in X_0$ implies $y = x$, thus $T^2 = T$. The analogous decomposition $x' = y' + z'$ shows $(x, T^*x') = (Tx, x') = (y, x') = (y, y'+z') = (y, y') = (y+z, y') = (x, Tx')$, hence $T = T^*$.
(b) Let T be an orthogonal projection with range X_0. Let $x = y + z$ be split as above. $T^2 = T$ shows $Ty = y$. For each $y' \in X_0$ holds $(Tz, y') = (z, T^*y') = (z, Ty') = (z, y') = 0$, thus $Tz \in X_0^{\perp}$. Together with $Tz \in X_0$ follows $Tz = 0$ so that $Tx = Ty + Tz = y$.
(c) $Tx = y$ and $\|x\|_X^2 = \|y\|_X^1 + \|z\|_X^2 \ge \|y\|_X^2$ prove $\|T\|_{X\leftarrow X} \le 1$. ■

6.3.3 Scales of Hilbert Spaces

We assume:

$$V \subset U \quad \text{are Hilbert spaces with a continuous and dense embedding.} \qquad (6.3.6)$$

Lemma 6.3.9. *Under assumption* (6) *U' is embedded continuously and densely in V'.*

PROOF. The continuity of the embedding $U' \subset V'$ was established in Lemma 1. For proof that U' is dense in V', we use Exercise 6.1.18d ($A := U', X := V'$). Let $0 \ne v' \in V'$ be arbitrary and $u := J_V^{-1}v' \in V \subset U$. According to the definition, $u' := J_U u \in U' \subset V'$ is characterised by $u'(x) = (x, u)_U$ for all $x \in U$. For $x := u = J_V^{-1}v' \in V$ follows

$$(v', u')_{V'} = (J_V^{-1}v', J_V^{-1}u')_V = (u, J_V^{-1}u')_V = u'(u) = (u, u)_U > 0. \quad ■$$

According to Corollary 7d , U and U' can be identified. By this one obtains the Gelfand triple

$$V \subset U \subset V' \quad (V \subset U \text{ continuously and densely embedded}). \qquad (6.3.7)$$

Corollary 6.3.10. *In a Gelfand triple* (7) *V and U are also continuously and densely embedded in V'.*

PROOF. For $U \subset V'$ see Lemma 9, for $V \subset V'$ see Exercise 6.2.25. ■

Attention. Likewise one could identify V with V' and one would obtain $U' \subset V' = V \subset U$. But it is not possible to identify U with U' and V with V' simultaneously. In the first case one interprets $x(y) = \langle y, x\rangle_{U\times U'}$ for $x, y \in U$ as $(y,x)_U$ (in particular for $x, y \in V \subset U$), in the second case as $(y,x)_V$.

Exercise 6.3.11. Let (7) be true. Set $W := \{J_V^{-1}u : u \in U\}$ and define $(x,y)_W := (J_V x, J_V y)_U$ as the scalar product on W. Show that
(a) W is a Hilbert space;
(b) $W \subset V$ is a continuous and dense embedding;
(c) $(v,w)_V = (v, J_V w)_U$ for all $v \in V$, $w \in W$;
(d) $|(x,y)_V| \le \|x\|_U \|y\|_W$ for all $x, y \in W$.

Because $U = U'$ the scalar product $(x,y)_U$ can also be written in the form $y(x) = \langle x, y\rangle_{U\times U'}$. If $x \in V$, then $y(x) = \langle x,y\rangle_{V\times V'}$ also holds. That means that $(x,y)_U = \langle x,y\rangle_{V\times V'}$ for all $x \in V, y \in U \subset V'$. Likewise one obtains $(x,y)_U = \langle x,y\rangle_{V'\times V}$ for all $x \in U$ and $y \in V$. The dense and continuous embedding $U \subset V'$ proves

Remark 6.3.12. Let $V \subset U \subset V'$ be a Gelfand triple. The continuous extension of the scalar product $(\cdot,\cdot)_U$ to $V \times V'$ [$V' \times V$] results in the dual form $\langle\cdot,\cdot\rangle_{V\times V'}$ [$\langle\cdot,\cdot\rangle_{V'\times V}$]. Therefore the following notation is practical:

$$\langle x,y\rangle_{V\times V'} = (x,y)_U \quad \text{for } x \in V, y \in V',$$
$$\langle x,y\rangle_{V'\times V} = (x,y)_U \quad \text{for } x \in V', y \in V.$$

In connection with Sobolev spaces one always chooses $U := L^2(\Omega)$ so that the embeddings read as follows:

$$H_0^s(\Omega) \subset L^2(\Omega) \subset (H_0^s(\Omega))' \quad (s \ge 0), \tag{6.3.8a}$$

$$H^s(\Omega) \subset L^2(\Omega) \subset (H^s(\Omega))' \quad (s \ge 0). \tag{6.3.8b}$$

Exercise 6.3.13. Show that (8a) and (8b) are Gelfand triples.

The dual space of $H_0^s(\Omega)$ is also denoted by $H^{-s}(\Omega)$ or $H_0^{-s}(\Omega)$:

$$H_0^{-s}(\Omega) := H^{-s}(\Omega) := (H_0^s(\Omega))' \quad (s \ge 0).$$

The norm of $H^{-s}(\Omega)$ according to (1) reads:

$$|u|_{-s} := \sup\{|(u,v)_{L^2(\Omega)}|/|v|_s : 0 \ne v \in H_0^s(\Omega)\}, \quad \text{for } s \ge 0$$

where $(u,v)_{L^2(\Omega)}$ is the dual form on $H_0^s(\Omega) \times H^{-s}(\Omega)$ (cf. Remark 12).

Remark 6.3.14. (a) Let $\Omega = \mathbb{R}^n$. The norm dual to $|\cdot|\hat{}_s$,

$$|u|\hat{}_{-s} := \sup\{|(u,v)_0|/|v|\hat{}_s : 0 \ne v \in H^s(\mathbb{R}^n)\},$$

is equivalent to $|\cdot|_{-s}$ and has the representation (2.16b) with $-s$ instead of s.
(b) The Fourier transform shows $D^\alpha \in L(H^s(\mathbb{R}^n), H^{s-|\alpha|}(\mathbb{R}^n))$ for all $s \in \mathbb{R}$.
(c) $au \in H^s(\Omega)$, if $u \in H^s(\Omega), a \in C^t(\overline{\Omega})$, where $t = |s| \in \mathbb{N} \cup \{0\}$ or $t > |s|$.

6.4 Compact Operators

Definition 6.4.1. A subset K of a Banach space is said to be precompact [compact] if each sequence $x_i \in K$ $(i \in \mathbb{N})$ contains a convergent subsequence x_{i_k} [and $\lim_{k\to\infty} x_{i_k} \in K$].

Another definition of compactness reads: Each open covering of K already contains a finite covering of K. Both definitions are equivalent in metric spaces (cf. Dieudonné [1, (3.16.1)]). For the terms "relatively compact" and "precompact" see also Dieudonné [1, (3.17.5)].

Remark 6.4.2. (a) $K \subset \mathbb{R}^n$ is precompact [compact] if and only if K is bounded [and complete].
(b) Let X be a Banach space. The unit sphere $\{x \in X: \|x\| \le 1\}$ is compact if and only if $\dim(X) < \infty$.

Definition 6.4.3. Let X and Y be Banach spaces. The mapping $T \in L(X, Y)$ is said to be compact if $\{Tx: x \in X, \|x\|_X \le 1\}$, the image of the unit sphere in X, is precompact in Y.

Exercise 6.4.4. When is the identity $I \in L(X, X)$ compact?

Lemma 6.4.5. *Let X, Y, and Z be Banach spaces.* (a) *Let one of the mappings $T_1 \in L(X, Y)$, $T_2 \in L(Y, Z)$ be compact. Then $T_2T_1 \in L(X, Z)$ is also compact.*
(b) *$T \in L(X, Y)$ is compact if and only if $T' \in L(Y', X')$ is compact.*

PROOF. (a) Let $K_1 := \{x \in X: \|x\|_X \le 1\}$. If T_1 is compact, i.e., $T_1(K_1)$ is precompact, then $T_2(T_1(K_1))$ is also precompact and thus T_2T_1 is compact. If, however, T_2 is compact, one proves the assertion as follows. Since scaling does not change compactness, $\|T_1\|_{Y\leftarrow X} \le 1$ can be assumed without loss of generality. Hence $T_1(K_1)$ is a subset of the unit sphere in Y and therefore $T_2(T_1(K_1))$ is precompact. (b) cf. Yosida [1, §X]. ■

A special type of compact mapping is a compact embedding:

Definition 6.4.6. Let $X \subset Y$ be a continuous embedding. X is said to be compactly embedded in Y if the inclusion $I \in L(X, Y)$, $Ix = x$ is compact.

Together with Definitions 1 and 3 one obtains: $X \subset Y$ is compactly embedded if every sequence $x_1 \in X$ with $\|x_i\|_X \le 1$ contains a subsequence convergent in Y.

Example 6.4.7. Let Ω be bounded. $C^s(\overline{\Omega}) \subset C^0(\overline{\Omega})$ is a compact embedding for $s > 0$.

PROOF. Functions $u_i \in C^s(\overline{\Omega})$ with $\|u_i\|_{C^s(\overline{\Omega})} \le 1$ are equi-continuous and uniformly bounded. The assertion is based on the theorem of Arzelà-Ascoli (cf. Yosida [1, §III.3]). ∎

Analogous results can be obtained for Sobolev spaces (cf. Adams [1, p. 144], Wloka [1, §7]):

Theorem 6.4.8. *Let $\Omega \subset \mathbb{R}^n$ be open and bounded.* (a) *The embeddings $H_0^s(\Omega) \subset H_0^t(\Omega)$ $(s, t \in \mathbb{R},\ s > t)$ are compact.*
(b) *Further, let $\Omega \in C^{0,1}$. The embeddings $H^k(\Omega) \subset H^l(\Omega)$ $(k, l \in \mathbb{N} \cup \{0\},$ $k > l)$ are compact.*
(c) *Let $0 \le t < s$ and $\Omega \in C^r$ $(r > t, r > 1)$ or $\Omega \in C^{k,1}$ $(k + 1 > t)$. Then the embedding $H^s(\Omega) \subset H^t(\Omega)$ is compact.*

Remark 6.4.9. In Theorem 8b one can replace $\Omega \in C^{0,1}$ by the "uniform cone property" (cf. Wloka [1, §2.1]). To ensure $\Omega \in C^{0,1}$ it is sufficient that the boundary $\partial\Omega$ is piecewise smooth and the inside angles of possible corners are smaller than 2π.

Indented corners (cf. Figure 2.1.1) are thus permitted while a cut domain (cf. Figure 5.2.1b) is excluded.

In Section 6.5 the following situation will arise:

$$V \subset U \subset V' \quad \text{Gelfand triple}\,, \quad T \in L(V', V). \tag{6.4.1}$$

Because of the continuous embeddings, T also belongs to $L(V', V')$, $L(U, U)$, $L(V, V)$, and $L(U, V)$.

Theorem 6.4.10. *Let* (1) *hold. Let $V \subset U$ be a compact embedding. Then $T \in L(V', V')$, $T \in L(U, U)$, $T \in L(V, V)$, $T \in L(V', U)$, and $T \in L(U, V)$ are compact.*

PROOF. As an example, let us do $T \in L(U, V)$. Since the inclusion $I \in L(V, U)$ is compact we see $I \in L(U, V')$ is also compact (cf. Lemma 5b). $T \in L(U, V)$, as the product of the compact mapping $I \in L(U, V')$ with $T \in L(V', V)$, is compact (cf. Lemma 5a). ∎

Exercise 6.4.11. Let $\dim X < \infty$ or $\dim Y < \infty$. Show that $T \in L(X, Y)$ is compact.

The significance of compact operators $T \in L(X, X)$ lies in the fact that the equation $Tx - \lambda x = y$ (with $x, y \in X$, y given, x sought) has properties analogous to the finite-dimensional case.

Theorem 6.4.12. (Riesz-Schauder theory) *Let X be a Banach space; let $T \in L(X, X)$ be compact.*
(a) *For each $\lambda \in \mathbb{C}\backslash\{0\}$ one of the following alternatives holds:*

$$\text{(i) } (T - \lambda I)^{-1} \in L(X, X) \quad \text{or} \quad \text{(ii) } \lambda \textit{ is an eigenvalue.}$$

In case (i) *the equation $Tx - \lambda x = y$ has a unique solution $x \in X$ for all $y \in X$. In case* (ii) *there exists a finite-dimensional eigenspace $E(\lambda, T) :=$ kernel $(T - \lambda I) \neq \{0\}$. All $x \in E(\lambda, T)\backslash\{0\}$ solve the eigenvalue problem $Tx = \lambda x$, $x \neq 0$.*
(b) *The spectrum $\sigma(T)$ of T consists by definition of all eigenvalues and, if not $T^{-1} \in L(X, X)$, $\lambda = 0$. There exist at most countably many eigenvalues which can only accumulate at zero. $\lambda \in \sigma(T)$ if and only if $\overline{\lambda} \in \sigma(T')$. Furthermore, $\dim(E(\lambda, T)) = \dim(E(\overline{\lambda}, T')) < \infty$.*
(c) *For $\lambda \in \sigma(T)\backslash\{0\}$, $Tx - \lambda x = y$ has at least one solution $x \in X$ if and only if $\langle y, x' \rangle_{X \times X'} = 0$ for all $x' \in E(\overline{\lambda}, T')$.*

In Lemma 6.5.18 we need

Lemma 6.4.13. *Let $X \subset Y \subset Z$ be continuously embedded Banach spaces and let $X \subset Y$ be compactly embedded. Then for every $\epsilon > 0$ there exists a C_ϵ such that*

$$||x||_Y \leq \epsilon ||x||_X + C_\epsilon ||x||_Z \qquad \textit{for all } x \in X. \tag{6.4.2}$$

PROOF. Let $\epsilon > 0$ be fixed. The negation of (2) reads: There exists $x_i \in X$ with $(||x_i||_Y - \epsilon ||x_i||_X)/||x_i||_Z \to \infty$. For $y_i := (\epsilon ||x_i||_X)^{-1} x_i \in X$ we thus have $(||y_i||_Y - 1)/||y_i||_Z \to \infty$. From this one infers $||y_i||_Z \to 0$ and $||y_i||_Y > 1$ for sufficiently large i. Since $||y_i||_X \leq 1/\epsilon$ and $X \subset Y$ is a compact embedding, a subsequence y_{i_k} converges to $y^* \in Y$. Now, $||y_i||_Y > 1$ implies $||y^*||_Y \geq 1$, i.e., $y^* \neq 0$. On the other hand, y_{i_k} also converges in Z to y^* since $Y \subset Z$ is continuously embedded. $||y_i||_Z \to 0$ gives the contradiction sought: $y^* = 0$. ∎

6.5 Bilinear Forms

In the following let us assume that V is a Hilbert space. The mapping $a(\cdot, \cdot): V \times V \to \mathbb{R}$ is called a bilinear form if

$$a(x, y + \lambda z) = a(x, y) + \lambda a(x, z), \qquad a(x + \lambda y, z) = a(x, z) + \lambda a(y, z)$$

for $(x, y, z \in V, \lambda \in \mathbb{R})$ (in the complex case one speaks of sesquilinear forms: $a(x, \lambda y) = \overline{\lambda} a(x, y)$). $a(\cdot, \cdot)$ is said to be continuous (or bounded) if there exists a C_S such that

$$|a(x, y)| \le C_S ||x||_V ||y||_V \quad \text{for all } x, y \in V. \tag{6.5.1}$$

Lemma 6.5.1. (a) *To a continuous bilinear form one can assign a unique operator $A \in L(V, V')$ such that*

$$a(x, y) = \langle Ax, y\rangle_{V' \times V} \quad \textit{for all } x, y \in V, \quad ||A||_{V' \leftarrow V} \le C_S. \tag{6.5.2}$$

(b) *Let V_1 and V_2 be dense in V. Let $a(\cdot, \cdot)$ be defined on $V_1 \times V_2$ and satisfy* (1) *with "$x \in V_1, y \in V_2$" instead of "$x, y \in V$". Then $a(\cdot, \cdot)$ can be extended uniquely to $V \times V$ so that* (1) *holds with the same C_S for all $x, y \in V$.*

PROOF. (a) Let $x \in V$ be fixed. $\varphi_x(y) := a(x, y)$ defines a functional $\varphi_x \in V'$ with $||\varphi_x||_{V'} \le C_S ||x||_V$. One sets $Ax := \varphi_x$ for $x \in V$. $||Ax||_{V'} \le C_S ||x||_V$ proves $||A||_{V' \leftarrow V} \le C_S$. The definitions show $\langle Ax, y\rangle_{V' \times V} = \langle \varphi_x, y\rangle_{V' \times V} = \varphi_x(y) = a(x, y)$. Conversely, for each $A \in L(V, V')$, $a(x, y) := \langle Ax, y\rangle_{V' \times V}$ is also a bilinear form with $\langle Ax, y\rangle_{V' \times V} \le ||Ax||_{V'} ||y||_V \le ||A||_{V' \leftarrow V} ||x||_V ||y||_V$.
(b) According to Theorem 6.1.11, A is also uniquely determined if $a(\cdot, \cdot)$ is only given on $V_1 \times V_2$. Then $\langle Ax, y\rangle_{V' \times V}$ represents the extension. ■

The proof shows

$$||A||_{V' \leftarrow V} = \sup\{|a(x, y)| : x, y \in V, ||x||_V = ||y||_V = 1\}. \tag{6.5.3}$$

A is called the operator that is associated to $a(\cdot, \cdot)$. The bilinear form $a^*(\cdot, \cdot)$ adjoint to $a(\cdot, \cdot)$ is given by $a^*(x, y) := a(y, x)$ $(x, y \in V)$. The bilinear form is said to be symmetric if $a(\cdot, \cdot) = a^*(\cdot, \cdot)$.

Exercise 6.5.2. Show that (a) If A belongs to $a(\cdot, \cdot)$, then A' belongs to $a^*(\cdot, \cdot)$. (b) If $a(\cdot, \cdot)$ is symmetric, then $A = A'$.

Lemma 6.5.3. *Let $A \in L(V, V')$ be the operator associated to a continuous bilinear form $a(\cdot, \cdot)$. Then the following statements* (i), (ii), (iii) *are equivalent:*
(i) *$A^{-1} \in L(V', V)$ exists;*
(ii) *$\epsilon, \epsilon' > 0$ exist such that*

$$\inf\{\sup\{|a(x, y)| : y \in V, ||y||_V = 1\} : x \in V, ||x||_V = 1\} = \epsilon > 0, \tag{6.5.4a}$$

$$\inf\{\sup\{|a(x, y)| : x \in V, ||x||_V = 1\} : y \in V, ||y||_V = 1\} = \epsilon' > 0; \tag{6.5.4b}$$

(iii) *the inequalities* (4a) *and* (4c) *hold:*

$$\sup\{|a(x, y)| : x \in V, ||x||_V = 1\} > 0. \tag{6.5.4c}$$

If one of the statements (i)–(iii) *holds, then*

$$\epsilon = \epsilon' = 1/\|A^{-1}\|_{V\leftarrow V'} \quad (\epsilon, \epsilon' \text{ from (4a,b)}). \tag{6.5.4d}$$

From (4a) *follows*

$$\inf\{\sup\{|a(x,y)|: y \in V, \|y\|_V = 1\}: x \in V, \|x\|_V = 1\} \geq \epsilon > 0. \tag{6.5.4e}$$

Conversely, (4a) *follows from* (4e) *with a possibly larger* $\epsilon > 0$. (4e) *and* (4c) *are also called the* *B a b u š k a c o n d i t i o n s.* (4e) *is equivalent to*

$$\sup\{|a(x,y)|: y \in V, \|y\|_V = 1\} \geq \epsilon \|x\|_V \quad \text{for all } x \in V, \tag{6.5.4e'}$$

because (4e) *is equal to* (4e′) *for all* $x \in V$, $\|x\|_V = 1$. *The scaling condition* $\|x\|_V = 1$ *can evidently be dropped. The left-hand side in* (4e′) *agrees with the definition of the dual norm of* Ax *so that* (4e) *and* (4e′) *are also equivalent to* (4e″)*:*

$$\|Ax\|_{V'} \geq \epsilon \|x\|_V \quad \text{for all } x \in V. \tag{6.5.4e''}$$

PROOF. (a) "(i) ⇒ (ii)": Let $A^{-1} \in L(V', V)$ exist. Then (4a) follows from

$$\begin{aligned}
\inf\{\ldots\} &= \inf_{\substack{x\in V\\x\neq 0}} \sup_{\substack{y\in V\\y\neq 0}} \frac{|a(x,y)|}{\|x\|_V\|y\|_V} = \inf_{\substack{x\in V\\x\neq 0}} \sup_{\substack{y\in V\\y\neq 0}} \frac{|\langle Ax,y\rangle|}{\|x\|_V\|y\|_V} \\
&= \inf_{\substack{x'\in V'\\x'\neq 0}} \sup_{\substack{y\in V\\y\neq 0}} \frac{|\langle AA^{-1}x',y\rangle|}{\|A^{-1}x'\|_V\|y\|_V} = \inf_{\substack{x'\in V'\\x'\neq 0}} \|A^{-1}x'\|_V^{-1} \sup_{\substack{y\in V\\y\neq 0}} \frac{|\langle x',y\rangle|}{\|y\|_V} \\
&= \inf_{\substack{x'\in V'\\x'\neq 0}} \|A^{-1}x'\|_V^{-1}\|x'\|_{V'} = 1/[\sup_{\substack{x'\in V'\\x'\neq 0}} \|A^{-1}x'\|_V/\|x'\|_{V'}] \\
&= 1/\|A^{-1}\|_{V\leftarrow V'} =: \epsilon.
\end{aligned}$$

In the same way one shows (4b) with $\epsilon' = 1/\|A'^{-1}\|_{V\leftarrow V'}$. Because $A'^{-1} = (A^{-1})'$, (3.3), and $V'' = V$, it follows that $\epsilon = \epsilon'$.
(b) "(ii) ⇒ (iii)": (4c) is a weakening of (4b).
(c) "(iii) ⇒ (i)": $\epsilon > 0$ in (4a) proves that A is injective. We wish to show that the image $W := \{Ax: x \in V\} \subset V'$ is closed. For a sequence $\{w_\nu\}$ with $\|w^* - w_\nu\|_{V'} \to 0$ we must therefore show that $w^* \in W$. According to the definition of W there exists $x_\nu \in V$ with $Ax_\nu = w_\nu$. From (4a) one infers via (4e) and (4e″) (with $x := x_\nu - x_\mu$) that $\|x_\nu - x_\mu\|_V \leq \|w_\nu - w_\mu\|_{V'}/\epsilon$. Since $\{w_\nu\}$ is Cauchy convergent, this property carries over to $\{x_\nu\}$. There exists an $x^* \in V$ with $x_\nu \to x^*$ in V. The continuity of $A \in L(V, V')$ proves $w_\nu = Ax_\nu \to Ax^*$ so that $w^* = Ax^* \in W$. According to Lemma 6.1.17 one can decompose V' into $W \oplus W^\perp$. If A were not surjective (thus $W \neq V'$), there would exist a $w \in W^\perp$ with $w \neq 0$. Then $y := J_{V'}w = J_V^{-1} \in V$ would satisfy $y \neq 0$ (cf. Theorem 6.3.6, Corollary 6.3.7). Since $a(x,y) = \langle Ax, y\rangle_{V'\times V} = (Ax, w)_{V'} = 0$ for all $x \in V$, a contradiction to (4c) would result. Therefore, A is also surjective, and Theorem 6.1.13 proves $A^{-1} \in L(V', V)$.
(d) Statement (4d) has already resulted from Part a of the proof. ■

It will be shown that for interesting cases conditions (4a) and (4b) are equivalent (cf. Lemma 17). A particularly simple case is in

Exercise 6.5.4. Show that if $\dim V < \infty$, then (4a) implies the statement (4b) with $\epsilon' = \epsilon$ and conversely.

Definition 6.5.5. A bilinear form is said to be V-elliptic if it is continuous on $V \times V$ and there is a constant C_E such that

$$a(x,x) \geq C_E \|x\|_V^2 \quad \text{for all } x \in V. \tag{6.5.5}$$

Exercise 6.5.6. Show that: (a) If $W \subset V$ is a Hilbert subspace with the norm equal (or equivalent) to V, then a V-elliptic bilinear form is also W-elliptic.
(b) Let $a(\cdot,\cdot): V \times V \to \mathbb{R}$ be continuous. If $a(x,x) \geq C_E \|x\|_V^2$ holds for all $x \in V_0$ where V_0 is dense in V, then (5) follows with the same C_E.
(c) Let $a(\cdot,\cdot): V \times V \to \mathbb{R}$ be continuous, symmetric, nonnegative ($a(v,v) \geq 0$) and satisfy (4a,c). Then $a(\cdot,\cdot)$ is V-elliptic with $C_E \geq \epsilon^2/C_S$ (ϵ from (4a,b), C_S from (1)). Hint: First prove that $|a(u,v)| \leq [a(u,u)a(v,v)]^{1/2}$ (cf. Exercise 6.1.16).

Lemma 6.5.7. *V-ellipticity* (5) *implies* (1) *and* (4a,b) *with* $\epsilon = \epsilon' \geq C_E$.

PROOF. Let $x \in V$, $\|x\|_V = 1$. $\sup\{|a(x,y)| : y \in V, \|y\|_V = 1\} \geq |a(x,x)| \geq C_E$ proves (4a) with $\epsilon \geq C_E$. (4b) follows analogously. ∎

The combination of Lemmata 1, 3, 7 together with $\|A'^{-1}\|_{V \leftarrow V'} = \|A^{-1}\|_{V \leftarrow V'}$ (cf. Lemma 6.3.2) proves

Theorem 6.5.8. *Let the bilinear form be V-elliptic [or satisfy* (1), (4a,c)*]. Then the corresponding operator A satisfies the conditions*

$$\begin{aligned} &A \in L(V,V'), && \|A'\|_{V' \leftarrow V} = \|A\|_{V' \leftarrow V} \leq C_S, \\ &A^{-1} \in L(V',V), && \|A'^{-1}\|_{V \leftarrow V'} = \|A^{-1}\|_{V \leftarrow V'} \leq C' \end{aligned} \tag{6.5.6}$$

with C_S from (1) *and $C' = 1/C_E$ [resp. $C' = 1/\epsilon = 1/\epsilon'$].*

With the help of the bilinear form $a(\cdot,\cdot)$ and a functional $f \in V'$ one can formulate the following problem:

$$\text{Find } x \in V \text{ with } a(x,y) = f(y) \text{ for all } y \in V. \tag{6.5.7}$$

According to Lemma 1 one can write (7) in the form $\langle Ax - f, y\rangle_{V' \times V} = 0$ for all $y \in V$, i.e., $Ax = f$ in V'. The equation $Ax = f$ is solvable for $f \in V'$ if and only if $A^{-1} \in L(V',V)$. Hence one obtains

Theorem 6.5.9. *Let the bilinear form be continuous* (*cf.* (1)) *and satisfy the stability condition* (4a,c) *[it is sufficient that $a(\cdot,\cdot)$ is V-elliptic]. Then Problem*

(7) has exactly one solution $x := A^{-1}f$. *This satisfies* $||x||_V \le C||f||_{V'}$ *with* $C = 1/\epsilon = 1/\epsilon'$ [*resp.* $C = 1/C_E$].

Corollary 6.5.10. *Under the assumptions of Theorem 9 the analogous statement holds with the same estimate for the* adjoint problem

$$\text{Find} \quad x^* \in V \quad \text{with} \quad a^*(x^*, y) = f(y) \quad \text{for all} \quad y \in V. \tag{6.5.8}$$

PROOF. $||A^{-1}||_{V \leftarrow V'} = ||A'^{-1}||_{V \leftarrow V'}$ (cf. Exercise 2a, Theorem 8). ∎

Exercise 6.5.11. Let $a(\cdot,\cdot)\colon V \times V \to \mathbb{R}$ be continuous. Let V_0 be dense in V. Show that the solution $x \in V$ of problem (7) is already uniquely determined by "$a(x, y) = f(y)$ for all $y \in V_0$". The same holds for (8).

Problem (7) may be equivalent to a variational problem:

Theorem 6.5.12. *Let* $a(\cdot,\cdot)$ *be* V*-elliptic and symmetric; furthermore, let* $f \in V'$. *Then*

$$J(x) := a(x, x) - 2f(x) \quad (x \in V) \tag{6.5.9}$$

assumes its unique minimum for the solution x *of Equation* (7).

PROOF. Let x be the solution of (7). For arbitrary $z \in V$ set $y := z - x$. From

$$\begin{aligned} J(z) &= J(x+y) = a(x+y, x+y) - 2f(x+y) \\ &= a(x,x) + a(x,y) + a(y,x) + a(y,y) - 2f(x) - 2f(y) \\ &= J(x) + a(y,y) + 2[a(x,y) - f(y)] \\ &= J(x) + a(y,y) \ge J(x) + C_E||y||_V^2 = J(x) + C_E||z-x||_V^2, \end{aligned}$$

one can read $J(z) > J(x)$ for all $z \neq x$. ∎

The term "V-elliptic" seems to indicate that to elliptic boundary value problems correspond V-elliptic bilinear forms. In general this is not the case. Rather, V-coercive forms will be assigned to the elliptic boundary value problems. Their definition necessitates the introduction of a Gelfand triple (cf. (6.3.7)):

$$V \subset U \subset V' \quad (U = U', V \subset U \text{ continuous and densely embedded}).$$

Definition 6.5.13. Let $V \subset U \subset V'$ be a Gelfand triple. A bilinear form $a(\cdot,\cdot)$ is said to be V-coercive if it is continuous and if there exists $C_K \in \mathbb{R}$ and $C_E > 0$ such that

$$a(x,x) \ge C_E||x||_V^2 - C_K||x||_U^2 \quad \text{for all } x \in V \quad \text{with } C_E > 0. \tag{6.5.10}$$

Exercise 6.5.14. Set $\tilde{a}(x,y) := a(x,y) + C_K(x,y)_U$ with C_K from (10). Let $I: V \to V'$ be the inclusion. Show that (a) the coercivity condition (10) is equivalent to the V-ellipticity of $\tilde{a}$.
(b) If $A \in L(V,V')$ is associated to $a(\cdot,\cdot)$, then so is $\tilde{A} := A + C_k I$ to $\tilde{a}(\cdot,\cdot)$. Why does $\tilde{A} \in L(V,V')$ hold?

The results of Riesz-Schauder theory (Theorem 6.4.12) transfer to A as soon as the embedding $V \subset U$ is not only continuous but also compact.

Theorem 6.5.15. *Let $V \subset U \subset V'$ be a Gelfand triple with compact embedding $V \subset U$. Let the bilinear form $a(\cdot,\cdot)$ be V-coercive with corresponding operator A. Let $I: V \to V'$ be the inclusion.*
(a) *For each $\lambda \in \mathbb{C}$ one of the following alternatives holds:*

$$\text{(i)}\quad (A - \lambda I)^{-1} \in L(V',V) \quad \textit{and} \quad (A' - \overline{\lambda} I)^{-1} \in L(V',V),$$
$$\text{(ii)}\quad \lambda \textit{ is an eigenvalue.}$$

In case (i) *$Ax - \lambda x = f$ and $A'x^* - \overline{\lambda}x^* = f$ are uniquely solvable for all $f \in V'$ (i.e., $a(x,y) - \lambda(x,y)_U = f(y)$ and $a^*(x^*,y) - \overline{\lambda}(x^*,y)_U = f(y)$ for all $y \in V$). In case* (ii) *there exist finite-dimensional eigenspaces $\{0\} \neq E(\lambda) :=$ kernel $(A - \lambda I)$ and $\{0\} \neq E'(\lambda) :=$ kernel $(A' - \overline{\lambda} I)$ such that*

$$Ax = \lambda x \quad \textit{for} \quad x \in E(\lambda), \quad \textit{i.e.,} \quad a(x,y) = \lambda(x,y)_U \quad \textit{for all } y \in V, \tag{6.5.11a}$$

$$A'x^* = \overline{\lambda}^* \quad \textit{for} \quad x^* \in E'(\overline{\lambda}), \quad \textit{i.e.,} \quad a^*(x^*,y) = \overline{\lambda}(x^*,y)_U \quad \textit{for all} \quad y \in V. \tag{6.5.11b}$$

(b) *The spectrum $\sigma(A)$ of A consists of at most countably many eigenvalues which cannot accumulate in $\mathbb{C}$. $\lambda \in \sigma(A)$ if and only if $\overline{\lambda} \in \sigma(A')$. Furthermore* $\dim E(\lambda) = \dim E'(\overline{\lambda}) < \infty$.
(c) *For $\lambda \in \sigma(A)$, $Ax - \lambda x = f \in V'$ has at least one solution $x \in V$ if and only $f \perp E'(\lambda)$, i.e., $\langle f, x^* \rangle_{V' \times V} = (f, x^*)_U = 0$ for all $x^* \in E'(\lambda)$.*

PROOF. With $V \subset U$, $V \subset V'$ is also a compact embedding, i.e., the inclusion $I: V \to V'$ is compact. $A + C_K I$ with C_K from (10) satisfies $A + C_K I \in L(V,V')$, $(A + C_K I)^{-1} \in L(V',V)$ (cf. Exercise 14 a). Lemma 6.5.4 shows that $K := (A + C_K I)^{-1} I: V \to V$ is compact. Hence the Riesz-Schauder theory is applicable to $K - \mu I$. Since

$$K - \mu I = -\mu(I - \mu^{-1}K) = -\mu(A + C_K I)^{-1}\{A + C_K I - \mu^{-1} I\}$$
$$= -\mu(A + C_K I)^{-1}(A - \lambda I)$$

with $\lambda = 1/\mu - C_K$, the statements of Theorem 6.4.12 transfer via $K - \mu I$ to $A - \lambda I = -\mu^{-1}(A + C_K I)\,(K - \mu I)$. ■

Remark 6.5.16. The spectrum $\sigma(A)$ has measure zero so that under the conditions of Theorem 15 the solvability of $Ax - \lambda x = f$ is guaranteed for almost all λ. Problem (7) is solvable if not "accidentally" $0 \in \sigma(A)$.

Lemma 6.5.17. *Under the conditions of Theorem* 15 *the inequalities* (4a,b) *are equivalent.*

PROOF. (4a) proves that A is injective, i.e., $0 \notin \sigma(A)$. Theorem 15a shows $A^{-1} \in L(V', V)$ so that (4b) follows from Lemma 3. ∎

Evidently $a(\cdot,\cdot)$ remains V-coercive if one adds a multiple of $(\cdot,\cdot)_U$. Generally, there holds

Lemma 6.5.18. *Let $a(\cdot,\cdot)$ be V-coercive where $V \subset U \subset V'$. Then $a(\cdot,\cdot) + b(\cdot,\cdot)$ is also V-coercive if the bilinear form $b(\cdot,\cdot)$ satisfies one of the following conditions:* (a) *for every $\epsilon > 0$ exists C_ϵ such that*

$$|b(x,x)| \le \epsilon \|x\|_V^2 + C_\epsilon \|x\|_U^2 \quad \text{for all } x \in V. \tag{6.5.12a}$$

(b) *Let the embeddings $V \subset X$ and $V \subset Y$ be continuous, with at least one of them compact. Let the following hold:*

$$|b(x,x)| \le C_B \|x\|_X \|x\|_Y \quad \text{for all } x \in V. \tag{6.5.12b}$$

(c) *Let the embeddings $V \subset X$, $V \subset Y$ be continuous. Let* (12b) *hold. For $\|\cdot\|_X$ or $\|\cdot\|_Y$ assume that for every $\epsilon > 0$ there exists a C'_ϵ such that*

$$\|x\|_X \le \epsilon \|x\|_V + C'_\epsilon \|x\|_U \quad \text{or} \quad \|x\|_Y \le \epsilon \|x\|_V + C'_\epsilon \|x\|_U \quad \text{for } x \in V. \tag{6.5.12c}$$

PROOF. (a) Select $\epsilon = C_E/2$ with C_E from (10). Then $a(\cdot,\cdot) + b(\cdot,\cdot)$ satisfies the V-coercivity condition with $C_E/2 > 0$ and $C_K + C_\epsilon$ instead of C_E and C_K.
(b) Lemma 6.4.13 proves (12c).
(c) Let the first inequality from (12c) hold, for example. Since the embedding $V \subset Y$ is continuous, C_Y exists with $\|x\|_Y \le C_Y \|x\|_V$. Choose $\epsilon' = \epsilon/(2C_B C_Y)$ in (12c):

$$|b(x,x)| \le C_B(\epsilon' \|x\|_V + C'_{\epsilon'} \|x\|_U) C_Y \|x\|_V \le \frac{\epsilon}{2} \|x\|_V^2 + K \|x\|_V \|x\|_U$$

with $K = C_B C_Y C'_{\epsilon'}$. Since $K \|x\|_V \|x\|_U \le \frac{\epsilon}{2} \|x\|_V^2 + \frac{1}{2} K^2 \epsilon^{-1} \|x\|_U^2$ (cf. (5.3.10)), (12a) follows. ∎

7 Variational Formulation

7.1 Historical Remarks

In the preceding chapters it was not possible to establish even for the Dirichlet problem of the potential equation (2.1.1a,b) whether, or under what conditions, a classical solution $u \in C^2(\Omega) \cap C^0(\overline{\Omega})$ exists. Green [1] took the view that his Green's function, described in 1828, always exists and that it provides the solution explicitly. This is not the case. Lebesgue proved in 1913 that for certain domains the Green function does not exist.

Thomson (1847), Kelvin (1847), and Dirichlet offered a different line of reasoning. The Dirichlet integral

$$I(u) := \int_\Omega |\nabla u(\mathrm{x})|^2 d\mathrm{x} = \int_\Omega \sum_{i=1}^n u_{x_i}^2(\mathrm{x}) d\mathrm{x} \tag{7.1.1}$$

describes the energy in physics. With boundary values $u = \varphi$ on Γ given, one seeks to minimise $I(u)$. This variational problem is equivalent to

$$I(u,v) := \int_\Omega \langle \nabla u(\mathrm{x}), \nabla v(\mathrm{x})\rangle d\mathrm{x} = 0 \quad \text{for all} \quad v \text{ with } v = 0 \text{ on } \Gamma. \tag{7.1.2}$$

The proof of the equivalence results from $I(u+v) = I(u) + 2I(u,v) + I(v)$ and $I(v) \geq 0$ for all v (cf. Theorem 6.5.12). Green's formula (2.2.5a) provides $I(u,v) = \int_\Omega v\Delta u \, d\mathrm{x} = 0$ for all v with $v = 0$ on Γ such that $\Delta u = 0$ follows. Thus, like (2), the variational problem $I(u) = \min$ is equivalent to the Dirichlet problem $\Delta u = 0$ in Ω, $u = \varphi$ on Γ.

The so-called Dirichlet principle states that $I(u)$, since it is bounded from below by $I(u) \geq 0$, must take a minimum for some u. According to the above considerations this would ensure the existence of a solution of the Dirichlet problem.

In 1870, Weierstraß argued against this line of reasoning, stating that while there may exist an infimum of $I(u)$ over $\{u \in C^2(\Omega) \cap C^0(\overline{\Omega}): u = \varphi \text{ on } \Gamma\}$ it need not necessarily be in this set. For example, the integral $J(u) := \int_0^1 u^2(x)dx$ in $\{u \in C^0([0,1]): u(0) = 0, u(1) = 1\}$ never takes the value $\inf J(u) = 0$.

Further, the following example due to Hadamard shows that no finite infimum of the Dirichlet integral need exist. Let r and φ be the polar coordinates in the circle $\Omega = K_1(0)$. The function $u(r,\varphi) = \sum_{n=1}^\infty r^{n!} n^{-2} \sin(n!\varphi)$ is harmonic in Ω but the integral $I(u)$ does not exist.

The above difficulties disappear if one seeks the solutions in the more suitable Sobolev spaces instead of in $C^2(\Omega) \cap C^0(\overline{\Omega})$.

7.2 Equations with Homogeneous Dirichlet Boundary Conditions

In the following we investigate the elliptic equation

$$Lu = g \quad \text{in } \Omega, \tag{7.2.1a}$$

$$L = \sum_{|\alpha|\le m} \sum_{|\beta|\le m} (-1)^{|\beta|} D^\beta a_{\alpha\beta}(\mathbf{x}) D^\alpha \tag{7.2.1b}$$

of order 2m (cf. Section 5.3; Exercise 5.3.5d). The principal part of L is

$$L_0 = (-1)^m \sum_{|\alpha|=|\beta|=m} D^\beta a_{\alpha\beta}(\mathbf{x}) D^\alpha. \tag{7.2.2}$$

According to Definition 5.3.4, L is uniformly elliptic in Ω if there exists $\epsilon > 0$ such that

$$\sum_{|\alpha|=|\beta|=m} a_{\alpha\beta}(\mathbf{x})\xi^{\alpha+\beta} \ge \epsilon|\xi|^{2m} \quad \text{for all } \mathbf{x} \in \Omega,\ \xi \in \mathbb{R}^n. \tag{7.2.3}$$

Attention. In the case that only $a_{\alpha\beta} \in L^\infty(\overline{\Omega})$ is assumed, one needs to replace "for all $\mathbf{x} \in \Omega$" by "for almost all $\mathbf{x} \in \Omega$".

We assume the homogeneous Dirichlet boundary conditions

$$u = 0,\ \frac{\partial u}{\partial n} = 0,\ \left(\frac{\partial}{\partial n}\right)^2 u = 0, \cdots, \left(\frac{\partial}{\partial n}\right)^{m-1} u = 0 \quad \text{on } \Gamma, \tag{7.2.4}$$

which are only meaningful if $\Gamma = \partial\Omega$ is sufficiently smooth. Note that in the standard case $m = 1$ (an equation of second order) condition (4) becomes $u = 0$.

Since with $u = 0$ on Γ the tangential derivatives also vanish, not only the kth normal derivatives ($k \le m-1$) but also all the derivatives of order $\le m-1$ are equal to zero:

$$D^\alpha u(\mathbf{x}) = 0 \quad \text{in } \mathbf{x} \in \Gamma \quad \text{for } |\alpha| \le m-1. \tag{7.2.4'}$$

Condition (4′) no longer requires the existence of a normal direction. According to Corollary 6.2.43, (4′) can also be formulated as

$$u \in H_0^m(\Omega). \tag{7.2.4''}$$

Let $u \in C^{2m}(\Omega) \cap H_0^m(\Omega)$ be a classical solution of (1a) and (4). To derive the variational formulation we take an arbitrary $v \in C_0^\infty(\Omega)$ and consider

$$(Lu, v)_0 = \sum_{\alpha,\beta} (-1)^{|\beta|} \int_\Omega D^\beta (v a_{\alpha\beta} D^\alpha u)\, d\mathbf{x}.$$

Since $v \in C_0^\infty(\Omega)$, the integrand vanishes in the proximity of Γ so that one can integrate by parts:

$$(-1)^{|\beta|} \int_\Omega D^\beta (v a_{\alpha\beta} D^\alpha u)\, d\mathbf{x} = \int_\Omega a_{\alpha\beta} (D^\alpha u)(D^\beta v)\, d\mathbf{x},$$

without boundary terms occurring. Thus we have found the variational formulation

$$\sum_{|\alpha|,|\beta| \le m} \int_\Omega a_{\alpha\beta}(\mathbf{x})(D^\alpha u(\mathbf{x}))(D^\beta v(\mathbf{x}))\, d\mathbf{x} = \int_\Omega g(\mathbf{x}) v(\mathbf{x})\, d\mathbf{x} \tag{7.2.5}$$

$$\text{for all } v \in C_0^\infty(\Omega),$$

since $Lu = g$. If conversely $u \in C^{2m}(\Omega)$ with boundary conditions (4) satisfies condition (5), then the partial integration can be reversed and $\int_\Omega (g - Lu) v dx = 0$ for all $v \in C_0^\infty(\Omega)$ proves $Lu = g$. This means that a classical solution of the variational problem (5) with boundary condition (4) is also a solution of the original boundary value problem. Hence the differential equation (1a,b) and the variational formulation (5) are equivalent with respect to classical solutions.

We introduce the bilinear form

$$a(u, v) := \sum_{|\alpha|,|\beta| \le m} \int_\Omega a_{\alpha\beta}(\mathbf{x})(D^\alpha u)(D^\beta v) d\mathbf{x} \tag{7.2.6}$$

and the functional

$$f(v) := \int_\Omega g(\mathbf{x}) v(\mathbf{x}) d\mathbf{x}. \tag{7.2.7}$$

As remarked above, the boundary condition (4) for classical solutions $u \in C^{2m}(\Omega) \cap H^m(\Omega)$ means that $u \in H_0^m(\Omega)$. Thus the "variational formulation" or "weak formulation" of the boundary value problem (1), (4) reads as follows:

$$\text{find} \quad u \in H_0^m(\Omega) \quad \text{with } a(u, v) = f(v) \quad \text{for all } v \in C_0^\infty(\Omega). \tag{7.2.8}$$

A solution of problem (8) which, according to the definition, lies in $H_0^m(\Omega)$ but not necessarily in $C^{2m}(\Omega)$, is called a weak solution.

Exercise 7.2.1. (a) Let Ω be bounded. Show that any classical solution $u \in C^{2m}(\Omega) \cap C^m(\overline{\Omega})$ is also a weak solution.
(b) With the aid of Example 2.4.2 show that this statement becomes false for unbounded domains.

Theorem 7.2.2. *Let $a_{\alpha\beta} \in L^\infty(\Omega)$. The bilinear form defined by (6) is bounded on $H_0^m(\Omega) \times H_0^m(\Omega)$.*

PROOF. Let $u, v \in C_0^\infty(\Omega)$. The inequality (6.2.5c) yields

$$|a(u,v)| \le \sum_{\alpha,\beta} \|a_{\alpha\beta}\|_{L^\infty(\Omega)} |D^\alpha u|_0 |D^\beta v|_0 \le \text{const } |u|_m |v|_m.$$

Since $C_0^\infty(\Omega)$ is dense in $H_0^m(\Omega)$ (cf. Theorem 6.2.10), $a(\cdot,\cdot)$ has an extension to $H_0^m(\Omega) \times H_0^m(\Omega))$ and is bounded by the same constant (cf. Lemma 6.5.1b). ∎

The function $f(v)$ is also defined and bounded for $v \in H_0^m(\Omega)$ if, for example, $g \in L^2(\Omega)$. According to Exercise 6.5.11, the variational formulation (8) is equivalent to the following one:

$$\text{find} \quad u \in H_0^m(\Omega) \quad \text{with} \quad a(u,v) = f(v) \quad \text{for all } v \in H_0^m(\Omega). \tag{7.2.9}$$

One can regain the form $Lu = f$ by applying Lemma 6.5.1. Let $L \in L(H_0^m(\Omega), H^{-m}(\Omega))$ and $f \in H^{-m}(\Omega) = (H_0^m(\Omega))'$ be defined by $a(u,v) = \langle Lu, v\rangle_{H^{-m}(\Omega)\times H_0^m(\Omega)}$ and $f(v) = \langle f, v\rangle_{H^{-m}(\Omega)\times H_0^m(\Omega)}$ for all $v \in H_0^m(\Omega)$. Equation (9) states that

$$Lu = f. \tag{7.2.9$'$}$$

While (1a) represents an equation $Lu = g$ in $C^0(\Omega)$ (i.e., for a classical solution), (9′) is an equation in $H^{-m}(\Omega)$.

Theorem 6.5.9 guarantees unique solvability of Equation (9) if $a(\cdot,\cdot)$ is $H_0^m(\Omega)$-elliptic. We first investigate the standard case $m = 1$ (equations of order $2m = 2$).

Theorem 7.2.3. *Let Ω be bounded, $m = 1$, $a_{\alpha\beta} \in L^\infty(\Omega)$. Let L satisfy* (3) *(uniform ellipticity) and be equal to the principal part L_0, i.e., $a_{\alpha\beta} = 0$ for $|\alpha| + |\beta| \le 1$. Then the form $a(\cdot,\cdot)$ is $H_0^1(\Omega)$-elliptic:*

$$a(u,u) \ge \epsilon' |u|_1^2, \quad \epsilon' > 0. \tag{7.2.10}$$

PROOF. Since $|\alpha| = |\beta| = 1$ one can identify α and β according to $D^\alpha = \partial/\partial x_i$, $D^\beta = \partial/\partial x_j$ with indices $i, j \in \{1, \cdots, n\}$. For fixed $x \in \Omega$ use (3) with $\xi = \nabla u(\mathbf{x})$:

$$\sum_{|\alpha|,|\beta|=1} a_{\alpha\beta}(\mathbf{x})\xi^{\alpha+\beta} = \sum_{i,j=1}^{n} a_{ij}(\mathbf{x})\xi_i\xi_j \ge \epsilon|\xi|^2 = \epsilon|\nabla u(\mathbf{x})|^2.$$

Integration over Ω yields $a(u,u) \ge \epsilon \int_\Omega |\nabla u|^2\, dx$. Since $\int_\Omega |\nabla u|^2\, dx \ge C_\Omega |u|_1^2$ (cf. Lemma 6.2.11), (10) follows with $\epsilon' = \epsilon C_\Omega$. ∎

Corollary 7.2.4. *The condition "Ω* bounded*" may be dropped if for $\alpha = \beta = 0$ one assumes $a_{00}(x) \ge \eta > 0$ (instead of $a_{00} = 0$).*

Example 7.2.5. The Helmholtz equation $-\Delta u + u = f$ in Ω leads to the bilinear form

$$a(u,v) := \int_\Omega [\sum_{i=1}^n u_{x_i}(\mathbf{x})v_{x_i}(\mathbf{x}) + u(\mathbf{x})v(\mathbf{x})]\, d\mathbf{x} = \int_\Omega [\langle \nabla u, \nabla v\rangle + uv]\, d\mathbf{x}.$$

$a(u,v)$ is the scalar product in $H_0^1(\Omega)$ (and $H^1(\Omega)$). The fact that $a(u,u) = |u|_1^2$ proves the $H_0^1(\Omega)$-ellipticity.

Exercise 7.2.6. Let the assumptions of Theorem 3 or Corollary 4 be satisfied, except for the fact that the coefficients $a_{\alpha 0}$ and $a_{0\beta}$ ($|\alpha| = |\beta| = 1$) of the first derivatives are arbitrary constants. Show that inequality (10) holds unchanged.

Theorem 3 cannot easily be extended to the case $m > 1$.

Theorem 7.2.7. *Let the coefficients of the principal part be constants:* $a_{\alpha\beta} = \text{const}$ *for* $|\alpha| = |\beta| = m$. *Furthermore, assume that* $a_{\alpha\beta} = 0$ *for* $0 < |\alpha|+|\beta| \le 2m-1$, $a_{00} \ge 0$ *for* $\alpha = \beta = 0$. *Let* L *be uniformly elliptic* (*cf.* (3)). *Further let either* Ω *be bounded or* $a_{00} \ge \eta > 0$. *Then* $a(\cdot,\cdot)$ *is* $H_0^m(\Omega)$*-elliptic.*

PROOF. We continue $u \in H_0^m(\Omega)$ through $u = 0$ onto $\mathbb{R}^n$. Theorem 6.2.21, Exercise 6.2.22, and inequality (3) show that

$$\begin{aligned}
a(u,u) - \int_\Omega a_{00}u^2\, d\mathbf{x} &= \sum_{|\alpha|,|\beta|=m} \int_\Omega a_{\alpha\beta} D^\alpha u D^\beta u\, d\mathbf{x} \\
&= \sum_{|\alpha|,|\beta|=m} a_{\alpha\beta} \int_{\mathbb{R}^n} D^\alpha u D^\beta u\, d\mathbf{x} \\
&= \sum_{|\alpha|,|\beta|=m} a_{\alpha\beta} \int_{\mathbb{R}^n} [(i\xi)^\alpha \hat{u}(\xi)]\overline{[(i\xi)^\beta \hat{u}(\xi)]}\, d\xi \\
&= \int_{\mathbb{R}^n} [\sum_{|\alpha|,|\beta|=m} a_{\alpha\beta}\xi^{\alpha+\beta}] |\hat{u}(\xi)|^2\, d\xi \\
&\ge \epsilon \int_{\mathbb{R}^n} |\xi|^{2m} |\hat{u}(\xi)|^2\, d\xi.
\end{aligned}$$

Let $a_{00} \ge \eta > 0$. There exists an $\epsilon' > 0$, so that $\epsilon|\xi|^{2m} \ge \epsilon' \sum_{|\alpha|\le m} |\xi^\alpha|^2 - \eta$ for all $\xi \in \mathbb{R}^n$. From this follows $\epsilon \int |\xi|^{2m}|\hat{u}(\xi)|^2 d\xi \ge \epsilon' |u|_m^2 - \eta|u|_0^2$ and $a(u,u) \ge \epsilon'|u|_m^2$ (cf. Lemma 6.2.23). If Ω is bounded, use Lemma 6.2.11. ■

Having shown the $H_0^m(\Omega)$-ellipticity of the form $a(\cdot,\cdot)$, we are now able to apply the general proofs after Theorem 6.5.9.

Theorem 7.2.8. (Existence and uniqueness of weak solutions)*If* $a(\cdot,\cdot)$ *is* $H_0^m(\Omega)$*-elliptic then there exists a solution* $u \in H_0^m(\Omega)$ *of Problem* (9) *which satisfies*

$$|u|_m \le \frac{1}{C_E}|f|_{-m} \quad (C_E \text{ from } (6.5.5)). \tag{7.2.11}$$

Since (11) *holds for all* $f \in H^{-m}(\Omega)$ *and* $u = L^{-1}f$ (*cf.* (9')), *inequality* (11) *is equivalent to*

$$\|L^{-1}\|_{H_0^m(\Omega)\leftarrow H^{-m}(\Omega)} \le C := 1/C_E. \tag{7.2.11'}$$

The term "variational problem" for (9) goes back to the following statement, inferred from Theorem 6.5.12:

Theorem 7.2.9. *Let* $a(\cdot,\cdot)$ *be an* $H_0^m(\Omega)$*-elliptic and symmetric bilinear form. Then* (9) *is equivalent to the variational problem*

$$\text{find} \quad u \in H_0^m(\Omega), \qquad \text{such that} \quad J(u) \le J(v) \quad \text{for all} \quad v \in H_0^m(\Omega), \tag{7.2.12a}$$

where

$$J(v) := \frac{1}{2}a(v,v) - f(v). \tag{7.2.12b}$$

Attention. If $a(\cdot,\cdot)$ is either not $H_0^m(\Omega)$-elliptic or not symmetric, Problem (9) remains meaningful although the solution does not minimize the functional $J(u)$.

Example 7.2.10. The Poisson equation $-\Delta u = f$ in Ω, $u = 0$ on Γ, leads to $a(u,v) = \int_\Omega \langle \nabla u(\mathbf{x}), \nabla v(\mathbf{x})\rangle d\mathbf{x}$. For a bounded domain Ω, $a(\cdot,\cdot)$ is $H_0^1(\Omega)$-elliptic (cf. Theorem 3), so that for any $f \in H^{-1}(\Omega)$ there exists exactly one (weak) solution $u \in H_0^1(\Omega)$ of the Poisson equation. This is also the solution of the variational problem $\frac{1}{2}\int_\Omega |\nabla u|^2 d\mathbf{x} - f(u) = \min$.

A weaker condition than $H_0^m(\Omega)$-ellipticity is the $H_0^m(\Omega)$-coercivity: $a(u,u) \ge \epsilon|u|_m^2 - C|u|_0^2$.

Theorem 7.2.11. *Let* $m = 1$, *and let the coefficients* $a_{\alpha\beta} \in L^\infty(\Omega)$ *satisfy condition* (3) *of uniform ellipticity. Then* $a(\cdot,\cdot)$ *is* $H_0^1(\Omega)$*-coercive.*

PROOF. We write L as $L = L_I + L_{II}$, with L_I satisfying the conditions of Theorem 3, resp. Corollary 4 if Ω is not bounded, and L_{II} containing only derivatives of order ≤ 1. Then we can apply the following lemma. ■

Lemma 7.2.12. *Let* $a(\cdot,\cdot) = a'(\cdot,\cdot) + a''(\cdot,\cdot)$ *be decomposed such that* $a'(\cdot,\cdot)$ *is* $H_0^m(\Omega)$*-elliptic, or perhaps only* $H_0^m(\Omega)$*-coercive, while*

$$a''(u,u) = \sum_{\substack{|\alpha|,|\beta|\le m \\ |\alpha|+|\beta|<2m}} \int_\Omega a_{\alpha\beta} D^\alpha u D^\beta u \, d\mathbf{x}$$

with $a_{\alpha\beta} \in L^\infty(\Omega)$ *contains only derivatives of order* $\le 2m-1$. *Then* $a(\cdot,\cdot)$ *is also* $H_0^m(\Omega)$*-coercive.*

PROOF. (6.5.12c) follows from (6.2.10b) with $X = H_0^{|\alpha|}(\Omega)$, $Y = H_0^{|\beta|}(\Omega)$, $V = H_0^m(\Omega)$, and $U = L^2(\Omega)$, so that Lemma 6.5.18c proves the assertion. ■

The generalisation of Theorem 11 to arbitrary $m \geq 1$ requires stronger conditions on the coefficients of the principal part.

Theorem 7.2.13. (Gårding's theorem) *Let L be uniformly elliptic (cf. (3)) and assume $a_{\alpha\beta} \in L^\infty(\Omega)$. Furthermore, let the coefficients $a_{\alpha\beta}$ with $|\alpha| = |\beta| = m$ be uniformly continuous in $\overline{\Omega}$. Then $a(\cdot,\cdot)$ is $H_0^m(\Omega)$-coercive. If conversely $a_{\alpha\beta} \in C(\Omega)$ holds for $|\alpha| = |\beta| = m$ and $a_{\alpha\beta} \in L^\infty(\Omega)$ otherwise, then from the $H_0^m(\Omega)$-coercivity follows uniform ellipticity* (3).

Details of the proof can be found in Wloka [1, Theorem 19.2]. The proof given there also holds for unbounded Ω, since the coefficients are *uniformly* continuous. For the first part of the theorem one uses a partition of unity.

The significance of coercivity lies in the following statement.

Theorem 7.2.14. *Let Ω be bounded and $a(\cdot,\cdot)$ be $H_0^m(\Omega)$-coercive. Then one of the following alternatives holds:*
(i) Problem (9) *has exactly one (weak) solution $u \in H_0^m(\Omega)$.*
(ii) The kernels $E = \mathrm{kernel}(L)$ and $E^* = \mathrm{kernel}(L')$ are k-dimensional for a $k \in \mathbb{N}$, i.e.,

$$a(e,v) = 0, \quad a(v,e^*) = 0 \quad \textit{for all } e \in E, e^* \in E^*, v \in H_0^m(\Omega).$$

Further, the eigenvalue problem

$$a(e,v) = \lambda(e,v)_{L^2(\Omega)} \quad \textit{for all } v \in H_0^m(\Omega) \tag{7.2.13}$$

has countably many eigenvalues which do not accumulate in $\mathbb{C}$.

PROOF. Since for bounded Ω the embedding $V := H_0^m(\Omega) \subset U := L^2(\Omega)$ is compact (cf. Theorem 6.4.8b), Theorem 6.5.15 is applicable. ■

7.3 Inhomogeneous Dirichlet Boundary Conditions

Next, we consider the boundary value problem

$$Lu = g \quad \text{in } \Omega, \quad u = \varphi \quad \text{on } \Gamma, \tag{7.3.1}$$

where L is a differential operator of second order (i.e., $m = 1$). The variational formulation of the boundary value problem reads:

$$\text{find} \quad u \in H^1(\Omega) \quad \text{with} \quad u = \varphi \quad \text{on } \Gamma \quad \text{such that} \tag{7.3.2a}$$

$$a(u,v) = f(v) \quad \text{for all } v \in H_0^1(\Omega). \tag{7.3.2b}$$

But according to Section 6.2.5 the restriction $u|_\Gamma$ of $u \in H^1(\Omega)$ on Γ is well defined as a function in $H^{1/2}(\Gamma)$. Thus "$u = \varphi$ on Γ" must be understood as the equality $u|_\Gamma = \varphi$ in $H^{1/2}(\Gamma)$. In contrast to the preceding section one uses $a(\cdot,\cdot)$ in (2b) as a bilinear form on $H^1(\Omega) \times H_0^1(\Omega)$. It is easy to see that $a(\cdot,\cdot)$ is well-defined and bounded on this product.

Remark 7.3.1. For the solvability of Problem (2a,b) it is necessary that:

$$\text{there exists a } u_0 \in H^1(\Omega) \quad \text{with} \quad u_0|_\Gamma = \varphi. \tag{7.3.3}$$

If a function u_0 with Property (3) is known, a second characterisation of the weak solution results:

$$\text{Let} \quad u_0 \quad \text{satisfy (3);} \qquad \text{find } w \in H_0^1(\Omega), \text{such that} \tag{7.3.4a}$$

$$a(w,v) = f'(v) := f(v) - a(u_0,v) \quad \text{for all } v \in H_0^1(\Omega). \tag{7.3.4b}$$

Remark 7.3.2. The variational problems (2a,b) and (4a,b) are equivalent. If u_0 and w are the solutions of (4a,b), then $u = u_0 + w$ is a solution of Problem (2a,b). If u is a solution of (2a,b), then, for example, $u_0 = u$ and $w = 0$ satisfy Problem (4a,b).

Exercise 7.3.3. Show that $f' \in H^{-1}(\Omega)$ for f' from (4b) and

$$|f'|_{-1} \le |f|_{-1} + C_S|u_0|_1 \tag{7.3.5}$$

with C_S from $|a(u,v)| \le C_S|u|_1|v|_1$ (cf. (6.5.1)).

Remark 7.3.4. The problem (1) and the variational formulation (2a,b) have the same classical solutions if such exist.

PROOF. It suffices to assume $v \in C_0^\infty(\Omega)$ in (2b). Integration by parts can be carried out as in Section 7.2 and proves the assertion. ■

Theorem 7.3.5. (Existence and uniqueness). *Let problem* (2.9) *(with homogeneous boundary values) be uniquely solvable for all $f \in H^{-1}(\Omega)$. Then Condition* (3) *is sufficient, and necessary, for the unique solvability of Problem* (2a,b).

PROOF. If there exists a solution $u \in H^1(\Omega)$ of (2a,b) then (3) is satisfied. However, if (3) holds, one obtains via (4a,b) a unique solution since (4b) agrees with (2.9). ■

Remark 7.3.6. Under the condition $\Omega \in C^{0,1}$, (3) is equivalent to

$$\varphi \in H^{1/2}(\Gamma). \tag{7.3.6}$$

PROOF. If u_0 satisfies Condition (3) then Theorem 6.2.40 shows that $\varphi \in H^{1/2}(\Gamma)$. If conversely $\varphi \in H^{1/2}(\Gamma)$, then the same theorem guarantees an extension $u_0 \in H^1(\Omega)$ to Ω with $u_0|_\Gamma = \varphi$ and

$$|u_0|_1 \leq C|\varphi|_{1/2}. \tag{7.3.7}$$

■

Let inequality (2.11′) hold in the case of homogeneous boundary values. Equation (5) shows $|u|_1 \leq |u_0|_1 + |w|_1 \leq |u_0|_1 + (|f|_{-1} + C'|u_0|_1)/\epsilon$ for the solution of Problem (4a,b). The estimate (7) proves

Theorem 7.3.7. *Let $\Omega \in C^{0,1}$. Let the bilinear form be restricted to $H^1(\Omega) \times H_0^1(\Omega)$ and let it satisfy* (2.11′)*. Then for every $f \in H^{-1}(\Omega)$ and $\varphi \in H^{1/2}(\Gamma)$ there exists exactly one solution $u \in H^1(\Omega)$ of Problem* (2a,b) *with*

$$|u|_1 \leq C(|f|_{-1} + |\varphi|_{1/2}). \tag{7.3.8}$$

Exercise 7.3.8. Let $a(\cdot,\cdot)$ be symmetric and $H_0^1(\Omega)$-elliptic. Show that Problem (2a,b) with $f = 0$ is equivalent to the variational problem: Find $u \in H^1(\Omega)$ with $u|_\Gamma = \varphi$ such that $a(u,u)$ becomes minimal (cf. (1.1)).

7.4 Natural Boundary Conditions

The bilinear form $a(\cdot,\cdot)$ defined in (2.6) is also well-defined on $H^m(\Omega) \times H^m(\Omega)$. In analogy to Theorem 7.2.2 there holds

Theorem 7.4.1. *Assume $a_{\alpha\beta} \in L^\infty(\Omega)$. The bilinear form defined by* (2.6) *is bounded on $H^m(\Omega) \times H^m(\Omega)$: $|a(u,v)| \leq \sum_{\alpha,\beta} ||a_{\alpha\beta}||_{L^\infty(\Omega)} |u|_m |v|_m$ for all $u, v \in H^m(\Omega)$.*

Now let f be a functional from $(H^m(\Omega))'$. Equation (2.7) with $g \in L^2(\Omega)$, for example, describes such a functional; but (2.7) is only a special case of the functional f subsequently defined in (1a), which we want to use as a foundation in the following.

Exercise 7.4.2. Let Γ be sufficiently smooth and let $g \in L^2(\Omega)$, $\varphi \in L^2(\Gamma)$ hold. Show that

$$f(v) := \int_\Omega g(\mathbf{x})v(\mathbf{x})\,d\mathbf{x} + \int_\Gamma \varphi(\mathbf{x})v(\mathbf{x})\,d\Gamma, \qquad v \in H^1(\Omega), \tag{7.4.1a}$$

defines a functional in $(H^1(\Omega))'$ with $||f||_{(H^1(\Omega))'} \leq C(||g||_{L^2(\Omega)} + ||\varphi||_{L^2(\Gamma)})$. This implies $f \in (H^m(\Omega))'$ for all $m > 1$. More precisely, the following also holds:

$$\|f\|_{(H^1(\Omega))'} \le C(\|g\|_{(H^1(\Omega))'} + \|\varphi\|_{H^{-1/2}(\Gamma)}). \tag{7.4.1b}$$

Frequently, variational problems with a physical background have the form:

$$\text{find} \quad u \in H^m(\Omega), \quad \text{such that} \tag{7.4.2a}$$

$$a(u,v) = f(v) \quad \text{for all } v \in H^m(\Omega). \tag{7.4.2b}$$

In contrast to the condition $u \in H_0^m(\Omega)$ from Section 7.2, $u \in H^m(\Omega)$ contains no boundary condition. Nevertheless, Problem (2a,b) has a unique solution if $a(\cdot,\cdot)$ is $H^m(\Omega)$-elliptic. This condition is easy to satisfy.

Theorem 7.4.3. *Under the conditions of Theorem* 7.2.3 *or Corollary* 7.2.4 *$a(\cdot,\cdot)$ is $H^1(\Omega)$-elliptic: $a(u,u) \ge \epsilon |u|_1^2$ for all $u \in H^1(\Omega)$. Problem* (2a,b) *(with $m = 1$) has exactly one solution which satisfies the estimate* (3)*:*

$$|u|_1 \le \frac{1}{\epsilon}\|f\|_{H^1(\Omega))'}. \tag{7.4.3}$$

PROOF. The same as for Theorem 7.2.3 or Corollary 7.2.4, and Theorem 7.2.8. ∎

Corollary 7.4.4. (a) *A unique solution which satisfies the estimate* (3), *also exists if instead of $H^1(\Omega)$-ellipticity one assumes: $a(\cdot,\cdot)$ is $H^1(\Omega)$-coercive, $\Omega \in C^{0,1}$ is bounded, $\lambda = 0$ is not an eigenvalue (i.e., $a(u,v) = 0$ for all $v \in H^1(\Omega)$ implies $u = 0$).*
(b) *The combination of inequalities* (3) *and* (1b) *results in*

$$|u|_1 \le C[\|g\|_{(H^1(\Omega))'} + \|\varphi\|_{H^{-1/2}(\Gamma)}] \tag{7.4.3'}$$

for the solution of (2a,b), *if f is defined by* (1a) *with $g \in (H^1(\Omega))'$, $\varphi \in H^{-1/2}(\Gamma)$.*

PROOF. (a). According to Theorem 6.4.8b, $H^1(\Omega)$ is compactly embedded in $L^2(\Omega)$ so that the statement of Theorem 7.2.14 can be transferred. If $\lambda = 0$ is not an eigenvalue then $L^{-1} \in L((H^1(\Omega))', H^1(\Omega))$ holds (cf. Theorem 6.4.12). ∎

To find out which classical boundary value problem corresponds to the variational formulation (2a,b), we assume that (2a,b) has a classical solution $u \in H^m(\Omega) \cap C^{2m}(\overline{\Omega})$. Further, $v \in C^\infty(\Omega)$ can be assumed also (cf. Lemma 6.5.1b). For reasons of simplicity we limit ourselves to the case $m = 1$. Under the assumption $a_{\alpha\beta} \in C^1(\overline{\Omega})$ and under suitable conditions on Ω the following general Green formula is applicable:

$$a(u,v) = \int_\Omega \left[\sum_{i,j=1}^n a_{ij}u_{x_i}v_{x_j} + \sum_{i=1}^n a_{0i}uv_{x_i} + \sum_{i=1}^n a_{i0}u_{x_i}v + a_{00}uv\right] \mathbf{d}\mathbf{x}$$

$$= \int_\Omega \left[-\sum_{i,j=1}^n (a_{ij}u_{x_i})_{x_j} - \sum_{i=1}^n (a_{0i}u)_{x_i} + \sum_{i=1}^n a_{i0}u_{x_i} + a_{00}u\right] v\,\mathbf{d}\mathbf{x}$$

$$+ \int_\Gamma \left[\sum_{i,j} n_j a_{ij}u_{x_i} + \sum_i n_i a_{0i}u\right] v\, d\Gamma. \tag{17.4.4}$$

Here, the n_i are the components of the normal direction $\mathbf{n} = \mathbf{n}(x)$, $\mathbf{x} \in \Gamma$. We define the boundary differential operator

$$B := \sum_{i,j=1}^n n_j a_{ij} \frac{\partial}{\partial x_i} + \sum_{i=1}^n n_i a_{0i}, \tag{7.4.5}$$

when L is described by (2.1b). Equation (4) becomes $a(u,v) = \int_\Omega vLudx + \int_\Gamma vBu\,d\Gamma$. By the formulation of the problem, $a(u,v)$ agrees with $f(v)$ from (1a). If we first choose $v \in H_0^1(\Omega) \subset H^1(\Omega)$ the boundary integrals drop out and we obtain $Lu = g$ as in Section 7.2. By this the identity $a(u,v) = f(v)$ reduces to $\int_\Gamma vBu\,d\Gamma = \int_\Gamma \varphi v\,d\Gamma$ for all $v \in H^1(\Omega)$. According to Theorem 6.2.40b, $v|_\Gamma$ runs over the set $H^{1/2}(\Gamma)$ if v runs over $H^1(\Omega)$, so that we have $\int_\Gamma \psi(Bu - \varphi)\,d\Gamma = 0$ for all $\psi \in H^{1/2}(\Gamma)$; thus $Bu = \varphi$. This proves

Theorem 7.4.5. *Let Γ be sufficiently smooth. A classical solution of the problem* (2a,b) *with f from* (1a) *is also the classical solution of the boundary value problem*

$$Lu = g \quad \text{in } \Omega, \quad Bu = \varphi \quad \text{on } \Gamma, \tag{7.4.6}$$

and conversely.

The condition $Bu = \varphi$ is called the natural boundary condition. This results from the fact that in (2b) (as distinct from (2.9)) the function v may assume arbitrary boundary values. Note that the bilinear form determines L as well as B.

Exercise 7.4.6. Show that the bilinear form from Example 7.2.10 for $-\Delta u = g$ has as the natural boundary condition the Neumann condition $-\partial u/\partial n = 0$.

Theorem 7.4.7. *Let $\Omega \in C^{0,1}$ be bounded and $a(\cdot,\cdot)$ be $H^m(\Omega)$-coercive. Then the statements of Theorem* 7.2.14 *holds with $H^m(\Omega)$ instead of $H_0^m(\Omega)$.*

Example 7.4.8. Let $\Omega \in C^{0,1}$ be a bounded domain. The bilinear form $a(u,v) = \int_\Omega (\langle \nabla u, \nabla v\rangle + cuv)dx$, associated to the Helmholtz equation

$$-\Delta u + cu = f \quad \text{in } \Omega \quad (c > 0), \quad \partial u/\partial n = 0 \quad \text{on } \Gamma,$$

is $H^1(\Omega)$-elliptic since $a(u,u) \geq \min(1,c)|u|_1^2$. For $c = 0$, however, $a(\cdot,\cdot)$ is only $H^1(\Omega)$-coercive. As is known from Theorem 3.4.1, the Neumann boundary value problem for the Poisson equation (i.e., for $c = 0$) is not uniquely solvable. According to alternative (ii) in Theorem 7, there exists a nontrivial eigenspace $E = \text{kernel}(L)$. $u \in E$ satisfies $a(u,u) = 0$, thus $\nabla u = 0$. Since Ω is connected, it follows that $u(\mathbf{x}) = \text{const}$ and therefore $\dim E = 1$. Since $a(\cdot,\cdot)$ is symmetric, $E^* := \text{kernel}(L')$ coincides with E. According to Theorem 7, the Neumann boundary value problem $a(u,v) = f(v)$ $(v \in H^1(\Omega))$ is solvable if and only if $f \perp E$, i.e., $f(1) = 0$. If $f(v) := \int_\Omega g(\mathbf{x})v(\mathbf{x})d\mathbf{x}$, $f(1) = 0$ reads as $\int_\Omega g dx = 0$. If, however, f is given by (1a), the integrability condition reads $f(1) = \int_\Omega g\, dx + \int_\Gamma \varphi\, d\Gamma = 0$ (this is Equation (3.4.2), in which f should be replaced by $-g$).

Remark 7.4.9. While the classical formulation of a boundary condition such as $\partial u/\partial n = 0$ requires conditions on the boundary Γ, the problem (2a,b) can be formulated for arbitrary measurable Ω.

In the following, we proceed in the opposite direction: does there exist, for a classically formulated boundary value problem $Lu = g$ in Ω, $Bu = \varphi$ on Γ, with given L and B, a bilinear form $a(\cdot,\cdot)$ such that (2a, b) is the corresponding variational formulation? This would mean that the freely prescribed boundary operator B represents the natural boundary condition.

For $m = 1$ the general form of the boundary operator reads

$$B = \sum_{i=1}^{n} b_i(\mathbf{x})\frac{\partial}{\partial x_i} + b_0(\mathbf{x}) \quad (\mathbf{x} \in \Gamma). \tag{7.4.7}$$

With $\mathbf{b}^\mathsf{T} = (b_1, \cdots, b_n)$ one can also write $B = \mathbf{b}^\mathsf{T}\nabla + b_0$ (cf. (5.2.1b,b′)). Here $\mathbf{b}^\mathsf{T}\nabla$ is not allowed to be a tangential derivative (cf. Remark 5.2.2):

$$\langle \mathbf{b}(\mathbf{x}), \mathbf{n}(\mathbf{x})\rangle \neq 0 \quad \text{for all } \mathbf{x} \in \Gamma. \tag{7.4.8}$$

Remark 7.4.10. Let $m = 1$. Let (8) hold. Let $\mathbf{A}(\mathbf{x})$ be the matrix $\mathbf{A} = (a_{ij})$ (cf. (5.1.1d)). By passing from $Bu = \varphi$ to the equivalent scaled equation $\sigma Bu = \sigma\varphi$ with $\sigma(\mathbf{x}) = \langle \mathbf{n}(\mathbf{x}), \mathbf{A}(\mathbf{x})\mathbf{n}(\mathbf{x})\rangle / \langle \mathbf{n}(\mathbf{x}), \mathbf{b}(\mathbf{x})\rangle$, one can ensure that $\langle \mathbf{n}, \sigma\mathbf{b}\rangle = \langle \mathbf{n}, \mathbf{A}\mathbf{n}\rangle$. Thus in the following it is always assumed that $\mathbf{b}$ already satisfies $\langle \mathbf{n}, \mathbf{b}\rangle = \langle \mathbf{n}, \mathbf{A}\mathbf{n}\rangle$.

PROOF. Because of (8) σ is well defined. For $Bu = \varphi$ and $\sigma Bu = \sigma\varphi$ to be equivalent, $\sigma \neq 0$ is required. This is guaranteed by the uniform ellipticity: $\langle \mathbf{n}, \mathbf{A}\mathbf{n}\rangle \geq \epsilon|\mathbf{n}|^2 = \epsilon$.

Theorem 7.4.11. (Construction of the bilinear form). *Let $m = 1$. Let L and B be given by* (2.1b) *and* (7), *with* $\mathbf{b}$ *satisfying condition* (8). *Then there exists a bilinear form $a(\cdot,\cdot)$ on $H^1(\Omega) \times H^1(\Omega)$ such that to the variational problem* (2a,b) *corresponds the classical formulation $Lu = g$ in Ω, $Bu = \varphi$ on Γ.*

PROOF. The bilinear form we seek is not uniquely determined . We will give two possibilities for its construction. First we discuss the absolute term in (7). On the basis of Remark 10 we assume $\langle \mathbf{n}, \mathbf{b}\rangle = \langle \mathbf{n}, \mathbf{A}\mathbf{n}\rangle$.
(a) Let the vector function $\beta(\mathbf{x}) := (\beta_1(\mathbf{x}), \cdots, \beta_n(\mathbf{x})) \in C^1(\overline{\Omega})$ be arbitrary. The differential operator

$$L_1 := -\sum_{i=1}^{n} \frac{\partial}{\partial x_i}\beta_i + \sum_{i=1}^{n} \beta_i \frac{\partial}{\partial x_i} + \sum_{i=1}^{n} (\beta_i)_{x_i}$$

maps every $u \in C^1(\Omega)$ into zero: $L_1 u = 0$. Thus the operator L can be replaced by $L + L_1$ without changing the boundary value problem. Let $a(\cdot,\cdot)$ be constructed according to (2.6) from the coefficients of $L + L_1$. Equation (5) shows that the boundary operator associated with $a(\cdot,\cdot)$ has the absolute term

$$\sum_{i=1}^{n} n_i(a_{0i} + \beta_i). \tag{7.4.9}$$

If $b_0 = 0$, the choice of $\beta_i = -a_{0i}$ is successful. Otherwise, two other options are available.
(aa) Select β_i such that on Γ the following holds: $\beta_i(\mathbf{x}) = b_0(\mathbf{x})n_i(\mathbf{x}) - a_{0i}(\mathbf{x})$. Since $|\mathbf{n}| = 1$, the term (9) then agrees with $b_0(\mathbf{x})$. The practical difficulty in this method consists in the need to construct a smooth continuation on Ω of the boundary values $\beta_i(\mathbf{x})$, $\mathbf{x} \in \Gamma$.
(ab) Set $\beta_i = -a_{0i}$ and add a suitable boundary integral:

$$a(u,v) := \int_\Omega \Big\{\sum_{i,j} a_{ij}u_{x_i}v_{x_j} + \sum_i (a_{i0} - a_{0i})u_{x_i}v + \Big[a_{00} - \sum_i (a_{0i})_{x_i}\Big]uv\Big\}\,d\mathbf{x} + \int_\Gamma b_0(\mathbf{x})uv\,d\Gamma. \tag{7.4.10}$$

The integration by parts described above shows that the boundary operator associated to (10) reads

$$\tilde{B} = \sum_{i,j} n_j a_{ij}\frac{\partial}{\partial x_i} + b_0. \tag{7.4.11}$$

(b) The operator (11) can be written in the form $\tilde{B} = \tilde{\mathbf{b}}^{\mathsf{T}}\nabla + b_0$ with $\tilde{\mathbf{b}} = \mathbf{A}\mathbf{n}$. Since $\langle \mathbf{n}, \tilde{\mathbf{b}}\rangle = \langle \mathbf{n}, \mathbf{A}\mathbf{n}\rangle = \langle \mathbf{n}, \mathbf{b}\rangle$ has been assumed already, $\mathbf{d} := \mathbf{b} - \tilde{\mathbf{b}}$ is orthogonal to $\mathbf{n}$. To change $\tilde{\mathbf{b}}$ to $\mathbf{b}$ there are again two options.
(ba) Define the $n \times n$-matrix $\mathbf{A}^s$ on Γ by $\mathbf{A}^s = \mathbf{d}\mathbf{n}^{\mathsf{T}} - \mathbf{n}\mathbf{d}^{\mathsf{T}}$, i.e., $a^s_{ij} = d_i n_j - n_i d_j$. This $\mathbf{A}^s(\mathbf{x})$ is skew-symmetric: $\mathbf{A}^{s^{\mathsf{T}}} = -\mathbf{A}^s$. Continue $\mathbf{A}^s(\mathbf{x})$, which is at first only defined on Γ, to a skew-symmetric matrix $\mathbf{A}^s \in C^1(\Omega)$. [Here we have the same practical difficulty as in step (aa).] The entries of $\mathbf{A}^s$ define

$$L_2 := \sum_{i,j=1}^{n} \Big[-\frac{\partial}{\partial x_j} a^s_{ij}\frac{\partial}{\partial x_i} + (a^s_{ij})_{x_j}\frac{\partial}{\partial x_i}\Big].$$

Again, $L_2u = 0$ holds for all $u \in C^2(\Omega)$, since

$$-(a^s_{ij}u_{x_i})_{x_j} - (a^s_{ji}u_{x_j})_{x_i} + (a^s_{ij})_{x_j}u_{x_i} + (a^s_{ji})_{x_i}u_{x_j} = -a^s_{ij}u_{x_ix_j} - a^s_{ji}u_{x_jx_i} = 0.$$

Thus L can be replaced by $L + L_2$ without changing the boundary value problem. The coefficients belonging to $L + L_2$ result in a boundary operator B whose derivative terms read $\sum_{i,j} n_j(a_{ij} + a^s_{ij})\partial/\partial x_i = [(\mathbf{A} + \mathbf{A}^s)\mathbf{n}]^\mathsf{T}\nabla$. By the construction of $\mathbf{A}^s$ we have $\mathbf{A}^s\mathbf{n} = (\mathbf{d}\mathbf{n}^\mathsf{T} - \mathbf{n}\mathbf{d}^\mathsf{T})\mathbf{n} = \mathbf{d} = \mathbf{b} - \tilde{\mathbf{b}}$, since $\langle \mathbf{n}, \mathbf{n}\rangle = 1$ and $\langle \mathbf{d}, \mathbf{n}\rangle = 0$. Since we also have $\mathbf{A}\mathbf{n} = \tilde{\mathbf{b}}$, the derivative term in B gives $\mathbf{b}^\mathsf{T}\nabla$ as desired. The transition $L \to L + L_2$ does not change the absolute term in $\tilde{B}$, so that from (11) follows $B = \mathbf{b}^\mathsf{T}\nabla + b_0$.
(bb) Let $\tilde{B} = \tilde{\mathbf{b}}^\mathsf{T}\nabla + b_0$ be given (cf. (11)) such that $\mathbf{d} = \mathbf{b} - \tilde{\mathbf{b}}$ is orthogonal to $\mathbf{n}$. From this the boundary operator $T := \mathbf{d}^\mathsf{T}\nabla$ is the derivative in a tangential direction if $n = 2$ [resp. in the tangent hyperplane if $n \ge 3$]. If Γ is sufficiently smooth then the restriction $v|_\Gamma$ of $v \in H^1(\Omega)$ is an element of $H^{1/2}(\Gamma)$. By Remark 6.3.14a, one can show that $T \in L(H^{1/2}(\Gamma), H^{-1/2}(\Gamma))$. Since $T(u|_\Gamma) \in H^{-1/2}(\Gamma)$, $\int_\Gamma \psi T(u|_\Gamma)\,d\Gamma)$ is well-defined for $\psi \in H^{1/2}(\Gamma)$, in particular, for $\psi = v|_\Gamma$ with $v \in H^1(\Omega)$. Thus

$$b(u,v) := \int_\Gamma (v|_\Gamma)\,T(u|_\Gamma)\,d\Gamma$$

is a bilinear form bounded on $H^1(\Omega) \times H^1(\Omega)$. We add $b(\cdot,\cdot)$ to $a(\cdot,\cdot)$ in (10). Integration by parts yields the boundary operator $\tilde{B} + T = \tilde{\mathbf{b}}^\mathsf{T}\nabla + b_0 + (\mathbf{b} - \tilde{\mathbf{b}})^\mathsf{T}\nabla = \mathbf{b}^\mathsf{T}\nabla + b_0 = B$. ■

Remark 7.4.12. Let the bilinear form (2.6) be $H^1(\Omega)$-coercive (cf. Theorem 3). Let its coefficients, as well as the boundary Γ, be sufficiently smooth ($\in C^1$). Then the constructions of the preceding proof again result in an B7j$H^1(\Omega)$-coercive form.

PROOF. We go through the steps.
(a) In step (aa) only terms of lower order are added so that Lemma 7.2.12 is applicable. As for step (ab), see step (bb).
(ba) In step (ba) one adds $b(u,v) := \int_\Omega \sum_{i,j}[a^s_{ij}u_{x_i}v_{x_j} + (a^s_{ij})_{x_j}vu_{x_i}]d\mathbf{x}$. Here Lemma 7.2.12 is also applicable, since the skew symmetry of $\mathbf{A}^s$ results in $b(u,u) = \int_\Omega \sum (a^s_{ij})_{x_j}uu_{x_i}d\mathbf{x}$.
(bb) In the construction in step (bb), Lemma 7.2.12 is applicable analogously. This is easiest to understand in the case $n = 2$. Let Γ be described by $\{(x_1(s), x_2(s))\colon\ 0 \le s \le 1\}$. If $x_1, x_2 \in C^1([0,1])$ and $x_i'(0) = x_i'(1)$, and if we have $\mathbf{d} \in C^1(\Gamma)$ for the $\mathbf{d}$ in $T = \mathbf{d}^\mathsf{T}\nabla$, then $b(\cdot,\cdot)$ has the representation $b(u,v) = \int_0^1 \overline{v}(s)\tau(s)\overline{u}'(s)\,ds$, where $\overline{u}(s) = u(x_1(s), x_2(s))$, $\overline{v}(s) = v(x_1(s), x_2(s))$, $\tau \in C^1([0,1])$. Thanks to periodicity, integration by parts yields in $b(u,v) = -\int_0^1 (\tau\overline{v})'\overline{u}\,ds$ without boundary terms, so that $b(u,u) = \frac{1}{2}\left[\int_0^1 \overline{u}\tau\overline{u}'\,ds - \int_0^1 (\tau\overline{u})'\overline{u}\,ds\right] = -\frac{1}{2}\int_0^1 \tau'\overline{u}^2\,ds$. This implies

$$|b(u,u)| \le \frac{1}{2}\|\tau\|_{C^1([0,1])}\,\|u|_\Gamma\|^2_{L^2(\Gamma)} \le C\|\tau\|_{C^1([0,1])}\,\|u\|^2_{H^s(\Omega)}$$

for all $s \in (1/2, 1)$ (cf. Theorem 6.2.40a). Since $|u|_s^2 \le \epsilon|u|_1^2 - C|u|_0^2$, one can apply Lemma 6.5.18c. ■

The case $m \ge 2$ has been excluded in this section (except for Theorem 1). Boundary value problems of order $2m$ require m boundary conditions $B_j u = \varphi_j$ on Γ $(j = 1, \cdots, m)$ (cf. Section 5.3). For $m \ge 2$ the proof of $H^m(\Omega)$-coercivity becomes more complicated.

In order to carry over Theorem 7.2.13, one needs in addition the so-called condition of Agmon (cf. Wloka [1, Theorem 19.3], Lions-Magenes [1, p. 210]).

The resulting complications can be seen with the aid of the biharmonic equation.

Example 7.4.13. (a) To the variational problem: find $u \in H^2(\Omega)$ with

$$\begin{aligned} a(u,v) &:= \int_\Omega \Delta u \Delta v\, dx = f(v) \\ &:= \int_\Omega gv\, \mathbf{dx} + \int_\Gamma \left(\varphi_1 \frac{\partial v}{\partial n} - \varphi_2 v\right) d\Gamma \end{aligned} \tag{7.4.12a}$$

for all $v \in H^2(\Omega)$, corresponds the classical formulation

$$\Delta^2 u = g \quad \text{in } \Omega, \quad \Delta u = \varphi_1 \quad \text{and} \quad \frac{\partial}{\partial n}\Delta u = \varphi_2 \quad \text{on } \Gamma. \tag{7.4.12b}$$

But the bilinear form $a(\cdot,\cdot)$ is not $H^2(\Omega)$-coercive.
(b) To the variational problem: find $u \in H^2(\Omega) \cap H_0^1(\Omega)$ with

$$a(u,v) = f(v) := \int_\Omega gv\, \mathbf{dx} + \int_\Gamma \varphi \frac{\partial v}{\partial n}\, d\Gamma \quad \text{for all } v \in H^2(\Omega) \cap H_0^1(\Omega) \tag{7.4.12c}$$

($a(\cdot,\cdot)$ as in (12a)) corresponds the classical formulation

$$\Delta^2 u = g \quad \text{in } \Omega, \quad u = 0 \quad \text{and } \Delta u = \varphi \quad \text{on } \Gamma. \tag{7.4.12d}$$

The bilinear form is $H^2(\Omega) \cap H_0^1(\Omega)$-coercive.
(c) The boundary conditions in

$$\Delta^2 u = g \quad \text{in } \Omega, \quad u = 0 \quad \text{and} \quad \frac{\partial}{\partial n}\Delta u = \varphi \quad \text{on } \Gamma \tag{7.4.12e}$$

are admissible. Nevertheless this boundary value problem cannot be written in the present form as a variational problem. A variational formulation for (12e) reads:

$$\text{Find} \quad u \in H^2(\Omega) \cap H_0^1(\Omega) \quad \text{such that}$$

$$a(u,v) = f(v) := \int_\Omega gv\, \mathbf{dx} - \int_\Gamma \varphi v\, d\Gamma \quad \text{for all } v \in H^2(\Omega) \text{ with } \frac{\partial v}{\partial n}|_\Gamma = 0 \tag{7.4.12f}$$

($a(\cdot,\cdot)$ as in (12a)). But this does not agree with the present concept since u and v belong to different spaces.

PROOF. The equivalence of the variational and the classical formulation can be shown via integration by parts:

$$a(u,v) = \int_\Omega v\Delta^2 u \, d\mathbf{x} + \int_\Gamma \left[\Delta u \frac{\partial v}{\partial n} - \frac{\partial \Delta u}{\partial n} v \right] d\Gamma.$$

The noncoercivity in part (a) results as follows. Let $\Omega \subset \mathbb{R}^n$ be bounded. For all $\alpha \in \mathbb{R}$, $u_\alpha(x_1, \cdots, x_n) = \sin(\alpha x_1)\exp(\alpha x_2)$ lies in $H^2(\Omega)$ and satisfies $\Delta u_\alpha = 0$, hence also $a(u_\alpha, u_\alpha) = 0$. If $a(\cdot,\cdot)$ were coercive, there would exist a C with $0 = a(u_\alpha, u_\alpha) \geq \epsilon |u_\alpha|_2 - C|u_\alpha|_0$ for all α, i.e., $|u_\alpha|_2 \leq (C/\epsilon)|u_\alpha|_0$. The contradiction results from $|u_\alpha|_2 \geq |\partial^2 u_\alpha / \partial x_1^2|_0 = \alpha^2 |u_\alpha|_0$ for sufficiently large α. ■

Natural and Dirichlet conditions can occur together. In Example 13b, $u = 0$ is a Dirichlet condition and $(\partial \Delta u/\partial n) = \varphi$ a natural boundary condition. Even in the case $m = 1$ both sorts of boundary conditions can occur.

Example 7.4.14. Let γ be a nonempty, proper subset of Γ. The boundary value problem

$$-\Delta u = g \quad \text{in } \Omega, \quad u = 0 \quad \text{on } \gamma, \quad \partial u/\partial n = \varphi \quad \text{on } \Gamma\backslash\gamma \tag{7.4.13}$$

in the variational formulation reads as follows: find $u \in H^1_\gamma(\Omega)$ such that

$$a(u,v) := \int_\Omega \langle \nabla u, \nabla v \rangle \, d\mathbf{x} = f(v) := \int_\Omega gv \, d\mathbf{x} + \int_{\Gamma\backslash\gamma} \varphi v \, d\Gamma$$

for all $v \in H^1_\gamma(\Omega)$, where $H^1_\gamma(\Omega) := \{u \in H^1(\Omega) \colon u = 0 \text{ on } \gamma\}$. Equation (13) is occasionally termed a Robin problem.

Exercise 7.4.15. Let $a(u,v) := \int_\Omega [\langle \nabla u, \nabla v \rangle + cuv] \, d\mathbf{x}$ with $c > 0$ on $V \times V$ with $V := \{u \in H^1(\Omega) \colon u \text{ constant on } \Gamma\}$ be defined. Show that
(a) $a(\cdot,\cdot)$ is V-elliptic.
(b) The weak formulation: $u \in V$, $a(u,v) = \int_\Omega gv \, d\mathbf{x} + \int_\Gamma \varphi v \, d\Gamma$ for all $v \in V$ corresponds to the problem

$$-\Delta u + cu = g \quad \text{in } \Omega, \quad u \text{ constant on } \Gamma, \quad \int_\Gamma \frac{\partial u}{\partial n} d\Gamma = \int_\Gamma \varphi \, d\Gamma. \tag{7.4.14}$$

which is also called an Adler problem.

Finally we want to point out the difficulty of classically interpreting a weak solution. In the variational formulation (2a,b) the right-hand sides g and φ of the differential equation and the boundary condition are combined in the functional f. In the variational formulation the components g and φ are indistinguishable! $u \in H^1(\Omega)$ has first derivatives in $L^2(\Omega)$, whose restrictions to Γ do not have to make sense. That is why Bu cannot be defined in general; $Bu = \varphi$ cannot be viewed as an equality in the space $H^{-1/2}(\Gamma)$ although $\varphi \in H^{-1/2}(\Gamma)$ (cf. Corollary 4b).

But even if there is a classical solution, the following paradox arises. Let u be a classical solution of $Lu = 0$ in Ω, $Bu = \varphi$ on Γ. According to (1a) define $f_\varphi \in (H^1(\Omega))'$ by $f_\varphi(v) := \int_\Gamma \varphi v \, d\Gamma$. One may also view u as a solution of $Lu = f_\varphi$ in Ω, $Bu = 0$ on Γ. These equations may even be interpreted classically in the following way: there exist $f_\nu \in C^\infty(\Omega)$ with $f_\nu \to f_\varphi$ in $(H^1(\Omega))'$. Let u_ν be the classical solution of $Lu_\nu = f_\nu$, $Bu_\nu = 0$. Then u_ν converges in $H^1(\Omega)$ to the above-mentioned classical solution u.

Incorporating the boundary values $Bu = \varphi$ in the differential equation $Lu = f_\varphi$ corresponds to a modification of the discretised problem as used in Section 4. The difference equations $D_h u_h = f_h$ in Ω_h and the boundary conditions $u_h = \varphi$ on Γ_h resulted in the system of equations $L_h u_h = q_h := f_h + \varphi_h$ (cf. (4.2.6b)). If one defines $\overline{u}_h$ by $\overline{u}_h = u_h$ in Ω_h, $\overline{u}_h = 0$ on Γ_h then $\overline{u}_h$ satisfies the equations $D_h \overline{u}_h = q_h$ in Ω_h, $\overline{u}_h = 0$ on Γ_h. Just as the functional f cannot be uniquely separated into g and φ, f_h and φ_h cannot be reconstructed from q_h. In contrast to the discrete case, the separation of f into g and φ is possible, however, provided stronger conditions than $g \in (H^1(\Omega))'$ are imposed on g (for example, $g \in L^2(\Omega)$).

8 The Method of Finite Elements

In Chapter 7 the variational formulation was introduced only for the purpose of proving the existence of a (weak) solution. It will now turn out that the variational formulation is the foundation of a new method of discretisation.

8.1 The Ritz-Galerkin Method

Suppose we have a boundary value problem in its variational formulation:

$$\text{Find } u \in V, \text{ so that } a(u,v) = f(v) \quad \text{for all } v \in V, \tag{8.1.1}$$

where we are thinking, in particular, of $V = H_0^m(\Omega)$ and $V = H^1(\Omega)$ (cf. Section 7.2, Section 7.4). Of course, it is assumed that $a(\cdot,\cdot)$ is a bounded bilinear form defined on $V \times V$, and that $f \in V'$:

$$|a(u,v)| \le C_S \|u\|_V \|v\|_V \quad \text{for } u,v \in V, \qquad f \in V'. \tag{8.1.2}$$

Difference methods arise through discretising the differential operators. Now we wish to leave the differential operator hidden in $a(\cdot,\cdot)$ unchanged. The Ritz-Galerkin discretisation consists in replacing the infinite-dimensional space V with a finite-dimensional space V_N:

$$V_N \subset V, \qquad \dim V_N = N < \infty. \tag{8.1.3}$$

V_N equipped with the norm $\|\cdot\|_V$ is still a Banach space. Since $V_N \subset V$, both $a(u,v)$ and $f(v)$ are defined for $u,v \in V_N$. Thus we may pose the problem (4):

$$\text{Find } u^N \in V_N, \text{ so that } a(u^N,v) = f(v) \quad \text{for all } v \in V_N. \tag{8.1.4}$$

The solution of (4), if it exists, is called the Ritz-Galerkin solution (belonging to V_N) of the boundary value problem (1).

To calculate a solution one needs a basis of V_N. Let $\{b_1,\dots,b_N\}$ be such a basis, i.e.,

$$V_N = \operatorname{span}\{b_1,\dots,b_N\}. \tag{8.1.5}$$

For each coefficient vector $\mathbf{v} = \{v_1,\dots,v_N\}^\mathsf{T}$ we define

$$\mathbf{P}: \mathbb{R}^n \to V_N \subset V, \qquad \mathbf{P}\mathbf{v} := \sum_{i=1}^{N} v_i b_i. \tag{8.1.6}$$

Remark 8.1.1. $\mathbf{P}$ is an isomorphism between $\mathbb{R}^N$ and V_N. The inverse $\mathbf{P}^{-1}: V_N \to \mathbb{R}^N$ is thus well-defined on V_N.

Lemma 8.1.2. *Assuming* (5) *the problem* (4) *is equivalent to*

$$\text{Find } u^N \in V_N, \text{ so that } a(u^N, b_i) = f(b_i) \quad \text{for all } i = 1, \ldots, N. \tag{8.1.7}$$

PROOF. Putting $v = b_i$ in (4) gives (7). On the other hand, suppose that $v = \sum_{i=1}^N v_i b_i \in V_N$ is arbitrary. Then (7) and the linearity of $a(u^N, \cdot)$ and f give (4):

$$a(u^N, v) - f(v) = a(u^N, \sum v_i b_i) - f(\sum v_i b_i) = \sum v_i (a(u^N, b_i) - f(b_i)) = 0.$$

■

We now seek $\mathbf{u} \in \mathbb{R}^N$ so that $u^N = \mathbf{Pu}$. The following theorem transforms the problem (4) [resp. (7)] into a system of linear equations.

Theorem 8.1.3. *Assume* (5). *The* $N \times N$*-matrix* $\mathbf{L} = (L_{ij})$ *and the* N*-vector* $\mathbf{f} = (f_1, \ldots, f_N)^\mathsf{T}$ *are defined by*

$$L_{ij} := a(b_j, b_i) \qquad (i, j = 1, \ldots, N), \tag{8.1.8a}$$

$$f_i := f(b_i) \qquad (i = 1, \ldots, N), \tag{8.1.8b}$$

Then the problems (4) *and*

$$\mathbf{Lu} = \mathbf{f} \tag{8.1.9}$$

are equivalent. If $\mathbf{u}$ *is a solution of* (9), *then* $u^N := \mathbf{Pu}$ *solves the problem* (4). *In the opposite direction, if* u^N *is a solution of* (4), *then* $\mathbf{u} := \mathbf{P}^{-1} u^N$ *is a solution of* (9) (*cf. Remark* 1).

PROOF. (4) is equivalent to (7). In (7) put $u^N = \mathbf{Pu} = \sum u_j b_j$; then $a(u^N, b_i) = \sum_j u_j a(b_j, b_i) = \sum_j L_{ij} u_j = f(b_i) = f_i$, and thus $\mathbf{Lu} = \mathbf{f}$. ■

In engineering applications where the boundary value problems arise from continuum mechanics (cf. the first paragraph of Section 5.3.1), one calls $\mathbf{L}$ the stiffness matrix.

The connections between $\mathbf{L}$ and $a(\cdot, \cdot)$, on the one hand, and between $\mathbf{f}$ and $f(\cdot)$, on the other, are clear from

Remark 8.1.4. If $\langle \mathbf{u}, \mathbf{v} \rangle := \sum_i u_i v_i$ is the usual scalar product, then $a(u, v) = \langle \mathbf{Lu}, \mathbf{v} \rangle$, and $f(v) = \langle \mathbf{f}, v \rangle$ with $u = \mathbf{Pu}, v = \mathbf{Pv}$.

A trivial consequence of Theorem 3 is:

Corollary 8.1.5. *The Ritz-Galerkin discretisation* (4) *has a unique solution* u^N *for each* $f \in V'$ *exactly when the matrix* $\mathbf{L}$ *in* (8a) *is nonsingular.*

The Gelfand triple $V \subset U \subset V'$ corresponds to the finite-dimensional situation $V_N \subset U_N \subset V'_N$, where the spaces are the same as sets:

$$V_N = U_N = V'_N.$$

However, the three spaces have different norms:

$$V_N := (V_N, ||\cdot||_V), \qquad U_N := (V_N, ||\cdot||_U), \qquad V'_N := (V_N, ||\cdot||_{V'_N}).$$

V_N has $||\cdot||_V$ and U_N has $||\cdot||_U$ as a norm, while the dual norm

$$||v||_{V'_N} := \sup\{|(v,u)_U|/||u||_V : 0 \neq u \in V_N\} \quad \text{for } v \in V'$$

(in particular for $v \in V'_N = V_N$) is defined on V', but is only a norm on V'_N, since $||v||_{V'_N} = 0$ for each v orthogonal to V_N.

By Lemma 6.5.1 there is an operator associated to $a(\cdot,\cdot): V_N \times V_N \to \mathbb{R}$

$$L_N : V_N \to V'_N,$$

so that

$$a(u,v) = (L_N u, v)_U \quad \text{for all } u, v \in V_N. \tag{8.1.10a}$$

Here we are writing $(\cdot,\cdot)_U$ for $\langle\cdot,\cdot\rangle_{V'_N \times V_N}$ (cf. Remark 6.3.12).

The map $\mathbf{P}: \mathbb{R}^N \to V_N$ can also be viewed as $\mathbf{P}: \mathbb{R}^N \to V$ since $V_N \subset V$. The map $\mathbf{P}^* \in L(V', \mathbb{R}^N)$ adjoint to $\mathbf{P}$ is defined by

$$\langle \mathbf{P}^* u, \mathbf{v}\rangle = (u, \mathbf{P}\mathbf{v})_U \quad \text{for } u \in V', \mathbf{v} \in \mathbb{R}^N. \tag{8.1.10b}$$

Exercise 8.1.6. (a) Show: The kernel of $\mathbf{P}^*$ is $V_N^\perp \subset U$ (orthogonal space relative to $||\cdot||_U$). The map $\mathbf{P}^*: V'_N \to \mathbb{R}^N$ is an isomorphism.
(b) Since $\mathbf{P}^{-1}: V_N \to \mathbb{R}^N$ exists, we can define

$$C_{\mathbf{P}} := ||\mathbf{P}^{-1}||_{\mathbb{R}^N \leftarrow U_N} := \max\{||\mathbf{P}^{-1}u||/||u||_U : 0 \neq u \in V_N\},$$

where $||\cdot||$ is the Euclidean norm. Show: The matrix $\mathbf{P}^*\mathbf{P}: \mathbb{R}^N \to \mathbb{R}^N$ has an inverse with spectral norm

$$||(\mathbf{P}^*\mathbf{P})^{-1}|| \le C_{\mathbf{P}}{}^2.$$

(c) Show:

$$Q_N := \mathbf{P}(\mathbf{P}^*\mathbf{P})^{-1}\mathbf{P}^* : U \to U$$

is the orthogonal projection (relative to $||\cdot||_U$) onto $U_N = V_N$. In addition we have $Q_N \in L(V', V)$.

Lemma 8.1.7. *Let $L_N \in L(V_N, V'_N)$ be the operator corresponding to $a(\cdot,\cdot): V_N \times V_N \to \mathbb{R}$, and $L \in L(V, V')$ that for $a(\cdot,\cdot): V \times V \to \mathbb{R}$. The following relationships hold between the operators L_n, L, and the stiffness matrix $\mathbf{L}$:*

$\mathbf{L} = \mathbf{P}^* L \mathbf{P} = \mathbf{P}^* L_N \mathbf{P}$,
$L_N = \mathbf{P}^{*-1} \mathbf{L} \mathbf{P}^{-1} : V_N \to V_N'$,
L_N is the restriction of $Q_N L$ and $Q_N L Q_N$ to V_N.

PROOF. For all $\mathbf{u}, \mathbf{v} \in \mathbb{R}^N$ we have

$$\begin{aligned}\langle \mathbf{L}\mathbf{u}, \mathbf{v}\rangle = a(\mathbf{P}\mathbf{u}, \mathbf{P}\mathbf{v}) &= (L_N \mathbf{P}\mathbf{u}, \mathbf{P}\mathbf{v})_U = \langle \mathbf{P}^* L_N \mathbf{P}\mathbf{u}, \mathbf{v}\rangle, \\ &= (L\mathbf{P}\mathbf{u}, \mathbf{P}\mathbf{v})_U = \langle \mathbf{P}^* L \mathbf{P}\mathbf{u}, \mathbf{v}\rangle.\end{aligned}$$ ■

By Remark 6.3.12 we can write $(f, v)_U$ for $f(v)$. If $v = \mathbf{P}\mathbf{v}$, it follows from (10b) that $f(v) = (f, \mathbf{P}\mathbf{v})_U = \langle \mathbf{P}^* f, \mathbf{v}\rangle$. Using Remark 4 we then have

$$\mathbf{f} = \mathbf{P}^* f \tag{8.1.10c}$$

for the $\mathbf{f}$ in the right side of Eq. (9).

In the case of a continuous variational problem the V-ellipticity guarantees its unique solvability. The same condition is also sufficient in the discrete case:

Theorem 8.1.8. *Assume* (3). *Suppose the bilinear form is V-elliptic: $a(u,u) \geq C_E \|u\|_V^2$ for all $u \in V$ with $C_E > 0$. Then the matrix $\mathbf{L}$ in* (9) *is nonsingular and the Ritz-Galerkin solution $u^N \in V_N$ satisfies*

$$\|U^N\|_V \leq \frac{1}{C_E}\|f\|_{V_N'} \leq \frac{1}{C_E}\|f\|_{V'}. \tag{8.1.11a}$$

PROOF. $\mathbf{L}$ is nonsingular since for each $\mathbf{u} \neq \mathbf{0}$ we also have $\mathbf{P}\mathbf{u} \neq \mathbf{0}$, and thus

$$\langle \mathbf{L}\mathbf{u}, \mathbf{u}\rangle = a(\mathbf{P}\mathbf{u}, \mathbf{P}\mathbf{u}) \geq C_E \|\mathbf{P}\mathbf{u}\|_V^2 > 0$$

and so, in particular, $\mathbf{L}\mathbf{u} \neq \mathbf{0}$. By Exercise 6.5.6a $a(\cdot,\cdot)$ is also V_N-elliptic with the same constant C_E. From Theorem 6.5.8 there holds $\|L_N^{-1}\|_{V_N' \leftarrow V'_N} \leq 1/C_E$, i.e., $\|U^N\|_V = \|L_N^{-1}\|_V \leq C_E^{-1}\|f\|_{V_N'}$. (11a) then results from

Exercise 8.1.9. Show $\|f\|_{V_N'} \leq \|f\|_{V'}$ for any $f \in V'$.

Exercise 8.1.10. Show: (a) If $a(\cdot,\cdot)$ is symmetric then so is $\mathbf{L}$.
(b) If $a(\cdot,\cdot)$ is symmetric and V-elliptic then $\mathbf{L}$ is positive definite. Under the same assumptions the Ritz-Galerkin solution u^N solves the following variational problem (cf. Theorem 6.5.12):

$$J(u^N) \leq J(u) := a(u,u) - 2f(u) \quad \text{for all } u \in V_N.$$

Example 8.1.11. (Dirichlet Problem) The boundary value problem is

$$-\Delta u(x,y) = 1 \quad \text{in } \Omega = (0,1)\times(0,1), \qquad u = 0 \quad \text{on } \Gamma.$$

The weak formulation is given by (1) with $V = H_0^1(\Omega)$,

$$a(u,v) := \int_\Omega \langle \nabla u, \nabla v\rangle \,dx\,dy = \int_\Omega (u_x v_x + u_y v_y)\,dx\,dy,$$
$$f(v) := \int_\Omega v\,dx\,dy.$$

The functions

$$b_1(x,y) = \sin(\pi x)\sin(\pi y), \qquad b_2(x,y) = \sin(3\pi x)\sin(\pi y),$$
$$b_3(x,y) = \sin(\pi x)\sin(3\pi y), \qquad b_4(x,y) = \sin(3\pi x)\sin(3\pi y),$$

fulfil the boundary conditions and so belong to $V = H_0^1(\Omega)$. They form a basis of $V_4 := \text{span}\,\{b_1,\dots,b_4\}$. The matrix elements $L_{ii} = a(b_i,b_i)$ can be worked out to be

$$L_{11} = \pi^2/2, \qquad L_{22} = L_{33} = 5\pi^2/2, \qquad L_{44} = 9\pi^2/2.$$

In addition the chosen basis is $a(\cdot,\cdot)$-orthogonal:

$$L_{ij} = 0 \quad \text{for } i \neq j,$$

so that the stiffness matrix $\mathbf{L}$ is diagonal. Furthermore one may calculate $f_i = f(b_i) = \int_\Omega b_i(x,y)\,dx\,dy$, getting

$$f_1 = 4/\pi^2, \qquad f_2 = f_3 = 4/(3\pi^2), \qquad f_4 = 4/(9\pi^2).$$

Hence $\mathbf{u} = \mathbf{L}^{-1}\mathbf{f}$ has the components

$$u_1 = 8/\pi^4, \qquad u_2 = u_3 = 8/(15\pi^4), \qquad u_4 = 8/(81\pi^4),$$

and the Ritz-Galerkin solution is then

$$u^N(x,y) = \frac{8}{\pi^4}\Big[\sin\pi x\sin\pi y + \frac{1}{15}(\sin 3\pi x\sin\pi y + \sin\pi x\sin 3\pi y) + \frac{1}{81}\sin 3\pi x\sin 3\pi y\Big].$$

The Ritz-Galerkin solution and the exact solution for $x = y = 1/2$ are

$$u^N(\tfrac{1}{2},\tfrac{1}{2}) = \frac{2848}{405}\pi^{-4} = 0.07219140\dots,$$
$$u(\tfrac{1}{2},\tfrac{1}{2}) = \sum_{\nu,\mu=0}^{\infty} \frac{16}{\pi^4}(-1)^{\nu+\mu}/[(1+2\nu)(1+2\mu)((1+2\nu)^2+(1+2\mu)^2)]$$
$$= 0.0736713\dots.$$

Example 8.1.12. (Natural boundary conditions) Let the boundary value problem be

$$-\Delta u(x,y) = \pi^2 \cos \pi x \text{ in } \Omega = (0,1)\times(0,1), \qquad \partial u/\partial n = 0 \text{ on } \Gamma.$$

The solution is given by $u = \cos \pi x + \text{const}$. The weak formulation is in terms of (1) with $V = H^1(\Omega)$,

$$a(u,v) := \int_\Omega (u_x v_x + u_y v_y)\,dx\,dy,$$

$$f(v) := \pi^2 \int_\Omega v(x,y) \cos \pi x\,dx\,dy.$$

The boundary value problem has a unique solution in

$$W := \{v \in V\colon \int_\Omega v\,dx\,dy = 0\}.$$

The basis functions

$$b_1(x,y) = x - 1/2, \qquad b_2(x,y) = (x-1/2)^3$$

are in W. The stiffness matrix $\mathbf{L}$ and the vector $\mathbf{f}$ are then

$$\mathbf{L} = \begin{bmatrix} 1 & 1/4 \\ 1/4 & 9/80 \end{bmatrix}, \quad \mathbf{f} = \begin{bmatrix} -2 \\ -(3/2) + 12/\pi^2 \end{bmatrix},$$

so that

$$\mathbf{u} = \mathbf{L}^{-1}\mathbf{f} = \begin{bmatrix} 3 - 60/\pi^2 \\ -20 + 240/\pi^2 \end{bmatrix}.$$

The solution is $u^N(x,y) = (3 - 60/\pi^2)(x - 1/2) - (20 - 240/\pi^2)(x-1/2)^3$. The Ritz-Galerkin solution satisfies the boundary condition $\partial u/\partial n = 0$ and the differential equation only approximately:

$$\partial u^N(0,y)/\partial n = 12 - 120/\pi^2 \approx -0.16\,.$$

For $x = 1/4$ the approximation is $u^N(1/4,y) = -7/16 + 45/(4\pi^2) = 0.70236\ldots$, whereas $u(1/4,y) = \cos \pi/4 = 0.7071\ldots$ is the exact value.

In the following we shall consider the case in which $a(\cdot,\cdot)$ is no longer V-elliptic, though it is V-coercive. That $a(\cdot,\cdot)$ is V-coercive guarantees that either problem (1) is solvable or $\lambda = 0$ is an eigenvalue. Even if one assumes V-coercivity and the solvability of the problem (1) one can n o t deduce the solvability of the discrete problem (4).

Example 8.1.13. $a(u,v) := \int_0^1 (u'v' - 10uv)\,dx$ is $H_0^1(0,1)$-coercive and $a(u,v) = f(v) := \int_0^1 gv\,dx$ ($v \in H_0^1(0,1)$) has a unique solution. Let V^N be spanned by $b_1(x) = x(1-x) \in V = H_0^1(0,1)$ (i.e., $N = 1$). Then the discrete problem (4) is not solvable since $\mathbf{L} = \mathbf{0}$.

If one replaces the space V in Lemma 6.5.3 with V_N, then there follows from Exercise 6.5.4

Theorem 8.1.14. *The problem* (4) *is solvable for all* $f \in V'$ *and has a unique solution,* u^N, *which satisfies the estimate*

$$||u^N||_V \le \frac{1}{\epsilon_N}||f||_{V_N'} \le \frac{1}{\epsilon_N}||f||_{V'}, \tag{8.1.11b}$$

if and only if

$$\inf\{\sup\{|a(u,v)|: v \in V_N, ||v||_V = 1\}: u \in V_N, ||u||_V = 1\} = \epsilon_N > 0. \tag{8.1.12}$$

Since (4) is equivalent to the system of equations (9), one has the

Corollary 8.1.15. *The matrix* $\mathbf{L}$ *is nonsingular if and only if* (12) *holds.*

Exercise 8.1.16. The requirement (12) is equivalent to

$$||u||_V \le \frac{1}{\epsilon_N}\sup\{|a(u,v)|: v \in V_N, ||v||_V = 1\} \quad \text{for all } u \in V_N. \tag{8.1.12$'$}$$

and

$$||L_N^{-1}||_{V_N \leftarrow V_N'} = 1/\epsilon_N \qquad (L_N \text{ as in (10a)}). \tag{8.1.12$''$}$$

Warning: The condition (12) for V_N does not follow from the analogous condition (6.4.5a) for V. See, however, Theorem 8.2.8.

The requirement (12″) guarantees the existence of $\mathbf{L}^{-1}$, but it does not say anything about its condition, $\text{cond}(\mathbf{L}) = ||\mathbf{L}||||\mathbf{L}^{-1}||$, which is the deciding factor for the sensitivity of the system of equations $\mathbf{Lu} = \mathbf{f}$. For example, if one chooses for $a(u,v) := \int_0^1 u'v'\,dx$ the basis $b_i = x^i$ $(i = 1, \ldots, N)$ then one obtains the very badly conditioned matrix $l_{ij} = ij/(i+j-1)$. The conditioning of $\mathbf{L}$ is optimal if one chooses the basis to be $a(\cdot,\cdot)$-orthogonal: $a(b_i, b_j) = \delta_{ij}$, as is the case in Example 11, up to a scaling factor.

8.2 Error Estimates

For difference methods the solution u and the grid function u_h are defined on different sets. The Ritz-Galerkin solution, u^N, is, on the other hand, directly comparable with u. One can measure the error due to discretisation with $||u - u^N||_V$ or with $||u - u^N||_U$.

Let u be the solution of (1.1): $a(u,v) = f(v)$ for $v \in V$. Suppose — by chance or because of a clever choice of V_N — that u also belongs to V_N; then $u^N := u$ also satisfies (1.4). That means: The discretisation error is zero if $u \in V_N$. We shall now show: The "closer" that u is to V_N the smaller is the discretisation error.

Theorem 8.2.1. (*Céa*) *Assume* (1.2), (1.3), *and* (1.12) *hold. Let* $u \in V$ *be a solution of the problem* (1.1), *and let* $u^N \in V_N$ *be the Ritz-Galerkin solution of* (1.4). *Then the following estimate holds:*

$$\|u - u^N\|_V \le (1 + C_S/\epsilon_N) \inf_{w \in V_N} \|u - w\|_V \tag{8.2.1}$$

with C_S *from* (1.2) *and* ϵ_N *from* (1.12). *Note* $\inf_{w \in V_N} \|u - w\|_V$ *is the distance of the function* u *from* V_N; *it will be abbreviated in the following to*

$$d(u, V_N) := \inf_{w \in V_N} \|u - w\|_V. \tag{8.2.2}$$

PROOF. If u satisfies $a(u, v) = f(v)$ for all $v \in V$ then it does in particular for all $v \in V_N$. Since we also have $a(u^N, v) = f(v)$ for $v \in V_N$, it follows that

$$a(u^N - u, v) = 0 \quad \text{for all} \quad v \in V_N. \tag{8.2.3}$$

For arbitrary $v, w \in V_N$ with $\|v\|_V = 1$ we can therefore conclude

$$a(u^N - w, v) = a([u^N - u] + [u - w], v) = a(u - w, v)$$

and

$$|a(u^N - w, v)| \le C_S \|u - w\|_V \|v\|_V = C_S \|u - w\|_V.$$

From (1.12′) we then obtain

$$\|u^N - w\|_V \le \frac{1}{\epsilon_N} \sup\{|a(u^N - w, v)| : v \in V_N, \|v\|_V = 1\} \le (C_S/\epsilon_N)\|u - w\|_V.$$

The triangle inequality then gives

$$\|u - u^N\|_V \le \|u - w\|_V + \|w - u^N\|_V \le (1 + C_S/\epsilon_N)\|u - w\|_V.$$

Since $w \in V_N$ is arbitrary we deduce the assertion (1). ■

In Theorem 1 the unique solvability of the problem (1.1) was not assumed, but only the existence of at least one solution.

If the discretisation error is supposed to converge to zero one makes use of a sequence of subspaces $V_{N_i} \subset V$ that converge to V in the following sense:

Theorem 8.2.2. *Let* $V_i := V_{N_i} \subset V (i \in \mathbb{N})$ *be a sequence of subspaces with*

$$\lim_{i \to \infty} d(u, V_i) = 0 \quad \text{for all} \quad u \in V. \tag{8.2.4a}$$

Assume that (1.12) *holds with* $\epsilon_{N_i} \ge \tilde{\epsilon} > 0$ *for all* $i \in \mathbb{N}$; *in addition assume* (1.2) (*continuity of* $a(\cdot, \cdot)$). *Then there exists a unique solution* u *of the problem* (1.1) *and the Ritz-Galerkin solution* $u^i := u^{N_i}$ *converges to* u:

$$\|u - u^i\|_V \to 0 \qquad \text{for} \quad i \to \infty.$$

Sufficient to ensure (4a) *is*

$$V_1 \subset V_2 \subset \ldots \subset V_{i-1} \subset V_i \subset \ldots \subset V, \qquad \bigcup_{i=1}^{\infty} V_i \textit{ dense in } V. \tag{8.2.4b}$$

PROOF. (a) Assume first the existence of a solution u.
(aa) The estimate (1) implies the convergence:

$$||u - u^i||_V \le (1 + C_S/\epsilon_{N_i})\, d(u, V_i) \to 0.$$

(ab) We now wish to show that (4a) follows from (4b). The inclusion $V_{i-1} \subset V_i$ implies $d(u, V_i) \le d(u, V_{i-1})$. Now $d(u, V_i)$ will be a null sequence if for each $\epsilon > 0$ there exists an i such that $d(u, V_i) \le \epsilon$. From the assumption (4b) there is, for each $u \in V$ and $\epsilon > 0$, a $w \in \bigcup_i V_i$ with $||u - w||_V \le \epsilon$. Therefore we have $w \in V_i$ for an $i \in \mathbb{N}$. That $d(u, V_i) \le ||u - w||_V \le \epsilon$ then proves (4a). The convergence $u^i \to u$ implies the uniqueness of the solution u.
(b) The next thing to show is that the image $W := \{Lv: v \in V\} \subset V'$ of the operator $L: V \to V'$ associated to $a(\cdot, \cdot)$ is closed. For each $f \in W$ there is a $u \in V$ with $Lu = f$, so that part (a) of this proof suffices to show the convergence $u^i \to u$. Since $||u^i||_V \le ||f||_{V'}/\tilde{\epsilon}$, it follows that $||u||_V = \lim ||u^i||_V \le ||f||_{v'}/\tilde{\epsilon}$. Let $f_\nu \in W$ be a sequence with $f_\nu \to f^*$ in V' and $f_\nu = Lu_\nu$. The Cauchy convergence $||f_\nu - f_\mu||_{V'} \to 0$ shows $||u_\nu - u_\mu||_V \le ||f_\nu - f_\mu||_{V'}/\tilde{\epsilon} \to 0$ so that the limit $u^* = \lim u_\nu \in V$ exists. The continuity of $L \in L(V, V')$ shows that $f^* = \lim f_\nu = \lim Lu_\nu = Lu^*$, and thus that $f^* \in W$. Hence W is closed.
(c) In order to demonstrate the existence of a solution to the problem (1.1) we have to show the surjectivity of $L: V \to V'$. If L were not surjective (i.e., $W \ne V'$), there would be an $f \in W^\perp$ with $||f||_{V'} = 1$. Let $J_V: V \to V'$ be the Riesz isomorphism (cf. Corollary 6.3.7). Set $v := -J_V^{-1} f \in V$. It follows that

$$f(v) = \langle f, v\rangle_{V' \times V} = -(f, f)_{V'} = -1, \qquad a(u, v) = \langle Lu, v\rangle_{V' \times V} = 0$$

for all $u \in V$. Let $u^i \in V_i$ be the Ritz-Galerkin solutions. They must also satisfy $a(u^i, v) = 0$, i.e.,

$$a(u^i, v) - f(v) = 1.$$

We split up v into $v^i + w^i$, where $v^i \in V_i$. By (4a), with v in place of u, one may guarantee that $||w^i||_V \to 0$. This shows

$$1 = a(u^i, v) - f(v) = a(u^i, v^i) - f(v^i) + a(u^i, w^i) - f(w^i)$$
$$= a(u^i, w^i) - f(w^i)$$

and

$$1 = |a(u^i, v) - f(v)| \le \left[C_S ||u^i||_V + ||f||_{V'}\right] ||w^i||_V.$$

Since $||u^i||_V \le ||f||_{V'}/\tilde{\epsilon}$ is uniformly bounded and $||w^i||_V \to 0$ this amounts to a contradiction. Therefore L must be surjective, so that for each $f \in V'$ there exists a solution u to the equation $Lu = f$, i.e., to the problem (1.1). ■

Corollary 8.2.3. *The requirement* (1.12) *with* $\epsilon_{N_i} \geq \tilde{\epsilon} > 0$ *from Theorem* 2 *is satisfied with* $\tilde{\epsilon} := C_E$ *if* $a(\cdot,\cdot)$ *is* V*-elliptic:* $a(u,u) \geq C_E\|u\|_V^2$.

Exercise 8.2.4. Show that (4a) implies that $\bigcup_{i=1}^{\infty} V_i$ is dense in V.

Let Q_N be the orthogonal projection onto V_N (cf. Exercise 8.1.6c). The factors $L: V \to V'$, $Q_N: V' \to V_N' = V_N$, $L_N^{-1}: V_N' = V_N \to V_N \subset V$ can be composed to give

$$S_N := L_N^{-1} Q_N L: V \to V_N \subset V. \tag{8.2.5a}$$

Exercise 8.2.5. Show there is also the representation

$$S_N = \mathbf{P}L^{-1}\mathbf{P}^* L. \tag{8.2.5a$'$}$$

Lemma 8.2.6. S_N *is the projection onto* V_N *and is called the* R i t z p r o j e c t i o n. *It sends the solution* u *of the problem* (1.1) *to the Ritz-Galerkin solution* $u^N \in V_N$*:* $u^N = S_N u$. *Assuming* (1.2) *and* (1.12) *we have*

$$\|S_N\|_{V \leftarrow V} \leq C_S/\epsilon_N. \tag{8.2.5b}$$

A definition of S_N *equivalent to that in* (5a) *is*

$$S_N u \in V_N \quad \text{and} \quad a(S_N u, v) = a(u,v) \quad \text{for all } v \in V_N,\ u \in V. \tag{8.2.5c}$$

PROOF. (a) Since

$$a(u,v) = \langle Lu, v\rangle_{V' \times V} = (Lu, v)_U = (Lu, Q_N v)_U = (Q_N Lu, v)_U$$

for all $v \in V_N$, it follows that $S_N u$ in (5c) is the Ritz-Galerkin solution for the right-hand side $f := Q_N Lu \in V'$, i.e., $S_N u = L_N^{-1} Q_N Lu$. Conversely one may argue similarly, and so show the equivalence of the definitions (5a) and (5c).
(b) (5c) shows that $u \in V_N$ leads to $S_N u = u$. Thus $S_N^2 = S_N$, i.e., S_N is a projection. Inequality (5b) follows from $\|S_N u\|_V \leq \|Q_N Lu\|_{V_N'}/\epsilon_N$ (cf. the proof of (1.11)) and $\|Q_N Lu\|_{V_N'} = \|Lu\|_{V_N'} \leq \|Lu\|_V' \leq C_S\|u\|_V$ with C_S from (1.2) (cf. Exercise 8.1.9). ∎

Remark 8.2.7. Let $a(\cdot,\cdot)$ be V-elliptic and symmetric. $|||v|||_V := a(v,v)^{1/2}$ is a norm equivalent to $\|\cdot\|_V$. The Ritz projection S_N is, with respect to $|||\cdot|||_V$, an orthogonal projection onto V_N. Thus, in particular, we have

$$|||S_N|||_{V \leftarrow V} \leq 1. \tag{8.2.5d}$$

PROOF. The scalar product associated to $|||\cdot|||_V$ is $a(\cdot,\cdot)$, so that it is to be shown that $a(S_N v, w) = a(v, S_N w)$. (5c) implies $a(S_N v, S_N w) = a(v, S_N w)$, since $S_N w \in V_N$. The symmetry of $a(\cdot,\cdot)$ and exchanging v and w give

$a(S_N v, w) = a(w, S_N v) = a(S_N w, S_N v) = a(S_N v, S_N w)$, so that $a(S_N v, w) = a(v, S_N w)$. (5d) results from Remark 6.3.8. ■

Remark 7 shows once again that the Ritz-Galerkin solution $u^N = S_N u$ is the best approximation to u in V_N in the sense of the norm $|||\cdot|||_V$. This assertion is equivalent to the variational formulation $J(u^N) \le J(v)$ for all $v \in V_N$ (cf. Exercise 8.1.10b).

The condition (1.12), with $\epsilon_n \ge \epsilon > 0$, is difficult to prove, except for V-elliptic bilinear forms. However, the following theorem shows that this condition does hold for subspaces approximating well enough.

Theorem 8.2.8. *Let the bilinear form $a(\cdot,\cdot)$ be V-coercive, where $V \subset U \subset V'$ is a continuous, dense, and compact embedding. Let Problem* (1.1) *be solvable for all $f \in V'$. Assume that* (4a) *holds for the subspaces $V_i \subset V$. For large enough i the stability condition* (1.12) *is then satisfied with $\epsilon_{N_i} \ge \epsilon > 0$.*

The proof of this will be postponed to a supplement to Lemma 11.2.7.

8.3 Finite Elements

8.3.1 Introduction: Linear Elements for $\Omega = (a, b)$

As soon as the dimension $N = \dim V_N$ becomes larger the essential disadvantage of the general Ritz-Galerkin method becomes apparent. The matrix $\mathbf{L}$ is in general full, i.e., $L_{ij} \neq 0$ for all $i, j = 1, \ldots, N$. Therefore one needs N^2 integrations to obtain the values of $L_{ij} = a(b_j, b_i) = \int_\Omega \ldots$, whether exactly or approximately. The final solution of the system of equations $\mathbf{Lu} = \mathbf{f}$ requires $O(N^3)$ operations. As soon as N is no longer small the general Ritz-Galerkin method therefore turns out to be unusable.

A glance at the difference method shows that the matrices L_h which occur there are sparse. Thus it is natural to wonder if it is possible to choose the basis $\{b_1, \ldots, b_N\}$ so that the stiffness matrix $L_{ij} = a(b_j, b_i)$ is also sparse. The best situation would be that the b_i were orthogonal with respect to $a(\cdot,\cdot)$: $a(b_j, b_i)$ for $i \neq j$. However, such a basis can be found only for special model problems such as the one in Example 8.1.11. Instead we shall base our further considerations on

Remark 8.3.1. Let the bilinear form $a(\cdot,\cdot)$ be given by (7.2.6). Let B_i be the interior of the support $\operatorname{supp}(b_i)$ of the basis function b_i, i.e., $B_i := \operatorname{supp}(b_i) \backslash \partial \operatorname{supp}(b_i)$. A sufficient condition that ensures $L_{ij} = a(b_j, b_i) = 0$ is $B_i \cap B_j = \emptyset$.

PROOF. The integration $a(b_j, b_i) = \int_\Omega \ldots$ can be restricted to $B_i \cap B_j$. ■

In order to be able to apply Remark 1 the basis functions should have as small supports as possible. In constructing them in general one goes about it from the desired goal: one defines partitions of Ω into small pieces, the so-called finite elements, from which the supports of b_i are pieced together.

As an introduction let us investigate the one-dimensional boundary value problem

$$-u''(x) = g(x) \quad \text{for } a < x < b, \qquad u(a) = u(b) = 0. \tag{8.3.1}$$

Assume we have a partition of the interval $[a, b]$ given by $a = x_0 < x_1 < \ldots < x_{N+1} = b$. Denote the interval pieces by $I_i := (x_{i-1}, x_i)$ $(1 \le i \le N+1)$. For the subspace $V_N \subset H_0^1(a, b)$ let us choose the piecewise linear functions:

$$\begin{aligned} V_N = \{u \in C^0([a,b])\colon & \text{the restriction of } u \text{ to } I_i \ (1 \le i \le N+1) \\ & \text{is linear; } u(a) = u(b) = 0\} \end{aligned} \tag{8.3.2}$$

See the figure below.

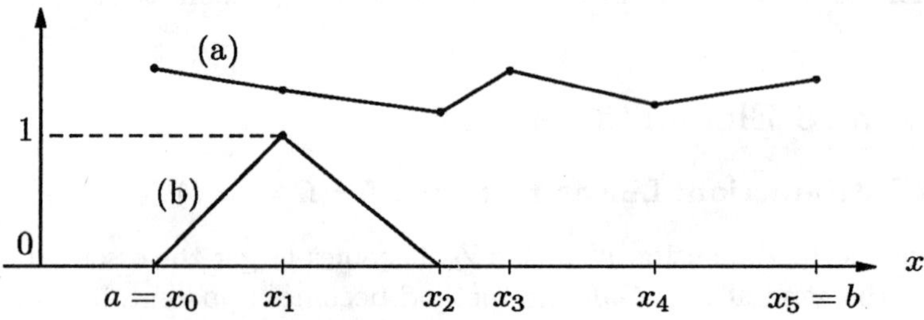

Figure 8.3.1. (a) a piecewise linear function, (b) a basis function

The continuity "$u \in C^0([a, b])$" is equivalent to continuity at the nodes x_i $(1 \le i \le N)$: $u(x_i + 0) = u(x_i - 0)$.

Remark 8.3.2. $u \in V_N$ is uniquely determined by its node values $u(x_i)$ $(1 \le i \le N)$:

$$u(x) = [u(x_i)(x_{i+1} - x) + u(x_{i+1})(x - x_i)]/[x_{i+1} - x_i] \quad \text{for } x \in I_{i+1}, \tag{8.3.3a}$$

where $u(x_0) = u(x_{N+1}) = 0$. From Theorem 6.2.42 we conclude that $V_N \subset H_0^1(a, b)$. The (weak) derivative $u' \in L^2(a, b)$ is piecewise constant:

$$u'(x) = [u(x_{i+1}) - u(x_i)]/[x_{i+1} - x_i] \quad \text{for } x \in I_{i+1}. \tag{8.3.3b}$$

The basis functions can be defined (see Figure 1b) by

$$b_i(x) = \begin{cases} (x - x_{i-1})/(x_i - x_{i-1}) & \text{for } x_{i-1} < x \le x_i \\ (x_{i+1} - x)/(x_{i+1} - x_i) & \text{for } x_i < x < x_{i+1}, \\ 0 & \text{otherwise.} \end{cases} \quad 1 \le i \le N \tag{8.3.4}$$

Remark 2 then has as a consequence

Remark 8.3.3. One has the representation $u = \sum_{i=1}^{N} u(x_i)b_i$. The supports of the functions b_i are $\overline{I}_{i-1} \cup \overline{I}_i = [x_{i-1}, x_{i+1}]$. The B_i from Remark 1 is now $B_i = (x_{i-1}, x_{i+1})$.

The weak formulation of the boundary value problem (1) is $a(u,v) = f(v)$ with

$$a(u,v) := \int_a^b u'v'\,dx, \qquad f(v) := \int_a^b gv\,dx \quad \text{for} \quad u,v \in H_0^1(a,b).$$

By Remark 1 we have, for the matrix elements, $L_{ij} = 0$ as soon as $|i-j| \geq 2$, since then $B_i \cap B_j = \emptyset$. For $|i-j| \leq 1$ one obtains

$$\begin{aligned} L_{i,i-1} &= a(b_{i-1}, b_i) = \int_{x_{i-1}}^{x_i} \frac{-1}{x_i - x_{i-1}} \frac{1}{x_i - x_{i-1}}\,dx = -1/(x_i - x_{i-1}), \\ L_{ii} &= a(b_i, b_i) = \int_{x_{i-1}}^{x_i} (x_i - x_{i-1})^{-2}\,dx + \int_{x_i}^{x_{i+1}} (x_{i+1} - x_i)^{-2}\,dx \qquad (8.3.5a) \\ &= 1/(x_i - x_{i-1}) + 1/(x_{i+1} - x_i), \\ L_{i,i+1} &= -1/(x_{i+1} - x_i). \end{aligned}$$

The right-hand side $\mathbf{f} = (f_1, \ldots, f_N)^\mathsf{T}$ is given by (5b):

$$\begin{aligned} f_i &= f(b_i) = \int_{x_{i-1}}^{x_{i+1}} gb_i\,dx \\ &= \frac{1}{x_i - x_{i-1}} \int_{x_{i-1}}^{x_i} g(x)(x - x_{i-1})\,dx + \frac{1}{x_{i+1} - x_i} \int_{x_i}^{x_{i+1}} g(x)(x_{i+1} - x)\,dx. \end{aligned} \qquad (8.3.5b)$$

Remark 8.3.4. The system of equations $\mathbf{Lu} = \mathbf{f}$, given in (5a,b), is tridiagonal, thus, in particular, sparse. For an equidistant partitioning $x_i := a + ih$ with $h := (b-a)/(N+1)$ the coefficients are $L_{i,i\pm1} = -1/h$, $L_{i,i} = 2/h$, $f_i = h\int_0^1 [g(x_i + th) + g(x_i - th)](1-t)\,dt$. If $g \in C^0(a,b)$, then by the intermediate value theorem $f_i = hg(x_i + \Theta_i h)$, with $|\Theta_i| < 1$. Therefore the system of equations $\mathbf{Lu} = \mathbf{f}$ is, up to a scaling, identical with the difference equation $L_h u_h = f_h$ in Section 4.1, if one defines f_h by $f_h(x_i) := f_i = g(x_i + \Theta_i h)$ instead of by $f_h(x_i) = g(x_i)$: then $\mathbf{L} = hL_h, \mathbf{f} = hf_h$.

This example shows that it is possible to find basis functions with small supports so that $\mathbf{L}$ is relatively easy to calculate. In addition, the similarity is apparent between the resulting discretisation method and difference methods. Finally consider the numerical evaluation of the integral (5b).

Exercise 8.3.5. Show that the Gaussian quadrature formula for $\int_{x_{i-1}}^{x_{i+1}} gb_i\,dx$ with $b_i(x)$ as weight function and one support point is:

$$f_i := \frac{x_{i+1} - x_{i-1}}{2} g\left(\frac{x_{i-1} + x_i + x_{i+1}}{3}\right). \tag{8.3.5c}$$

What is the formula for a partition of equal intervals?

If one replaces the Dirichlet boundary conditions in (1) by the Neumann condition then $V = H^1(a,b)$. The subspace $V_N \subset H^1(a,b)$ results from (2) after removal of the condition $u(a) = u(b) = 0$. In order that $\dim V_N = N$ the numbering has to be changed: The partition of (a,b) is given by $a = x_1 < x_2 < \ldots < x_N = b$.

Exercise 8.3.6. Let $\partial u(a)/\partial n = g_a, \partial u(b)/\partial n = g_b$, with $a(\cdot,\cdot)$ as before. Suppose the partition of (a,b) is equidistant: $x_i = a + (b-a)(i-1)/(N-1)$. What is the form of the equation $\mathbf{Lu} = \mathbf{f}$? Show that $\mathbf{Lu} = \mathbf{f}$ has at least one solution if $\int_a^b g(x)\,dx + g_a + g_b = 0$.

8.3.2 Linear Elements for $\Omega \subset \mathbb{R}^2$

We shall assume:

$$\Omega \subset \mathbb{R}^2 \text{ is a polygon.} \tag{8.3.6}$$

As shown in Fig. 2 we divide Ω into triangles.

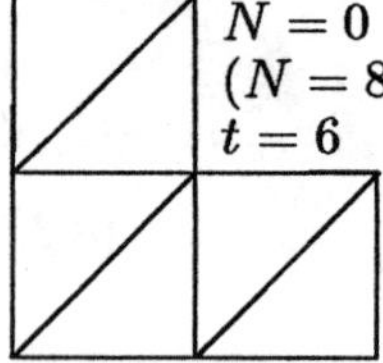

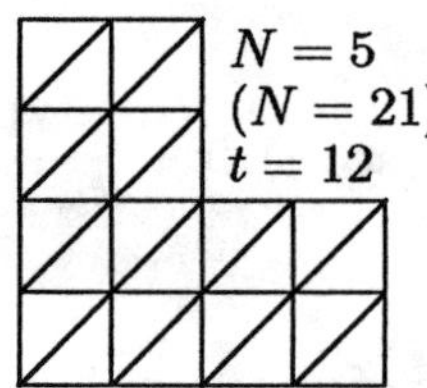

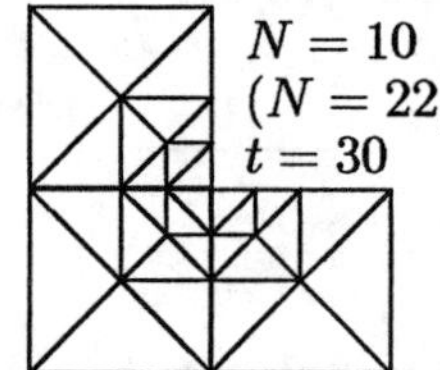

Figure 8.3.2. Triangulations of Ω, $t :=$ number of triangles, $N :=$ number of inner nodes (number of all nodes)

Definition 8.3.7. $\tau := \{T_1, \ldots, T_t\}$ is called an admissible triangulation of Ω if the following conditions are fulfilled:

$$T_i \ (1 \le i \le t) \text{ are open triangles ("finite elements")} \tag{8.3.7a}$$

$$T_i \text{ are disjoint, i.e., } T_i \cap T_j = \emptyset \text{ for } i \ne j, \tag{8.3.7b}$$

$$\bigcup_{1 \le i \le t} \overline{T}_i = \overline{\Omega}, \tag{8.3.7c}$$

for $i \ne j$ the set $\overline{T}_i \cap \overline{T}_j$ is either

i) empty, or

ii) a common side of the elements T_i and T_j, or

iii) a common edge of the elements T_i and T_j. (8.3.7d)

Remark 8.3.8. (a) The conditions (7a) and (7c) imply the polygonal shape of Ω, i.e., the assumption (6). (b) Figure 2 shows only admissible triangulations. An example of an inadmissible triangulation would be, e.g., a square partitioned as follows: ⊠

Let τ be an admissible triangulation. The point $\mathbf{x}$ is called a node (of τ) if $\mathbf{x}$ is a vertex of one of the $T_i \in \tau$. One distinguishes interior and boundary nodes, according to whether $\mathbf{x} \in \Omega$ or $\mathbf{x} \in \partial\Omega$. Let N be the total number of interior nodes. We define V_N as the subspace of the piecewise linear functions:

$$V_N := \{u \in C^0(\overline{\Omega}) : u = 0 \text{ on } \partial\Omega;\ \text{on each } T_i \in \tau \text{ the function } u \text{ agrees with a linear function, i.e., } u(x,y) = a_{i1} + a_{i2}x + a_{i3}y \text{ on } T_i\}. \tag{8.3.8}$$

Remark 2 can be applied to the two-dimensional case under consideration:

Remark 8.3.9. (a) It is true that $V_N \subset H_0^1(\Omega)$. (b) Each function is uniquely determined by its node values $u(\mathbf{x}_i)$ at the interior nodes $\mathbf{x}_i$, $1 \le i \le N$.

PROOF. (a) Example 6.2.5 shows that $V_N \subset H^1(\Omega)$. Since $u = 0$ on $\partial\Omega$, Theorem 6.2.42 proves that $V_N \subset H_0^1(\Omega)$.
(b) Let $\mathbf{x}$, $\mathbf{x}'$, $\mathbf{x}''$ be the three vertices of $T_i \in \tau$. The linear function $u(x,y) = a_{i1}+a_{i2}x+a_{i3}y$ is uniquely determined on T_i by the values $u(\mathbf{x})$, $u(\mathbf{x}')$, $u(\mathbf{x}'')$. ■

The converse to Remark 9b is as follows:

Remark 8.3.10. Let $\mathbf{x}^i$ $(1 \le i \le N)$ be the interior nodes of τ. For an arbitrary u_i $(1 \le i \le N)$ there exists exactly one $u \in V_N$ with $u(\mathbf{x}^i) = u_i$. It may be written as $u = \sum_{i=1}^N u_i b_i$, where the basis functions b_i are characterised by

$$b_i(\mathbf{x}^i) = 1, \qquad b_i(\mathbf{x}^j) = 0 \quad \text{for } j \ne i. \tag{8.3.9a}$$

If $T \in \tau$ is a triangle with the vertices $\mathbf{x}^i = (x_i, y_i)$ [$\mathbf{x}^i$ as in (9a)] and $\mathbf{x}' = (x', y')$, $\mathbf{x}'' = (x'', y'')$, then

$$b_i(x,y) = \frac{(x-x')(y''-y') - (y-y')(x''-x')}{(x_i-x')(y''-y') - (y_i-y')(x''-x')} \quad \text{on } T. \tag{8.3.9b}$$

On all $T \in \tau$ that do not have an $\mathbf{x}^i$ as a vertex, $b_i = 0$.

PROOF. Obviously, using (9b) we get a linear function defined on all $T \in \tau$, which satisfies (9a). In addition, (9a) forces continuity at all nodes. If the vertices $\mathbf{x}^j = (x_j, y_j)$ and $x^k = (x_k, y_k)$ are directly connected by the side of a triangle, then (9b) provides the representation $b_i(\mathbf{x}^j + s(\mathbf{x}^k - \mathbf{x}^j)) = sb_i(\mathbf{x}^k) + (1-s)b_i(\mathbf{x}^j)$ with $s \in [0,1]$ for both triangles that have this side in common. Thus b_i is also continuous on the common edges, so that $b_i \in C^0(\Omega)$. Applying this consideration to two boundary nodes $\mathbf{x}', \mathbf{x}''$ with $b_i(\mathbf{x}') = b_i(\mathbf{x}'') = 0$ we derive $b_i = 0$ on $\partial\Omega$. Thus b_i belongs to V_N. ■

Remark 8.3.11. (a) The dimension of the subspace determined by (8) is the number of interior nodes of τ.
(b) The support of the basis function b_i is

$$\operatorname{supp}(b_i) = \bigcup\{\overline{T}\colon\ T \in \tau \text{ has } \mathbf{x}^i \text{ as a corner}\}.$$

(c) Let B_i be the interior of $\operatorname{supp}(b_i)$. There holds $B_i \cap B_j = \emptyset$ if and only if the nodes $\mathbf{x}^i$ and $\mathbf{x}^j$ are directly connected by a side.

Corollary 8.3.12. *$a(u,v) = \int_\Omega \langle \nabla u, \nabla v \rangle\, dx$ is the bilinear form associated to the Poisson equation. The integrals $L_{ij} = a(b_j, b_i) = \sum_k \int_{T_k} \langle \nabla b_j, \nabla b_i \rangle\, dx$ are to be taken over the following T_k:*
(i) all T_k with $\mathbf{x}^i$ as a vertex, if $i = j$;
(ii) all T_k with $\mathbf{x}^i$ and $\mathbf{x}^j$ as vertices, if $i \neq j$.
(iii) We have $L_{ij} = 0$ if $\mathbf{x}^i$ and $\mathbf{x}^j$ are not directly connected by the side of a triangle.

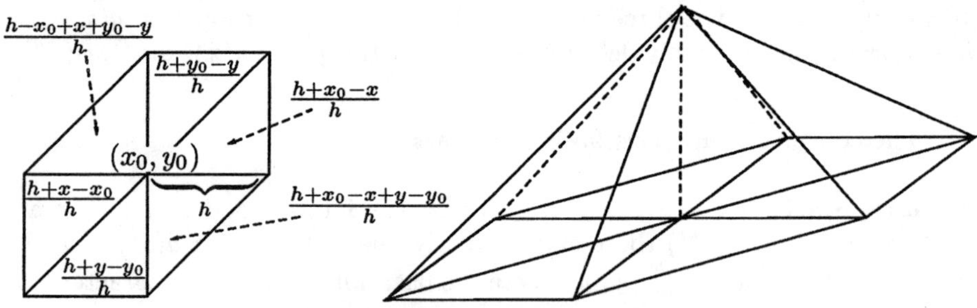

Figure 8.3.3. Basis functions for the node (x_0, y_0)

We get an especially regular triangulation when we first divide into squares with sides of length h, and then divide these into two triangles (⧄). The first and second triangulations in Figure 2 are of this sort. We call them "square grid triangulations". The corresponding basis function is depicted in Figure 3. One therefore expects that the matrix $\mathbf{L}$ corresponds to a 7-point formula. For the Laplace operator, however, one finds the well-known 5-point formula (4.2.11) from Section 4.2:

Exercise 8.3.13. Let τ be a square grid triangulation. Further let $a(u,v) = \int_\Omega \langle \nabla u, \nabla v\rangle\, d\mathbf{x}$. The basis functions are described by Figure 3. Show that one has for the entries of the stiffness matrix $\mathbf{L}$

$$L_{ii} = 4, \quad L_{ij} = -1 \text{ if } \mathbf{x}^i - \mathbf{x}^j = (0, \pm h) \text{ or } (\pm h, 0), \quad L_{ij} = 0 \text{ otherwise };$$

i.e., $\mathbf{L}$ agrees with $h^2 L_h$ from (4.2.8).

Although $\mathbf{L} = h^2 L_h$ holds in the case of the Poisson equation $-\Delta u = g$, the finite-element discretisation and the difference method still do not coincide, since $h^2 f_h$ has $h^2 g(\mathbf{x}^i)$ as components and so differs from $f_i = \int_\Omega g b_i\, d\mathbf{x}$.

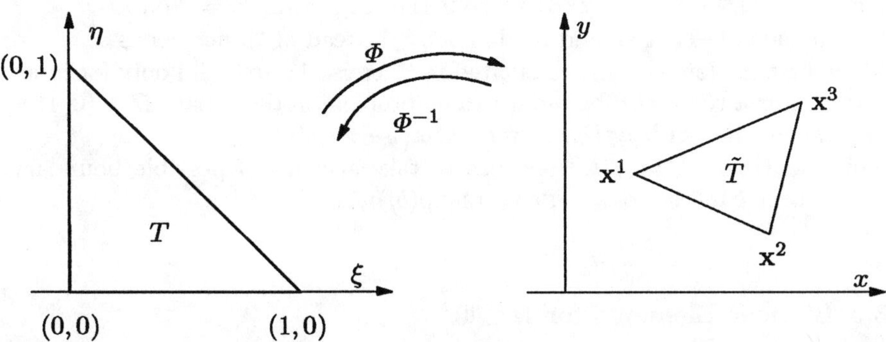

Figure 8.3.4. Reference triangle T

The integration $\int_{T_i} \ldots d\mathbf{x}$ over the triangle $T_i \in \tau$ seems at first difficult. However, for each i one can express $\int_{T_i} \ldots d\mathbf{x}$ as an integral over the reference triangle T in Figure 4. The details are in

Exercise 8.3.14. Let $\mathbf{x}^i = (x^i, y^i)$ $(i = 1, 2, 3)$ be the vertices of $\tilde{T} \in \tau$, and let T be the unit triangle in Figure 4. Show:
(a) $\phi : (\xi, \eta) \mapsto \mathbf{x}^1 + \xi(\mathbf{x}^2 - \mathbf{x}^1) + \eta(\mathbf{x}^3 - \mathbf{x}^1)$ maps T onto $\tilde{T}$.
(b) $\det \phi'(\xi, \eta) = (x^2 - x^1)(y^3 - y^1) - (y^2 - y^1)(x^3 - x^1)$ for all $\xi, \eta \in \mathbb{R}$.
(c) The substitution rule gives

$$\int_{\tilde{T}} \mathbf{v}(\mathbf{x}, \mathbf{y})\, d\mathbf{x}\, d\mathbf{y} = |(x^2 - x^1)(y^3 - y^1) - (y^2 - y^1)(x^3 - x^1)| \int_T \mathbf{v}(\phi(\xi, \eta))\, d\xi\, d\eta. \tag{8.3.10}$$

In general one evaluates the integral $\int_T \ldots d\xi\, d\eta$ over the unit triangle numerically. Examples of integration formulae can be found in Schwarz [1, §2.4.3] and Ciarlet [1, §4.1]. The necessary ordering of the quadrature formulae is discussed in the same place by Ciarlet [14] and by Witsch [1].

In contrast to the difference methods finite-element discretisation offers one the possibility of changing the size of the triangles locally. The third triangulation in Figure 2 contains triangles that become smaller as they are nearer to the intruding corner. This flexibility of the finite-element method is an essential advantage. On the other hand, one does obtain systems of equations $\mathbf{Lu} = \mathbf{f}$ with more complex structures, since (a) $\mathbf{u}$ can no longer be stored in a two-dimensional array, (b) $\mathbf{L}$ cannot be characterised by a star as in (4.2.12).

Remark 8.3.15. If one replaces the Dirichlet condition $u = 0$ on $\partial\Omega$ by natural boundary conditions, then the following changes take place:
(a) $N = \dim V_N$ is the number of all nodes (inner and boundary nodes).
(b) $V_N \subset H^1(\Omega)$ is given by (8) without the restriction "$u = 0$ on $\partial\Omega$".
(c) In Remarks 9–11 it should read "nodes" instead of "inner nodes".
(d) The matrix elements L_{ij} calculated in Exercise 13 are valid only for inner nodes $\mathbf{x}^i$. For a Neumann boundary value problem in the square $\Omega = (0,1) \times (0,1)$, $\mathbf{L}$ coincides with $h^2 D_h L_h$ from Exercise 4.7.8b.
(e) In calculating $f_i = f(b_i)$ one has to take account of possible boundary integrals over $\partial\Omega \cap \operatorname{supp}(b_i)$, if $\partial\Omega \cap \operatorname{supp}(b_i) \neq \emptyset$.

8.3.3 Bilinear Elements for $\Omega \subset \mathbb{R}^2$

The difference procedures in a square grid are near to partitioning Ω into squares of side h (cf. Figure 5a). If, more generally, one replaces the squares by parallelograms one obtains partitions like those in Figure 5a,b. An admissible partition by parallelograms is described by the conditions (7a–d), if in (7a) the expression "triangle" is replaced by "parallelogram".

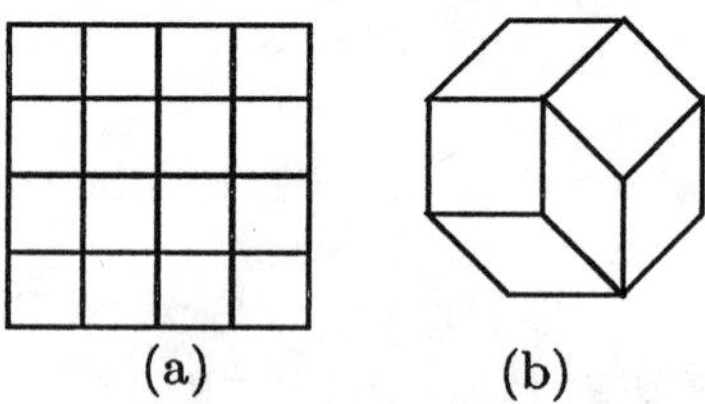

Figure 8.3.5. Partition of Ω into parallelograms

If one were to define the subspace V_N by the condition that u must be a linear function in each parallelogram, then there would be only three of the four vertex values that could be arbitrarily assigned. In the case of the partition in Figure 5b one can see that the only piecewise linear function u with $u = 0$ on $\partial\Omega$ is the null function. Thus in each parallelogram u must be a function that involves four free parameters. We next consider the case of a rectangle parallel to the axes $P = (x_1, x_2) \times (y_1, y_4)$ and define a bilinear function on P by

$$u(x, y) = (a_1 + a_2 x)(a_3 + a_4 y). \tag{8.3.11a}$$

Then u is linear in each direction parallel to the axes — thus, in particular, along the sides of the rectangle. For an arbitrary parallelogram $\tilde{P}$ such as in Figure 6a, the restriction of the function (11a) to the side of a parallelogram is in general a quadratic function. Therefore one generalises the definition as follows. Let

$$\phi\colon (\xi, \eta) \mapsto \mathbf{x}^1 + \xi(\mathbf{x}^2 - \mathbf{x}^1) + \eta(\mathbf{x}^4 - \mathbf{x}^1) \in \mathbb{R}^2 \tag{8.3.11b}$$

be the mapping taking the unit square $(0, 1) \times (0, 1)$ onto the parallelogram $\tilde{P}$ (cf. Figure 6a,b). A bilinear function is defined on $\tilde{P}$ by

$$u(x, y) := v(\phi^{-1}(x, y)), \qquad v(\xi, \eta) := (\alpha + \beta\xi)(\gamma + \delta\eta). \tag{8.3.11c}$$

It is not necessary to calculate $v(\phi^{-1}(x, y))$ explicitly, since all the integrations can be carried out over $(0, 1) \times (0, 1)$ as the reference parallelogram (cf. Exercise 14).

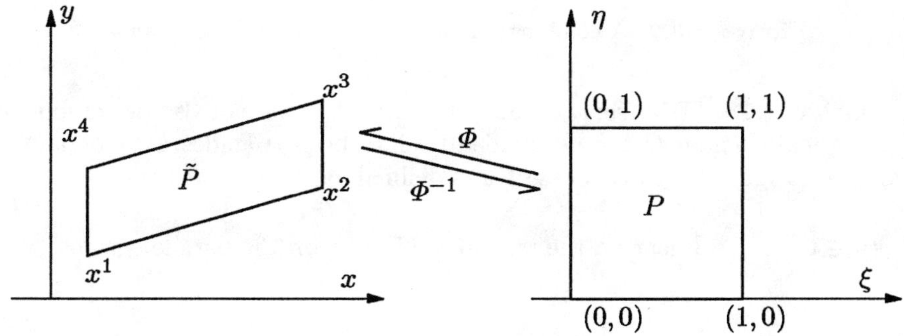

Figure 8.3.6. (a) Parallelogram (b) Unit square as reference parallelogram

If $\pi = \{P_1, \ldots, P_t\}$ is an admissible partition into parallelograms one defines $V_N \subset H^1(\Omega)$ [resp. $V_N \subset H_0^1(\Omega)$] by

$$V_N := \{u \in C^0(\overline{\Omega})\colon \text{ on all } P \in \pi,\ u \text{ coincides with a bilinear function}\} \tag{8.3.12a}$$

resp.

$$V_N := \{u \in C^0(\overline{\Omega})\colon\ u = 0 \text{ on } \partial\Omega;\ \text{ on all } P \in \pi,\ u \text{ coincides with a bilinear function}\} \tag{8.3.12b}$$

Here $N = \dim V_N$ is the number of nodes [resp. inner nodes].

A bilinear function on $P \in \pi$ is linear along each side of P. Continuity in the node points thus already implies continuity in Ω.

Remark 8.3.16. The Remarks 8a, 9, 10, and 11 hold with appropriate changes.

Exercise 8.3.17. Let the bilinear form associated to $-\Delta$ be $a(u,v) = \int_\Omega \langle \nabla u, \nabla v \rangle \, dx$. Assume $\Omega = (0,1) \times (0,1)$ is divided into squares of side h as in Figure 5a. What are the basis functions characterised by (9a)? Show that the matrix $\mathbf{L}$ coincides with the difference star $\frac{1}{3}\begin{bmatrix} -1 & -1 & -1 \\ -1 & 8 & -1 \\ -1 & -1 & -1 \end{bmatrix}$.

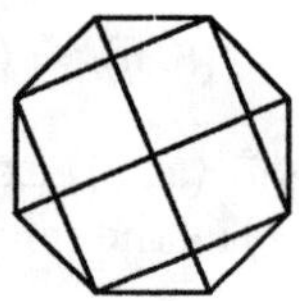

Figure 8.3.7. A combination of triangles and parallelograms

Remark 8.3.18. Triangle and parallelogram divisions can also be combined. A polygonal domain Ω can be divided up into both triangles and parallelograms (cf. Figure 7). In this case V_N is defined as

$$\{u \in C^0(\overline{\Omega}) \colon u \text{ linear on the triangles, bilinear on the parallelograms}\}.$$

8.3.4 Quadratic Elements for $\Omega \subset \mathbb{R}^2$

Let τ be an admissible triangulation of a polygonal domain Ω. We wish to increase the dimension of the finite-element subspace by allowing, instead of linear functions, quadratics

$$u(x,y) = a_{i1} + a_{i2}x + a_{i3}y + a_{i4}x^2 + a_{i5}xy + a_{i6}y^2 \text{ on } T_i \in \tau \tag{8.3.13}$$

so that

$$V_N = \{u \in C^0(\overline{\Omega}) \colon u = 0 \text{ on } \partial\Omega; \text{ on each } T_i \in \tau, \ u \text{ coincides with a quadratic function}\}. \tag{8.3.14}$$

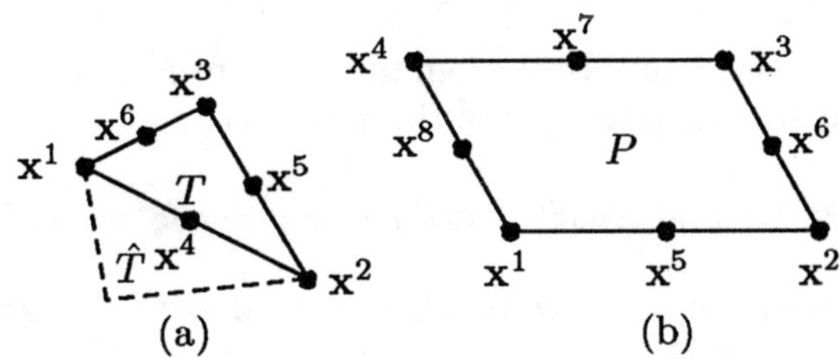

Figure 8.3.8. Nodes for
(a) a quadratic ansatz on a triangle,
(b) a quadratic ansatz of the serendipity class on a parallelogram

Lemma 8.3.19. (a) *Let* $\mathbf{x}^1, \mathbf{x}^2, \mathbf{x}^3$ *be the vertices of a triangle* $T \in \tau$, *while* $\mathbf{x}^4, \mathbf{x}^5, \mathbf{x}^6$ *are the midpoints of the sides (cf. Figure* 8a). *Each function quadratic on t is determined by the values* $\{u(\mathbf{x}^j): j = 1, \ldots, 6\}$.
(b) *The restriction of the function* (13) *to a side of* $T \in \tau$ *gives a one-dimensional quadratic function, which is uniquely determined by three of the nodes lying on this side (e.g.,* $u(\mathbf{x}^1), u(\mathbf{x}^4), u(\mathbf{x}^2)$ *in Figure* 8a).
(c) *If u is quadratic on each* $T_i \in \tau$ *and continuous at all nodes (i.e., vertices of triangles and mid-points of sides), then u is continuous on* $\overline{\Omega}$.

PROOF. (a) The quadratic function can be obtained as the uniquely defined interpolating polynomial of the form (13). Part (b) is also elementary.
(c) Let T and $\hat{T}$ be neighbouring triangles in Figure 8a. Since $u|_T$ and $u|_{\hat{T}}$ coincide on $\mathbf{x}^1, \mathbf{x}^4, \mathbf{x}^2$, by part (b) they represent the same quadratic function on the common side of T and $\hat{T}$. ■

We call all [inner] vertices of triangles and mid-points of sides [inner] nodes. By Lemma 18a we can find for each inner node $\mathbf{x}^i$ a basis function b_i which is quadratic on each $T \in \tau$ and satisfies $b_i(\mathbf{x}^i) = 1, b_i(\mathbf{x}^j) = 0 \ (j \neq i)$. By Lemma 18c b_i belongs to V_N. This proves

Remark 8.3.20. The number of inner nodes is the dimension of V_N in 14. Each $u \in V_N$ admits the expression $u = \sum_{i=1}^{N} u(\mathbf{x}^i) b_i$, where the basis function belonging to the node is characterised by (9a).

In the sequel we wish to assume that, as in Figure 7, both triangles and parallelograms are used in the partition. On the triangles the function $u \in V_N$ are quadratic. The functions on the parallelograms P must satisfy the following conditions:
(α) $u(x, y)|_P$ must be uniquely determined by the values at the 4 vertices and 4 mid-points of the sides (cf. Figure 8b).
(β) The restriction to a side of P gives a (one-dimensional) quadratic function (of the arc length).

By condition (α) the ansatz must contain exactly 8 coefficients. The quadratic (13) has only 6 coefficients, while the biquadratic $\sum_{0 \le i,j \le 2} a_{ij} \xi^i \eta^j$

has one parameter too many. If one omits the term $\xi^2\eta^2$ in the biquadratic ansatz one obtains for the unit square the function

$$v(\xi,\eta) = a_1 + a_2\xi + a_3\eta + a_4\xi^2 + a_5\xi\eta + a_6\eta^2 + a_7\xi^2\eta + a_8\xi\eta^2, \quad (8.3.15)$$

the so-called quadratic ansatz of the serendipity class. The restrictions to the sides $\xi = 0,1$ [resp. $\eta = 0,1$] give a quadratic function in η [resp. ξ]. The mapping ϕ in (11b) (cf. Figure 6a,b) yields the function $u(x,y) = v(\phi^{-1}(x,y))$ defined on the parallelogram P, which still satisfies the conditions (α) and (β).

If one were to use the full biquadratic ansatz one would one additional node which one could choose as the barycentre of the parallelogram. Cubic ansatzes can be carried out in the same way in triangles and parallelograms (cf. Schwarz [1]).

8.3.5 Elements for $\Omega \subset \mathbb{R}^3$

In the three-dimensional case assume

$$\Omega \subset \mathbb{R}^3 \quad \text{is a polyhedron.} \quad (8.3.16)$$

The triangulation from Section 8.3.2 corresponds now to a division of Ω into tetrahedra. It is called admissible, if (7a–d) hold in the appropriate sense: (7a) becomes "T_i $(1 \le i \le t)$ are open tetrahedra"; in (7d) it now should read "For $i \ne j$, $\overline{T}_i \cap \overline{T}_j$ is either empty, or a common vertex, side, or face of T_i and T_j".

Each linear function $a_1 + a_2x + a_3y + a_4z$ is uniquely determined by its values at the 4 vertices of the tetrahedron. As a basis for the space

$$V_N = \{u \in C^0(\overline{\Omega}):\ u = 0 \ \text{ on } \ \partial\Omega;\ u \ \text{ is linear on each tetrahedron } \ T_i \ (1 \le i \le t)\}$$

one chooses b_i with the property (9a): $b_i(\mathbf{x}^i) = 1,\ b_i(\mathbf{x}^j) = 0\ (j \ne i)$. The support of b_i consists in all tetrahedra that share $\mathbf{x}^i$ as a vertex. The dimension $N = \dim V_N$ is again the number of inner nodes (i.e., vertices of tetrahedra).

As in the two-dimensional case the linear ansatz may be replaced by a quadratic one. Instead of a tetrahedron one can use a parallelipiped or a triangular prism with corresponding ansatzes for the functions (cf. Schwarz [1]).

8.3.6 Handling of Side Conditions

The space V which lies at the foundation of the whole matter may be a subspace of a simply discretisable space $W \supset V$. A given function $w \in W$ belongs to V if certain side conditions are satisfied. Before we describe this situation in full generality, two examples will be provided as illustrations.

Example 8.3.21. If one wishes to make the Neumann boundary-value problem $-\Delta u = g$ in Ω and $\partial u/\partial n = \varphi$ on Γ uniquely solvable by the addition of the side condition $\int_\Omega u(\mathbf{x})\,d\mathbf{x} = 0$, one can choose the space V to be

$$V = \{u \in H^1(\Omega) : \int_\Omega u(\mathbf{x})\,d\mathbf{x} = 0\}. \tag{8.3.17a}$$

For a bounded domain Ω the form $a(u,v) := \int_\Omega \langle \nabla u, \nabla v \rangle\,d\mathbf{x}$ is V-elliptic. The weak formulation (1.1) with $f(v) := \int_\Omega gv\,d\mathbf{x} + \int_\Gamma \varphi v\,d\Gamma$ corresponds to the equation $-\Delta u = g$, $\partial u/\partial n = \varphi$, if g and φ satisfy the integrability condition $f(1) = 0$ (cf. (3.4.2)). But even when $f(1) \neq 0$ there exists a weak solution of the corrected equation $-\Delta u = \tilde{g}$, $\tilde{g}(\mathbf{x}) := g(\mathbf{x}) - [\int_\Omega g\,d\mathbf{x} + \int_\Gamma \varphi\,d\Gamma]/\int_\Omega d\mathbf{x}$.

Example 8.3.22. The Adler problem of Exercise 7.4.15 uses

$$V = \{u \in H^1(\Omega) : u \text{ constant on } \Gamma\}. \tag{8.3.17b}$$

In both cases $W = H^1(\Omega)$ is a proper superset of V. Let τ be a triangulation of Ω with I_{in} inner nodes and I_{bd} boundary nodes. Let $W_h \subset W$ be the space of linear triangular elements (cf. Remark 15):

$$N_h := \dim W_h = I_{\text{in}} + I_{\text{bd}}. \tag{8.3.18a}$$

The basis functions $\{b_i : 1 \le i \le N_h\} \subset W_h$ will be described as usual by $b_i(\mathbf{x}^j) = \delta_{ij}$ (cf. (9a)).

As the finite-element subspace of V we define $V_h := W_h \cap V$, which is of smaller dimension:

$$V_h := W_h \cap V; \qquad M_h := N_h - \dim V_h. \tag{8.3.18b}$$

In the Examples 21 and 22 we have $M_h = 1$, resp. $M_h = I_{\text{bd}} - 1$. The difficulty in the numerical solution of the discrete problem (19),

$$\text{Find } u^h \in V_h \text{ with } a(u^h, v) = f(v) \text{ for all } v \in V_h, \tag{8.3.19}$$

begins with the choice of a basis for V_h. It is not possible to use a subset of the functions $b_1, \ldots, b_{N_h}$. Thus, for example, in Example 21 no b_i belongs to V, and thus also none to V_h, since $\int_\Omega b_i\,d\mathbf{x} > 0$.

In principle, it is possible in the case of (17a) to construct as new basis functions linear combinations $b_i := \alpha b_{i_1} + \beta b_{i_2} \in V_h$ which have again localised (but a bit larger) supports. In the case of (17b) it is still relatively simple to find a practical basis for V_h: The basis functions $b_i \in W_h$ which belong to inner nodes are also in V_h, since $b_i \in H_0^1(\Omega) \subset V$. As further elements one uses $b_0(\mathbf{x}) := \sum' b_j(\mathbf{x})$, where $\sum'$ denotes the sum over all boundary nodes. In spite of this, even in this case, it would simply be easier if one could work with the standard basis of W_h.

In order to treat the problem (19) with the aid of $b_i \in W_h$, we reintroduce the notation $w^h = \mathbf{P}\mathbf{w}$ ($\mathbf{w} \in \mathbb{R}^{N_h}$ a coefficient vector, $w^h \in W_h$, cf. (1.6)).

Each $v \in V_h \subset W_h$ can be written as $\mathbf{Pv}$ with $\mathbf{v}$ in $\mathbf{V} := \{\mathbf{v} \in \mathbb{R}^{N_h} : \mathbf{Pv} \in V_h\} = \mathbf{P}^{-1}V_h$. Thus problem (19) is equivalent to:

$$\text{Find } \mathbf{u} \in \mathbf{V} \text{ with } a(\mathbf{Pu}, \mathbf{Pv}) = f(\mathbf{Pv}) \text{ for all } \mathbf{v} \in \mathbf{V}. \tag{8.3.19'}$$

From (18a,b) we have $\dim \mathbf{V} = \dim V_h = N_h - M_h = \dim W_h - M_h$. The spaces V_h and $\mathbf{V}$ are described by M_h linear conditions

$$\sum_{j=1}^{N_h} c_{ij} w_j = 0 \qquad (1 \le i \le M_h); \tag{8.3.20}$$

$$\mathbf{V} = \text{ Ker } \mathbf{C} = \{\mathbf{w} \in \mathbb{R}^{N_h} : \mathbf{Cw} = \mathbf{0}\}, \tag{8.3.21}$$

where $\mathbf{C} = (c_{ij})$ is an $M_h \times N_h$ matrix. In the case of Example 21 we have

$$M_h = 1, \qquad c_{1j} = \int_\Omega b_j(\mathbf{x})\,d\mathbf{x} \qquad (1 \le j \le N_h).$$

For Example 22 let $\mathbf{x}^0, \ldots, \mathbf{x}^{M_h}$, with $M_h = I_{\text{bd}} - 1$, be the boundary nodes. Then $\mathbf{C}$ can be defined as follows:

$$\begin{aligned} &M_h = I_{\text{bd}} - 1. \qquad c_{ii} = 1 \ (1 \le i \le M_h), \qquad c_{i,i+1} = -1 \ (1 \le i < M_H), \\ &c_{M_h,0} = -1, \qquad c_{ij} = 0 \text{ otherwise.} \end{aligned}$$

The variation of $\mathbf{v}$ over $\mathbf{V}$ in (19′) can be replaced by the variation of $\mathbf{w}$ over $\mathbb{R}^{N_h}$, if one couples the conditions (20) by using Lagrange multipliers λ_i $(1 \le i \le M_h)$, which can be put together to form a vector $\lambda = (\lambda_1, \ldots, \lambda_{M_h})$. The resulting formulation of the problem is:

$$\text{find } \mathbf{u} \in \mathbb{R}^{N_h} \text{ and } \lambda \in \mathbb{R}^{M_h} \text{ with } \mathbf{Cu} = \mathbf{0} \text{ and} \tag{8.3.22a}$$

$$a(\mathbf{Pu}, \mathbf{Pw}) + \langle \lambda, \mathbf{Cw} \rangle = f(\mathbf{Pw}) \text{ for all } \mathbf{w} \in \mathbb{R}^{N_h}, \tag{8.3.22b}$$

where $\langle \lambda, \mu \rangle = \sum \lambda_i \mu_i$ is the scalar product in $\mathbb{R}^{M_h}$.

Theorem 8.3.23. *The problems* (19) *and* (22a,b) *are equivalent in the following sense. If* $\mathbf{u}, \lambda$ *is a solution pair for* (22a,b) *then* $\mathbf{u}$ *is a solution of* (19′) *and* $\mathbf{Pu}$ *is a solution of* (19). *Conversely, if* $u^h = \mathbf{Pu}$ *is a solution of* (19) *then there exists precisely one* $\lambda \in \mathbb{R}^{M_h}$ *such that* $\mathbf{u}$ *and* λ *solve* (22a,b).

Before we prove this theorem, we give a matrix formulation which is equivalent to (22a,b)

Remark 8.3.24. Let $\mathbf{L}$ and $\mathbf{f}$ be defined as in (1.8a,b). Further, suppose that $\mathbf{B} := \mathbf{C}^\mathsf{T}$ with $\mathbf{C}$ from (21). Then (22a,b) is equivalent to the system of equations

$$\begin{bmatrix} \mathbf{L} & \mathbf{B} \\ \mathbf{B}^\mathsf{T} & \mathbf{O} \end{bmatrix} \begin{bmatrix} \mathbf{u} \\ \lambda \end{bmatrix} = \begin{bmatrix} \mathbf{f} \\ \mathbf{0} \end{bmatrix}. \tag{8.3.22'}$$

PROOF. $\mathbf{Lu} + \mathbf{B}\lambda = \mathbf{f}$ is equivalent to $a(\mathbf{Pu}, \mathbf{Pw}) + \langle \lambda, \mathbf{Cw} \rangle = \langle \mathbf{Lu}, \mathbf{w} \rangle + \langle \mathbf{B}\lambda, \mathbf{w} \rangle = \langle \mathbf{f}, \mathbf{w} \rangle = f(\mathbf{Pw})$ for all $\mathbf{w} \in \mathbb{R}^{N_h}$. Note $\mathbf{B}^\mathsf{T}\mathbf{u} = \mathbf{0}$ is the same as $\mathbf{Cu} = \mathbf{0}$. ■

PROOF *(of Theorem 23).*
(a) Assume a solution of (22a,b) is given in terms of $\mathbf{u}, \lambda$. Then $\mathbf{Cu} = \mathbf{0}$ implies $\mathbf{u} \in \mathbf{V}$ and $u^h := \mathbf{Pu} \in V_h$. Equation (22b) holds in particular for all $\mathbf{w} \in \mathbf{V}$, so that (19′) follows since $\mathbf{Cw} = \mathbf{0}$.
(b) (19′) implies $\mathbf{f} - \mathbf{Lu} \in \mathbf{V}^\perp$. Since $\mathbf{V}^\perp = (\operatorname{Ker} \mathbf{C})^\perp = (\operatorname{Ker} \mathbf{B}^\mathsf{T})^\perp = \operatorname{Im}(\mathbf{B})$, there exists a $\lambda \in \mathbb{R}^{M_h}$ with $\mathbf{f} - \mathbf{Lu} = \mathbf{B}\lambda$, so that (22′), and thus (22a,b) are satisfied. For the $N_h \times M_h$ matrix $\mathbf{B}$ with rank M_h we have $\operatorname{Ker} \mathbf{B} = \{\mathbf{0}\}$, so that the solution is unique. ■

Note that in the case of Example 21 the system of equations (22′) is essentially identical to (4.7.10a,b).

8.4 Error Estimates for Finite Element Methods

8.4.1 H^1-Estimates for Linear Elements

In this section we shall restrict ourselves to the consideration of the linear elements from Section 8.3.2 and therefore assume:

$$\mathcal{T} \text{ is an admissible triangulation of } \Omega \subset \mathbb{R}^2, \tag{8.4.1a}$$
$$V_N \text{ is defined by (3.8), if } V = H_0^1(\Omega), \tag{8.4.1b}$$
$$V_N \text{ is as in Remark 8.3.15b, if } V = H^1(\Omega). \tag{8.4.1c}$$

By Theorem 8.2.1 one has to determine $d(u, V_N) = \inf\{|u - w|_1 : w \in V_N\}$. To do this one begins by looking at the reference triangle from Figure 8.3.4.

Lemma 8.4.1. *Let $T = \{(\xi, \eta) : \xi, \eta \ge 0, \xi + \eta \le 1\}$. For all $u \in H^2(T)$ there holds*

$$\|u\|^2_{H^2(T)} \le C[\,|u(0,0)|^2 + |u(0,1)|^2 + |u(1,0)|^2 + \sum_{|\alpha|=2} \|D^\alpha u\|^2_{L^2(T)}\,]. \tag{8.4.2}$$

PROOF. (a) First it has to be shown that the bilinear form

$$a(u,v) := u(0,0)v(0,0) + u(0,1)v(0,1) + u(1,0)v(1,0) + \sum_{|\alpha|=2} (D^\alpha u, D^\alpha v)_{L^2(T)}$$

is continuous on $H^2(T) \times H^2(T)$ and is $H^2(T)$-coercive. The continuity follows from the continuous embedding $H^2(T) \subset C^0(T)$, which implies $|u(\mathbf{x})| \le \hat{C}|u|_2$ for all $\mathbf{x} \in \overline{T}$ (cf. Theorem 6.2.30). Thus we have $|a(u,v)| \le (1 + 3\hat{C})|u|_2|v|_2$.

Since $T \in C^{0,1}$, one may apply Theorem 6.4.8b: $H^2(T)$ is compactly embedded in $H^1(T)$. By Lemma 6.4.13, there exists a constant $C_{1/2}$ such that

$$|u|_1^2 \le \left(\frac{1}{2}|u|_2 + C_{1/2}|u|_0\right)^2 \le \frac{1}{2}|u|_2^2 + 2C_{1/2}^2|u|_0^2 \text{ for all } u \in H^2(T).$$

Since $\sum_{|\alpha|=2} |D^\alpha u|_0^2 = |u|_2^2 - |u|_1^2$, the estimate

$$a(u,u) \ge |u|_2^2 - |u|_1^2 \ge \frac{1}{2}|u|_2^2 - 2C_{1/2}^2|u|_0^2,$$

shows that $a(\cdot,\cdot)$ is $H^2(T)$-coercive.
(b) Since the embedding $H^2(T) \subset L^2(T)$ is also compact one may apply Theorem 6.5.15. The operator $A \in L(H^2(T), H^2(T)')$ which is associated with $a(\cdot,\cdot)$ either has an inverse $A^{-1} \in L(H^2(T)', H^2(T))$ or has $\lambda = 0$ as an eigenvalue with an eigenfunction $0 \ne e \in H^2(T)$. In the latter case e must, in particular, satisfy the equation $a(e,e) = 0$. From this follows that $D^\alpha e = 0$ for all $|\alpha| = 2$, so that e must be linear: $e(x,y) = \alpha + \beta x + \gamma y$. Furthermore, from $a(e,e) = 0$ one may conclude that $e(0,0) = e(1,0) = e(0,1) = 0$, so that $e = 0$ in contradiction to what was just assumed. From Lemma 6.5.3 and Exercise 6.5.6c one may show the $H^2(T)$-ellipticity: $a(u,u) \ge C_E|u|_2^2$ with $C_E := 1/[\,\|A^{-1}\|_{H^2(T)\leftarrow H^2(T)'}^2(1 + 3\hat{C})\,] > 0$ and $\hat{C}$ from part (a). Thus we have assertion (2) with $C := 1/C_E$. ■

Lemma 8.4.2. *Let $T_h := hT := \{(x,y)\colon\ x,y \ge 0,\ x + y \le h\}$. For each $u \in H^2(T_h)$ there holds with $|\beta| \le 2$*

$$\|D^\beta u\|_{L^2(T_h)}^2 \le C\{h^{2-2|\beta|}[u^2(0,0) + u^2(0,h) + u^2(h,0)] + h^{4-2|\beta|}\sum_{|\alpha|=2}\|D^\alpha u\|_{L^2(T_h)}^2\}, \tag{8.4.3}$$

with C independent of u, β, and h.

PROOF. Let $v(\xi,\eta) := u(\xi h, \eta h) \in H^2(T)$. The derivatives with respect to (x,y) and to $(\xi,\eta) = (x,y)/h$ are related by $D_{x,y}^\beta = h^{-|\beta|}D_{\xi,\eta}^\beta$. For each $|\beta| \le 2$ Lemma 1 gives

$$\begin{aligned}\|D^\beta u\|_{L^2(T_h)}^2 &= \iint_{T_h} |D_{x,y}^\beta u|^2\,dx\,dy = \iint_T |h^{-|\beta|}D_{\xi,\eta}^\beta v|^2 h^2\,d\xi\,d\eta \\ &= h^{2-2|\beta|}\|D^\beta v\|_{L^2(T)}^2 \le h^{2-2|\beta|}\|v\|_{H^2(T)}^2 \\ &\le Ch^{2-2|\beta|}[v^2(0,0) + v^2(1,0) + v^2(0,1) + \sum_{|\alpha|=2}\|D^\alpha v\|_{L^2(T)}^2].\end{aligned}$$

Since $v(0,0) = u(0,0)$, $v(1,0) = u(h,0)$, $v(0,1) = u(0,h)$, and $\|D_{\xi,\eta}^\alpha v\|_{L^2(T)}^2 = h^2\|D_{x,y}^\alpha u\|_{L^2(T_h)}^2$, one may conclude assertion (3). ■

Lemma 8.4.3. *Let $\tilde{T}$ be an arbitrary triangle with*

$$\textit{Side lengths} \le h_{\max}, \qquad \textit{Interior angles} \ge \alpha_0 > 0. \tag{8.4.4}$$

For each $u \in H^2(\tilde{T})$ and $|\beta| \le 2$, there holds

$$||D^\beta u||^2_{L^2(\tilde{T})} \le C(\alpha_0) \left[h_{\max}^{2-2|\beta|} \sum_{\substack{\mathbf{x}\text{ vertex}\\ \text{of } \tilde{T}}} |u(\mathbf{x})|^2 + h_{\max}^{4-2|\beta|} \sum_{|\alpha|=2} ||D^\alpha u||^2_{L^2(\tilde{T})} \right], \tag{8.4.5}$$

where $C(\alpha_0)$ depends only on α_0 and not on u, β, or h.

PROOF. Let $h \le h_{\max}$ be one of the lengths of the sides of $\tilde{T}$. In a way similar to that in Exercise 8.3.14a, let $\varphi: T_h \to \tilde{T}$ be the linear transformation that maps T_h onto $\tilde{T}$. Then $v(\xi,\eta) := u(\varphi(\xi,\eta))$ belongs to $H^2(T_h)$. Under the conditon (4) we know that the determinant $|\det \varphi'| \in [1/K(\alpha_0), K(\alpha_0)]$ is bounded both above and below. From

$$||D^\beta_{x,y} u||^2_{L^2(\tilde{T})} \le C_1(\alpha_0) \sum_{|\beta'|=|\beta|} ||D^{\beta'}_{\xi,\eta} v||^2_{L^2(T_h)},$$

(3), and

$$\sum_{|\alpha|=2} ||D^\alpha_{\xi,\eta} v||^2_{L^2(T_h)} \le C_2(\alpha_0) \sum_{|\alpha|=2} ||D^\alpha_{x,y} u||^2_{L^2(\tilde{T})},$$

there then follows (5). ■

From Lemma 3 follows the important result:

Theorem 8.4.4. *Assume that conditions* (1a–c) *hold for τ, V_N, and V. Let α_0 be the smallest interior angle of all $T_i \in \tau$, while h is the maximum length of the sides of all $T_i \in \tau$. Then*

$$\inf_{v \in V_N} ||u - v||_{H^k(\Omega)} \le C'(\alpha_0) h^{2-k} ||u||_{H^2(\Omega)} \text{ for } k = 0, 1 \text{ and all } u \in H^2(\Omega) \cap V. \tag{8.4.6}$$

PROOF. For a $u \in H^2(\Omega)$ [resp. $u \in H^2(\Omega) \cap H^1_0(\Omega)$, if $V = H^1_0(\Omega)$] one chooses $v := \sum_i u(\mathbf{x}^i) b_i \in V_N$, i.e., $v \in V_N$ with $v(\mathbf{x}^i) = u(\mathbf{x}^i)$ at the [inner] nodes $\mathbf{x}^i$. Since $w := u - v$ vanishes at the vertices of each $T_i \in \tau$ and $D^\alpha w = D^\alpha u$ for $|\alpha| = 2$ (because $D^\alpha v = 0$ for all $v \in V_N$), inequality (5) implies the estimate

$$\begin{aligned} ||D^\beta w||^2_{L^2(\Omega)} &= \sum_{T_i \in \tau} ||D^\beta w||^2_{L^2(T_i)} \\ &\le \sum_{T_i \in \tau} [C(\alpha_0) h^{4-2|\beta|} \sum_{|\alpha|=2} ||D^\alpha u||^2_{L^2(T_i)}] \\ &= h^{4-2|\beta|} C(\alpha_0) \sum_{|\alpha|=2} ||D^\alpha u||^2_{L^2(\Omega)} \\ &\le h^{4-2|\beta|} C(\alpha_0) ||u||^2_{H^2(\Omega)}. \end{aligned}$$

Summation over $|\beta| \le k$ now proves the inequality (6). ■

Let h_T be the longest side of $T \in \tau$, then $h := \max\{h_T : T \in \tau\}$ is the longest side-length which occurs in τ. The parameter h is the essential one and will be used as an index in the sequel: $\tau = \tau_h$. Denote by ρ_T the radius of the inscribed circle in $T \in \tau$. The ratio h_T/ρ_T tends to infinity exactly when the smallest interior angle tends to zero. If one has the bound $\max\{h_T/\rho_T : T \in \tau\} \le \text{const}$ for a family of triangulations, then one calls the family quasi-uniform. In this condition the constant should be the same for all τ in the family $\{\tau_h\}$. A family $\{\tau_h\}$ is called uniform if it fulfils the stronger requirement $h/\min\{\rho_T : T \in \tau\} \le \text{const}$. A family $\{\tau_h\}$ which is both unform and for which $h \to 0$ is called regular by Ciarlet [1, §3.1].

In the construction of triangulations it should be noted that with increasing refinement ($h_\nu \to 0$) the interior angle may not also be reduced. Strategies for the systematic construction of quasi-uniform sequences V_{h_ν} can be found, for example, in Hackbusch [1, §3.8.2].

The case $k = 1$ of Theorem 4 may now be written as follows:

Theorem 8.4.5. *Assume conditions* (1a–c) *hold for a sequence of quasi-uniform triangulations τ_ν. Then there exists a constant C, such that for all $h = h_\nu$ and $V_h = V_{h_\nu}$ there holds the following estimate:*

$$\inf_{v \in V_h} |u - v|_1 \le Ch|u|_2 \quad \text{for all } u \in H^2(\Omega) \cap V. \tag{8.4.6'}$$

The combination of this theorem with Theorem 8.2.1 gives

Theorem 8.4.6. *Assume conditions* (1a–c) *hold for a sequence of quasi-uniform triangulations τ_ν. Let the bilinear form fulfil conditions* (1.2) *and* (1.12). *Suppose the constants $\epsilon_N := \epsilon_{h_\nu} > 0$ from* (1.12) *are bounded below by $\epsilon_{h_\nu} \ge \tilde{\epsilon} > 0$ (cf. Corollary* 8.2.3). *Let Problem* (1.1) *have the solution $u \in H^2(\Omega) \cap V$. Let $u^h \in V_h$ be the finite-element solution. Then there exists a constant C, which does not depend on u, $h = h_\nu$ or ν, such that*

$$|u - u^h|_1 \le Ch|u|_2. \tag{8.4.7}$$

The combination of the Theorems 5 and 8.2.2 can be written:

Theorem 8.4.7. *Assume conditions* (1a–c) *hold for a sequence of quasi-uniform triangulations τ_ν with $h_\nu \to 0$. Let the bilinear form fulfil* (1.2) *and* (1.12) *with $\epsilon_{h_\nu} \ge \tilde{\epsilon} > 0$. Then the problem* (1.1) *has a unique solution $u \in V$, and the finite-element solution $u^{h_\nu} \in V_{h_\nu}$ converges to u:*

$$|u - u^{h_\nu}|_1 \to 0 \quad (\nu \to \infty). \tag{8.4.8}$$

PROOF. Let $u \in V$ and $\epsilon > 0$ be arbitrary. Since $H^2(\Omega) \cap V$ is dense in V, there exists $u_\epsilon \in H^2(\Omega) \cap V$ with $|u - u_\epsilon|_1 \le \epsilon/2$. From (6) and $h_\nu \to 0$ we see there are ν and $v_\epsilon \in V_{h_\nu}$ with $|u_\epsilon - v_\epsilon|_1 \le \epsilon/2$; thus $|u - v_\epsilon|_1 \le \epsilon$. This proves (2.4a). ■

Theorem 7 proves convergence without restrictive conditions on u. On the other hand the order of convergence $O(h)$ in the estimate (7) requires the assumption that the solution $u \in V$ lies in $H^2(\Omega)$. As will later become clear, this assumption is hardly to be taken as fulfilled in every situation. A weaker assumption is $u \in V \cap H^s(\Omega)$ with $s \in (1,2)$. The corresponding result is as follows.

Theorem 8.4.8. *Assume that in the assumptions of Theorem* 6 $u \in V \cap H^2(\Omega)$ *is replaced by* $u \in V \cap H^s(\Omega)$ *with* $1 \le s \le 2$. *Let* Ω *be sufficiently smooth. Then there holds*

$$|u - u^h|_1 \le Ch^{s-1}|u|_s. \tag{8.4.9}$$

The proof uses a generalisation of Theorem 4:

Lemma 8.4.9. *Under the assumptions of Theorem* 4 *and suitable conditions on* Ω, *there holds*

$$\inf_{v \in V_h} |u - v|_1 \le C''(\alpha_0)h^{s-1}|u|_s \quad \text{for } s \in [1,2], u \in H^s(\Omega) \cap V. \tag{8.4.10}$$

PROOF. The proof is based on an interpolation argument that will not be further explained here. Equation (10) holds for $s = 2$ (cf. (7)) and for $s = 1$, since $\inf |u - v|_1 \le |u - 0|_1 = |u|_1$. From this follows (10) for $s \in (1,2)$ with the norm $|\cdot|_s$ of the interpolating space $[H^1(\Omega) \cap V, H^2(\Omega) \cap V]_{s-1}$ (cf. Lions-Magenes [1]), which under suitable assumptions on Ω coincides with $H^s(\Omega) \cap V$. ■

For $s = 1$ the right-hand side of (10) becomes const $\cdot |u|_1$. In fact the estimate $\inf\{|u - v|_1 : v \in V_h\} \le |u|_1$ is the best possible. To prove this ,choose $u \perp V_h$ (orthogonal with respect to $|\cdot|_1$). On the other side $s = 2$ is the maximal value for which the estimates (9) and (10) can hold. Even $u \in C^\infty(\Omega)$ permits no better order of approximation than $O(h)$!

The Ritz projections $S_N : V \to V_N$ introduced in (2.5a) will now be written $S_h : V \to V_h$. The inequality (9) becomes $|u - S_h u|_1 / |u|_s \le Ch^{s-1}$, and this proves

Corollary 8.4.10. *Assume conditions* (1a–c) *for* τ, V_h *and* V. *Let the bilinear form fulfil* (1.2) *and* (1.12). *The Ritz projection* S_h *satisfies the estimate*

$$||I - S_h||_{H^1(\Omega) \leftarrow H^2(\Omega) \cap V} \le Ch^1. \tag{8.4.11}$$

Under the asumptions of Lemma 9 there holds

$$||I - S_h||_{H^1(\Omega) \leftarrow H^s(\Omega) \cap V} \le Ch^{s-1} \quad \text{for all } 1 \le s \le 2. \tag{8.4.11'}$$

8.4.2 L^2 and H^s Estimates for Linear Elements

According to Theorem 6 $O(h)$, is the optimal order of convergence. This result seems to contradict the $O(h^2)$ convergence of the five-point formula (cf. Section 4.5), since the finite-element method with a particular triangulation is almost identical to the five-point formula (cf. Exercise 8.3.13). The reason for this is that the error $u - u^h$ is measured in the $|\cdot|_1$ norm. The estimate $\inf\{|u-v|_0 : v \in V_h\} \le Ch^2|u|_2$ in Theorem 4 suggests the conjecture that the $|\cdot|_0$ norm of the error is of the order $O(h^2)$: $|u-u^h|_0 \le Ch^2|u|_2$. However, this statement is false without further assumptions.

Up to this point only the existence of a weak solution $u \in V$ (e.g., $V = H_0^1(\Omega)$ or $V = H^1(\Omega)$) has been guaranteed. But in Theorem 6 we needed the stronger assumption $u \in H^2(\Omega) \cap V$. A similar regularity condition will also be imposed on the adjoint problem to (1.1)

$$\text{Find } u \in V \text{ with } a^*(u,v) = f(v) \quad \text{for all } v \in V, \tag{8.4.12}$$

which uses the adjoint bilinear form $a^*(u,v) := a(v,u)$. For $f \in L^2(\Omega) \subset V'$, the value $f(v)$ becomes $(f,v)_{L^2(\Omega)}$. The regularity condition is:

$$\begin{aligned}&\text{For each } f \in L^2(\Omega) \text{ the problem (12)}\\ &\text{has a solution } u \in H^2(\Omega) \cap V \text{ with } |u|_2 \le C_R|f|_0.\end{aligned} \tag{8.4.13}$$

In Section 9 we shall see that this statement holds for sufficiently smooth domains Ω.

Theorem 8.4.11. *(Aubin–Nitsche) Assume* (13), (1.2), *and* (1.12) *with* $\epsilon_N = \epsilon_h \ge \tilde{\epsilon} > 0$, *and*

$$\inf_{v \in V_h} |u - v|_1 \le C_0 h |u|_2 \quad \textit{for all } u \in H^2(\Omega) \cap V. \tag{8.4.14}$$

Let problem (1.1) *have the solution* $u \in V$. *Let* $u^h \in V_h \subset V$ *be the finite-element solution. Then, with a constant* C_1 *independent of* u *and* h,

$$|u - u^h|_0 \le C_1 h |u|_1. \tag{8.4.15a}$$

If the solution u *also belongs to* $H^2(\Omega) \cap V$, *then there is a constant* C_2, *independent of* u *and* h, *such that*

$$|u - u^h|_0 \le C_2 h^2 |u|_2. \tag{8.4.15b}$$

From Theorem 4, it is sufficent to ensure (14) *that* V_h *be the space of finite elements of an admissible and quasi-uniform triangulation.*

Proof (cf. Nitsche [1]). For each $e := u - u^h \in L^2(\Omega)$ define $w \in H^2(\Omega) \cap V$ as the solution of (12) for $f = e$:

$$a(v,w) = (e,v)_{L^2(\Omega)} \qquad \text{for all } v \in V. \tag{8.4.16a}$$

Corresponding to w there is, by (14), a $w^h \in V_h$ with

$$|w - w^h|_1 \le C_0 h |w|_2 \le C_0 C_R h |e|_0. \tag{8.4.16b}$$

Equation (2.3) is $a(e, v) = 0$ for all $v \in V_h$; therefore, in particular, we have

$$a(e, w^h) = 0. \tag{8.4.16c}$$

From (16a–c) we obtain

$$\begin{aligned} |e|_0^2 &= (e,e)_{L^2(\Omega)} = a(e,w) = a(e, w - w^h) \\ &\le C_s |e|_1 |w - w^h|_1 \le C_S |e|_1 C_0 C_R h |e|_0, \end{aligned}$$

and thus

$$|e|_0 \le C_S C_0 C_R h |e|_1. \tag{8.4.16d}$$

From (2.1) one deduces

$$|e|_1 = |u - u^h|_V \le (1 + C_S/\epsilon_h) \inf_{v \in V_h} |u - v|_V \le (1 + C_S/\tilde{\epsilon})|u|_V$$

so that (15a) follows with $C_1 := C_S C_0 C_R (1 + C_S/\tilde{\epsilon})$. If $u \in H^2(\Omega)$, one may use the inequality (7), $|e|_1 \le Ch|u|_2$, and deduce (15b) with $C_2 := C_S C_0 C_R C$. ■

Corollary 8.4.12. *The inequalities* (15a) *and* (15b) *are equivalent to the properties* (17a) *and* (17b) *of the Ritz projection* S_h*:*

$$\|I - S_h\|_{L^2(\Omega) \leftarrow V} \le C_1 h, \tag{8.4.17a}$$

$$\|I - S_h\|_{L^2(\Omega) \leftarrow H^2(\Omega) \cap V} \le C_2 h^2. \tag{8.4.17b}$$

The estimates (17a) and (17b) may also be proved directly. The definition of $\hat{S}_h$ and the connection between S_h and $\hat{S}_h$ may be found in

Exercise 8.4.13. To $a(\cdot,\cdot)$ let the operator $L : V \to V'$, the Ritz projection S_h, and the stiffness matrix $\mathbf{L}$ be associated. Show:
(a) To the adjoint bilinear form $a^*(\cdot,\cdot)$ belong the stiffness matrix $\mathbf{L}^\mathsf{T}$ and the Ritz projection

$$\hat{S}_h = \mathbf{P}(\mathbf{L}^\mathsf{T})^{-1}\mathbf{P}^* L^*. \tag{8.4.18a}$$

(b) There holds

$$S_h = L^{-1} \hat{S}_h^* L. \tag{8.4.14b}$$

PROOF. Second proof of Theorem 11: Let

$$H_*^2(\Omega) := \{u \in H^2(\Omega) : u \text{ is a solution of (12) for an } f \in L^2(\Omega)\} \subset V$$

be the range of $L^{*-1} \in L(L^2(\Omega), H^2(\Omega))$. Equip the space $H^2(\Omega)$ with the norm $|\cdot|_2$. Since $L^2(\Omega)' = L^2(\Omega)$ we have that $L^{*-1} \in L(L^2(\Omega), H_*^2(\Omega))$ is equivalent to $L^{-1} \in L(H^2_*(\Omega)', L^2(\Omega))$ (cf. Lemma 6.3.2), so that there is a C_α with

$$||L^{-1}||_{L^2(\Omega)\leftarrow H^2_*(\Omega)'} \le C_\alpha. \tag{8.4.19a}$$

The assumptions (1.2) and (1.12) of Theorem 11 can be brought over without change of the constants to the adjoint problem, so that the statement of Theorem 8.2.1 can be written in the form $||I - \hat{S}_h||_{V\leftarrow H^2_*(\Omega)} \le C_\beta h$. For the adjoint operator we have the estimate

$$||(I - \hat{S}_h)^*||_{H^2_*(\Omega)'\leftarrow V'} \le C_\beta h. \tag{8.4.19b}$$

The basic assumption (1.2) may be written

$$||L||_{V'\leftarrow V} \le C_S. \tag{8.4.19c}$$

From Equation (18b) one may read the expression

$$I - S_h = I - L^{-1}\hat{S}_h^* L = L^{-1}(I - \hat{S}_h^*)L = L^{-1}(I - \hat{S}_h)^* L.$$

From (19a–c) one has

$$\begin{aligned} ||I - S_h||_{L^2(\Omega)\leftarrow V} &\le ||L^{-1}||_{L^2(\Omega)\leftarrow H^2_*(\Omega)'} ||(I - \hat{S}_h)^*||_{H^2_*(\Omega)'\leftarrow V'} ||L||_{V'\leftarrow V} \\ &\le C_\alpha C_\beta C_S h. \end{aligned} \tag{8.4.19d}$$

Therefore we have proved (17a) with $C_1 = C_\alpha C_\beta C_S$. Since S_h is a projection, so is $I - S_h$, so that

$$\begin{aligned} ||I - S_h||_{L^2(\Omega)\leftarrow H^2(\Omega)\cap V} &= ||(I - S_h)^2||_{L^2(\Omega)\leftarrow H^2(\Omega)\cap V} \\ &\le ||I - S_h||_{L^2(\Omega)\leftarrow V} ||I - S_h||_{V\leftarrow H^2(\Omega)\cap V}. \end{aligned}$$

From (17a) and (11) we deduce (17b). ■

The regularity condition (13) is weakened in the following theorem:

Theorem 8.4.14. *Assume* (1.2), *and* (1.12) *with* $\epsilon_N = \epsilon_h \ge \tilde{\epsilon} > 0$, *and inequality* (10) *for all* $1 \le s \le 2$. *In the place of* (13) *we assume* H^{2-t}*-regularity for some* $t \in [0, 1]$. *If* $V = H^1(\Omega)$ *then assume* $L^{*-1} \in L(H^t(\Omega)', H^{2-t}(\Omega))$; *if* $V = H_0^1(\Omega)$, *then assume* $L^{*-1} \in L(H^{-t}(\Omega), H^{2-t}(\Omega) \cap V)$. *Then the finite-element solution satisfies the inequality*

$$||u - u^h||_t \le C_{t,s} h^{s-t} ||u||_s \qquad (0 \le t \le 1 \le s \le 2). \tag{8.4.20}$$

The Ritz projection satisfies

$$||I - S_h||_{H^t(\Omega)\leftarrow H^s(\Omega)\cap V} \le C_{t,s} h^{s-t}. \tag{8.4.21}$$

PROOF. (For the case $V = H^1(\Omega)$) In (19a–c) replace $L^2(\Omega)$ by $H^t(\Omega)$, and $H^2_*(\Omega)'$ by $H^{2-t}_*(\Omega)'$ with $H^{2-t}_*(\Omega) := \{u \in H^{2-t}(\Omega)\colon u$ is the solution to $f \in H^t(\Omega)'\}$. Because of (11′), (19b) becomes [with s replaced by $2 - t$]

$$||(I - \hat{S}_h)^*||_{H^{2-t}_*(\Omega)' \leftarrow V'} = ||I - \hat{S}_h||_{V \leftarrow H^{2-t}_*(\Omega)} \leq C_\beta h^{1-t}.$$

As in (19d) one shows that $||I - S_h||_{H^t(\Omega) \leftarrow V} \leq Ch^{1-t}$. Combining this with $||I - S_h||_{V \leftarrow H^s(\Omega) \cap V} \leq Ch^{s-1}$ (cf. (11′)) there follows (21), and thus (20). ■

For nonlinear boundary value problems, estimates of the errors in terms of other norms (e.g., $L^\infty(\Omega), L^p(\Omega)$) are of interest. For this we refer to Ciarlet [1. §3.3] and Schatz [1].

8.5 Generalisations

8.5.1 Error Estimates for Other Elements

The estimates for the errors for linear functions on triangular elements proved in Section 8.4 are also true for bilinear functions on parallelograms and for combinations of both sorts of elements (cf. Section 8.3.7). The proofs are along analogous lines. Even for tetrahedral elements in a three-dimensional region $\Omega \subset \mathbb{R}^3$ the same results can be carried over. For the proof one notices that $u \in H^2(\Omega)$ has well-defined nodal values $u(\mathbf{x}^i)$ even for $\Omega \subset \mathbb{R}^3$, since $2 > n/2$ ($n = 3$, dimension of Ω; cf. Theorem 6.2.30).

For quadratic elements (cf. Section 8.3.4) one expects a correspondingly improved order of convergence. In general one can show the following: If the ansatz function is, in each $T_i \in \tau$, a polynomial of degree $k \geq 1$ (i.e., $u(\mathbf{x}) = \sum_{|\nu| \leq k} a_\nu \mathbf{x}^\nu$ on T_i), then

$$d(u, V_h) = \inf\{\,|u - v|_1 : v \in V_h\} \leq Ch^k |u|_{k+1} \quad \text{for all } u \in H^{k+1}(\Omega) \cap V \tag{8.5.1}$$

(cf. Ciarlet [1, Theorem 3.2.1]). If one uses parallelograms as a basis and uses ansatz functions that are polynomials of degree at least k, then (1) also holds. For example, biquadratic elements and quadratic ansatz functions of the serendipity class fulfil this requirement for $k = 2$. The result corresponding to Theorem 8.4.6 follows from Theorem 8.2.1:

Theorem 8.5.1. *Let V_h fulfil* (1), *and let the bilinear form satisfy* (1.2) *and* (1.12) *with $\epsilon_n =: \epsilon_h \geq \tilde{\epsilon} > 0$. Assume problem* (1.1) *has a solution $u \in V \cap H^{k+1}(\Omega)$. Then the finite-element solution $u^h \in V_h$ satisfies the inequality*

$$|u - u^h|_1 \leq Ch^k |u|_{k+1}. \tag{8.5.2}$$

The Ritz projection satisfies $||I - S_h||_{V \leftarrow H^{k+1}(\Omega) \cap V} \leq Ch^k$.

The estimate (4.9) now holds for $s \in [1, k]$ if $u \in H^s(\Omega) \cap V$. With suitable regularity conditions, there are, as in Theorem 8.4.11, the error estimates

$$|u - u^h|_0 \leq Ch^{k+1} |u|_{k+1}, \tag{8.5.3a}$$

$$||u - u^h||_{H^{k-1}(\Omega)'} \leq Ch^{2k} |u|_{k+1}. \tag{8.5.3b}$$

For (3b) one needs, for instance, H^{k+1}-regularity: For each $f \in H^{k-1}(\Omega)$ the adjoint problem (4.12) has a solution $u \in H^{k+1}(\Omega)$.

Remark 8.5.2. Under suitable assumptions the inequality (4.20) holds for $1-k \le t \le 1 \le s \le k+1$. For negative t the norm $|\cdot|_t$ should be understood as the dual norm of $H^{-t}(\Omega)$.

8.5.2 Finite Elements for Equations of Higher Order

8.5.2.1 Introduction: The One-Dimensional Biharmonic Equation

All the spaces of finite elements, V_h, constructed so far are useless for equations of order $2m > 2$, since $V_h \not\subset H^m(\Omega)$. According to Example 6.2.5, in order to have $V_h \subset H^2(\Omega)$ it is necessary that not only the function u but also its derivatives $u_{\mathbf{x}_i}$ change continuously between elements. The ansatz functions must therefore be piecewise smooth and globally in $C^1(\Omega)$.

As a model problem we introduce the one-dimensional biharmonic equation

$$u''''(x) = g(x) \text{ for } 0 < x < 1, \quad u(0) = u'(0) = u(1) = u'(1) = 0$$

which becomes in its weak formulation

$$\begin{aligned} &a(u,v) = f(v) \quad \text{for all } v \in H_0^2(0,1), \text{ where} \\ &a(u,v) = \int_0^1 u''v''\,dx, \qquad f(v) = \int_0^1 gv\,dx. \end{aligned} \tag{8.5.4}$$

Divide the interval $\Omega = (0,1)$ into equal subintervals of length h. Piecewise linear functions (cf. Figure 8.3.1) can be viewed as linear spline functions, so that it is natural to define V_h as the space of c u b i c s p l i n e s (with $u = u' = 0$ for $x = 0$ and $x = 1$) (cf. Stoer [1, §2.4] and Stoer-Bulirsch [1, §2.4]). We can take as basis functions B-splines, whose supports in general consist of four subintervals (cf. Figure 1). Since cubic spline functions belong to $C^2(0,1)$, they are not just in $H_0^2(0,1)$, but even in $H^3(0,1) \cap H_0^2(0,1)$.

Figure 8.5.1. B-spline

Simpler yet than working with spline functions is to use c u b i c H e r m i t e i n t e r p o l a t i o n:

$$\begin{aligned} V_h := \{u \in C^1(0,1)\colon & u \text{ cubic on each subinterval}, \\ & u(0) = u'(0) = u(1) = u'(1) = 0\}. \end{aligned} \tag{8.5.5}$$

To each of the inner nodes $x_j = jh$ there are two basis functions b_{1i} and b_{2i} with $b_{1i}(x_i) = 1, b'_{1i}(x_i) = 0, b_{2i}(x_i) = 0, b'_{2i}(x_i) = 1, b_{ki}(x_j) = b'_{ki}(x_j) = 0$ for $k = 1, 2$ $(j \neq i)$ (cf. Figure 2).

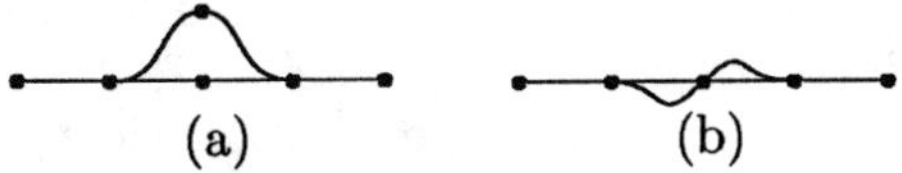

Figure 8.5.2. Hermite basis functions

The support consist of only two subintervals. The expressions are

$$b_{1i}(x) = (h - |x - x_i|)^2(2|x - x_i| + h)/h^3 \quad \text{for } x_{i-1} \le x \le x_{i+1},$$
$$b_{2i}(x) = (h - |x - x_i|)^2(x - x_i)/h^2 \quad \text{for } x_{i-1} \le x \le x_{i+1},$$
$$b_{ji}(x) = 0 \quad \text{for } x \notin [x_{i-1}, x_{i+1}].$$

Exercise 8.5.3. Let u_{1i} and u_{2i} $(0 < i < 1/h)$ be the coefficients of the expression $u(x) = \sum_i [u_{1i}b_{1i}(x) + u_{2i}b_{2i}(x)] \in V_h$. Show that the coefficients of the finite-element solution of the problem (4) are given by the equations

$$\begin{aligned} &h^{-3}[-12u_{1,i-1} + 24u_{1i} - 12u_{1,i+1}] + h^{-2}[-6u_{2,i-1} + 6u_{2,i+1}] = f_{1i}, \\ &h^{-2}[6u_{1,i-1} - 6u_{1,i+1}] + h^{-1}[2u_{2,i-1} + 8u_{2i} + 2u_{2,i+1}] = f_{2i} \end{aligned} \tag{8.5.6}$$

where $f_{1i} = \int_{x_{i-1}}^{x_{i+1}} b_{1i}g\,dx \approx hg(x_i)$, $f_{2i} = \int_{x_{i-1}}^{x_{i+1}} b_{2i}g\,dx$, $x_i = ih$.

The system of equations (6) differs completely from the difference equations (cf. Section 5.3.3), since in Equation (6) there appear values u_{1i} of the functions at the nodes together with the values u_{2i} of the derivatives at the nodes.

8.5.2.2 The Two-Dimensional Case

The ansatz (5) can be carried over to $\Omega \subset \mathbb{R}^2$ if one starts with a partition into rectangular elements. The ansatz function is bicubic:

$$V_h := \{u \in C^1(\Omega) \colon u = u_n = 0 \text{ on } \Gamma,\ u = \sum_{\nu,\mu=0}^{3} \alpha_{\nu\mu} x^\nu y^\mu$$
$$\text{on each rectangle of the partition.}\}.$$

At each inner node one may prescribe the four values u, u_x, u_y, u_{xy}. Consequently there are four unknowns and four basis functions $b_{1i}(x, y), \ldots, b_{4i}(x, y)$ belonging to each node. The latter are products $b_{ji}(x)b_{ki}(y)$ $(j, k = 1, 2)$ of the one-dimensional basis functions described in Section 8.5.2.1 (cf. Meis-Marcowitz [1, Figure 15.22], Schwarz [1, §2.6.1]).

If one starts from a triangulation τ, a fifth-order ansatz gives what is wanted: $u(\mathbf{x}) = \sum_{|\nu| \le 5} a_\nu \mathbf{x}^\nu$ on $T_i \in \tau$. The number of degrees of freedom is

21 (the number of $\nu \in \mathbb{Z}^2$ with $\nu \geq 0, |\nu| \leq 5$). For the nodes one chooses the points $\mathbf{x}^1, \ldots, \mathbf{x}^6$ in Figure 8.3.8a. At the vertices $\mathbf{x}^1, \mathbf{x}^2, \mathbf{x}^3$ one prescribes 6 values $\{D^\mu u(\mathbf{x}^j): |\mu| \leq 2\}$. The 3 remaining degrees of freedom result from giving the normal derivatives $\partial u(\mathbf{x}^i)/\partial n$ at the midpoints of the sides $\mathbf{x}^4, \mathbf{x}^5, \mathbf{x}^6$. Notice that if two neighbouring triangles (T and $\hat{T}$ in Figure 8.3.8a) have the same vertex values $\{D^\mu u(\mathbf{x}^j): |\mu| \leq 2,\ j = 1, 2\}$ and the same normal derivative at $\mathbf{x}^4$, then both u and ∇u are continuous on the shared side, i.e., $V_h \subset H^2(\Omega)$. According to whether $V = H_0^2(\Omega)$, $V = H^2(\Omega) \cap H_0^1(\Omega)$ or $V = H^2(\Omega)$ one has to set u, or the first derivatives, to be zero at the boundary nodes $\mathbf{x}^i$.

Remark 8.5.4. The finite-element space $V_h \subset H^2(\Omega)$ described here can of course be used for differential equations of the order $2m = 2$.

8.5.2.3 Estimating Errors

Instead of (5.1) one obtains

$$\inf\{|u - v|_m : v \in V_h\} \leq Ch^{k-m+1}|u|_{k+1} \quad \text{for all } u \in H^{k+1}(\Omega) \cap V, \quad (8.5.7)$$

where $k \geq m$ depends on the order of the polynomial ansatz (e.g., $k = 3$ for cubic splines, cubic [resp. bicubic] Hermite interpolation). Here $m = 2$ for the biharmonic equation. As in Theorem 8.5.1, there follows from (7) the error estimate

$$|u - u^h|_m \leq Ch^{k+1-m}|u|_{k+1} \quad (8.5.8a)$$

for the finite-element solution $u^h \in V_h$. Under suitable regularity assumptions one gets

$$|u - u^h|_t \leq Ch^{s-t}|u|_s \qquad (2m - k - 1 \leq t \leq m \leq s \leq k + 1) \quad (8.5.8b)$$

(cf. Remark 8.5.2). The maximal order of convergence $2(k - m) + 2$ results for $s = k + 1$, $t = 2m - k - 1$ and requires $u \in H^{k+1}(\Omega) \cap V$. In addition each solution of the adjoint problem (12) with $f \in H^{k+1-2m}(\Omega)$ must belong to $H^{k+1}(\Omega)$.

8.6 Finite Elements for Non-Polygonal Regions

Since the union of triangles and parallelograms only generates polygonal regions a polygonal shape was assumed in (3.6). The finite-element method is, however, in no way restricted to just such regions. On the contrary finite elements can readily be adapted to curved boundaries.

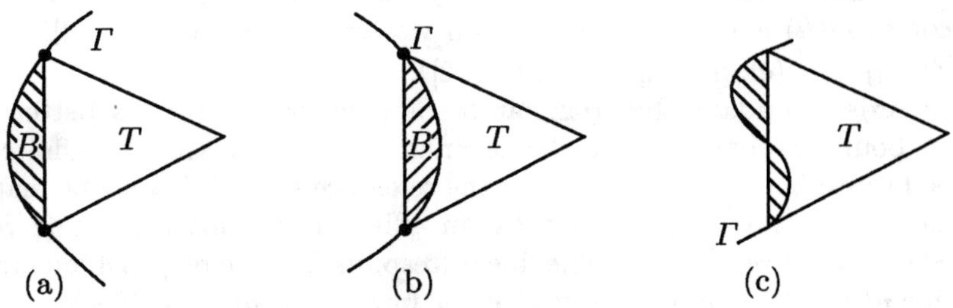

Figure 8.6.1. Curvilinear boundaries

Let $V = H^1(\Omega)$ and let Ω be arbitrary. The triangulation τ can be so chosen that (3.7a,b,d) hold and the "outer" triangles have two vertices on $\Gamma = \partial\Omega$ as in Figure 1. In the convex case of Figure 1a one extends the linear functions defined on T onto $\tilde{T} := \overline{T} \cup B$. In the case that the boundary is concave (see Figure 1b) one should replace T by $\tilde{T} := T\backslash B$. One copes correspondingly in the situation of Figure 1c. The nodes and the expressions for the basis functions remain undisturbed by these changes. $\overline{\Omega}$ is the union of the closures of all the inner triangles $T_i \in \tau$ and all modified triangles $\tilde{T}$ which lie on the boundary. All the properties and results of Sections 8.3–8.4 extend to the new situation. The only difficulty is of a practical sort: To calculate $\mathbf{L}$ and $\mathbf{f}$ one has to work out integrals over the triangle $\tilde{T}$ which involves arcs.

Assume now $V = H_0^1(\Omega)$. The previous construction will not be a success, since the extensions of the linear functions to $\tilde{T}$ will not vanish on the boundary piece $\Gamma \cap \partial\tilde{T}$. Thus $V_h \subset H_0^1(\Omega)$ will not be satisfied. As long as the region Ω is convex, only the situation shown in Figure 1a occurs, and u^h may be extended to $B \subset \tilde{T}$ by $u^h = 0$. In the case of Figure 1b, one must set the values at the inner nodes to zero, so that $u^h = 0$ on $\tilde{T} = T\backslash B$, and in particular $u^h = 0$ on $\Gamma \cap \partial(T\backslash B)$. All in all, what results is that the support of any $u^h \in V_h$ is in a polygonal region inscribed in Ω. One interpretation is the following: In the boundary value problem, Ω should be replaced by an approximating region $\Omega_h \subset \Omega$ (cf. Theorem 2.4.6). The finite-element solution described agrees with that which would result from the smaller region. However, the error estimate in Theorem 8.4.4 only holds for the smaller region Ω_h: $\inf_{v\in V_h} \|u - v\|_{H^1(\Omega_h)} \le Ch\|u\|_{H^2(\Omega)}$. Since $v = 0$ on $\Omega\backslash\Omega_h$ for any $v \in V_h$ one should also estimate $\|u\|_{H^1(\Omega\backslash\Omega_h)}$. Now $\Omega\backslash\Omega_h$ is contained in a strip $S_\delta = \{\mathbf{x} \in \Omega_h : \operatorname{dist}(\mathbf{x}, \Gamma) \le \delta\}$ of width $\delta := \max\{\operatorname{dist}(\mathbf{x}, \Gamma) : \mathbf{x} \in \Omega\backslash\Omega_h\}$. One can estimate as follows:

$$\inf_{v\in V_h} \|u - v\|_{H^1(\Omega\backslash\Omega_h)} = \|u\|_{H^1(\Omega\backslash\Omega_h)} \le \|u\|_{H^1(S_\delta)} \le C\sqrt{\delta}\|u\|_{H^2(\Omega)}.$$

If $\Omega\backslash\Omega_h$ consists only of arc segments B as in Figure 1a (for instance, in the case of a convex region), and if $\Omega \in C^{1,1}$, then $\delta = O(h^2)$. From this follows the estimate $\inf_{v\in V_h} \|u - v\|_{H^1(\Omega)} \le Ch^2\|u\|_{H^2(\Omega)}$, as for a polygonal region. If, however, as in Figure 1b the whole triangle is part of $\Omega\backslash\Omega_h$, then δ becomes $O(h)$ and the approximation worsens to $\inf_{v\in V_h}\|u-v\|_{H^1(\Omega)}\le Ch^{1/2}\|u\|_{H^2(\Omega)}$ (cf. Strang-Fix [1, p. 192]).

In order to adapt the trianglar or parallelogram elements better to a curved boundary one can use the technique of isoparametric finite elements. From Figure 8.3.4 we see that the reference triangle T may be mapped into an arbitrary triangle $\tilde{T}\in\tau$ by an affine transformation $\phi:(\xi,\eta)\mapsto \mathbf{x}^1+\xi(\mathbf{x}^2-\mathbf{x}^1)+\eta(\mathbf{x}^3-\mathbf{x}^1)$. The linear [resp. in the case of §8.3.4, quadratic] function $u(\mathbf{x})$ on $\tilde{T}$ can be expressed as the image, $v(\xi,\eta)=u(\phi(\xi,\eta))$, of a linear [resp. quadratic] function v on T. We now replace the affine transformation of triangles ϕ by a more general quadratic mapping

$$\phi(\xi,\eta):=\begin{pmatrix} a_1+a_2\xi+a_3\eta+a_4\xi^2+a_5\xi\eta+a_6\eta^2\\ a_7+a_8\xi+a_9\eta+a_{10}\xi^2+a_{11}\xi\eta+a_{12}\eta^2\end{pmatrix}:T\to\tilde{T}\subset\mathbb{R}^2.$$

The image $\tilde{T}$ is a curvilinear triangle. The coefficients $a_1,\ldots,a_{12}$ are uniquely determined by the nodes $\mathbf{x}^1,\ldots,\mathbf{x}^6$ of $\tilde{T}$ which are the images of the vertices and midpoints of the sides of the reference triangle T (cf. Figure 2). The triangulation τ used so far can be replaced by a "triangulation" by triangles with curved sides, if neighbouring elements have the same arcs as common boundaries and the midpoints also coincide. The ansatz functions on $\tilde{T}\in\tilde{\tau}$ have the form $u(\mathbf{x})=v(\phi^{-1}(\mathbf{x}))$, where $v(\xi,\eta)$ is linear [resp. quadratic] in ξ and η. The resulting finite-element space is the space of isoparametric linear [resp. quadratic] elements (cf. Strang-Fix [1], Ciarlet [1], Schwarz [1], Zienkiewicz[1]).

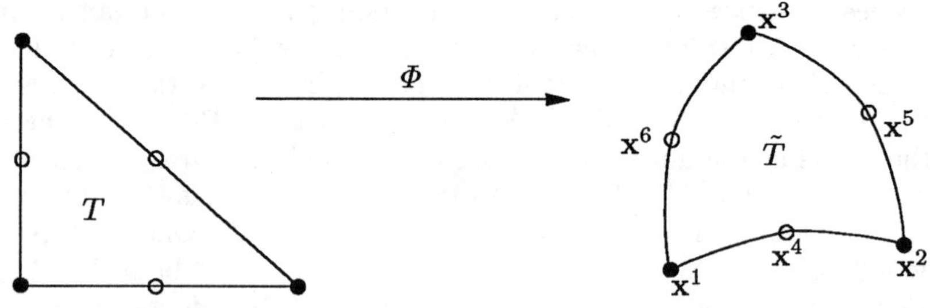

Figure 8.6.2. Mapping of the reference triangle T to the curvilinear triangle $\tilde{T}$

In general, there is no reason to use curvilinear triangles in the interior of Ω. As in Figure 3, one chooses ordinary triangles in the interior (i.e., the quadratic transformation ϕ is again taken to be linear). A boundary triangle, like $\tilde{T}$ in Figure 3, is, on the other hand, defined as follows: $\mathbf{x}^1$ and $\mathbf{x}^2$ are the vertices of $\tilde{T}$ which lie on Γ. One chooses another boundary point $\mathbf{x}^4$ between $\mathbf{x}^1$ and $\mathbf{x}^2$ and requires that the boundary $\partial\tilde{T}$ of the triangle cuts the boundary Γ of the region at $\mathbf{x}^1,\mathbf{x}^4$, and $\mathbf{x}^2$.

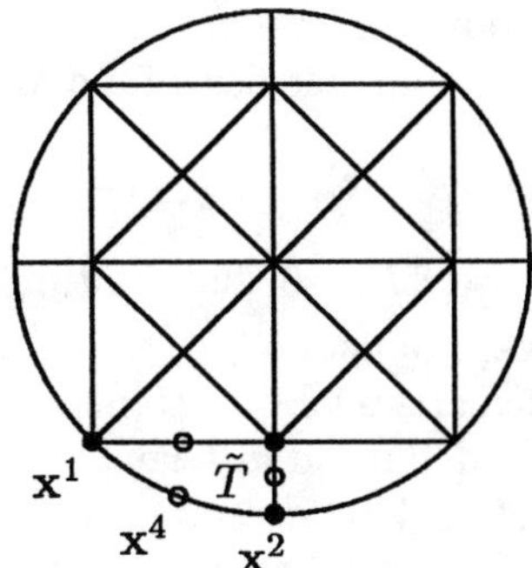

Figure 8.6.3. Isoparametric triangular dissection

Remark 8.6.1. Section 5.2.2 shows that, in the case of other than Dirichlet boundary conditions, the construction of difference schemes is increasingly complicated. Instead one may restrict the usual difference methods to the inner grid points, and near the boundary discretise by using (e.g., isoparametric) finite elements.

8.7 Additional Remarks

It is not possible to consider here many of the details about, and modifications of, finite-element methods. In the following, individual cases are just treated summarily.

8.7.1 Non-Conformal Elements

Condition (1.3), $V_h \subset V$, characterises conformal finite-element methods. Discretisations for which $V_h \not\subset V$ are called non-conformal. An example of a non-conformal element is the "Wilson rectangle". For this method one starts with a dissection into rectangles $R_i = (x_{1i}, x_{2i}) \times (y_{1i}, y_{2i})$ and assumes the function to be quadratic on R_i:

$$V_h := \{u \text{ quadratic on } R_i, u \text{ continuous at vertices of the rectangle } R_i\}.$$

Notice that $u \in V_h$ is required to be continuous only at the nodes, and not in the interior of Ω. A function which is not continuous along a side of R_i cannot belong to $H^1(\Omega)$: $V_h \not\subset V = H^1(\Omega)$. A possible basis for V_h is the following. For each node $\mathbf{x}^j$, let b_j be the basis function belonging to the bilinear elements (cf. Section 8.3.3). The quadratic ansatz chosen here has two additional degrees of freedom. Because of this one chooses for each rectangle $R_i = (x_{1i}, x_{2i}) \times (y_{1i}, y_{2i})$ two further basis functions

$$b_i^{(1)}(x,y) := (x - x_{1i})(x_{2i} - x), \qquad b_i^{(2)}(x,y) := (y - y_{1i})(y_{2i} - y),$$

which vanish at the vertices of R_i, so that continuity is ensured there if $b_i^{(1)}$ and $b_i^{(2)}$ are extended by zero outside R_i.

The first problem which arises for non-conformal elements is the definition of the stiffness matrix $\mathbf{L}$. A definition in terms of $L_{ij} = a(b_j, b_i)$, as in (1.8a), is not possible because $b_j, b_i \notin V$! Therefore, in addition to discretising V to V_h, one must replace $a(\cdot,\cdot)$ by a bilinear form $a_h(\cdot,\cdot): V_h \times V_h \to \mathbb{R}$. For a partition of Ω into rectangles R_i one defines $a_h(\cdot,\cdot)$ by

$$a_h(u,v) := \sum_i \int_{R_i} \sum_{|\alpha|,|\beta| \le m} a_{\alpha\beta}(\mathbf{x})(D^\alpha u)(D^\beta u)\, d\mathbf{x},$$

if $a(\cdot,\cdot)$ is given by (7.2.6). Notice that $a(u,v) = a_h(u,v)$ for $u, v \in V$. Conformal finite-element methods always give consistent discretisations. For non-conformal methods one does not always have consistency. For this problem, in particular for the so-called "patch test" see Strang-Fix [1, p. 174], Ciarlet [1, §4.2], Stummel [2], Gladwell-Wait [1, p. 83–92], and Thomasset [1].

That an ansatz which seems completely reasonable can possibly yield completely false discretisations is shown by

Exercise 8.7.1. Let τ be an admissible triangulation. Assume that $V_h := \{u \text{ constant on each } T_i \in \tau\}$. Show that
(a) $V_h \not\subset V := H^1(\Omega)$.
(b) $\inf\{|u - v|_0 : v \in V_h\} \le Ch|u|_1$ for all $u \in V = H^1(\Omega)$.
(c) Let $a(u,v) := \int_\Omega \langle \nabla u, \nabla v\rangle\, dx$ and $a_h(u,v) := \sum_{T_i \in \tau} \int_{T_i} \langle \nabla u, \nabla v\rangle\, dx$. What is the result for $L_{ij} = a_h(b_j, b_i)$?

Non-conformal elements are, in particular, of importance for equations of higher order, since they are less complicated than conformal elements.

8.7.2 The Trefftz Method

In Theorem 6.5.12 it was shown that for symmetric and V-elliptic bilinear forms the weak formulation (1.1) of the boundary value problem is equivalent to the variational problem

$$\text{Find } u \in V \text{ with } J(u) = \min\{J(v) : v \in V\} \tag{8.7.1}$$

(cf. also Theorem 7.2.9, Exercise 7.3.8). Another variational problem

$$\text{Find } w \in W \text{ with } K(w) = \max\{K(\tilde{w}) : \tilde{w} \in W\} \tag{8.7.2}$$

is called the dual or complementary variational problem to (1), if both have the same solutions $u = w \in V \cap W$.

For example, the Poisson problem, $-\Delta u = f \in L^2(\Omega)$ in Ω, $u = 0$ on Γ, leads to $J(v) = \int_\Omega |\nabla v|^2 \, d\mathbf{x}$ for $v \in V := H_0^1(\Omega)$. A special dual variational problem is due to Trefftz (1926):

$$K(w) := -\int_\Omega |\nabla w|^2 \, d\mathbf{x} \text{ for } w \in W := \{w \in H^1(\Omega): -\Delta w = f\}$$

(cf. Velte [1, p. 91]). Notice that $v \in V$ satisfies the boundary condition $v = 0$ on Γ, while for $w \in W$ no boundary condition is required though it must satisfy the differential equation $Lw = f$.

8.7.3 Finite-Element Methods for Singular Solutions

The error estimates provided so far, such as for instance (4.7), depend on the assumption $u \in H^2(\Omega) \cap V$, which will be discussed more carefully in Section 9.1. That this assumption does not always hold is shown by

Example 8.7.2. (a) The Laplace equation $\Delta u = 0$ in the L-shaped region of Example 2.1.4 has the solution $u = r^{2/3} \sin((2\varphi - \pi)/3)$. This function only belongs to $H^s(\Omega)$ for $s < 1 + \frac{2}{3}$.
(b) The solution $u = r^{1/2} \sin(\varphi/2)$ in Example 5.2.3 only belongs to $H^s(\Omega)$ for $s < 3/2$.

If one replaces, for example in the L-shaped region, the Laplace equation by the more general Poisson equation, $\Delta u = f$ in Ω, $u = 0$ on Γ, then one may still show that for $f \in L^2(\Omega)$ the solution u may be split up as $u = u_1 + \alpha\chi(r) r^{2/3} \sin((2\varphi - \pi)/3)$ with $u_1 \in H^2(\Omega) \cap H_0^1(\Omega)$ and $\alpha \in \mathbb{R}$ (cf. Strang-Fix [1, p. 275ff.]). In this equation $\chi(r) \in C^\infty(\Omega)$ is a specifiable function with $\chi(r) = 1$ in $0 \le r \le 1/4$, $\chi(r) = 0$ for $r \ge 1/2$, such that $\chi(r) r^{2/3} \sin(\ldots) \in H_0^1(\Omega)$ is ensured. An ordinary discretisation by linear triangular elements can only attain $|u - u^h|_1 = O(h^s)$, $s < 2/3$. However, if one includes the function $\chi(r) r^{2/3} \sin(\ldots)$ or similar functions in V_h, then the approximation of u in V_h can be improved: $d(u, V_h) = O(h)$, so that $|u - u^h|_1 = O(h)$. For such matters refer to Babuška-Rosenzweig [1] and Blum-Dobrowolski [1]. See also Gladwell-Wait [1, p. 119] and Hackbusch [4].

A completely different approach using finite elements which does not need explicit knowledge of the singularity is the following. Let τ be a triangulation whose triangles become finer near where the singularity is expected, for example, at a re-entrant corner (cf. Figure 2.1.1, 5.2.1). For analysis of the errors see Schatz-Wahlbin [1].

8.7.4 Adaptive Triangulation

The sort of triangulation mentioned in the last paragraph can be most easily obtained by adaptive methods. So far $\tau = \tau_h$ has been given depending on a given "step size" h. In an adaptive procedure one starts with a triangulation

τ_0 and uses a-posteriori error estimates. Those triangles of τ_0 that lead to the largest errors are then divided further so that a new triangulation τ is formed. One repeats this procedure until the errors are satisfactory. For the literature on this we refer to, for example, Rheinboldt's work [1] and the paper of Babuška and Rheinboldt referred to therein, as well as to Eriksson–Johnson [1].

8.7.5 Hierarchical Bases

Let us give the fundamental idea in the one-dimensional case. Let $V_h \subset H_0^1(0,1)$ be, as in (3.2), the space of piecewise linear functions for an equidistant division of the interval $\Omega = (0,1)$ into parts of length h. For $h = 1/2$ we have $\dim(V_{1/2}) = 1$, and b_1 as in Figure 1a is the only basis function. Clearly $V_{1/4} \subset V_{1/2}$. For $V_{1/4}$ one takes over $b_1 \in V_{1/2}$ as a basis function, and adds the functions b_2, b_3 as in Figure 1b; $\{b_1, b_2, b_3\}$ is a basis for $V_{1/4}$. Similarly $V_{1/8} \subset V_{1/4}$. Thus b_1, b_2, b_3 supplemented by $b_4, \ldots, b_7$ (cf. Figure 1c) gives a basis — the so-called hierarchical basis — of $V_{1/8}$. In contrast to the basis functions that have as yet been used, the supports of the b_i are not necessarily of length $O(h)$; e.g., $\operatorname{supp}(b_1) = [0,1]$. From this it follows that the stiffness matrix has more entries than is otherwise usual. In spite of this, the above choice of basis does have important advantages:
(i) A matrix $\mathbf{L} = \mathbf{L}^{(2h)}$ already calculated for V_{2h} is a submatrix of the stiffness matrix $\mathbf{L} = \mathbf{L}^{(h)}$ for V_h, so that (a global or local) refinement of the triangulation requires only a minimum of additional effort.
(ii) The condition $\operatorname{cond}(\mathbf{L}) = ||\mathbf{L}||\,||\mathbf{L}^{-1}||$ of the stiffness matrix is essentially better than in the standard case (for example, $\operatorname{cond}(\mathbf{L}) = O(\log h)$ instead of $O(h^{-2})$; cf. Section 8.8).
(iii) There are suitable methods available for solving the system of equations $\mathbf{Lu} = \mathbf{f}$. Further details on this can be found in Yserentant [1], Bank–Dupont–Yserentant [1], and Hackbusch [9].

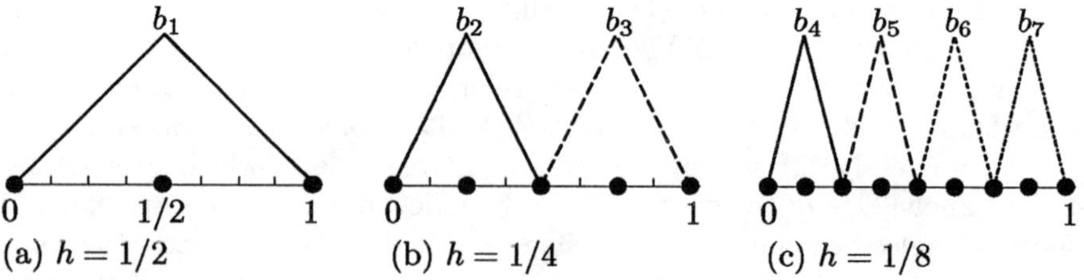

Figure 8.7.1. A hierarchical basis

8.7.6 Superconvergence

In difference methods the error is given by a mesh function $u_h - R_h u$ (u_h: the difference solution, u: the exact solution), while for finite-element methods the solution u^h is defined on the whole region Ω, and therefore $u^h - u$ can be used as the error.

A grid function u_h could in principle be extended to a function u^h defined on all Ω, for example, by choosing the values $u_h(\mathbf{x}^i)$ ($\mathbf{x}^i$: a grid point of Ω_h) as the nodal values of the piecewise linear or bilinear finite elements, and thus interpolating u_h with $u^h := \mathbf{P}u_h$ (for $\mathbf{P}$ see (1.6)). In addition to the errors $u_h - R_h u$ at the grid points, there are now the interpolation errors $u - \mathbf{P}R_h u$ as well: $u^h - u = \mathbf{P}u_h - u = \mathbf{P}(u_h - R_h u) - (u - \mathbf{P}R_h u)$. As soon as the interpolation errors are of the same order of magnitude as $u_h - R_h u$, one has for $\mathbf{P}u_h - u$ the same convergence assertions, otherwise $\mathbf{P}u_h - u$ displays worse convergence than $u_h - R_h u$.

Conversely, for finite-element methods, one can ask if one has the same convergence properties (e.g., $\|u_h - R_h u\| = O(h^2)$, R_h: restriction to the nodes) for the values at the nodes $u_h = \{u^h(\mathbf{x}^i) : \mathbf{x}^i \text{ nodes}\}$ or for suitable averaged values. If one has indeed better convergence than for $u^h - u$, one speaks of superconvergence. This notion can also be applied to difference quotients of node values, if these are more precise than $|D^\alpha u^h - D^\alpha u|_0 \le |u^h - u|_1$ ($|\alpha| = 1$). For this see Lesaint-Zlámal [1], Bramble-Schatz [1], Thomée [2], and Louis [1]. Greater precision can be achieved with extrapolation techniques which, under suitable conditions, can be applied to finite-element problems (cf. Blum-Rannacher [2]).

8.8 Properties of the Stiffness Matrix

The properties of the matrix L_h have been carefully studied for difference procedures since the solvability of the difference equations and the analysis of convergence properties depend on them. In the case of finite-element discretisation we obtain the corresponding assertions in another way. The factors that control the solvability and convergence are the subspace V_h and the operator $L_h : V_h \to V_h'$, which is associated to the bilinear form $a(\cdot,\cdot) : V_h \times V_h \to \mathbb{R}$, and which was introduced as L_N in (1.10a).

The stiffness matrix $\mathbf{L}$, for a given V_h, depends on the basis $\{b_1, \ldots\}$ chosen. Let

$$N_h := \dim(V_h).$$

The isomorphism $\mathbf{P} : \mathbb{R}^{N_h} \to V_h$ defined in (1.6) maps the coefficient vector $\mathbf{v}$ to $\mathbf{P}\mathbf{v} = \sum_{i=1}^{N_h} v_i b_i \in V_h$. For finite elements, in general, the coefficients coincide with the nodal values: $v_i = (\mathbf{P}\mathbf{v})(\mathbf{x}^i)$, $(1 \le i \le N_h)$. The connection between the matrix $\mathbf{L}$ and the operator $L_h : V_h \to V_h'$ is, according to Lemma 8.1.7,

$$\mathbf{L} = \mathbf{P}^* L_h \mathbf{P}, \qquad L_h = \mathbf{P}^{*-1}\mathbf{L}\mathbf{P}^{-1}. \tag{8.8.1}$$

The definition of $\mathbf{P}^*$ in (1.10b), $\langle \mathbf{P}^* u, \mathbf{v}\rangle = (u, \mathbf{P}\mathbf{v})_{L^2(\Omega)}$, also depends on the choice of the scalar product $\langle\cdot,\cdot\rangle$ in $\mathbb{R}^{N_h}$. Let us choose here the usual

$$\langle \mathbf{u}, \mathbf{v} \rangle = \sum_{i=1}^{N_h} u_i v_i. \tag{8.8.2a}$$

In Exercise 8.1.10 we have already established the following properties of $\mathbf{L}$: If $a(\cdot,\cdot)$ is symmetric [and V-elliptic] then $\mathbf{L}$ is symmetric [and positive definite]. Notice that the difference methods do not always have these properties. For the Poisson problem there is a symmetric form, but nonetheless the matrix L_h in Section 4.8.1 is not symmetric (cf. Theorem 4.8.4).

The condition cond($\mathbf{L}$) plays an important role in the solution of the equation $\mathbf{Lu} = \mathbf{f}$. We would like to show that, under standard conditions, we have cond($\mathbf{L}$) $= O(h^{-2m})$ (cf. Remark 5.3.10). First we have to decide on the fundamental norm of $\mathbb{R}^{N_h}$. Up to a scalar factor

$$||\mathbf{u}||_h := \left(h^n \sum_{i=1}^{N_h} |u_i|^2 \right)^{1/2} \qquad (n: \text{ dimension of } \Omega \subset \mathbb{R}^n) \tag{8.8.2b}$$

is the Euclidean norm coming from the scalar product (2a). The corresponding matrix norm

$$||\mathbf{L}|| := \sup\{||\mathbf{Lv}||_h/||\mathbf{v}||_h : \mathbf{0} \neq \mathbf{v} \in \mathbb{R}^{N_h}\} \tag{8.8.2c}$$

is the usual spectral norm of $\mathbf{L}$. As an alternative to $||\cdot||_h$ let us introduce the norm

$$||\mathbf{u}||_P := ||\mathbf{Pu}||_{L^2(\Omega)} \quad \text{for } u \in \mathbb{R}^{N_h}. \tag{8.8.3}$$

In some important cases $||\cdot||_P$ and $||\cdot||_h$ are equivalent (uniformly in h). As an example consider the linear elements.

Theorem 8.8.1. *Let τ_h, $h \to 0$ be a sequence of quasiuniform triangulations. Let V_h be the space of linear elements defined in* (3.8) *with the usual basis (see* (3.9a)*). Then there is a constant C_P, which does not depend on h, such that*

$$\frac{1}{C_P}||u||_P \le ||u||_h \le C_P||u||_P. \tag{8.8.4}$$

The basis for the proof of this is

Lemma 8.8.2. *Let $T = \{(\xi,\eta): \xi,\eta \ge 0, \xi+\eta \le 1\}$ be the unit triangle (see Figure* 8.3.4*). If u is linear on T, then*

$$\begin{aligned} &\frac{1}{24}[u^2(0,0)+u^2(1,0)+u^2(0,1)] \le \int\!\!\int_T u^2(\xi,\eta)\,d\xi\,d\eta \\ &\le \frac{1}{6}[u^2(0,0)+u^2(1,0)+u^2(0,1)]. \end{aligned} \tag{8.8.5}$$

PROOF. Put $u_1 := u(0,0)$, $u_2 := u(1,0)$, $u_3 := u(0,1)$. From $u(\xi,\eta) = u_1 + (u_2-u_1)\xi + (u_3-u_1)\eta$, it follows that

$$\iint_T u^2(\xi,\eta)\,d\xi\,d\eta = \frac{1}{24}\begin{pmatrix} u_1 & u_2 & u_3 \end{pmatrix}\begin{bmatrix} 2 & 1 & 1 \\ 1 & 2 & 1 \\ 1 & 1 & 2 \end{bmatrix}\begin{pmatrix} u_1 \\ u_2 \\ u_3 \end{pmatrix}.$$

The eigenvalues of the symmetric 3×3 matrix are $\lambda_1 = \lambda_2 = 1$ and $\lambda_3 = 4$. Therefore the right-hand side is $\geq \sum_{i=1}^3 u_i^2/24$ and $\leq 4\sum u_i^2/24$. ■

To prove Theorem 1 write $||\mathbf{u}||_P^2$ as $\int_\Omega (\mathbf{Pu})^2\,d\mathbf{x} = \sum_{T_i \in \mathcal{T}_h} \int_{T_i} (\mathbf{Pu})^2\,d\mathbf{x}$. Let $\phi_i : T \to T_i$ be the linear maps from T to T_i as in Exercise 8.3.14a. Let u', u'', u''' be the values of $u = \mathbf{Pu}$ at the vertices of T_i. The inequality $0 < C_1 h^2 \leq |\det \phi_i'| \leq C_2 h^2$ and Lemma 2 imply

$$\begin{aligned} \frac{C_1}{24}h^2(u'^2 + u''^2 + u'''^2) &\leq \int_{T_i} |\mathbf{Pu}|^2\,d\mathbf{x} = |\det \phi_i'| \iint_T |(\mathbf{Pu})(\phi_i(\xi,\eta))|^2\,d\xi\,d\eta \\ &\leq \frac{C_2}{6}h^2(u'^2 + u''^2 + u'''^2). \end{aligned} \tag{8.8.6}$$

By definition of a quasiuniform triangulation all the angles of the triangles are $\geq \alpha_0 > 0$. Let $M > 2\pi/\alpha_0$. Each node belongs to at least one triangle, and to at most M triangles. Since u', u'', u''' are the components of the vector $\mathbf{u}$, the summation in Equation (6) over all $T_i \in \mathcal{T}_h$ gives

$$\frac{C_1}{24}||\mathbf{u}||_h^2 \leq ||\mathbf{u}||_P^2 = \sum_{T_i \in \mathcal{T}} \int_{T_i} (\mathbf{Pu})^2\,d\mathbf{x} \leq \frac{MC_2}{6}||\mathbf{u}||_h^2,$$

so that inequality (4) holds with $C_P := \max(MC_2/6, 24/C_1)^{1/2}$. ■

Exercise 8.8.3. Let $V_h \subset H_0^1(\Omega)$ be derived from the square grid triangulation (cf. Exercise 8.3.13). Show that $\sqrt{4/3}||\mathbf{u}||_P \leq ||\mathbf{u}||_h \leq \sqrt{8}||\mathbf{u}||_P$.

The matrix

$$\mathbf{M} := \mathbf{P}^*\mathbf{P}, \text{ i.e., } M_{ij} := \int_\Omega b_i b_j\,d\mathbf{x} \tag{8.8.7}$$

is called the mass matrix when it occurs in engineering applications. Concrete bounds in the inequality (4) have been given by Wathen [1] for various specific elements.

Remark 8.8.4. (a) Inequality (4) is equivalent to

$$||\mathbf{M}|| \leq C_P^2 h^n, \qquad ||\mathbf{M}^{-1}|| \leq C_P^2 h^{-n}.$$

(b) We have $||\mathbf{u}||_P \leq ||h^{-n}\mathbf{M}||^{1/2}||\mathbf{u}||_h$, $||\mathbf{u}||_h \leq ||h^n\mathbf{M}^{-1}||^{1/2}||\mathbf{u}||_P$.

PROOF. (a) The calculation $||\mathbf{u}||_P^2 = \int_\Omega (\mathbf{Pu})^2\,d\mathbf{x} = (\mathbf{Pu}, \mathbf{Pu})_0 = \langle \mathbf{P}^*\mathbf{Pu}, \mathbf{u}\rangle = \langle \mathbf{Mu}, \mathbf{u}\rangle \leq ||\mathbf{M}||\langle \mathbf{u}, \mathbf{u}\rangle = ||\mathbf{M}||h^{-n}||\mathbf{u}||_h^2$ proves the first inequality in (b). The next follows from $||\mathbf{u}||_h^2 = h^n\langle \mathbf{u}, \mathbf{u}\rangle = h^n\langle \mathbf{M}^{1/2}\mathbf{u}, \mathbf{M}^{-1}\mathbf{M}^{1/2}\mathbf{u}\rangle \leq ||h^n\mathbf{M}^{-1}||\langle \mathbf{M}^{1/2}\mathbf{u}, \mathbf{M}^{1/2}\mathbf{u}\rangle = ||h^n\mathbf{M}^{-1}||\langle \mathbf{Mu}, \mathbf{u}\rangle = ||h^n\mathbf{M}^{-1}||\,||\mathbf{u}||_P^2$, since $\mathbf{M}$ is positive definite.

(b) From $h^{-n}||\mathbf{M}|| \le C_P^2$ and $h^n||\mathbf{M}^{-1}|| \le C_P^2$, using (b), we deduce the inequality (4). Conversely $\langle \mathbf{Mu}, \mathbf{u}\rangle^{1/2} = ||\mathbf{u}||_P \le C_P||\mathbf{u}||_h = C_P h^{n/2}\langle \mathbf{u}, \mathbf{u}\rangle^{1/2}$ implies that $h^{-n}||\mathbf{M}|| \le C_P^2$, since M is positive definite (cf. Lemma 4.3.25). Similarly, $h^n||\mathbf{M}^{-1}|| \le C_P^2$ follows from $||\mathbf{u}||_h \le C_P||\mathbf{u}||_P$. ■

Since V_h is finite-dimensional $C_h := \sup\{|u|_1/|u|_0 : 0 \ne u \in V_h\}$ is finite. One says that V_h satisfies the inverse estimate if $C_h = O(h^{-1})$, that is,

$$|u|_1 \le C_1 h^{-1}|u|_0 \quad \text{for all } u \in V_h \tag{8.8.8}$$

where C_1 does not depend on h as $h \to 0$.

Theorem 8.8.5. *Let τ_h be a sequence of quasiuniform triangulations. Then the space of finite elements introduced in* (3.8) *satisfies the inverse estimate.*

PROOF. As in Theorem 1, the proof follows by transforming the integrals $\int_{T_i} |D^\alpha(P\mathbf{u})|^2\, d\mathbf{x}$ for $T_i \in \tau_h$ and $|\alpha| \le 1$ into integrals over T. Here we have in addition to transform $D = D_{\mathbf{x}}$ into $D_{\xi,\eta}$ (cf. the proof of Lemma 8.4.2). ■

Theorem 8.8.6. *Suppose $\Omega \subset \mathbb{R}^n$. Assume that* (1.2), (8), *and* $||\mathbf{u}||_P \le C_P||\mathbf{u}||_h$ *hold. Then we have*

$$||\mathbf{L}|| \le h^{n-2} C_S C_P^2 C_I^2. \tag{8.8.9}$$

PROOF. Multiply the inequality

$$\begin{aligned}\langle \mathbf{u}, \mathbf{Lv}\rangle = a(\mathbf{Pu}, \mathbf{Pv}) &\le C_S|\mathbf{Pu}|_1|\mathbf{Pv}|_1 \le C_S C_I^2 h^{-2}|\mathbf{Pu}|_0|\mathbf{Pv}|_0 \\ &\le C_S C_I^2 h^{-2} C_P^2 ||\mathbf{u}||_h||\mathbf{v}||_h\end{aligned}$$

by h^n and set $\mathbf{u} := \mathbf{Lv}$:

$$||\mathbf{Lv}||_h^2 = h^n\langle \mathbf{Lv}, \mathbf{Lv}\rangle \le C_S C_I^2 C_P^2 h^{n-2}||\mathbf{Lv}||_h||\mathbf{v}||_h,$$

and thus $||\mathbf{Lv}||_h \le C_S C_I^2 C_P^2 h^{n-2}||\mathbf{v}||_h$, that is (9). ■

Theorem 8.8.7. *Assume $\Omega \subset \mathbb{R}^n$. Let* (1.12) *hold with $\epsilon_h \ge \epsilon > 0$ and $||\cdot||_h \le C_P||\cdot||_P$. Then there holds*

$$||\mathbf{L}^{-1}|| \le C_P^2 h^{-n}/\epsilon. \tag{8.8.10}$$

PROOF. We know $\mathbf{L}^{-1} = \mathbf{P}^{-1}L_h^{-1}\mathbf{P}^{*-1}$ and $||L_h^{-1}||_{V_h \leftarrow V_h'} \le 1/\epsilon_h \le 1/\epsilon$ (cf. Section 8.1), so that

$$\begin{aligned}||\mathbf{L}^{-1}\mathbf{f}||_h \le C_P||\mathbf{L}^{-1}\mathbf{f}||_P &= C_P|\mathbf{PL}^{-1}\mathbf{f}|_0 = C_P|L_h^{-1}\mathbf{P}^{*-1}\mathbf{f}|_0 \\ &\le C_P||L_h^{-1}\mathbf{P}^{*-1}\mathbf{f}||_{V_h} \le (C_P/\epsilon)||\mathbf{P}^{*-1}\mathbf{f}||_{V_h'}.\end{aligned}$$

Since

$$\begin{aligned}||\mathbf{P}^{*-1}\mathbf{f}||_{V_h'} &= \sup_{0\neq v\in V_h} |(v,\mathbf{P}^{*-1}\mathbf{f})_0|/|v|_1 = \sup_{0\neq v\in V_h} |\langle \mathbf{P}^{-1}v,\mathbf{f}\rangle|/|v|_1 \\ &= \sup_{\mathbf{v}\neq\mathbf{0}} |\langle \mathbf{v},\mathbf{f}\rangle|/|\mathbf{P}\mathbf{v}|_1 = \sup_{\mathbf{v}\neq\mathbf{0}} h^{-n}||\mathbf{v}||_h||\mathbf{f}||_h/|\mathbf{P}\mathbf{v}|_0 \\ &\leq C_P \sup_{\mathbf{v}\neq\mathbf{0}} h^{-n}||\mathbf{v}||_h||\mathbf{f}||_h/||\mathbf{v}||_h = C_P h^{-n}||\mathbf{f}||_h,\end{aligned}$$

we have that $||\mathbf{L}^{-1}\mathbf{f}||_h \leq C_P^2 h^{-n}||\mathbf{f}||_h/\epsilon$ for all $\mathbf{f}$; thus we conclude (10). ■

The combination of Theorems 6 and 7 leads to

Theorem 8.8.8. *Assume* (1.2), (1.12) *with* $\epsilon_h \geq \epsilon > 0$, (4), $m = 1$, *and* (8). *Then we have*

$$\operatorname{cond}(\mathbf{L}) \leq h^{-2} C_S C_P^4 C_I^2/\epsilon. \tag{8.8.11}$$

The ideas involved in the proof can, in principle, be carried over to the case $2m > 2$, i.e., to boundary value problems of higher order. The inverse estimate becomes

$$|u|_m \leq C_I h^{-m}|u|_0 \quad \text{for all } u \in V_h. \tag{8.8.12}$$

In order that inequality (4) hold one must be careful in defining the norm $||\cdot||_h$. If all the components u_i of $\mathbf{u}$ are nodal values $(\mathbf{Pu})(\mathbf{x}^i)$ (as for instance is the case for the spline ansatz for $V_h \subset H^2(\Omega)$), then $||\cdot||_h$ can be defined as in (2b). However as soon as the components u_i of $\mathbf{u}$ involve derivatives $(D^\alpha\mathbf{Pu})(\mathbf{x}^i)$, the u_i^2 in (2b) must be replaced by $(h^{|\alpha|}u_i)^2$. For example, when the Hermite functions of (5.5) are used then $\mathbf{u}$ has the components u_{1i} and u_{2i} (cf. Exercise 8.5.3), where $u_{1i} = (\mathbf{Pu})(\mathbf{x}^i)$ and $u_{2i} = \frac{d}{dx}(\mathbf{Pu})(\mathbf{x}^i)$. The appropriate definition of $||\cdot||_h$ is $||\mathbf{u}||_h^2 := h^n\sum_i(u_{1i}^2 + h^2u_{2i}^2)$ with $n = 1$, since $(0,1) \subset \mathbb{R}^1$.

Exercise 8.8.9. Let $V \subset U \subset V'$ be a Gelfand triple and assume $V_h \subset V$. Show that the inverse estimate

$$||u||_V \leq C_I h^{-m}||u||_U \quad \text{for all } u \in V_h \tag{8.8.13a}$$

implies

$$||u||_U \leq C_I h^{-m}||u||_{V'} \quad \text{and} \quad ||u||_V \leq C_I^2 h^{-2m}||u||_{V'} \quad \text{for all } u \in V_h. \tag{8.8.13b}$$

9 Regularity

9.1 Solutions of the Boundary Value Problem in $H^s(\Omega)$, $s > m$

9.1.1 The Regularity Problem

The weak formulation of a boundary value problem

$$Lu = g \quad \text{in } \Omega, \qquad Bu = \varphi \quad \text{on } \Gamma \tag{9.1.1}$$

as

$$u \in V, \quad a(u, v) = f(v) \quad \text{for all } v \in V \tag{9.1.2}$$

was, in Section 7, the basis upon which we were able to answer the questions of existence and uniqueness of the solution. Here, by existence of a solution we understand the existence of a weak solution $u \in V$.

The error estimates in Section 8.4 made it clear that the statement $u \in V$ is not enough. Under this assumption we can only show $\|u - u^h\|_V \to 0$. The more interesting quantitative estimate $|u - u^h|_1 = O(h)$ as in, for example, Theorem 8.4.6 for $V = H_0^1(\Omega)$ or $V = H^1(\Omega)$, requires the assumption $u \in H^2(\Omega) \cap V$. The assertion $u \in H^2(\Omega)$ or, more generally $u \in H^s(\Omega)$, is a statement of regularity, i.e., a statement about the smoothness of the solution, which will be examined in greater detail in this section.

The regularity proofs in the following sections are very technical. To make the proof ideas clearer, let us sketch the proof of inequality (4) below for the Helmholtz equation $-\Delta u + u = f$ in $\Omega \subset \mathbb{R}^2$, $u = 0$ on Γ.

Step 1: $\Omega = \mathbb{R}^2$. Since the bilinear form for this situation $a(u, v) = \int_\Omega \langle \nabla u, \nabla v \rangle \, dx + \int_\Omega uv \, dx$ is $H^1(\mathbb{R}^2)$-elliptic, (4) holds for $s = m = 1$. We shall prove (4) by induction for $s = 1, 2, 3, \cdots$. To this end we take the derivative of the differential equation with respect to x, $-\Delta u_x + u_x = f_x$. If $f \in H^{s-2}(\Omega)$, then $f_x \in H^{s-3}(\Omega)$ and the equation $-\Delta v + v = f_x$, by the induction assumption, has a unique solution $v \in H^{s-1}(\Omega)$ with $|v|_{s-1} \le C'_{s-1} |f_x|_{s-3} \le C'_{s-1} |f|_{s-2}$. If one sets up this inequality for $v = u_x$ and likewise for $v = u_y$, the result is $|u|_s \le |u|_{s-1} + |u_x|_{s-1} + |u_y|_{s-1} \le 3C'_{s-1} |f|_{s-2}$. Thus, (4) has been shown for s.

Step 2: $\Omega = \mathbb{R}^2_+ = \mathbb{R} \times (0, \infty)$. As above, we can obtain the estimate $|u_x|_{s-1} \le C'_{s-1} |f|_{s-2}$, since u_x also satisfies $-\Delta u_x + u_x = f_x$ and the boundary condition

$u_x = 0$ on Γ. This is not the case for u_y. But $u_x \in H^{s-1}(\mathbb{R}^2_+)$ implies $u_{xx} \in H^{s-2}(\mathbb{R}^2_+)$ and $u_{xy} \in H^{s-2}(\mathbb{R}^2_+)$. We would have $u \in H^s(\mathbb{R}^2_+)$ if we could also show $u_{yy} \in H^{s-2}(\mathbb{R}^2_+)$. This property, however, results from the differential equation, $u_{yy} = \Delta u - u_{xx} = u - f - u_{xx} \in H^{s-2}(\mathbb{R}^2_+)$.

Step 3: Let Ω be arbitrary, but sufficiently smooth. As in Section 6.2.1, Ω is decomposed into (overlapping) pieces Ω_i, which can be mapped into $\mathbb{R}^2$ or $\mathbb{R}^2_+$. Correspondingly one splits the solution u into $\sum \chi_i u$ (χ_i is a partition of unity). Then the arguments from steps 1 and 2 prove inequalities for $|\chi_i u|_s$ which together result in (4).

Note that only a sketch of the proof was given. Some of the steps of the proof are incomplete. For example, might not the equation $-\Delta u_x + u_x = f_x$ have a solution in $\mathbb{R}^2$, $u_x \in L^2(\mathbb{R}^2)$, which does not belong to $H^1(\mathbb{R}^2)$ and hence does not coincide with the solution $v \in H^1(\mathbb{R}^2)$ of $-\Delta v + v = f_x$?

In the following, always let $s \geq m$. The boundary value problem (2) with $V = H_0^m(\Omega)$ is said to be H^s-regular if each solution $u \in H_0^m(\Omega)$ of problem (2) with $f \in H^{s-2m}(\Omega)$ belongs to $H^s(\Omega) \cap H_0^m(\Omega)$ and satisfies the estimate

$$|u|_s \leq C_s[\,|f|_{s-2m} + |u|_m]. \tag{9.1.3}$$

If L is the operator associated with $a(\cdot,\cdot)$, then it is also said that L is H^s-regular.

Remark 9.1.1. (a) H^m-regularity always holds.
(b) Let the variational problem (2) have a unique solution $u \in H_0^m(\Omega)$ with $|u|_m \leq C_0|f|_{-m}$ for all $f \in H^{-m}(\Omega)$. If the boundary value problem is H^s-regular, then the weak solution of (2) with $f \in H^{s-2m}$ satisfies the inequality

$$|u|_s \leq C_s'|f|_{s-2m}. \tag{9.1.4}$$

(c) Let L be the operator associated with $a(\cdot,\cdot)$. (4) is equivalent to $L^{-1} \in L(H^{s-2m}(\Omega), H^s(\Omega))$ and (4′):

$$||L^{-1}||_{H^s(\Omega)\leftarrow H^{s-2m}(\Omega)} \leq C_s'. \tag{9.1.4′}$$

PROOF. (a) Set $C_m = 1$. (b) In (3) estimate $|u|_m$ by $C_0|f|_{-m}$ and use the fact that because of $s \geq m$ the embedding $H^{s-2m}(\Omega) \subset H^{-m}(\Omega)$ is continuous such that $|f|_{-m} \leq C'|f|_{s-2m}$. Thus we obtain (4) with $C_s' = C_s(1+C_0C')$. ■

The following remark shows that a perturbation of $a(\cdot,\cdot)$ by a smooth term of order $< 2m$ does not change the H^s-regularity.

Remark 9.1.2. Let L be H^t-regular for all $t = m, m+1, \cdots, s \in \mathbb{N}$. Let δL be an operator of order $\leq 2m-1$: $\delta L \in L(H^r(\Omega) \cap H_0^m(\Omega), H^{r+1-2m}(\Omega))$ for all $r = m, m+1, \cdots, s-1$. Then $L+\delta L$ is also H^t-regular for $t = m, m+1, \cdots, s$. The assumption on δL holds in particular if δL belongs to the bilinear form $a''(\cdot,\cdot)$ in Lemma 7.2.12 and its coefficients $a_{\alpha\beta}$ are sufficiently smooth: for example, $a_{\alpha\beta} \in C^{\max\{0,s-2m+|\beta|\}}$.

PROOF. (By induction) For $s = m$ the statement follows from Remark 1a. Assume the assertion for $s-1$. Let u be the weak solution of $(L + \delta L)u = f \in H^{s-2m}(\Omega)$, hence also the solution of $Lu = \tilde{f} := f - \delta Lu$. Under the induction assumption the following already holds:

$$|u|_{s-1} \le C'_{s-1}[|f|_{s-2m-1} + |u|_m]$$

such that

$$|\tilde{f}|_{s-2m} \le |f|_{s-2m} + ||\delta L||_{s-2m\leftarrow s-1}|u|_{s-1} \le C''_{s-1}[|f|_{s-2m} + |u|_m].$$

Because of the H^s-regularity of L, u belongs to $H^s(\Omega)$ and satisfies $|u|_s \le C'_s[\,|\tilde{f}|_{s-2m} + |u|_m]$. Together these inequalities result in statement (4). ■

9.1.2 Regularity Theorems for $\Omega = \mathbb{R}^n$

The domain $\Omega = \mathbb{R}^n$ is distinguished by the fact that it has no boundary, and hence no boundary conditions either.

Theorem 9.1.3. *Let $m \in \mathbb{N}$, $\Omega = \mathbb{R}^n$. Let the bilinear form*

$$a(u,v) := \sum_{|\alpha|,|\beta|\le m} \int_\Omega a_{\alpha\beta}(\mathbf{x})(D^\alpha u)(D^\beta v)d\mathbf{x} \tag{9.1.5}$$

be $H^m(\mathbb{R}^n)$-coercive. For some $k \in \mathbb{N}$ let the following hold:

$$D^\gamma a_{\alpha\beta} \in L^\infty(\Omega) \quad \text{for all } \alpha,\beta,\gamma \text{ with } |\gamma| \le \max(0, k + |\beta| - m). \tag{9.1.6}$$

Then every weak solution $u \in H^m(\mathbb{R}^n)$ of the problem

$$a(u,v) = (f,v)_0 \quad \text{for all } v \in H^m(\mathbb{R}^n) \tag{9.1.7}$$

with $f \in H^{-m+k}(\mathbb{R}^n)$ belongs to $H^{m+k}(\mathbb{R}^n)$ and satisfies the estimate

$$|u|_{m+k} \le C_k[\,|f|_{-m+k} + |u|_m]. \tag{9.1.8}$$

PROOF. (a) First we must investigate the start of the induction, $k = 1$. Let $\partial = \partial_{h,i}$ be the difference operator

$$\partial_{h,i}u(\mathbf{x}) := \left[u\left(\mathbf{x} + \frac{h}{2}\mathbf{e}_i\right) - u\left(\mathbf{x} - \frac{h}{2}\mathbf{e}_i\right)\right]/h, \mathbf{e}^i: \ i\text{-th unit vector}, 1 \le i \le n.$$

Let u be the solution of (7). For $v \in H^m(\mathbb{R}^n)$ set

$$d(u,v) := a(u, \partial_{h,i}v) + a_P(\partial_{h,i}u, v), \tag{9.1.9a}$$

where $a_P(u,v) := \sum_{\substack{|\alpha|\le m\\|\beta|=m}} \int_{\mathbb{R}^n} a_{\alpha\beta}(D^\alpha u)(D^\beta v)\,d\mathbf{x}$.

Let $u^{\pm}(\mathbf{x}) := u(\mathbf{x} \pm h\mathbf{e}_i/2)$. From $h\int_{\mathbb{R}^n} au\partial v\,d\mathbf{x} = \int_{\mathbb{R}^n} au[v^+ - v^-]\,d\mathbf{x} = \int_{\mathbb{R}^n}[a^-u^- - a^+u^+]v\,d\mathbf{x} = -h\int_{\mathbb{R}^n} av\,\partial u\,d\mathbf{x} + \int_{\mathbb{R}^n}\{(a^- - a)u^- + (a - a^+)u^+\}v\,d\mathbf{x}$ we see that

$$d(u,v) = \sum_{\substack{|\alpha|\le m\\|\beta|\le m-1}} \int_{\mathbb{R}^n} a_{\alpha\beta}(D^\alpha u)(D^\beta \partial v)\,d\mathbf{x}$$
$$-\sum_{\substack{|\alpha|\le m\\|\beta|=m}} \int_{\mathbb{R}^n} \frac{1}{h}\{(a_{\alpha\beta} - a^-_{\alpha\beta})D^\alpha u^- + (a^+_{\alpha\beta} - a_{\alpha\beta})D^\alpha u^+\}D^\beta v\,d\mathbf{x}. \tag{9.1.9b}$$

Since $|\partial v|_s \le |v|_{s+1}$ (cf. Lemma 6.2.24), one bounds the first sum in (9b) with $C|u|_m|\partial v|_{m-1} \le C|u|_m|v|_m$. One obtains the same bound for the second sum:

$$|d(u,v)| \le C|u|_m|v|_m \quad \text{for} \quad v \in H^m(\mathbb{R}^n). \tag{9.1.9c}$$

However, if $a(\cdot,\cdot)$ is $H^m(\mathbb{R}^n)$-coercive $a_P(\cdot,\cdot)$ is also (cf. Lemma 7.2.12). For $C_E > 0$ and C_K from (6.5.10) we thus obtain

$$C_E|\partial_{h,i}u|^2_m \le a_P(\partial u, \partial u) + C_K|\partial u|^2_0 = -a(u, \partial^2 u) + d(u, \partial u) + C_K|\partial u|^2_0 \tag{9.1.10a}$$

The first summand can be transformed according to (7): $a(u,\partial^2 u)_0 = (f, \partial^2 u)_0$ and bounded by $|f|_{-m+1}|\partial^2 u|_{m-i}$. The inequality $|\partial v|_s \le |v|_{s+1}$ for $s = m-1$ and $v = \partial u$ yields

$$|a(u, \partial^2_{h,i}u)| \le |f|_{-m+1}|\partial_{h,i}u|_m. \tag{9.1.10b}$$

By (9c), the bound for the second summand in (10a) reads:

$$d(u, \partial_{h,i}u) \le C_d|u|_m|\partial_{h,i}u|_m. \tag{9.1.10c}$$

Finally we have

$$|\partial_{h,i}u|^2_0 \le |\partial_{h,i}u|_{m-1}|\partial_{h,i}u|_m \le |u|_m|\partial_{h,i}u|_m. \tag{9.1.10d}$$

(10a-d) yields

$$|\partial_{h,i}u|_m \le [\,|f|_{-m+1} + (C_d + C_K)|u|_m]/C_E \quad \text{for all } h > 0,\ 1 \le i \le n.$$

Lemma 6.2.24 shows that $u \in H^{m+1}(\mathbb{R}^n)$ and inequality (8) for $k = 1$.
(b) Now let $k = 2$. In (9a) substitute $D^\gamma \partial v$ with $|\gamma| = 1$ for ∂v and integrate by parts:

$$
\begin{aligned}
d(u,v) &:= a(u, D^\gamma \partial v) - a_P(D^\gamma u, v) \\
&= \sum_{\substack{|\alpha|\le m\\ |\beta|\le m-2}} \int_{\mathbb{R}^n} a_{\alpha\beta}(D^\alpha u)(D^{\beta+\gamma}\partial v)\,d\mathbf{x} \\
&\qquad - \sum_{\substack{|\alpha|\le m\\ |\beta|\le m-1}} \int_{\mathbb{R}^n} [D^\gamma(a_{\alpha\beta}D^\alpha u)](D^\beta \partial v)\,d\mathbf{x} \\
&+ \sum_{\substack{|\alpha|\le m\\ |\beta|= m}} \int_{\mathbb{R}^n} \big[(D^\gamma \partial a_{\alpha\beta})\,(D^\alpha u(\mathbf{x}-\tfrac{h}{2}\mathbf{e}_i)) \\
&\qquad + (\partial a_{\alpha\beta})\,(D^{\alpha+\gamma} u(\mathbf{x}-\tfrac{h}{2}\mathbf{e}_i)) \\
&\qquad + (D^\gamma a_{\alpha\beta}(\mathbf{x}+\tfrac{h}{2}\mathbf{e}_i) D^\alpha \partial u) \\
&\qquad + (a_{\alpha\beta}(\mathbf{x}+\tfrac{h}{2}\mathbf{e}_i) - a_{\alpha\beta}(\mathbf{x}))\, D^{\alpha+\gamma}\partial u\big]\,(D^\beta v)\,d\mathbf{x}.
\end{aligned}
$$

As in the first part of the proof, from $|d(u,v)| \le C_d |u|_{m+1}|v|_m$ and

$$|(f, D^\gamma \partial v)_0| \le |f|_{2-m}|D^\gamma \partial v|_{m-2} \le |f|_{2-m}|v|_m$$

follows

$$|D^\gamma \partial_{h,i} u|_m \le [\,|f|_{-m+2} + (C_d + C_K)|u|_{m+1}]/C_E$$

for all $|\gamma| = 1$, $h > 0$, $1 \le i \le n$. Thus, the differences $|\partial_{h,i}u|_{m+1}$ are uniformly bounded so that Lemma 6.2.24 proves $u \in H^{m+2}(\mathbb{R}^n)$ and $|u|_{m+2} \le C[|f|_{-m+2} + |u|_{m+1}]$. If one uses the estimate $|u|_{m+1} \le C[|f|_{-m+1} + |u|_m] \le C[|f|_{-m+2} + |u|_m]$ proved in (a), (8) follows for $k = 2$.
(c) An induction argument is carried out analogously for $k \ge 2$. ■

Corollary 9.1.4. *The coercivity in Theorem* 3 *can be replaced* (a) *by the sufficient conditions from Theorem* 7.2.11; (b) *by the assumption that for some* λ *the bilinear form* $a_P(u,v) + \lambda(u,v)_0$ *satisfies the assumption* (6.5.4a,b).

Corollary 9.1.5. *If, in addition,* $a(\cdot,\cdot)$ *is* $H^m(\mathbb{R}^n)$*-elliptic or if* $a(\cdot,\cdot)$ *satisfies condition* (6.5.4a,b), *then in Theorem* 3 *the estimate* (8) *should be replaced by*

$$|u|_{m+k} \le C_k |f|_{-m+k}. \tag{9.1.11}$$

PROOF. By Remark 2. ■

Corollary 9.1.6. *Let* $a(\cdot,\cdot)$ *be* $H^m(\mathbb{R}^n)$*-coercive. Let the conditions* (6) *and* $f \in H^{k-m}(\mathbb{R}^n)$ *be satisfied for* $k \in \mathbb{N}$ *with* $k > s + n/2 \ge n/2$. *Then the weak solution* u *of* (7) *belongs to* $C^s(\mathbb{R}^n)$. *Hence for* $s \ge 2m$, *the weak solution is also a classical solution.*

PROOF. The statement results from Sobolev's lemma (Theorem 6.2.30). ■

Corollary 9.1.7. *Let $a(\cdot,\cdot)$ be $H^m(\mathbb{R}^n)$-coercive. Let the conditions (6) and $f \in H^{k-m}(\mathbb{R}^n)$ be satisfied for all $k \in \mathbb{N}$. Then the weak solution of problem (7) belongs to $C^\infty(\mathbb{R}^n)$. The conditions are satisfied in particular if f belongs to $C_0^\infty(\mathbb{R}^n)$ and the coefficients $a_{\alpha\beta}$ are constant.*

The generalisation of $u \in H^{m+k}(\mathbb{R}^n)$ with $k \in \mathbb{N}$ to $u \in H^{m+s}(\mathbb{R}^n)$ with real $s > 0$ reads as follows:

Theorem 9.1.8. *Let $a(\cdot,\cdot)$ in (5) be $H^m(\mathbb{R}^n)$-coercive. Let*

$$s = k + \Theta, \quad k \in \mathbb{N} \cup \{0\}, \quad 0 < \Theta < \vartheta < 1, \quad t := k + \vartheta.$$

For the coefficients ($\Omega = \mathbb{R}^n$) let

$$a_{\alpha\beta} \in C^{t+|\beta|-m}(\Omega) \quad \text{for } k + |\beta| \ge m, \quad a_{\alpha\beta} \in L^\infty(\Omega) \quad \textit{otherwise.} \tag{9.1.12}$$

Then each weak solution of (7) with $f \in H^{-m+s}(\mathbb{R}^n)$ belongs to $H^{m+s}(\mathbb{R}^n)$, and satisfies the estimate

$$|u|_{m+s} \le C_s[\,|f|_{-m+s} + |u|_m]. \tag{9.1.13}$$

PROOF. The beginning of the induction is given by $k = 0$, i.e., $0 < s < t < 1$. Replace the difference quotient $\partial = \partial_{h,j}$ by an approximation to its power ∂^s:

$$Ru(\mathbf{x}) := R_{h,j}u(\mathbf{x}) := h^{-s} \sum_{\mu=0}^{\infty} e^{-\mu h}(-1)^\mu \binom{s}{\mu} u(\mathbf{x} + \mu h \mathbf{e}_j),$$

where $\mathbf{e}_j$ is the jth unit vector. Here $\binom{s}{0} = 1$, and

$$\binom{s}{\mu}(-1)^\mu = -s(1-s)(2-s)\dots(\mu-1-s)/\mu!$$

is the binomial coefficient.

Exercise 9.1.9. Let $0 < s < 1$. Show that (a) The operator adjoint to $R_{h,j}$ is

$$R^*u(x) = R^*_{h,j}u(\mathbf{x}) = h^{-s} \sum_{\mu=0}^{\infty} e^{-\mu h}(-1)^\mu \binom{s}{\mu} u(\mathbf{x} - \mu h \mathbf{e}_j).$$

(b) For any $z \in \mathbb{C}$ with $|z| < 1$ we have $(1-z)^s = \sum_{\mu=o}^{\infty} \binom{s}{\mu}(-z)^\mu$.
(c) For the Fourier transform we have

$$\mathcal{F}(R_{h,j}u)(\xi) = [(1 - e^{-h+i\xi_j h})/h]^s \hat{u}(\xi),$$
$$\mathcal{F}(R^*_{h,j}u)(\xi) = [(1 - e^{-h-i\xi_j h})/h]^s \hat{u}(\xi).$$

Hint: $\mathcal{F}(u(\cdot + \delta \mathbf{e}_j))(\xi) = e^{i\xi_j h}\hat{u}(\xi)$.

(d) $|R_{h,j}u|_\tau \le C_{|\tau|}|u|_{\tau+s}$ for all $\tau \in \mathbb{R}$, $h > 0$, $1 \le j \le n$, $u \in H^{\tau+s}(\mathbb{R}^n)$; similarly $|R^*_{h,j}u|_\tau \le |U|_{\tau+s}$. Hint: Make use of the norms $|\cdot|^\wedge_\tau$, and $|\cdot|^\wedge_{\tau+s}$ (cf. (6.2.15)), and show $|\mathcal{F}(Ru)(\xi)| \le (1+|\xi|^2)^{s/2}|\hat{u}(\xi)|$.

By analogy with (9) one obtains

$$\begin{aligned}
d(u,v) &:= a(u, R_{h,j}v) - a_P(R^*_{h,j}u, v)\\
&= \sum_{\substack{|\alpha|\le m\\ |\beta|\le m-1}} \int_{\mathbb{R}^n} a_{\alpha\beta}(D^\alpha u)(R_{h,j}D^\beta v)\,d\mathbf{x}\\
&\quad + \sum_{\substack{|\alpha|\le m\\ |\beta|= m}} \sum_{\mu=1}^{\infty} h^{-s}e^{-\mu h}(-1)^\mu \binom{s}{\mu}\\
&\quad \int_{\mathbb{R}^n} [a_{\alpha\beta}(\mathbf{x}-\mu h\mathbf{e}_j) - a_{\alpha\beta}(\mathbf{x})][D^\alpha u(\mathbf{x}-\mu h\mathbf{e}_j)](D^\beta v)\,d\mathbf{x},
\end{aligned}$$

since the first summand in $(a_{\alpha\beta}D^\alpha u)(\mathbf{x}-\mu h\mathbf{e}_j) = a_{\alpha\beta}(\mathbf{x})D^\alpha u(\mathbf{x}-\mu h\mathbf{e}_j) + [a_{\alpha\beta}(\mathbf{x}-\mu h\mathbf{e}_j) - a_{\alpha\beta}(\mathbf{x})]D^\alpha u(\mathbf{x}-\mu h e_j)$ belongs to $a_P(R^*u, v)$. Since we have $|a_{\alpha\beta}(\mathbf{x}-\mu h\mathbf{e}_j) - a_{\alpha\beta}(\mathbf{x})| \le C(\mu h)^t$, it follows that

$$|d(u,v)| \le C|u|_m|v|_m\left[1 + h^{t-s}\sum_{\mu=1}^{\infty} e^{-\mu h}(-1)^\mu\binom{s}{\mu}\mu^t\right] \le C'|u|_m|v|_m,$$

for it is true that $\binom{s}{\mu} = O(\mu^{-s-1})$ and $\sum_{\mu=1}^{\infty} e^{-\mu h}\mu^{t-s-1} = O(h^{s-t})$. The same consideration as in Theorem 3 yields

$$|R^*_{h,j}u|_m \le C[|R_{h,j}f|_{-m}+|u|_m] \le C'[|f|_{-m+s}+|u|_m] \quad \text{for all } h>0,\ 1\le j\le n$$

(cf. Exercise 9d). To obtain this estimate for $|u|_{m+s}$ we write $(|u|^\wedge_{m+s})^2$ as

$$\sum_{|\xi|\le 1/h}(1+|\xi|^2)^{m+s}|\hat{u}(\xi)|^2 d\xi + \int_{|\xi|\ge 1/h}(1+|\xi|^2)^{m+s}|\hat{u}(\xi)|^2 d\xi.$$

The second integral converges to zero for $h \to 0$. The first one can be estimated by $|u|^2_m$ plus the sum

$$\begin{aligned}
C\sum_{j=1}^{n}|R^*_{h,j}u|^2_m &\ge C'\int_{|\xi|\le 1/h}(1+|\xi|^2)^m\sum_j|\mathcal{F}(R^*_{h,j}u)(\xi)|^2 d\xi\\
&= C''\int_{|\xi|\le 1/h}(1+|\xi|^2)^m\sum_j|(1-e^{-h-i\xi_j h})/h|^{2s}|\hat{u}(\xi)|^2 d\xi\\
&= C''\int_{|\xi|\le 1/h}(1+|\xi|^2)^m\sum_j\left|\frac{\sin\xi_j h}{h}\right|^{2s}|\hat{u}(\xi)|^2 d\xi\\
&\ge \tilde{C}\int_{|\xi|\le 1/h}(1+|\xi|^2)^m|\xi|^{2s}|\hat{u}(\xi)|^2 d\xi.
\end{aligned}$$

As in Lemma 6.2.24, $u \in H^{m+s}(\mathbb{R}^n)$ and (13) follow.
(2) As in the proof of Theorem 3, one carries out induction over k and investigates $d(u,v) := a(u, RD^\gamma v) - (-1)^{|\gamma|} a_P(R^* D^\gamma u, v)$, $|\gamma| = k$. ■

Exercise 9.1.10. Generalise Corollary 6 with the aid of Theorem 8.

9.1.3 Regularity Theorems for $\Omega = \mathbb{R}^n_+$.

The halfspace $\mathbb{R}^n_+$ in (6.2.18) is characterised by $x_n > 0$. As in Section 7, we limit ourselves to the following two cases: either a Dirichlet problem is given for arbitrary $m \geq 1$, or the natural boundary condition is posed for $m = 1$.

Theorem 9.1.11. (Homogeneous Dirichlet problem). *An analogue to Theorem 3 holds for the Dirichlet problem:*

$$u \in H_0^m(\mathbb{R}^n_+), \quad a(u,v) = (f,v)_0 \quad \text{for all } v \in H_0^m(\mathbb{R}^n_+). \tag{9.1.14}$$

Theorem 8 can also be carried over if one excludes the values $s = 1/2, 3/2, \cdots, m - 1/2$.

For the proof we need the following highly technical lemma.

Lemma 9.1.12. *Let $s > 0$, $s \notin \{\frac{1}{2}, \frac{3}{2}, \cdots, m - \frac{1}{2}\}$. The norm $|\cdot|_s$ of $H^s(\mathbb{R}^n_+)$ is equivalent to*

$$|||u|||_s := (|u|_0^2 + \sum_{|\alpha|=m} |D^\alpha u|_{s-m}^2)^{1/2}. \tag{9.1.15}$$

PROOF. The relatively elementary case $s \geq m$ is left to the reader. For $0 < s < m$ too, the proof would be considerably simpler if in (15) one were to replace the dual norm $|\cdot|_{s-m}$ of $(H_0^{m-s}(\mathbf{R}^n_+))'$ by that of $(H^{m-s}(\mathbb{R}^n_+))'$.
Step 1. First we prove the statement for $\Omega = \mathbb{R}^n$ instead of $\Omega = \mathbb{R}^n_+$. According to Theorem 6.2.25a, for $\Omega = \mathbb{R}^n$ the norm $|||\cdot|||_s$ is equivalent to

$$|||u|||_s^\wedge := (|u|_0^2 + \sum_{|\alpha|=m} (|D^\alpha u|_{s-m}^\wedge)^2)^{1/2}.$$

Since $(|||u|||_s^\wedge)^2 = \int_{\mathbb{R}^n} [1 + (\sum_{|\alpha|=m} |\xi^\alpha|^2)(1 + |\xi|^2)^{s-m}] |\hat{u}(\xi)|^2 d\xi$ and

$$0 < C_0(1+|\xi|^2)^s \leq 1 + (\sum_{|\alpha|=m} |\xi^\alpha|^2)(1+|\xi|^2)^{s-m} \leq C_1(1+|\xi|^2)^s,$$

$|||\cdot|||_s^\wedge$ and $|\cdot|_s^\wedge$ are also equivalent.
Step 2. For the transition to $\Omega = \mathbb{R}^n_+$ the following extension ϕ: $H^s(\mathbb{R}^n_+) \to H^s(\mathbb{R}^n)$ must be investigated, where $\mathbf{x} = (\mathbf{x}', x_n) \in \mathbb{R}^n$:

$$(\phi u)(\mathbf{x}) := u(\mathbf{x}) \quad \text{for } \mathbf{x} \in \mathbb{R}^n_+,$$
$$(\phi u)(\mathbf{x}', x_n) := \sum_{\nu=1}^{L} a_\nu [u(\mathbf{x}', -\nu x_n) + u(\mathbf{x}', -x_n/\nu)] \quad \text{for } x_n < 0.$$

We list the properties of ϕ as

Exercise 9.1.13. Let the coefficients a_ν of ϕ be selected as the solution of the system of equations $\sum_{\nu=1}^{L} a_\nu(\nu^k + \nu^{-k}) = (-1)^k$ $(0 \le k \le L-1)$. Show that
(a) $u \in C^{L-1}(\mathbb{R}^n_+)$ yields $\phi u \in C^{L-1}(\mathbb{R}^n)$.
(b) $\phi \in L(H^k(\mathbb{R}^n_+), H^k(\mathbb{R}^n))$ for $k = 0, 1, \ldots, L$.
(c) The operator adjoint to ϕ reads

$$(\phi^* u)(\mathbf{x}', x_n) = u(\mathbf{x}', x_n) + \sum_{\nu=1}^{L} a_\nu \left[\frac{1}{\nu} u(\mathbf{x}', -x_n/\nu) + \nu u(\mathbf{x}', -\nu x_n)\right] \quad \text{for } x_n > 0.$$

(d) $(\partial/\partial x_n)^k (\phi^* u)(\mathbf{x}', 0) = 0$ for $k = 0, 1, \ldots, L-2$ and $u \in C^k(\mathbb{R}^n)$.
(e) $\phi^* \in L(H^k(\mathbb{R}^n), H^k_0(\mathbb{R}^n_+))$ for $k = 0, 1, \ldots, L-1$. Hint: cf. Corollary 6.2.43.
(f) $\phi \in L(H^s(\mathbb{R}^n_+), H^s_0(\mathbb{R}^n))$ for $s = 1-L, 2-L, \ldots, 0, 1, \ldots, L$.

Step 3. $|||\cdot|||_s \le C|\cdot|_s$ results from $D^\alpha \in L(H^s_0(\mathbb{R}^n_+), H^{s-m}(\mathbb{R}^n_+))$ (provable via continuation arguments from Remark 6.3.14b) so that $|\cdot|_s \le |||\cdot|||_s$ remains to be shown.
Step 4. $|u|_s \le |\phi u|_s \le |||\phi u|||_s$ is true according to Step 1 of the proof. The inequality $|||\phi u|||_s \le C|||u|||_s$, which would finish the proof, reduces to

$$|D^\alpha \phi u|_{s-m} \le C |D^\alpha u|_{s-m} \quad \text{for } |\alpha| = m, \ u \in H^s(\mathbb{R}^n_+). \tag{9.1.16}$$

Let $|\alpha| = m$. For

$$(\phi_\alpha u)(\mathbf{x}', x_n) := \begin{cases} u(\mathbf{x}', x_n) & \text{for } x_n > 0, \\ \sum_{\nu=1}^{L} a_\nu \left[(-\nu)^{\alpha_n} u(\mathbf{x}', -\nu x_n) + \left(-\frac{1}{\nu}\right)^{\alpha_n} u(\mathbf{x}', -x_n/\nu)\right] & \text{otherwise,} \end{cases}$$

one verifies $D^\alpha \phi = \phi_\alpha D^\alpha$. As in Exercise 13f one shows that $\phi_\alpha \in L(H^s(\mathbb{R}^n_+), H^s(\mathbb{R}^n))$ for $s = 1+m-L, 2+m-L, \ldots, L-m$. This result can be carried over to real $s \in [1+m-L, L-m]$ except for the cases $\frac{1}{2} - s \in \mathbb{N}$ (i.e., $s = -1/2, -3/2, \cdots$) (cf. Lions-Magenes [1, p. 54 ff]). Let $L \ge 2m+1$ and $v \in H^{m-s}(\mathbb{R}^n)$. Since $\phi^*_\alpha v \in H^{m-s}_0(\mathbb{R}^n_+)$, one infers from $(D^\alpha \phi u, v)_{L^2(\mathbb{R}^n)} = (D^\alpha u, \phi^*_\alpha v)_{L^2(\mathbb{R}^n_+)}$ the estimate

$$|(D^\alpha \phi u, v)_0| = |D^\alpha u|_{s-m} \|\phi^*_\alpha\|_{H^{m-s}_0(\mathbb{R}^n_+) \leftarrow H^{m-s}(\mathbb{R}^n)} |v|_{m-s},$$

for all $v \in H^{m-s}(\mathbb{R}^n)$, and therefore (16) with

$$C := \|\phi^*_\alpha\|_{H^{m-s}_0(\mathbb{R}^n_+) \leftarrow H^{m-s}(\mathbb{R}^n)} = \|\phi_\alpha\|_{H^{s-m}(\mathbb{R}^n) \leftarrow H^{s-m}(\mathbb{R}^n_+)}.$$

■

PROOF. (of Theorem 11) (a) First let $k = s = 1$. The proof of Theorem 3 can be repeated for the differences $\partial_{h,j}u$ $(j = 1, \cdots, n-1)$ and implies the existence of the derivatives $\partial u/\partial x_j \in H_0^m(\mathbb{R}_+^n)$, $j \neq n$. Thus one has $D^\alpha u \in H^1(\mathbb{R}_+^n)$ for all $|\alpha| = m$ except for $\alpha = (0, \cdots, 0, m)$.
(b) We set

$$\hat{\alpha} := (0, \cdots, 0, m) \in \mathbb{Z}^n, \quad w := a_{\hat{\alpha}\hat{\alpha}} D^{\hat{\alpha}} u,$$

$$F_\alpha(v) := \int_{\mathbb{R}_+^n} w(\mathbf{x}) D^\alpha v(\mathbf{x})\, d\mathbf{x} \quad \text{for} \quad |\alpha| = m,$$

where $a_{\hat{\alpha}\hat{\alpha}}$ is the coefficient from the bilinear form (5). The remainder of the proof runs as follows. In part (c) we will show that

$$|F_\alpha(v)| \leq C|v|_{m-1} \quad \text{for } |\alpha| = m, \ v \in H_0^{m-1}(\mathbb{R}_+^n). \tag{9.1.17}$$

Since $F_\alpha(v) = (w, D^\alpha v)_0 = (-1)^m (D^\alpha w, v)_0$, (17) means that $D^\alpha w \in H^{1-m}(\mathbb{R}_+^n)$ and $|D^\alpha w|_{1-m} \leq C$ for $|\alpha| = m$. According to Lemma 12 it follows that $w \in H^1(\mathbb{R}_+^n)$. The coercivity of $a(\cdot,\cdot)$ implies uniform ellipticity of $L = \sum_{|\alpha|=|\beta|=m} (-1)^m D^\beta a_{\alpha\beta} D^\alpha$, i.e., $\sum a_{\alpha\beta}\xi^{\alpha+\beta} \geq \epsilon|\xi|^{2m}$ (cf. Theorem 7.2.13). For $\xi = (0, \cdots, 0, 1)$ one obtains $a_{\hat{\alpha}\hat{\alpha}}(\mathbf{x}) \geq \epsilon$. Hence it follows from $w \in H^1(\mathbb{R}_+^n)$ and $D^\gamma a_{\hat{\alpha}\hat{\alpha}} \in L^\infty(\mathbb{R}_+^n)$, $|\gamma| = 1$, that $D^{\hat{\alpha}} u \in H^1(\mathbb{R}_+^n)$. According to part (a) all other derivatives $D^\alpha u$ $(|\alpha| \leq m, \alpha \neq \hat{\alpha})$ belong to $H^1(\mathbb{R}_+^n)$ anyway so that $u \in H^{m-1}(\mathbb{R}_+^n)$ has been proved.
(c) Proof of (17). For each $\alpha \neq \hat{\alpha}$ there exists a γ with $|\gamma| = 1$, $\gamma_n = 0$, $0 \leq \gamma \leq \alpha$ (component-wise inequalities). Integration by parts yields

$$F_\alpha(v) = -\int_{\mathbb{R}_+^n} [(D^\gamma a_{\hat{\alpha}\hat{\alpha}})(D^{\hat{\alpha}} u)(D^{\alpha-\gamma} v) + a_{\hat{\alpha}\hat{\alpha}}(D^{\hat{\alpha}+\gamma} u)(D^{\alpha-\gamma} v)]\, d\mathbf{x}$$

for $v \in C_0^\infty(\mathbb{R}_+^n)$; and thus

$$|F_\alpha(v)| \leq C_\alpha |v|_{m-1} \quad \text{with } C_\alpha := C[|u|_m + |D^\gamma u|_m] \quad \text{for all } v \in C_0^\infty(\mathbb{R}_+^n).$$

Here we used the fact that $D^\gamma u \in H^m(\mathbb{R}_+^n)$ according to part (a). Since $C_0^\infty(\mathbb{R}_+^n)$ is dense in $H_0^{m-1}(\mathbb{R}_+^n)$, (17) follows for $\alpha \neq \hat{\alpha}$.

There remains to investigate $\alpha = \hat{\alpha}$. We write

$$F_{\hat{\alpha}}(v) = a(u, v) - \hat{a}(u, v) \quad \text{with } \hat{a}(u, v) = \sum{}' \int_{\mathbb{R}_+^n} a_{\alpha\beta}(D^\alpha u)(D^\beta v)\, dx,$$

where $\sum'$ represents the summation over all pairs $(\alpha, \beta) \neq (\hat{\alpha}, \hat{\alpha})$. For each $(\alpha, \beta) \neq (\hat{\alpha}, \hat{\alpha})$ there exists a γ with

$$|\gamma| = 1, \quad 0 \leq \gamma \leq \beta, \quad \alpha + \gamma \neq (0, \cdots, 0, m+1).$$

As above, one integrates each summand with $|\beta| = m$ by parts:

$$\int_{\mathbb{R}^n_+} a_{\alpha\beta} D^\alpha u D^\beta v \, \mathbf{dx}$$
$$= -\int_{\mathbb{R}^n_+} (D^\gamma a_{\alpha\beta})(D^\alpha u)(D^{\beta-\gamma} v)\, \mathbf{dx} - \int_{\mathbb{R}^n_+} a_{\alpha\beta}(D^{\alpha+\gamma} u)(D^{\beta-\gamma} v)\, \mathbf{dx}$$

for $v \in C_0^\infty(\mathbb{R}^n_+)$, and estimates using $C|v|_{m-1}$ with $C = C(u)$. Altogether one obtains $|\hat{a}(u,v)| \le C|v|_{m-1}$. Together with $|a(u,v)| = |(f,v)_0| \le |f|_{-m+1}|v|_{m-1}$, (17) also follows for $\alpha = \hat{\alpha}$.
(d) By induction for $(k = 2, \ldots)$ one proves in the same way $|F_\alpha(v)| \le C|v|_{m-k}$, and from this $u \in H^{k+m}(\mathbb{R}^n_+)$. For real $s > 0$, $s \ne 1/2, \cdots, m-1/2$, one proves correspondingly $|F_\alpha(v)| \le C|v|_{m-s}$ and hence $u \in H^{s+m}(\mathbb{R}^n_+)$. ■

The generalisation of Theorem 11 to inhomogeneous boundary values reads as follows.

Theorem 9.1.14. *Let the bilinear form $a(\cdot,\cdot)$ from (5) be $H_0^m(\mathbb{R}^n_+)$-coercive. For an $s > 0$, $s \notin \{1/2, \ldots, m-1/2\}$ either let (6) hold if $s = k \in \mathbb{N}$, or (12), if $s \notin \mathbb{N}$. Let $u \in H^m(\mathbb{R}^n_+)$ be the weak solution of the inhomogeneous Dirichlet problem*

$$a(u,v) = (f,v)_0 \quad \textit{for all} \;\; v \in H_0^m(\mathbb{R}^n_+), \tag{9.1.18a}$$
$$\partial^l u/\partial n^l = \varphi_l \qquad \textit{on } \Gamma = \partial\mathbb{R}^n_+ \textit{ for } l = 0, 1, \ldots, m-1, \tag{9.1.18b}$$

where

$$f \in H^{-m+s}(\mathbb{R}^n_+), \quad \varphi_l \in H^{m+s-l-1/2}(\Gamma) \quad (0 \le l \le m-1). \tag{9.1.19}$$

Then u belongs to $H^{m+s}(\mathbb{R}^n_+)$ and satisfies the inequality

$$|u|_{m+s} \le C_s \left[|f|_{-m+s} + \sum_{l=0}^{m-1} |\varphi_l|_{m+s-l-1/2} + |u|_m \right]. \tag{9.1.20}$$

PROOF. For $m = 1$ Theorem 6.2.32 guarantees the existence of $u_0 \in H^{m+s}(\mathbb{R}^n_+)$ which satisfies the boundary conditions (18b) (for $m > 1$ cf. Wloka [1, Theorem 8.8]). $w := u - u_0$ is the solution of the homogeneous problem $a(w,v) = F(v) := (f,v)_0 - a(u_0,v)$ (cf. Remark 7.3.2). Theorem 11 may also be carried over to the right-hand side $F(v)$ under discussion here (instead of $(f,v)_0$) and yields $w \in H^{m+s}(\mathbb{R}^n_+)$. ■

By similar means one proves

Theorem 9.1.15. (Natural boundary conditions) *Let the bilinear form $a(\cdot,\cdot)$ from (5) be $H^1(\mathbb{R}^n_+)$-coercive. For $s > 0$ either let (6) hold if $s = k \in \mathbb{N}$, or (12) if $s \notin \mathbb{N}$. Let $u \in H^1(\mathbb{R}^n_+)$ be the weak solution of the problem*

$$a(u,v) = f(v) := \int_{\mathbb{R}^n_+} g(\mathbf{x})v(\mathbf{x})\, \mathbf{dx} + \int_\Gamma \varphi(\mathbf{x})v(\mathbf{x})\, d\Gamma \quad \textit{for all } v \in H^1(\mathbb{R}^n_+),$$

where

$$g \in H^{s-1} := \begin{Bmatrix} H^{s-1}(\mathbb{R}^n_+) & \text{for } s \geq 1 \\ (H^{1-s}(\mathbb{R}^n_+))' & \text{for } s < 1 \end{Bmatrix}, \quad \varphi \in H^{s-1/2}(\Gamma).$$

Then u belongs to $H^{1+s}(\mathbb{R}^n_+)$ and satisfies the estimate

$$|u|_{1+s} \leq C_s[\|g\|_{H^{s-1}} + |\varphi|_{s-1/2} + |u|_1]. \tag{9.1.21}$$

If a boundary value problem is in the form (1): $Lu = g$, $Bu = \varphi$ with $B = b^{\mathsf{T}}\nabla + b_0$, $|b_n(\mathbf{x})| \geq \epsilon > 0$ on $\Gamma = \partial\mathbb{R}^n_+$, then according to Theorem 7.4.11 one can find an associated variational formulation and apply Theorem 15.

9.1.4 Regularity Theorems for General $\Omega \subset \mathbb{R}^n$

The following theorems show that the above regularity statements also hold for $\Omega \subset \mathbb{R}^n$ if Ω is bounded sufficiently smoothly.

Theorem 9.1.16. *Let $\Omega \in C^{t+m}$ for some $t \geq 0$. Let the bilinear form* (5) *be $H_0^m(\Omega)$-coercive. Let $s \geq 0$ satisfy*

$$s + 1/2 \notin \{1, 2, \cdots, m\}; \; 0 \leq s \leq t, \;\; \textit{if } t \in \mathbb{N}; \; 0 \leq s < t, \;\; \textit{if } \; t \notin \mathbb{N}.$$

For the coefficients let the following hold:

$$D^\gamma a_{\alpha\beta} \in L^\infty(\Omega) \;\; \textit{for all } \alpha, \beta, \gamma \;\; \textit{with } |\gamma| \leq \max(0, t + |\beta| - m), \;\; \textit{if } t \in \mathbb{N},$$
$$\tag{9.1.22}$$
$$a_{\alpha\beta} \in C^{t+|\beta|-m}(\overline{\Omega}) \;\; \textit{for} \;\; |\beta| > m-1, \; a_{\alpha\beta} \in L^\infty(\Omega) \;\; \textit{otherwise}, \;\; \textit{if} \;\; t \notin \mathbb{N}.$$

Then each weak solution $u \in H_0^m(\Omega)$ of the problem

$$a(u, v) = \int_\Omega f(\mathbf{x}) v(\mathbf{x})\, d\mathbf{x} \quad \textit{for all } v \in H_0^m(\Omega)$$

with $f \in H^{-m+s}(\Omega)$ belongs to $H^{m+s}(\Omega) \cap H_0^m(\Omega)$ and satisfies the estimate

$$|u|_{m+s} \leq C_s[|f|_{-m+s} + |u|_m]. \tag{9.1.23}$$

For inhomogeneous boundary conditions

$$\partial^l u / \partial n^l = \varphi_l \quad \textit{with } \varphi_l \in H^{m+s-l-1/2}(\Gamma) \quad (l = 0, \cdots, m-1)$$

instead of $u \in H_0^m(\Omega)$, the statement $u \in H^m(\Omega)$ implies the statement $u \in H^{m+s}(\Omega)$ and the estimate (20).

PROOF. (a) Let $\{U^i : i = 0, 1, \cdots, N\}$ with $U^i \subset \Omega$ be a covering of Ω as in Lemma 6.2.36. Let $\{\chi_i\}$, with $\chi_i = \sigma_i^2 \in C^\infty(\Omega)$ and $\operatorname{supp}(\chi_i) \subset U^i$, be the associated partition of unity from Lemma 6.2.37. There exist derivatives $\alpha^i \in C^t(U^i)$ which map U^i $(i \geq 1)$ into $\mathbb{R}^n_+$ such that $\alpha^i(\partial U^i \cap \Gamma) \subset \partial\mathbb{R}^n_+$. By contrast, U^0 lies in the interior of Ω so that $\Gamma \cap \partial U^0 = \emptyset$. The solution u

can be written as $\sum_i \chi_i u$. In part (b) of the proof we will treat $\chi_0 u$, and in part (c) $\chi_i u$ for $i \geq 1$.

(b) We set

$$d_i(u,v) := a(\chi_i u, v) - a(u, \chi_i v) \qquad (i = 0, 1, \ldots, N)$$

and wish to show the estimate

$$|d_0(u,v)| \leq C_d |u|_m |v|_{m-s} \qquad (u \in H_0^m(\Omega), v \in H_0^{m-s}(\Omega), s \geq 1) \quad (9.1.24)$$

for $s = 1$. Each summand of $d_0(u,v)$ has the form

$$\int_{U^0} a_{\alpha\beta}[\,(D^\alpha(\chi_0 u))(D^\beta v) - (D^\alpha u)(D^\beta(\chi_0 v))\,]\, \mathbf{dx}.$$

Since $D^\alpha(\chi_0 u) = \chi_0 D^\alpha u +$ lower derivatives of u, one has

$$[\,(D^\alpha(\chi_0 u))(D^\beta v) - (D^\alpha u)(D^\beta(\chi_0 v))\,] = \sum_{\gamma,\delta} c_{\gamma\delta} D^\gamma u D^\delta v$$

with $|\gamma|, |\delta| \leq m, |\gamma| + |\delta| \leq 2m-1$. One integrates $\int_{U^0} a_{\alpha\beta} c_{\gamma\delta} D^\gamma u D^\delta v\, \mathbf{dx}$ with $|\delta| = m$ by parts, and obtains a bound $C|u|_m |v|_{m-1}$, whence (24) follows.

The coefficients $a_{\alpha\beta}$ can be extended to $\mathbb{R}^n$ in such a way that the corresponding condition (22) is satisfied on $\mathbb{R}^n$. Denote the resulting bilinear form by $\overline{a}_0(u,v)$. Since $\chi_0 \in C^\infty(\Omega)$ has a support $\operatorname{supp}(\chi_0) \subset U^0$, the extension of $\chi_0 u$ through $\chi_0 u(\mathbf{x}) = 0$ for $\mathbf{x} \in \mathbb{R}^n \backslash U^0$ poses no problems. We may formally define $d_0(u,v)$ for $v \in H^{m-s}(\mathbb{R}^n)$ since only the restriction of v to U^0 is of any consequence. By (24) $d_0(u,v)$ can be written, for a fixed $u \in H_0^m(\Omega)$, in the form

$$d_0(u,v) = (d_0, v)_{L^2(\mathbb{R}^n)} \;\text{ with } d_0 \in H^{s-m}(\mathbb{R}^n), |d_0|_{s-m} \leq C_d |u|_m.$$

$\chi_0 u$ is the weak solution of

$$\begin{aligned} \overline{a}_0(\chi_0 u, v) &= a(\chi_0 u, v) = a(u, \chi_0 v) + d_0(u,v) = (f, \chi_0 v)_0 + (d_0, v)_0 \\ &= (\chi_0 f + d_0, v)_{L^2(\mathbb{R}^n)} \;\text{ for all } v \in H^m(\mathbb{R}^n). \end{aligned}$$

Theorem 3 [resp. 8] proves $\chi_0 u \in H^{m+s}(\mathbb{R}^n)$ (thus too $\chi_0 u \in H^{m+s}(\Omega)$) and

$$\begin{aligned} |\chi_0 u|_{m+s} &\leq C[|\chi_0 f|_{-m+s} + |d_0|_{-m+s} + |\chi_0 u|_m] \\ &\leq C'[|f|_{-m+s} + C_d|u|_m + C|u|_m] \leq C''[|f|_{-m+s} + |u|_m], \end{aligned} \quad (9.1.25a)$$

where s is still restricted to $s \leq 1$.

(c) The same reasoning as for $\chi_i u$ $(i = 1, \ldots, N)$ shows

$$a(\chi_i u, v) = (\chi_i f + d_i, v)_0 \;\text{ for } v \in H_0^m(\Omega), \; |d_i|_{-m+s} \leq C_d |u|_m.$$

By assumption the maps $\alpha^i : U^i \to \mathbb{R}^n_+$ and their inverses $(\alpha^i)^{-1}$ belong to $C^{t+m}(U^i)$ [resp. $C^{t+m}(\alpha^i(U^i))$]. Put $\tilde{u}(\tilde{\mathbf{x}}) := u(\mathbf{x})$ for $\tilde{\mathbf{x}} = \alpha^i(\mathbf{x})$, that is $\tilde{u} = u \circ (\alpha^i)^{-1}$. In a similar way define $\tilde{a}_{\alpha\beta}, \tilde{\chi}_i, \tilde{f}, \tilde{d}_i$. In

$$a(\chi_i u, v) = \sum_{\alpha,\beta} \int_{\alpha^i(U^i)} \tilde{a}_{\alpha\beta} D^\alpha_{\mathbf{x}}(\tilde{\chi}_i\tilde{u})(D^\beta_{\mathbf{x}}\tilde{v})|\det(\alpha^i)'|^{-1}\, d\tilde{\mathbf{x}}$$

one can replace the derivatives $D^\alpha_{\mathbf{x}}, D^\beta_{\mathbf{x}}$ with derivatives with respect to the new coordinates, since $t \ge 0$, and thus obtain the form

$$a_i(\tilde{\chi}_i\tilde{u}, \tilde{v}) = \sum_{\alpha,\beta} \int_{\alpha^i(U^i)} \hat{a}_{\alpha\beta}(D^\alpha(\tilde{\chi}_i\tilde{u})(D^\beta \tilde{v})\, d\tilde{\mathbf{x}},$$

which again is $H^m_0(\alpha^i(U^i))$-coercive and whose new coefficients $\hat{a}_{\alpha\beta}$ satisfy the conditions corresponding to (22). As in (b), $\hat{a}_{\alpha\beta}$ can be continued to $\mathbb{R}^n_+ \supset \alpha^i(U^i)$ such that the resulting bilinear form $\overline{a}_i(\cdot,\cdot)$ is $H^m_0(\mathbb{R}^n_+)$-coercive. One can apply Theorem 11 to

$$\overline{a}_i(\tilde{\chi}_i\tilde{u}, \tilde{v}) = a_i(\tilde{\chi}_i\tilde{u}, \tilde{v}) = a(\chi_i u, v) = (\chi_i f + d_i, v)_0$$

$$= ((\tilde{\chi}_i\tilde{f} + \tilde{d}_i)/|\det(\alpha^i)'|, \tilde{v})_{L^2(\mathbb{R}^n_+)} \quad \text{for all } \tilde{v} \in H^m_0(\mathbb{R}^n_+)$$

which yields $\tilde{\chi}_i\tilde{u} \in H^{m+s}(\mathbb{R}^n_+)$ and

$$|\tilde{\chi}_i\tilde{u}|_{m+s} \le \tilde{C}_s[|\tilde{f}|_{-m+s} + |\tilde{d}_i|_{-m+s} + |\tilde{\chi}_i\tilde{u}|_m].$$

Transforming back (cf. Theorems 6.2.17, 6.2.25g) yields

$$|\chi_i u|_{m+s} \le C_s[|f|_{-m+s} + |d_i|_{-m+s} + |\chi_i u|_m] \le C_s[|f|_{-m+s} + C_d|u|_m + C|u|_m],$$

and thus

$$|\chi_i u|_{m+s} \le C'_s[|f|_{-m+s} + |u|_m] \quad \text{for } 1 \le i \le N,\ s \le 1. \tag{9.1.25b}$$

(d) (25a,b) hold for $s \le 1$. Since $|u|_{m+s} = |\sum \chi_i u|_{m+s} \le \sum |\chi_i u|_{m+s}$, estimate (23) has been proved for $s \le 1$. If the conditions of the theorem allow an $s \in (1,2]$, one proves (23) as follows. Since $u \in H^{m+1}(\Omega)$ has been proved already, one can estimate the forms $d_i(u,v)$ after further integration by parts via $|d_i(u,v)| \le C_d|u|_{m+1}|v|_{m-s}$. Accordingly, $d_i(u,v) = (d_i, v)_0$ with $d_i \in H^{-m+s}(\Omega)$ and $|d_i|_{-m+s} \le C_d|u|_{m+1}$. If one inserts the above estimate (23) for $s = 1$, one obtains (25a,b) and hence also (23) for $1 < s \le 2$. Further induction yields (23) for admissible $s \in (k, k+1]$.

(e) The case of inhomogeneous boundary values is treated as in Theorem 14. ∎

Analogously one may prove

Theorem 9.1.17. (*Natural boundary conditions*) *Let $\Omega \in C^{t+1}$ with $t \ge 0$. If $0 \le s \le t \in \mathbb{N}$ or $0 \le s < t \notin \mathbb{N}$, $\mathbb{R}^n_+$ in Theorem 15 can be replaced by Ω.*

Exercise 9.1.18. Transfer Corollaries 5, 6, and 7 to the situation of Theorems 16 and 17. What conditions on Ω must be added?

Corollary 9.1.19. *Let Ω and the coefficients of $a(\cdot,\cdot)$ satisfy the conditions in Theorem* 16, *resp.* 17. *An eigenfunction, i.e., a solution $u \in V$ ($V = H_0^m(\Omega)$ in the case of Theorem* 16, *$V = H^1(\Omega)$ in the case of Theorem* 17) *of*

$$a(u,v) = 0 \quad \textit{for all } v \in V, \quad u \neq 0 \tag{9.1.26}$$

belongs to $H^{m+s}(\Omega)$ for all $0 \le s \le t \in \mathbb{N}$ or $0 \le s < t \notin \mathbb{N}$.

PROOF. Use Theorem 16 (17) for $f = 0$ (and $\varphi = 0$). ■

According to Theorem 6.2.30 (Sobolev's lemma) one obtains sufficient conditions via $C^{k+\lambda}(\overline{\Omega}) \supset H^{k+\lambda+n/2}(\Omega)$ for u to be a classical solution from $C^{k+\lambda}(\overline{\Omega})$. The minimal conditions for $u \in C^{k+\lambda}(\overline{\Omega})$ result from a different theoretical approach, which essentially goes back to Schauder. The following theorem, for example, can be found in Miranda [1, §V]. It shows the $C^{k+\lambda}$-regularity of the operator L in (5.1.1).

Theorem 9.1.20. *Let $k \ge 2$, $0 < \lambda < 1$. Let $\Omega \in C^{k+\lambda}$ be a bounded domain. Let the differential operator $L = \sum a_{ij}\partial^2/\partial x_i \partial x_j + \sum a_i \partial/\partial x_i + a$ be uniformly elliptic in Ω (i.e.,* (5.1.3a) *holds). Let $a_{ij}, a_i, a \in C^{k-2+\lambda}(\overline{\Omega})$ and $f \in C^{k-2+\lambda}(\overline{\Omega})$, $\varphi \in C^{k+\lambda}(\Gamma)$. Then the boundary value problem $Lu = f$ in Ω, $u = \varphi$ on Γ either has a unique (classical) solution $u \in C^{k+\lambda}(\overline{\Omega})$, or there exists a finite-dimensional eigenspace $\{0\} \neq E \subset C^{k+\lambda}(\overline{\Omega})$ such that for all $e \in E$ the following holds: $Le = 0$ in Ω, $e = 0$ on Γ. If $a \le 0$, the first alternative always holds.*

The condition $\Omega \in C^{t+m}$ in Theorem 16 is stronger than necessary. For the Dirichlet problem the Lipschitz continuity of Γ is already sufficient to obtain the following result.

Theorem 9.1.21. (*Nečas* [1]) *Let $\Omega \in C^{0,1}$ be a bounded domain. Let the bilinear form* (5) *be $H_0^m(\Omega)$-coercive. Let the following hold:*

$$1/2 \ge t > s > 0.$$

The coefficients $a_{\alpha\beta} \in L^\infty(\Omega)$ must belong to $C^t(\overline{\Omega})$ if $|\beta| = m$. Then the weak solution $u \in H_0^m(\Omega)$ of the problem

$$a(u,v) = \int_\Omega f(\mathbf{x})v(\mathbf{x})\,d\mathbf{x} \quad \textit{for all } v \in H_0^m(\Omega)$$

with $f \in H^{-m+s}(\Omega)$ belongs to $H_0^{m+s}(\Omega)$ and satisfies the estimate (23).

The condition of $H_0^m(\Omega)$-coercivity can be replaced by that of uniform ellipticity (7.2.3) (cf. Theorems 7.2.11, 7.2.13). The statement of Theorem 21 cannot be extended to $s \ge 1/2$ since then $u \in H_0^{m+s}(\Omega)$ would contain another boundary condition.

The proof of Theorem 21 uses an isomorphism $R = R^*$ related to $R_{h,j}$ (cf. proof of Theorem 8) between $H_0^{m+s}(\Omega)$ and $H_0^m(\Omega)$ and also between $H_0^m(\Omega)$ and $H_0^{m-s}(\Omega)$ such that the form $b(u,v) := a(Ru, Rv)$ is $H_0^{m+s}(\Omega)$-coercive. It is necessary to prove that $\tilde{b}(u,v) := a(u, R^2 v)$ is also $H_0^{m+s}(\Omega)$-coercive. We know $f \in H^{-m+s}(\Omega)$ implies $\tilde{f} := R^2 f \in H^{-m-s}(\Omega)$. Each solution of $a(u,v) = (f,v)_0$ is also a solution of $a(u, R^2\tilde{v}) = \tilde{b}(u,\tilde{v}) = (\tilde{f},\tilde{v})_0 = (f, R^2\tilde{v})_0$ so that $u \in H_0^{m+s}(\Omega)$ follows.

9.1.5 Regularity for Convex Domains and Domains with Corners

A domain Ω is convex if with $\mathbf{x}', \mathbf{x}'' \in \Omega$, $\mathbf{x}' + t(\mathbf{x}'' - \mathbf{x}')$ belongs to Ω for all $0 \le t \le 1$. Convex domains in particular belong to $C^{0,1}$, but permit stronger regularity statements than Theorem 21.

Theorem 9.1.22. (*Kadlec* [1]) *Let Ω be bounded and convex. Let the bilinear form* (5) *be $H_0^m(\Omega)$-coercive. Let the coefficients of the principal part be Lipschitz-continuous:*

$$a_{\alpha\beta} \in C^{0,1}(\overline{\Omega}) \quad \textit{for all } |\alpha| = |\beta| = 1;$$

for the remaining ones let the following hold:

$$D^\gamma a_{\alpha\beta} \in L^\infty(\Omega) \quad \textit{for all } \alpha, \beta, \gamma \ \textit{ with } \gamma \le |\beta|,\ |\alpha| + |\beta| < 1.$$

Then every weak solution $u \in H_0^1(\Omega)$ of the problem

$$a(u,v) = \int_\Omega f(\mathbf{x}) v(\mathbf{x})\, dx \quad \textit{for all } \ v \in H_0^1(\Omega)$$

with $f \in L^2(\Omega)$ belongs to $H^2(\Omega) \cap H_0^1(\Omega)$ and satisfies the estimate

$$|u|_2 \le C_1 [\, |f|_0 + |u|_1 \,]. \tag{9.1.27}$$

The constant C_1 depends only on the diameter of Ω.

In Section 8.4.2 H^2-regularity was required. A generalisation of (27) in the form of H^{m+1}-regularity for the biharmonic differential equation with $m = 2$ is known for convex polygons (cf. Blum–Rannacher [1]). For the Poisson equation the inequality (27) can be reformulated as follows.

Corollary 9.1.23. *For the solution of the Poisson equation $-\Delta u = f \in L^2(\Omega)$ in a convex domain Ω with $u = 0$ on Γ, the following holds:*

$$\left(\sum_{|\alpha|=2} |D^\alpha u|_0^2 \right)^{1/2} \le |f|_0. \tag{9.1.28}$$

As in Lemma 8.4.1 one shows that the left side of (28) is a norm of $H^2(\Omega) \cap H_0^1(\Omega)$ which is equivalent to $|\cdot|_2$. Thus, (27) follows. As a model for the proof of Theorem 22 we carry out the proof for Corollary 23 for the case $\Omega \subset \mathbb{R}^2$.

PROOF. (a) First let us assume that the convex domain is smooth: $\Omega \in C^\infty$. According to Theorem 16 $-\Delta u = f \in L^2(\Omega)$ has a solution $u \in H^2(\Omega) \cap H_0^1(\Omega)$. We want to show

$$|||u|||_2^2 := \sum_{|\alpha|=2} |D^\alpha u|_0^2 \le |\Delta u|_0^2 \quad \text{for all } u \in H^2(\Omega) \cap H_0^1(\Omega). \tag{9.1.29}$$

It suffices to prove (29) that for all u in the dense subset $\{u \in C^\infty(\Omega):$ $u = 0$ on $\Gamma\} = C^\infty(\Omega) \cap H_0^1(\Omega)$. Integration by parts yields

$$\begin{aligned}
\int_\Omega |\Delta u|^2\, dx\, dy &= \int_\Omega (u_{xx}^2 + u_{yy}^2 + 2u_{xx}u_{yy})\, dx\, dy \\
&= \int_\Omega (u_{xx}^2 + u_{yy}^2 - 2u_{xxy}u_y)\, dx\, dy + 2\int_\Gamma u_{xx}u_y n_y\, d\Gamma \\
&= \int_\Omega (u_{xx}^2 + u_{yy}^2 + 2u_{xy}^2)\, dx\, dy + 2\int_\Gamma (u_{xx}u_y n_y - u_{xy}u_y n_x)\, dx\, dy \\
&= |||u|||_2^2 + 2\int_\Gamma (u_{xx}n_y - u_{xy}n_x)u_y\, dx\, dy,
\end{aligned}$$

where $\mathbf{n} = (n_x, n_y)$ is the normal vector. The tangent direction is given by $\mathbf{t} = (-n_y, n_x)$. Then $u_{xx}n_y - u_{xy}n_x$ is the negative tangential derivative $-(u_x)_t$. Now u_x and u_y can be expressed in terms of u_t and u_n. Since both u and u_t vanish on Γ, $u_x = n_x u_n$ and $u_y = n_y u_n$. Hence the boundary integral becomes

$$\begin{aligned}
2\int_\Gamma (u_{xx}n_y - u_{xy}n_x)u_y d\Gamma &= -2\int_\Gamma (n_x u_n)_t n_y u_n d\Gamma \\
&= -\int_\Gamma [2u_n^2 n_y (n_x)_t + n_x n_y (u_n^2)_t] d\Gamma.
\end{aligned}$$

Integration by parts of the second summand yields

$$-\int_\Gamma [2u_n^2 n_y (n_x)_t - (n_x n_y)_t u_n^2] d\Gamma = \int_\Gamma u_n^2 [(n_y)_t n_x - (n_x)_t n_y] d\Gamma.$$

The bracketed expression in the last display is the curvature in $\mathbf{x} \in \Gamma$, which for a convex domain is always ≥ 0. (28) has thus been proved.

(b) Every convex domain Ω can be approximated monotonically by convex $\Omega_\nu \in C^\infty$: $\Omega_1 \subset \Omega_2 \subset \cdots \subset \Omega$, $\bigcup_\nu \Omega_\nu = \Omega$. We interpret $V_\nu := \{u \in H_0^1(\Omega) :$ $\mathrm{supp}(u) \subset \overline{\Omega}_\nu\}$ as Ritz-Galerkin space $V_\nu \subset H_0^1(\Omega)$. Each $u \in C_0^\infty(\Omega)$ lies in V_μ for sufficiently large μ. With $C_0^\infty(\Omega)$, $\bigcup_\nu V_\nu$ is therefore also a dense subset of $H_0^1(\Omega)$. For every ν, the Ritz-Galerkin problem provides the solution $u_\nu \in H_0^1(\Omega_\nu)$ of $-\Delta u_\nu = f$ in Ω_ν, $u_\nu = 0$ on $\partial\Omega_\nu$. In $\Omega\backslash\Omega_\nu$, u_ν may be continued by $u_\nu = 0$. Theorem 8.2.2, which is also applicable in the case $\dim V_\nu = \infty$, proves $|u_\nu - u|_1 \to 0$, where $u \in H_0^1(\Omega)$ is the solution of

$-\Delta u = f$ in Ω. Theorem 9.1.26 will show that for every μ the restriction of u on Ω_μ belongs to $H^2(\Omega_\mu)$. For each $v \in V_\mu \subset V_\nu$, $\nu \geq \mu$, we have

$$|\int_{\Omega_\mu} u_{xx} v\,dx\,dy| = |\int_{\Omega_\mu} u_x v_x\,dx\,dy| = |\lim_{\substack{\nu\to\infty\\ \nu\geq\mu}} \int_{\Omega_\mu} (u_\nu)_x v_x\,dx\,dy|$$

$$= |\lim_{\nu\to\infty} \int_{\Omega_\mu} (u_\nu)_{xx} v\,dx\,dy| \leq \sup_{\nu\geq\mu} ||(u_\nu)_{xx}||_{L^2(\Omega_\mu)} ||v||_{L^2(\Omega_\mu)}.$$

From this one infers $\sum_{|\alpha|=2} ||D^\alpha u||^2_{L^2(\Omega_\mu)} \leq \sup_{\nu\geq\mu} \sum_{|\alpha|=2} ||D^\alpha u_\nu||^2_{L^2(\Omega_\mu)} \leq ||f||^2_{L^2(\Omega_\mu)} \leq ||f||^2_{L^2(\Omega)}$ (cf. (29)), and obtains (28). ■

What role the H^2-regularity plays for the $H_0^1(\Omega)$-projection on a subspace $V_h \subset H_0^1(\Omega)$, is shown in

Exercise 9.1.24. Let the subspace $V_h \subset H_0^1(\Omega)$ satisfy

$$\inf\{|u - v|_1 : v \in V_h\} \leq C_0 h |u|_2 \quad \text{for all } u \in H^2(\Omega) \cap H_0^1(\Omega)$$

(cf. (8.4.14)). Let the Poisson problem be H^2-regular (according to Theorem 21 convexity of Ω is sufficient). Let $Q_V: H_0^1(\Omega) \to V_h \subset H_0^1(\Omega)$ be the orthogonal projection on V_h with respect to $|\cdot|_1$. Show that there exists a C_1 such that

$$|u - Q_V u|_0 \leq C_1 h |u|_1 \quad \text{for all } u \in H_0^1(\Omega).$$

Hints: (1) With the Poisson problem the boundary value problem $-\Delta u + u = f$ in Ω and $u = 0$ on Γ are also H^2-regular (cf. Remark 2). The corresponding bilinear form $a(\cdot,\cdot)$ is the scalar product in $H_0^1(\Omega)$.
(2) Q_V agrees with the Ritz projection S_h for $a(\cdot,\cdot)$.
(3) Use Corollary 8.4.12.

In connection with finite elements one often considers polygonal domains Ω. Since polygons belong to $C^{0,1}$, the Dirichlet problem, according to Theorem 21, is H^{m+s}-regular with $0 \leq s < 1/2$. If the polygon is convex (i.e., if the inner angles are $\leq \pi$) then as in Theorem 22 one has H^2-regularity ($m = 1$). One obtains results between $H^{3/2}$ and H^2 if the maximal inner angle of the polygon lies between π and 2π (cf. Schatz-Wahlbin [1]). If Ω has a reentrant corner the boundary value problem can no longer be H^2-regular (cf. Example 2.1.4; with an inner angle α the solution belongs to $H^{1+s}(\Omega)$ for $s < \pi/\alpha$). Stronger regularity properties may be obtained, however, if special compatibility conditions are satisfied in the corners (cf. Kondrat'ev [1]).

Example 9.1.25. Let u be the solution of the Poisson equation $-\Delta u = f$ in the rectangle $\Omega = (0,1) \times (0,1)$ with $u = 0$ on Γ. Only for $s < 3$ does $f \in H^{s-2}(\Omega)$ lead to $u \in H^s(\Omega)$. Under the additional compatibility condition that f vanish at all corners, $f(0,0) = f(0,1) = f(1,0) = f(1,1) = 0$, however, one can also conclude, for $f \in H^{s-2}(\Omega)$ with $3 < s < 4$, that $u \in H^s(\Omega)$.

9.1.6 Regularity in the Interior

Up to now all regularity statements have referred to the entire domain Ω. Instead one can also investigate the regularity of the solution u in $\Omega_0 \subset\subset \Omega$. The following theorem shows that the answer depends neither on the smoothness of the boundary nor on the type of boundary condition.

Theorem 9.1.26. *Let $\Omega_0 \subset\subset \Omega_1 \subset \Omega$ and $s \ge 0$. Let the bilinear form* (5) *be $H_0^m(\Omega)$-coercive. For the coefficients assume condition* (22) *with Ω be replaced by Ω_1 and with $t \ge s \in \mathbb{N}$ or $t > s$. Let $u \in V \subset H^m(\Omega)$ be a weak solution of the problem $a(u,v) = \int_\Omega fv\,d\mathbf{x}$ for all $v \in V$, where the restriction $f|_{\Omega_1}$ belongs to $H^{-m+s}(\Omega_1)$. Then the restriction of u to Ω_0 belongs to $H^{m+s}(\Omega_0)$ and satisfies*

$$||u||_{H^{m+s}(\Omega_0)} \le C(s, \Omega_0, \Omega_1, \Omega)[||f||_{H^{-m+s}(\Omega_1)} + ||u||_{H^m(\Omega)}].$$

PROOF. A special covering of Ω is given by $U^0 = \Omega_1$, $U^1 = \Omega\backslash\Omega_0$. Thus we obtain the assertion from Part b of the proof of Theorem 16. ■

9.2 Regularity Properties of Difference Equations

The convergence estimates for difference equations in Section 4.4 read, for example, $||u - u_h||_\infty \le Ch^2||u||_{C^4(\overline{\Omega})}$ under the condition that $u \in C^4(\overline{\Omega})$ or $u \in C^{3,1}(\overline{\Omega})$. This regularity assumption is frequently not satisfied (cf. Examples 2.1.3–4). A comparison with the error estimate $|u - u^h|_0 \le Ch^2|u|_2$ for the finite-element method suggests that similar estimates also exist for difference solutions. To obtain the latter, one needs to replace the stability estimate $||L_h^{-1}||_2 \le C$ (or $||L_h^{-1}||_\infty \le C$), which corresponds to $L^{-1} \in L(L^2(\Omega), L^2(\Omega))$, by stronger estimates which correspond to $L^{-1} \in L(H^{-1}(\Omega), H_0^1(\Omega))$ or $L^{-1} \in L(L^2(\Omega), H^2(\Omega))$.

9.2.1 Discrete H^1-Regularity

For $\Omega \subset \mathbb{R}^n$ the following grids are defined:

$$Q_h := \{\mathbf{x} \in \mathbb{R}^n : x_i = \nu_i h,\ \nu_i \in \mathbb{Z}\}, \quad \Omega_h := \Omega \cap Q_h. \tag{9.2.1}$$

A grid function v_h defined on Ω_h is extended to Q_h by $v_h = 0$:

$$v_h(\mathbf{x}) = 0 \quad \text{for } \mathbf{x} \in Q_h\backslash\Omega_h. \tag{9.2.2}$$

The Euclidean norm is now called the L_h^2-norm:

$$|v_h|_0 := ||v_h||_{L_h^2} := \left[h^n \sum_{\mathbf{x}\in Q_h} |v_h(\mathbf{x})|^2\right]^{1/2}. \tag{9.2.3a}$$

It comes from the scalar product

$$(v_h, w_h)_0 := (v_h, w_h)_{L_h^2} := h^n \sum_{\mathbf{x}\in Q_h} v_h(\mathbf{x}) w_h(\mathbf{x}). \tag{9.2.3b}$$

The discrete analogue of $H_0^1(\Omega)$ is H_h^1 with the norm

$$|v_h|_1 := ||v_h||_{H_h^1} := \left[|v_h|_0^2 + \sum_{i=1}^{n} |\partial_i^+ v_h|_0^2 \right]^{1/2} \qquad (v_h = 0 \text{ on } Q_h \backslash \Omega_h) \tag{9.2.3c}$$

where ∂_i^+ is the forward difference in the x_i direction. The dual norm reads

$$|v_h|_{-1} := ||v_h||_{H_h^{-1}} := \sup\{|(v_h, w_h)_0|/|w_h|_1 \colon w_h = 0 \text{ in } Q_h \backslash \Omega_h\}. \tag{9.2.3d}$$

The associated matrix norms $||L_h||_{H_h^1 \leftarrow H_h^{-1}} = |L_h|_{1\leftarrow -1}$, $||L_h||_{H_h^1 \leftarrow L_h^2} = |L_h|_{1\leftarrow 0}$, etc., are defined by

$$|L_h|_{i\leftarrow j} := \sup\{|L_h v_h|_i / |v_h|_j \colon 0 \neq v_h \text{ satisfies (2)}\} \quad \text{for } i, j \in \{-1, 0, 1\}. \tag{9.2.3d'}$$

Exercise 9.2.1. Show that (a) $|L_h|_{0\leftarrow 0}$ is the spectral norm of L_h (cf. §4.3). (b) The following inverse estimates hold:

$$|v_h|_i \le C_{ij} h^{j-i} |v_h|_j \quad \text{for } 1 \ge i \ge j \ge -1. \tag{9.2.3e}$$

The matrix L_h yields the bilinear form

$$a_h(u_h, v_h) := (L_h u_h, v_h)_{L_h^2}. \tag{9.2.4a}$$

$a_h(\cdot,\cdot)$ is said to be H_h^1-elliptic if a $C_E > 0$ exists such that

$$a_h(u_h, u_h) \ge C_E |u_h|_1^2 \quad \text{for all } u_h \text{ and all } h > 0. \tag{9.2.4b}$$

Correspondingly, $a_h(\cdot,\cdot)$ is said to be H_h^1-coercive if there exist $C_E > 0$ and $C_K \in \mathbb{R}$ with

$$a_h(u_h, u_h) \ge C_E |u_h|_1^2 - C_K |u_h|_0^2 \quad \text{for all } u_h \text{ and all } h > 0. \tag{9.2.4c}$$

As defined in Sect. 4.5, L_h (resp. $a_h(\cdot,\cdot)$) is said to be L_h^2-stable if

$$|L_h^{-1}|_{0\leftarrow 0} \le C_0 \quad \text{for all } h > 0. \tag{9.2.4d}$$

We call L_h H_h^1-regular if

$$|L_h^{-1}|_{1\leftarrow -1} \le C_1 \quad \text{for all } h > 0. \tag{9.2.4e}$$

Exercise 9.2.2. (a) H_h^1-regularity implies L_h^2-stability.
(b) $|A_h|_{i\leftarrow j} = |A_h^{\mathsf{T}}|_{-j\leftarrow -i}$ for $i, j \in \{-1, 0, 1\}$.
(c) If L_h is H_h^1-regular, then so is L_h^{T}.

(d) If L_h and L_h^{T} are stable with respect to $|\cdot|_\infty$, i.e., $||L_h||_\infty \le C_\infty$, $||L_h^{\mathsf{T}}||_\infty \le C_1$, then L_h^2-stability (4d) follows with $C_0 := (C_1 C_\infty)^{1/2}$.
(e) H_h^1-ellipticity implies H_h^1-regularity.

The following statement resembles the alternative in Theorem 6.5.15.

Theorem 9.2.3. *If $a_h(\cdot,\cdot)$ is H_h^1-coercive and if L_h is L_h^2-stable, then L_h is also H_h^1-regular.*

PROOF. (a) Let $L_h^{\mathsf{T}} u_h = f_h$ such that $a_h(u_h, u_h) = (u_h, L_h^{\mathsf{T}} u_h)_0 = (u_h, f_h)_0$. Coercivity provides

$$|u_h|_1^2 \le [a_h(u_h,u_h) + C|u_h|_0^2]/C_E = [(u_h, f_h)_0 + C|u_h|_0^2]/C_E$$

$$\le C'[|f_h|_0 + C|u_h|_0]|u_h|_0.$$

On the basis of the stability estimate $|u_h|_0 \le C_0 |f_h|_0$ we obtain $|u_h|_1^2 \le C''|f_h|_0^2$. From this one infers $|L_h^{\mathsf{T}-1}|_{1\leftarrow 0} \le \sqrt{C''} := C^*$ and hence $|L_h^{-1}|_{0\leftarrow -1} \le C^*$ (cf. Exercise 2b).
(b) Now let $L_h u_h = f_h$. According to Part a, one has $|u_h|_0 \le C^*|f_h|_{-1}$. By estimating $|a_h(u_h,u_h)| = |(f_h,u_h)_0| \le |f_h|_{-1}|u_h|_1$ through $\frac{1}{2}C_E|u_h|_1^2 + \frac{1}{2}C_E^{-1}|f_h|_{-1}^2$, one obtains the coercivity

$$|u_h|_1^2 \le [a_h(u_h,u_h) + C|u_h|_0^2]/C_E \le \frac{1}{2}|u_h|_1^2 + \left(\frac{1}{2}C_E^{-2} + CC^{*2}C_E^{-1}\right)|f_h|_{-1}^2$$

and hence $|L_h^{-1}|_{1\leftarrow -1} \le C_1$ with $C_1^2 := C_E^{-2} + 2CC^{*2}C_E^{-1}$. ∎

Instead of the L_h^2-stability in Theorem 3 one can also assume the solvability of the continuous problem and a consistency condition (cf. Corollary 11.3.5).

By analogy with Lemma 7.2.12 the following may be proved.

Exercise 9.2.4. If L_h is H_h^1-coercive, and if $\delta L_h := a_0 + \sum_i b_i \partial_i^\pm + \sum_i \partial_i^\pm c_i$, with $|a_0|, |b_i|, |c_i| \le$ const, contains at most first differences, then $L_h + \delta L_h$ is also H_h^1-coercive. According to Exercise 4 it is sufficient to investigate the principal part of a difference operator as to its coercivity.

Example 9.2.5. Let $\Omega \subset \mathbb{R}^2$ be bounded. On Ω_h let L_h be given by the difference method (5.1.18) (with $a = 0$):

$$(L_h u_h)(x,y) = \partial_x^- a_{11}\left(x + \frac{h}{2}, y\right)\partial_x^+ u_h(x,y) + \partial_y^- a_{22}\left(x, y + \frac{h}{2}\right)\partial_y^+ u_h(x,y) \tag{9.2.5}$$

for $(x,y) \in \Omega_h$, where $u_h(x,y) = 0$ for $(x,y) \in Q_h \backslash \Omega_h$. Here, let $-a_{11}$, $-a_{22} \ge \epsilon > 0$ in $\overline{\Omega}$. Then $a_h(\cdot,\cdot)$ is H_h^1-elliptic and L_h is H_h^1-regular.

PROOF. For arbitrary v_h, w_h defined on Q_h the following rules of summation by parts hold:

$$(v_h, \partial_{x_i}^+ w_h)_0 = -(\partial_{x_i}^- v_h, w_h)_0, \quad (v_h, \partial_{x_i}^- w_h)_0 = -(\partial_{x_i}^+ v_h, w_h)_0. \qquad (9.2.6)$$

Hence it follows that

$$\begin{aligned} a_h(u_h, u_h) &= (L_h u_h, u_h)_0 \\ &= -\left(a_{11}\left(\cdot + \frac{h}{2}, \cdot\right) \partial_x^+ u_h, \partial_x^+ u_h\right)_0 - \left(a_{22}\left(\cdot, \cdot + \frac{h}{2}\right) \partial_y^+ u_h, \partial_y^+ u_h\right)_0 \\ &\geq \epsilon[|\partial_x^+ u_h|_0^2 + |\partial_y^+ u_h|_0^2]. \end{aligned}$$

As in Lemma 6.2.11 one proves that for bounded Ω the norms $|\cdot|_1$ and $|u_h|_{1,0} := [|\partial_x^+ u_h|_0^2 + \partial_y^+ u_h|_0^2]^{1/2}$ are equivalent (uniformly with respect to h) so that (4b) follows. The H_h^1-regularity results from Exercise 2e. ■

Exercise 9.2.6. Let $\Omega \subset \mathbb{R}^2$ be bounded. Let the equation

$$(a_{11}u_x)_x + (a_{12}u_y)_x + (a_{12}u_x)_y + (a_{22}u_y)_y = f$$

be given on Ω_h by the difference stars (5.1.18/19), where $u_h = 0$ in $Q_h \backslash \Omega_h$. Let the differential equation be uniformly elliptic in $\overline{\Omega}$: $a_{ii} < 0$, $a_{11}a_{22} - a_{12}a_{21} \geq -\epsilon(a_{11} + a_{22}) > 0$. Further, let $a_{ij} \in C^0(\overline{\Omega})$ hold. Show that for sufficiently small h the associated matrix L_h is H_h^1-regular; for all $h > 0$, L_h is H_h^1-coercive. Hint: For sufficiently small h the following holds:

$$\begin{aligned} -a_{11}\left(x + \frac{h}{2}, y\right) d_1^2 - a_{22}\left(x, y + \frac{h}{2}\right) d_2^2 \\ - \left[a_{12}\left(x + \frac{h}{2}, y + h\right) + a_{12}\left(x + \frac{h}{2}, y\right)\right] d_1 d_2 \\ \geq \frac{\epsilon}{2}(d_1^2 + d_2^2), \quad \text{for all } d_1, d_2 \in \mathbb{R}. \end{aligned}$$

The H_h^1-coercive difference methods constructed so far remain H_h^1-coercive if differences of lower order are added (cf. Exercise 4) or if the principal term $(a_{11}u_x)_x + \cdots$ is replaced by $a_{11}u_{xx} + \cdots$ with $a_{11} \in C^1(\overline{\Omega})$. The above difference methods are described by the same difference operator regardless whether the grid points are close to or far from the boundary. The homogeneous Dirichlet boundary condition is discretised by (2): "$u_h = 0$ on $Q_h \backslash \Omega_h$". If one wants to approximate the boundary condition more accurately, one needs to select special discretisations in the points near the boundary of Ω_h (cf. Section 4.8.1, 4.8.2). One thus obtains an irregularity which makes a proof of H_h^1-regularity, as in Example 5, difficult. We begin with the one-dimensional case.

Lemma 9.2.7. *Let L_h be the matrix of the one-dimensional Shortley-Weller discretisations of $-u'' = f$ on Ω, $u = 0$ on $\partial\Omega$:*

$$h^{-2}\left\{\frac{2}{s_l s_r}u_h(x)-\frac{2}{s_r(s_r+s_l)}u_h(x+s_r h)-\frac{2}{s_l(s_r+s_l)}u_h(x-s_r h)\right\}=f(x) \tag{9.2.7}$$

for $x\in\Omega_h$, where $0<s_l,s_r<1$ (cf. (4.8.7)), $u(\xi)=0$ for $\xi\in\partial\Omega$. For arbitrary $\Omega\subset\mathbb{R}$ there holds $(v_h,L_hv_h)_0\geq\frac{1}{2}|\partial^+v_h|_0^2$ if $v_h=0$ on $Q_h\backslash\Omega_h$. Thus for bounded Ω, L_h is H_h^1-regular.

PROOF. (a) First, let Ω be assumed to be connected. Let the grid points of Ω_h be $x_l:=x_0+lh\in\Omega$ for $l=0,\cdots,k>0$. Let the boundary points of Ω be $x_0-s_{l,0}h$ and $x_k+s_{r,k}h$ with $s_{l,0},s_{r,k}\in(0,1]$. The other factors of Equation (7) in $x=x_i$ are $s_{l,i}=s_{r,i}=1$. If one takes into account $v_h=0$ on $\mathbb{R}\backslash\Omega$ one obtains the following identity:

$$\begin{aligned}(v_h,L_hv_h)_0 &:= h\sum_{l=0}^{k}v_h(x_l)(L_hv_h)(x_l)\\ &=|\partial^+v_h|_0^2+h^{-1}\left\{v_h(x_0)\left[\left(\frac{2}{s_l}-2\right)v_h(x_0)+\left(1-\frac{2}{1+s_l}\right)v_h(x_1)\right]\right.\\ &\quad\left.+v_h(x_k)\left[\left(\frac{2}{s_r}-2\right)v_h(x_k)+\left(1-\frac{2}{1+s_r}\right)v_h(x_{k-1})\right]\right\},\end{aligned} \tag{9.2.8a}$$

where $s_l:=s_{l,0}$ and $s_r:=s_{r,k}$. Since $v_h(x_0)=h\partial^+v_h(x_{-1})$ and $v_h(x_1)=h[\partial^+v_h(x_{-1})+\partial^+v_h(x_0)]$ because of $v_h(x_{-1})=0$, the first summand inside the braces can be written as

$$\begin{aligned}&h^{-1}v_h(x_0)\left[\left(\frac{2}{s_l}-2\right)v_h(x_0)+\left(1-\frac{2}{1+s_l}\right)v_h(x_1)\right]\\ &=h\frac{1-s_l}{1+s_l}\left[\frac{s_l+2}{s_l}\partial^+v_h(x_{-1})^2-\partial^+v_h(x_{-1})\partial^+v_h(x_0)\right]\end{aligned} \tag{9.2.8b}$$

For $\alpha:=\partial^+v_h(x_{-1})$, $\beta:=\partial^+v_h(x_0)$ use the inequality $-\alpha\beta\geq-\frac{\lambda}{2}\alpha^2-\frac{1}{2\lambda}\beta^2$ with $\frac{\lambda}{2}=(s_l+2)/s_l$ and note that the function $s(1-s)/[4(1+s)(2+s)]$ is bounded by 0.018 (maximum at $s=(\sqrt{3}-1)/2$). The second summand inside the braces is treated analogously and yields

$$(v_h,L_hv_h)_0\geq|\partial^+v_h|_0^2-0.018h[\partial^+v_h(x_0)^2+\partial^+v_h(x_{k-1})^2]. \tag{9.2.8c}$$

(b) In Part (a) it was assumed that $k>0$. For $k=0$ one obtains

$$(v_h,L_hv_h)_0=h^{-1}\frac{2}{s_ls_r}v_h(x_0)^2=\frac{h}{s_ls_r}[\partial^+v_h(x_0)^2+\partial^+v_h(x_{-1})^2\geq|\partial^+v_h|_0^2,$$

so that (8c) also holds.

(c) Let Ω be arbitrary. Let the components of connection be $I_i=(a_i,b_i)$ $(i\in\mathbb{Z})$ with $b_i\leq a_{i+1}$. Let $L_h^{(i)}$ be the discretisation matrix associated with I_i. Each v_h with support in $\Omega_h=Q_h\cap\Omega$ can be written as $\sum_i v_h^{(i)}$ where

$v_h^{(i)} = 0$ outside I_i. Let $x_0^{(i)}$ and $x_{k(i)}^{(i)}$ be the first and last grid points of $\Omega_h^{(i)} := \Omega_h \cap I_i$. The statement follows from

$$|\partial^+ v_h|_0^2 \le \sum_i \{|\partial^+ v_h^{(i)}|_0^2 + [h\partial^+ v_h^{(i)}(x_{k(i)}^{(i)})]^2\} = 2\sum_i |\partial^+ v_h^{(i)}|_0^2$$

$$\le 2\sum_i (v_h^{(i)}, L_h^{(i)} v_h^{(i)})_0 = 2(v_h, L_h v_h)_0. \qquad \blacksquare$$

Theorem 9.2.8. *Let $\Omega \subset \mathbb{R}^2$ be bounded. Let L_h be the matrix associated with the Shortley-Weller discretisation of the Poisson equation* (*cf.* *Section* 4.8.1). *Then L_h is H_h^1-regular.*

PROOF. Let L_h^x $[L_h^y]$ be the portion of the x differences [y differences] such that $L_h = L_h^x + L_h^y$. The restriction of L_h^x to a "grid row" $\{(\nu h, y) \in \Omega : \nu \in \mathbb{Z}\}$ (y is fixed) corresponds to the matrix L_h in Lemma 7. Thus one obtains $(v_h, L_h^x v_h)_0 \ge \frac{1}{2}|\partial_x^+ v_h|_0^2$. With the analogous inequality $(v_h, L_h^y v_h)_0 \ge \frac{1}{2}|\partial_y^+ v_h|_0^2$ one obtains $2(v_h, L_h v_h)_0 \ge |\partial_x^+ v_h|_0^2 + |\partial_y^+ v_h|_0^2 = |v_h|_1^2 - |v_h|_0^2 \ge \epsilon_\Omega |v_h|_1^2$, thus $\|L_h^{-1}\|_{1\leftarrow -1} \le 1/(2\epsilon_\Omega)$.

Theorem 9.2.9. *Let $\Omega \subset \mathbb{R}^2$ be bounded. Let the Poisson equation be discretised with the aid of the five-point formula* (4.8.14c) *at points far from the boundary and by interpolation* (4.8.16) *at points near the boundary. The associated matrix is H_h^1-regular.*

PROOF. The corresponding, one-dimensional formulae, except for the scaling factor $s_r + s_l \le 2$, agree with (7) so that the proof of Theorem 8 can easily be carried over.

Exercise 9.2.10. Show that for a convex domain $\Omega \subset \mathbb{R}^2$ the matrix L_h of the Shortley-Weller discretisation satisfies the inequality

$$(v_h, L_h v_h)_0 \ge 0.982(|\partial_x^+ v_h|_0^2 + |\partial_y^+ v_h|_0^2).$$

Theorems 8 and 9 can be strengthened in the following way:

Corollary 9.2.11. *Let L_h be as in Theorem* 8 *or* 9. *Let $D_h = \operatorname{diag}\{d(\mathbf{x}) : \mathbf{x} \in \Omega_h\}$ be a diagonal matrix with*

$$d(\mathbf{x}) := \min\{2s_l s_r, 2s_o s_u, 1\} \quad (\mathbf{x} \in \Omega_h),$$

where s_l, s_r, s_o, s_u come from (4.8.7), *respectively* (4.8.16). *For points far from the boundary, $\mathbf{x} \in \Omega_h$, evidently $d(\mathbf{x}) = 1$ holds. The matrix*

$$L_h' := D_h L_h,$$

which belongs to the re-scaled system of equations $L'_h u_h = f'_h := D_h f_h$, *is also* H^1_h*-regular:* $|L'^{-1}_h|_{1\leftarrow -1} \le C$.

PROOF. In the case of the Shortley-Weller method one needs to combine the correction terms (8b) for the x and y directions. The result is that they remain $\ge -\frac{1}{2}[|\partial_x^+ v_h|_0^2 + |\partial_y^+ v_h|_0^2]$ so that $(v_h, L'_h v_h)_0 \ge \frac{1}{2}[|\partial_x^+ v_h|_0^2 + |\partial_y^+ v_h|_0^2] \ge \epsilon |v_h|_1^2$ with $\epsilon > 0$. One obtains an analogous estimate for the difference method in Section 4.8.2. ■

9.2.2 Consistency

In the following we carry over the estimate $|u^h - u|_1 \le Ch|u|_2 \le C'h|f|_0$, which holds for finite-element solutions, to difference methods. To this end one needs to prove the consistency condition

$$|L_h R_h - \tilde{R}_h L|_{-1\leftarrow 2} := \|L_h R_h - \tilde{R}_h L\|_{H_h^{-1}\leftarrow H^2(\Omega)} \le C_K h \tag{9.2.9}$$

for suitable restrictions

$$R_h: H^2(\Omega) \cap H_0^1(\Omega) \to H_h^2, \quad \tilde{R}_h: L^2(\Omega) \to L_h^2.$$

To construct the restrictions we first continue $u \in H^2(\Omega)$ in $\overline{u} := E_2 u \in H^2(\mathbb{R}^2)$. According to Theorem 6.2.40c one assumes

$$\overline{u} := E_2 u = u \quad \text{on } \Omega, \quad \|E_2 u\|_{H^2(\mathbb{R}^2)} \le C\|u\|_{H^2(\Omega)} \tag{9.2.10a}$$

for all $u \in H^2(\Omega) \cap H_0^1(\Omega)$. An analogous continuation $E_0: L^2(\Omega) \to L^2(\mathbb{R}^2)$ with

$$\overline{f} := E_0 f = f \quad \text{on } \Omega, \quad \|E_0 f\|_{L^2(\mathbb{R}^2)} \le C\|f\|_{L^2(\Omega)} \quad \text{for all } f \in L^2(\Omega) \tag{9.2.10b}$$

is given, for example, by $\overline{f} = 0$ on $\mathbb{R}^2 \backslash \Omega$. Let the averaging operators $\sigma_h^x, \sigma_h^y: C_0^\infty(\mathbb{R}^2) \to C_0(\mathbb{R}^2)$ be defined by

$$(\sigma_h^x u)(x,y) = h^{-1} \int_{-h/2}^{h/2} u(x+\xi, y)\, d\xi, \quad (\sigma_h^y u)(x,y) = h^{-1} \int_{-h/2}^{h/2} u(x, y+\eta)\, d\eta. \tag{9.2.11}$$

The restrictions $R_h, \tilde{R}_h$ are chosen as follows:

$$R_h := \sigma_h^x \sigma_h^y E_2, \quad \text{i.e., } (R_h u)(x,y) = h^{-2} \int_{-h/2}^{h/2} \int_{-h/2}^{h/2} \overline{u}(x+\xi, y+\eta)\, d\xi\, d\eta$$
$$\text{with } \overline{u} = E_2 u \quad \text{for } (x,y) \in \Omega_h, \tag{9.2.12a}$$

$$\tilde{R}_h := (\sigma_h^x \sigma_h^y)^2 E_0, \quad \text{i.e., } (\tilde{R}_h f)(x,y)$$
$$= h^{-4} \int_{-h/2}^{h/2} \int_{-h/2}^{h/2} \int_{-h/2}^{h/2} \int_{-h/2}^{h/2} \overline{f}(x+\xi+\xi', y+\eta+\eta')\, d\xi\, d\xi'\, d\eta\, d\eta'$$
$$\text{with } \overline{f} = E_0 f \quad \text{for } (x,y) \in \Omega_h. \tag{9.2.12b}$$

The characteristic properties of the convolutions σ_h^x, σ_h^y are the subject of

Exercise 9.2.12. Let $\hat{\partial}_x$ be the symmetric difference operator $(\hat{\partial}_x u)(x,y) := [u(x+h/2,y) - u(x-h/2,y)]/h$; $\hat{\partial}_y$ is defined analogously. Show that
(a) $\sigma_h^x\sigma_h^y = \sigma_h^y\sigma_h^x$, $\quad \hat{\partial}_x = \frac{\partial}{\partial x}\sigma_h^x = \sigma_h^x\frac{\partial}{\partial x}$, $\hat{\partial}_y = \frac{\partial}{\partial y}\sigma_h^y = \sigma_h^y\frac{\partial}{\partial y}$.
(b) $\|\sigma_h^x\|_{H^k(\mathbb{R}^2)\leftarrow H^k(\mathbb{R}^2)} \le C, \|\sigma_h^y\|_{H^k(\mathbb{R}^2)\leftarrow H^k(\mathbb{R}^2)} \le C$ (in particular for $k = 0, \pm 1, 2$).
(c) $\|\sigma_h^x\sigma_h^y v\|_{L_h^2} := [\sum_{\mathbf{x}\in\Omega_h} |(\sigma_h^x\sigma_h^y v)(\mathbf{x})|^2]^{1/2} \le \|v\|_{L^2(\mathbb{R}^2)}$ for all $v \in L^2(\mathbb{R}^2)$.
(d) For $a \in C^{0,1}(\mathbb{R}^2)$ holds $\|a\sigma_h^x\sigma_h^y - \sigma_h^x\sigma_h^y a\|_{L_h^2\leftarrow L^2(\mathbb{R}^2)} \le Ch\|a\|_{C^{0,1}(\mathbb{R}^2)}$. Here $(a\sigma_h^x u)(\mathbf{x}) = (a(\mathbf{x})\sigma_h^x u)(\mathbf{x})$ and $(\sigma_h^x a u)(\mathbf{x}) = (\sigma_h^x(au))(\mathbf{x})$.
(e) $\|a(\sigma_h^x)^\nu(\sigma_h)^\mu - (\sigma_h^x)^\nu(\sigma_h)^\mu a\|_{L_h^2\leftarrow L^2(\mathbb{R}^2)} \le C_{\nu\mu}h\|a\|_{C^{0,1}(\mathbb{R}^2)}$ for $\nu, \mu \in \mathbb{N}$.
(f) $\|u - (\sigma_h^x)^\nu(\sigma_h)^\mu u\|_{H^k(\mathbb{R}^2)} \le C_{\nu\mu}h\|u\|_{H^{k+1}(\mathbb{R}^2)}$ for $u \in H^{k+1}(\mathbb{R}^2)$.

Consider the differential operator

$$L = \sum_{i,j=1}^{n} a_{ij}(\mathbf{x})\partial^2/\partial x_i\partial x_j + \sum_{i=1}^{n} a_i(\mathbf{x})\partial/\partial x_i + a(\mathbf{x}). \tag{9.2.13a}$$

First we assume that $\Omega = \mathbb{R}^n$ and discretise L with the regular difference operator

$$L_h = \sum a_{ij}(\mathbf{x})\partial_{x_i}^{\pm}\partial_{x_j}^{\pm} + a_i(\mathbf{x})\partial_{x_i}^{\pm} + a(\mathbf{x}) \tag{9.2.13b}$$

with an arbitrary combination of the $\pm$-signs.

Lemma 9.2.13. *Let $\Omega = \mathbb{R}^n$. Let $a_{ij}, a_i, a \in C^{0,1}(\mathbb{R}^n)$. Let L and L_h be given by* (13a,b). *Then the consistency estimate* (9) *holds.*

PROOF. To simplify the notation let us assume that $n = 2$. Let

$$\tilde{R}_h a_{11} u_{xx} = (\sigma_h^x\sigma_h^y)^2 a_{11} u_{xx} = a_{11}(\mathbf{x})(\sigma_h^x\sigma_h^y)^2 u_{xx} - \delta_1$$

with $\|\delta_1\|_{L_h^2} \le Ch\|a_{11}\|_{C^{0,1}(\mathbb{R}^2)}|u|_2$ (cf. Exercise 12e,b). Let the term corresponding to $a_{11}u_{xx}$ in (13b) be, for example, $a_{11}(\mathbf{x};h)\partial_x^+\partial_h^+$. Exercise 12a shows that

$$\begin{aligned} a_{11}(\sigma_h^x\sigma_h^y)^2 u_{xx} &= a_{11}\hat{\partial}_x\hat{\partial}_x\sigma_h^y\sigma_h^y u(x,y) \\ &= a_{11}\partial_x^+\partial_x^+\sigma_h^y\sigma_h^y u(x-h,y) = a_{11}\partial_x^+\partial_x^+\sigma_h^y\sigma_h^y u(x,y) - a_{11}\partial_x^+\delta_2 \end{aligned}$$

with $\delta_2(x,y) := -\partial_x^+\sigma_h^y\sigma_h^y[u(x-h,y) - u(x,y)] = -\sigma_h^x\sigma_h^y\sigma_h^y[u_x(x-h,y) - u_x(x,y)]$ and $\|\delta_2\|_2 \le \|\sigma_h^y[u_x(x-h,y) - u_x(x,y)]\|_{L^2(\mathbb{R}^2)} \le h|u|_2$ (cf. Exercise 12c,b). Since $\hat{\partial}_x\hat{\partial}_x = \partial_x^+\partial_x^-$, the error term δ_2 does not appear if one also approximates $a_{11}u_{xx}$ by $a_{11}(\mathbf{x})\partial_x^+\partial_x^-$. Finally one obtains

$$\begin{aligned} a_{11}\partial_x^+\partial_x^+\sigma_h^y\sigma_h^y u &= a_{11}\partial_x^+\partial_x^+ R_h u + a_{11}\partial_x^+\partial_x^+[\sigma_h^y\sigma_h^y - \sigma_h^x\sigma_h^y]u \\ &= a_{11}\partial_x^+\partial_x^+ R_h u + a_{11}\partial_x^+\sigma_h^x\sigma_h^y[\sigma_h^y - \sigma_h^x]u_x = a_{11}\partial_x^+\partial_x^+ R_h u - a_{11}\partial_x^+\delta_3 \end{aligned}$$

with $||\delta_3||_{L_h^2} = ||\sigma_h^x\sigma_h^y[\sigma_h^y - \sigma_h^x]u_x||_{L_h^2} \le ||[\sigma_h^y - \sigma_h^x]u_x||_{L^2(\mathbb{R}^2)} \le Ch|u|_2$ (cf. Exercise 12c,f). Putting this altogether one obtains

$$\left[a_{11}\partial_x^\pm\partial_x^\pm R_h - \tilde{R}_h a_{11}\frac{\partial^2}{\partial x^2}\right]u = \delta_1 + a_{11}\partial_x^\pm(\delta_2 + \delta_3).$$

For the first error term the following holds:

$$||\delta_1||_{H_h^{-1}} \le ||\delta_1||_{L_h^2} \le C'h|u|_2. \tag{9.2.14a}$$

For arbitrary $v_h \in H_h^1$ we have

$$(v_h, a_{11}\partial_x^\pm(\delta_2 + \delta_3))_{L_h^2} = -(\partial_x^\pm[a_{11}v_h], \delta_2 + \delta_3)_{L_h^2}. \tag{9.2.14b}$$

From $a \in C^{0,1}(\mathbb{R}^2)$ it follows that $||\partial_x^\pm[av_h]||_{L_h^2} \le ||a||_{C^{0,1}(\mathbb{R}^2)}||v_h||_{L_h^2} + ||a||_{C^0(\mathbb{R}^2)}||v_h||_{H_h^1} \le C||v_h||_{H_h^1}$, so that

$$\begin{aligned} ||a_{11}\partial_x^\pm(\delta_2 + \delta_3)||_{H_h^{-1}} &= \sup_{v_h \in H_h^1} |(v_h, a_{11}\partial_x^\pm(\delta_2 + \delta_3))_{L_h^2}| \\ &\le C||\delta_2 + \delta_3||_{L_h^2} \le C'h|u|_2. \end{aligned} \tag{9.2.14c}$$

From (14a,c) we have

$$||a_{11}\partial_x^\pm\partial_x^\pm R_h - \tilde{R}_h a_{11}\frac{\partial^2}{\partial x^2}||_{H_h^{-1}\leftarrow H^2(\mathbb{R}^2)} \le Ch. \tag{9.2.14d}$$

Analogously, one shows

$$||a_{ij}\partial_{x_i}^\pm\partial_{x_j}^\pm R_h - \tilde{R}_h a_{ij}\partial^2/\partial x_i\partial x_j||_{H_h^{-1}\leftarrow H^2(\mathbb{R}^2)} \le Ch. \tag{9.2.14e}$$

For $a_i\partial_{x_i}^\pm R_h - \tilde{R}_h a_i\partial/\partial x_i$ similar reasoning results in an $O(h)$-estimate for the norms $||\cdot||_{H_h^{-1}\leftarrow H^1(\mathbb{R}^2)}$ and $||\cdot||_{L_h^2\leftarrow H^2(\mathbb{R}^2)}$. Both are upper bounds for the larger norm $||\cdot||_{H_h^{-1}\leftarrow H^2(\mathbb{R}^2)}$ such that

$$||a_i\partial_{x_i}^\pm R_h - \tilde{R}_h a_i\partial/\partial x_i||_{H_h^{-1}\leftarrow H^2(\mathbb{R}^2)} \le Ch. \tag{9.2.14f}$$

Likewise

$$||aR_h - \tilde{R}_h a||_{H_h^{-1}\leftarrow H^2(\mathbb{R}^2)} \le Ch. \tag{9.2.14g}$$

Statement (9) follows from (14e,f,g). ■

When generalising the consistency estimate to more general domains $\Omega \subset \mathbb{R}^2$, the following difficulty arises. The entries of the matrix L_h according to (4.8.7) or (14.8.16) are not bounded by Ch^{-2}. Rather, at points near the boundary the inverse of the distance to the boundary point enters, and this distance may be arbitrarily small. One way around this, would be to formulate the discretisation so that the distances between boundary points and points near near the boundary remain, for example, $\ge h/2$. A second possibility would be a suitable definition of R_h so that the product L_hR_h appearing in

(9) can be estimated (cf. Hackbusch [2]). Here we choose a third option: L_h is replaced by the re-scaled matrix $L'_h = D_h L_h$ from Corollary 11.

Theorem 9.2.14. *Let $\Omega \in C^2$ (or convex) and bounded. For the discretisation of $Lu = f$ for $L = -\Delta$ on Ω with $u = 0$ on Γ use the discretisation $L_h u_h$ according to* (4.8.7) *or* (4.8.16). *Let $L'_h = D_h L_h$ be defined as in Corollary* 11. *Then the consistency estimate*

$$|L'_h R_h - D_h \tilde{R}_h L|_{-1\leftarrow 2} \le Ch \tag{9.2.15}$$

holds.

Here, the matrix L_h from (4.8.7/16) is only taken as an example. The proof will show that the estimate (9), respectively (15), also holds for other L_h if $(L_h u_h)(\mathbf{x})$, $\mathbf{x}$ near the boundary, represents a second difference. First, two lemmas are needed.

Let $\gamma_h \subset \Omega_h$ be the set of points near the boundary. If v_h is a grid function defined on Ω_h then we denote by $v_h|_{\gamma_h}$ the function

$$(v_h|_{\gamma_h})(\mathbf{x}) = v_h(\mathbf{x}) \quad \text{for } \mathbf{x} \in \gamma_h, \quad (v_h|_{\gamma_h})(\mathbf{x}) = 0 \quad \text{for } \mathbf{x} \in \Omega_h \backslash \gamma_h.$$

Lemma 9.2.15. *Let $\Omega \in C^{0,1}$ be bounded. Then there exists a $C = C(\Omega)$ independent of h, such that*

$$|\, v_h|_{\gamma_h}|_0 \le Ch|v_h|_1. \tag{9.2.16}$$

PROOF. From $\Omega \in C^{0,1}$ follows: there exist numbers $K \in \mathbb{N}$ and $h_0 > 0$ such that for all $\mathbf{x} \in \gamma_h$, with $h \le h_0$, not all grid points $\{\mathbf{x} + (\nu h, \mu h) \colon -K \le \nu, \mu \le K\}$ lie in Ω. For each $\mathbf{x} \in \Omega_h$ select a pair $(\nu_0, \mu_o) \in \mathbb{Z}^2$ with $-K \le \nu_0, \mu_0 \le K$ and $\mathbf{x} + (\nu_0 h, \mu_0 h) \notin \Omega$. At this grid point $\mathbf{x}$ the functions $w_h^{\nu\mu}$ $(-K \le \nu, \mu \le K)$ are defined by $w_h^{\nu_0\mu_0}(\mathbf{x}) := v_h(\mathbf{x})$, $w_h^{\nu\mu}(\mathbf{x}) := 0$ for $(\nu, \mu) \ne (\nu_0, \mu_0)$. Therefore $\sum_{\nu,\mu=-K}^{K} w_h^{\nu\mu} = v_h|_{\gamma_h}$ is a decomposition with $w_h^{\nu\mu}(\mathbf{x}) = v_h(\mathbf{x})$ or $w_h^{\nu\mu}(\mathbf{x}) = 0$.

Without loss of generality let us assume that $\nu > 0$ and $\mu > 0$. For $\mathbf{x} \in \Omega_h$ we define the chain

$$\mathbf{x}^0 = \mathbf{x},\ \mathbf{x}^1 = \mathbf{x} + (h, 0)\ ,\ \cdots\ ,\ \mathbf{x}^\nu = \mathbf{x} + (\nu h, 0),$$
$$\mathbf{x}^{\nu+1} = \mathbf{x} + (\nu h, h)\ ,\ \cdots\ ,\ \mathbf{x}^{\nu+\mu} = \mathbf{x} + (\nu h, \mu h).$$

According to the definition, either $w_h^{\nu\mu}(\mathbf{x}) = 0$ holds or $\mathbf{x}^{\nu+\mu} \notin \Omega$, i.e., $w_h^{\nu\mu}(\mathbf{x}^{\nu+\mu}) = 0$ (cf. (2)). In both cases we have

$$\begin{aligned}
|w_h^{\nu\mu}(\mathbf{x})| &\le |w_h^{\nu\mu}(\mathbf{x}^0) - w_h^{\nu\mu}(\mathbf{x}^{\nu+\mu}| \\
&= h \left| \frac{1}{h}[w_h^{\nu\mu}(\mathbf{x}^0) - w_h^{\nu\mu}(\mathbf{x}^1)] + \frac{1}{h}[w_h^{\nu\mu}(\mathbf{x}^1) - w_h^{\nu\mu}(\mathbf{x}^2)] + \ldots \right| \\
&\le h[\,|\partial_x^- w_h^{\nu\mu}(\mathbf{x}^1)| + |\partial_x^- w_h^{\nu\mu}(\mathbf{x}^2)| + \ldots + |\partial_x^- w_h(\mathbf{x}^\nu)| \\
&\qquad + |\partial_y^- w_h^{\nu\mu}(\mathbf{x}^{\nu+1})| + \ldots + |\partial_y^- w_h^{\nu\mu}(\mathbf{x}^{\nu+\mu})|\,]
\end{aligned}$$

and thus $|w_h^{\nu\mu}|_0 \le h[|\nu|\,|\partial_x^+ w_h^{\nu\mu}|_0 + |\mu|\,|\partial_y^+ w_h^{\nu\mu}|_0] \le 2Kh|w_h^{\nu\mu}|_1$. Summation over ν, μ yields estimate (16) for $v_h|_{\gamma_h} = \sum w_h^{\nu\mu}$. ■

In most cases $\mathbf{x} \in \gamma_h$ already has a direct neighbour in $\mathbb{R}^2\backslash\Omega$ of $\mathbf{x} \in \gamma_h$ such that (16) follows with $C = \sqrt{2}$. This holds in particular for convex domains.

Lemma 9.2.16. *Let* R_h *be defined by* (12a). *Let* E_2 *satisfy* (10a). *Then we have*

$$|(R_h u)(\xi)| \le Ch\|E_2 u\|_{H^2(K_{h/2}(\xi))} \quad \text{for all } \xi \in \Gamma, u \in H_0^1(\Omega) \cap H^2(\Omega), \tag{9.2.17}$$

where $K_{h/2}(\xi) := \{\xi + \mathbf{x} \in \mathbb{R}^2 : \mathbf{x} \in (-\frac{h}{2}, \frac{h}{2}) \times (-\frac{h}{2}, \frac{h}{2})\}$.

PROOF. (a) Let $Q = (-\frac{1}{2}, \frac{1}{2}) \times (-\frac{1}{2}, \frac{1}{2})$. For $v \in H^2(Q)$ one shows

$$|v(\mathbf{0}) - \int_Q v(\mathbf{x})\, d\mathbf{x}| \le C\sqrt{\int\!\!\int_Q (v_{xx}^2 + 2v_{xy}^2 + v_{yy}^2)\, dx\, dy},$$

since the left-hand side vanishes for linear functions $u(x,y) = \alpha + \beta x + \gamma y$. The proof is similar to the one for (8.4.2). Transforming from Q to $Q_h := (-\frac{h}{2}, \frac{h}{2}) \times (-\frac{h}{2}, \frac{h}{2})$ gives

$$|h^2 v(\mathbf{0}) - \int_{Q_h} v(\mathbf{x})\, d\mathbf{x}| \le Ch^3\sqrt{\int\!\!\int_{Q_h} (v_{xx}^2 + 2v_{xy}^2 + v_{yy}^2)\, dx\, dy}. \tag{9.2.18}$$

(b) Let $\xi \in \Gamma$. Statement (17) follows from (18) with

$$v(\mathbf{x}) := \overline{u}(\xi + h\mathbf{x}) = (E_2 u)(\xi + h\mathbf{x}), \quad \text{since } u(\xi) = 0.$$ ■

PROOF. (of Theorem 14) (a) Inequality (15) is proved if

$$|(v_h, [L_h' R_h - D_h \tilde{R}_h L]u)_0| \le Ch \tag{9.2.19a}$$

for all $v_h \in H_h^1$ and $u \in H^2(\Omega) \cap H_0^1(\Omega)$ with $|v_h|_1 = |u|_2 = 1$. To this end v_h is split into

$$v_h = v_h' + v_h'' \quad \text{with } v_h' := v_h|_{\gamma_h}.$$

In part (b) we show

$$|(v_h'', [L_h' R_h - D_h \tilde{R}_h L]u)_0| \le C_1 h, \tag{9.2.19b}$$

The other steps of the proof, (c) and (d), yield

$$|(v_h', [L_h' R_h - D_h \tilde{R}_h L]u)_0| \le C_2 h, \tag{9.2.19c}$$

such that (19a) with $C = C_1 + C_2$ follows.
(b) Lemma 15 shows

$$|v_h'|_0 \le C_3 h |v_h|_1 = C_3 h. \tag{9.2.19d}$$

For v_h'' one obtains $|v_h''|_1 \le |v_h|_1 + |v_h'|_1 = 1 + |v_h'|_1$. The inverse estimate (3e) yields $|v_h'|_1 \le Ch^{-1}|v_h'|_0 \le CC_3$, thus

$$|v_h''|_1 \le C_4. \tag{9.2.19e}$$

Let $\hat{L}_h$ be the (regular) difference operator on the infinite grid

$$Q_h = \{(\nu h, \mu h) \colon \nu, \mu \in \mathbb{Z}\}.$$

Since the support of v_h'' is $\Omega_h \backslash \gamma_h$, $(v_h'', L_h' w_h)_0 = (v_h'', \hat{L}_h w_h)_0$ holds for all w_h. Furthermore, $(v_h'', D_h w_h)_0 = (v_h'', w_h)_0$. This proves the first equality in

$$\begin{aligned} |(v_h'', [L_h' R_h - D_h \tilde{R}_h L]u)_0| &= |(v_h'', [\hat{L}_h R_h - \tilde{R}_h L]\overline{u})_0| \\ &\le C_5 h |v_h''|_1 |\overline{u}|_2 \le C_5 h C_4 C_6 := C_1 h, \end{aligned}$$

where $\overline{u} := E_2 u$ is the continuation of u to $\mathbb{R}^2$. Further inequalities result from Lemma 13, (19e), and

$$\|\overline{u}\|_{H^2(\mathbb{R}^2)} \le C_6 \|u\|_{H^2(\Omega)} = C_6 \tag{9.2.19f}$$

(cf. (10a)). Theorem 6.2.40c guarantees the existence of an extension $\overline{u}$ with (19f) if $\Omega \in C^2$. Another sufficient condition for (19f) is the convexity of Ω.

(c) The left side of (19c) splits into $(v_h', L_h' R_h u)_0$ and $(v_h', D_h \tilde{R}_h L u)_0$. The first part is estimated in part (d). Exercise 12c and (19d) yield the second term

$$|(v_h', D_h \tilde{R}_h L u)_0| \le |v_h'|_0 |D_h \tilde{R}_h L u|_0 \le C_3 h C' |Lu|_0 \le C_6 h |u|_2 = C_6 h. \tag{9.2.19g}$$

(d) We set $w_h := (L_h' R_h u)|_{\gamma_h}$. Since the support of v_h' is contained in γ_h, we have $(v_h', L_h' R_h u)_0 = (v_h', w_h)_0$. L_h' contains differences with respect to the x and y directions. Accordingly we write $w_h = w_h^x + w_h^y$. In the following we limit ourselves (i) to the Shortley-Weller discretisation, (ii) to the term w_h^x and (ii) to the case that

$$\mathbf{x} \in \gamma_h, \quad \mathbf{x}^r = \mathbf{x} + (s_r h, 0) \in \Omega_h \quad \text{(i.e., } s_r = 1), \quad \mathbf{x}^l := \mathbf{x} - (s_l h, 0) \in \Gamma.$$

The other cases should be treated analogously. We set

$$\hat{u} := R_h u = \sigma_h^x \sigma_h^y E_2 u \in H^2(\mathbb{R}^2).$$

The Shortley-Weller difference in the x direction reads

$$\hat{w}_h^x(\mathbf{x}) = d_{\mathbf{x}} \left[\frac{\hat{u}(\mathbf{x}) - \hat{u}(\mathbf{x}^r)}{h s_r} - \frac{\hat{u}(\mathbf{x}^l) - \hat{u}(\mathbf{x})}{h s_l} \right] \Big/ \frac{h(s_r + s_l)}{2},$$

where $d_{\mathbf{x}}$ is the diagonal element of D_h. Since in the equation $L_h' u_h = f_h$ the variables $u_h(\xi)$, $\xi \in \Gamma$ (for example, $\xi = \mathbf{x}^l$), have already been eliminated, $w_h^x(\mathbf{x})$ has the form

$$w_h^x(\mathbf{x}) = \hat{w}_h^x(\mathbf{x}) + \frac{2 d_{\mathbf{x}}}{h^2 s_l (s_l + s_r)} \hat{u}(\mathbf{x}^l).$$

The factor $2d_{\mathbf{x}}/[h^2 s_l(s_l+s_r)]$, due to the definition of D_h, remains bounded by $4h^{-2}$. From Lemma 16 and $K_{h/2}(\mathbf{x}^l) \subset K_{3h/2}(\mathbf{x})$ one infers

$$|w_h^x(\mathbf{x}) - \hat{w}_h^x(\mathbf{x})| \le 4h^{-2}|\hat{u}(\mathbf{x}^l)| \le Ch^{-1}\|\overline{u}\|_{H^2(K_{3h/2}(\mathbf{x}))}. \tag{9.2.19h}$$

The second divided difference $\hat{w}_h^x$ in $\mathbf{x} = (x,y)$ can be represented by

$$\hat{w}_h^x(x,y) = d_{\mathbf{x}} \int_{-hs_r}^{h} g(t)\hat{u}_{xx}(x+t,y)\,dt$$

with

$$g(t) = \begin{cases} 2(t-h)/(h^2+h^2 s_l), & \text{for } 0 \le t \le h, \\ -2(s_l h + t)/(h^2 s_l(1+s_l)), & \text{for } -s_l h \le t \le 0, \\ 0 & \text{otherwise.} \end{cases}$$

From this one infers

$$\begin{aligned} |\hat{w}_h^x(x,y)| &\le d_{\mathbf{x}} \left[\int_{-hs_l}^{h} g^2(t)\,dt\right]^{1/2} \left[\int_{-hs_l}^{h} \hat{u}_{xx}^2(x+t,y)\,dt\right]^{1/2} \\ &\le 2h^{-1}\left[\int_{-h}^{h} \hat{u}_{xx}^2(x+t,y)\,dt\right]^{1/2} \le Ch^{-1}\|\overline{u}\|_{H^2(K_{3h/2}(\mathbf{x}))} \end{aligned} \tag{9.2.19i}$$

(cf. (6.2.5a)). From (19h,i) and the corresponding estimate for w_h^y one obtains $|w_h(\mathbf{x})| \le C_7 h^{-1}\|\overline{u}\|_{H^2(K_{3h/2}(\mathbf{x}))}$, such that

$$\begin{aligned} |w_h|_0^2 &= h^2 \sum_{\mathbf{x}\in\gamma_h} |w_h(\mathbf{x})|^2 \le C_7^2 \sum_{\mathbf{x}\in\gamma_h} \|\overline{u}\|^2_{H^2(K_{3h/2}(\mathbf{x}))} \le 9C_7^2\|\overline{u}\|^2_{H^2(\mathbb{R}^2)} \\ &\le (3C_7C_6)^2 := C_8^2. \end{aligned}$$

From this follows

$$|(v_h', L_h' R_h u)_0| \le |(v_h', w_h)_0| \le |v_h'|_0 |w_h|_0 \le C_3 h C_8 := C_9 h. \tag{9.2.19j}$$

(19g) and (19j) yield the required inequality (19c). ■

Remark 9.2.17. The proof steps for Theorem 14 can be carried out in the same manner for more general difference equations (for example, with variable coefficients, as in Lemma 13).

9.2.3 Optimal Error Estimates

In the following we compare the discrete solution $u_h = L_h^{-1} f_h$ with the restriction $u_h^* := R_h u$ of the exact solution $u = L^{-1}f$. From the representation

$$\begin{aligned} u_h - u_h^* &= L_h^{-1} f_h - R_h u = L_h^{-1}(f_h - \tilde{R}_h f) + L_h^{-1}\tilde{R}_h f - R_h u \\ &= L_h^{-1}(f_h - \tilde{R}_h f) - L_h^{-1}(L_h R_h - \tilde{R}_h L)u \end{aligned} \tag{9.2.20}$$

one immediately obtains

Theorem 9.2.18. *Let $u \in H^2(\Omega)$ hold for the solution of $Lu = f$. Let the right-hand side f_h of the discrete equation $L_h u_h = f_h$ be chosen so that*

$$|f_h - \tilde{R}_h f|_{-1} \le C_f h. \tag{9.2.21}$$

If, furthermore, L_h is H_h^1-regular, and if the consistency condition (9) *holds, then u_h satisfies the error estimate*

$$|u_h - u_h^*|_1 \le C_1(C_f + C_K |u|_2)h. \tag{9.2.22}$$

PROOF. $|u_h - u_h^*|_1 \le |L_h^{-1}|_{1\leftarrow -1}(|f_h - \tilde{R}_h f|_{-1} + |L_h R_h - \tilde{R}_h L|_{-1\leftarrow 2}|u|_2)$. ∎

Corollary 9.2.19. (*a*) *Inequality* (21) *is satisfied in particular if one chooses $f_h := \tilde{R}_h f$.*
(*b*) *The choice $f_h(\mathbf{x}) := f(\mathbf{x})$ for $\mathbf{x} \in \Omega_h$ (cf.* (4.2.6b)) *leads to* (21) *if $f \in C^{0,1}(\overline{\Omega})$ or $f \in H^2(\Omega)$. In these cases the following even holds:*

$$|f_h - \tilde{R}_h f|_0 \le C\|f_h - \tilde{R}_h f\|_\infty \le C' h \|f\|_{C^{0,1}(\overline{\Omega})},$$
$$\text{resp. } |f_h - \tilde{R}_h f|_0 \le C h^2 |f|_2. \tag{9.2.23}$$

(*c*) *In Theorem* 18 *one can replace the H_h^1-regularity of L_h by that of $L_h' = D_h L_h$ (cf. Corollary* 11), (9) *by* (15), *and* (21) *by*

$$|D_h f_h - D_h \tilde{R}_h f|_{-1} \le C_f h. \tag{9.2.21'}$$

The proof of (23) is based on the inequality (18). ∎

Error estimates of order $O(h^2)$ can be derived in the same way if one has consistency conditions of second order. These are, for example, (24a) or (24b):

$$|L_h R_h - \tilde{R}_h L|_{-2\leftarrow 2} \le C h^2, \tag{9.2.24a}$$

$$|L_h R_h - \tilde{R}_h L|_{-1\leftarrow 3} \le C h^2. \tag{9.2.24b}$$

Remark 9.2.20. If $\Omega_h = Q_h$ (i.e., $\Omega = \mathbb{R}^2$) or $\Omega = (x', x'') \times (y', y'')$, the inequalities (24a,b) can be shown in a way similar to Lemma 13.

Example 9.2.21. The difference method in Example 4.5.8 shows quadratic convergence. This case can be analysed as follows. Using Remark 20 one shows (24a). In the following section we prove the H_h^2-regularity $|L_h^{-1}|_{2\leftarrow 0} \le C$ which for the symmetric matrix under discussion, L_h, is equivalent with $|L_h^{-1}|_{0\leftarrow -2} \le C$ (cf. (6.3.3)). Expression (20) leads to

$$|u_h - u_h^*|_0 \le C h^2 |u|_2. \tag{9.2.25a}$$

The corresponding estimate

$$|u_h - u_h^*|_1 \le Ch^2 |u|_3, \tag{9.2.25b}$$

which is based on (24b), fails due to the fact that the solution in Example 4.5.8 does not belong to $H^3(\Omega)$ (cf. Example 9.1.25).

The verification of (24b) in the presence of irregular discretisations at the boundary becomes more complicated.

9.2.4 H_h^2-Regularity

Under suitable conditions on Ω, $Lu = f \in L^2(\Omega)$ has a solution $u \in H^2(\Omega) \cap H_0^1(\Omega)$: $|u|_2 \le C|f|_0$ (cf., for example, Theorem 9.1.22). For the discrete solution of $L_h u_h = f_h$ the corresponding question arises: does $|u_h|_2 \le C|f_h|_0$ hold? First one has to define the norm $|\cdot|_2$ of H_0^2.

If $\Omega = \mathbb{R}^2$ or if the boundary Γ coincides with grid lines, one can define

$$|u_h|_2 := \left\{ |u_h|_1^2 + h^2 \sum_{x \in \Omega_h} [|\partial_x^+ \partial_x^- u_h(\mathbf{x})|^2 + |\partial_y^+ \partial_y^- u_h(\mathbf{x})|^2] \right\}^{1/2}.$$

In general, however, one has to use irregular differences at the boundary. At points far from the boundary $\mathbf{x} \in \Omega_h \backslash \gamma_h$, let $D_{xx} u_h(\mathbf{x}) = \partial_x^+ \partial_x^- u_h(\mathbf{x})$. At a grid point near the boundary $\mathbf{x} = (x, y) \in \gamma_h$, for example, with $\mathbf{x}^l := (x - s_l h, y) \in \Gamma$, $\mathbf{x}^r := (x + h, y) \in \Omega_h$ it appears practical to use the divided difference (4.8.5). In general, however, this would violate the inverse estimate $|u_h|_2 \le Ch^{-1} |u_h|_1$, since s_l may become arbitrarily small. Therefore one defines $D_{xx} u_h(\mathbf{x})$ with the aid of the grid points $(x - s_l h, y)$, $(x + h, y)$, $(x + 2h, y)$:

$$D_{xx} u_h(x, y) := \frac{2}{(2 + s_l)h} \left[\frac{u_h(x + 2h, y) - u_h(x + h, y)}{h} - \frac{u_h(x + h, y) - u_h(x - s_l h, y)}{s_l h} \right],$$

where $u_h(x - s_l h, y) = 0$.

If, in addition, Γ passes between $(x + h, y)$ and $(x + 2h, y)$, one needs to replace $x + 2h$ by $x + h + s_r h$. If $(x + h, y)$ also no longer belongs to Ω, set $D_{xx} u_h(\mathbf{x}) := h^{-2} u_h(\mathbf{x})$. The second difference D_{yy} is defined accordingly. The mixed difference in the interior reads $D_{xy} := \partial_x^- \partial_y^-$. Near the boundary several choices for D_{xy} are possible. The definition of $|\cdot|_2$ reads

$$|u_h|_2 := \left\{ |u_h|_1^2 + h^2 \sum_{\mathbf{x} \in \Omega_h} [|D_{xx} u_h(\mathbf{x})|^2 + 2|D_{xy} u_h(\mathbf{x})|^2 + |D_{yy} u_h(\mathbf{x})|^2] \right\}^{1/2}.$$

Exercise 9.2.22. Prove the inverse estimate $|\cdot|_2 \le Ch^{-1} |\cdot|_1$.

We call L_h H_h^2-regular if $|L_h^{-1}|_{2 \leftarrow 0} \le C$. An equivalent formulation is $|u_h|_2 \le C|f_h|_0$ for $u_h = L_h^{-1} f_h$, $f_h \in L_h^2$.

Exercise 9.2.23. Let L_h be H_h^2-regular, let $L_h + \delta L_h$ be H_h^1-regular and $|\delta L_h|_{0\leftarrow 1} \le C$ for all h. Show that $L_h + \delta L_h$ is also H_h^2-regular. Hint: Remark 9.1.2.

For the proof of H_h^2-regularity one can —in analogy to the proof of Corollary 9.1.23— partially sum the scalar product $(L_h u_h, L_h u_h)_0$ in order to show the equivalence of $|f_h|_0^2 = |L_h u_h|_0^2$ and $|u_h|_2^2$. This technique, however, can only be applied to a rectangle Ω. Here we use a simpler proof which is also applicable to general domains.

Let $P_h: L_h^2 \to L^2(\Omega)$ be the following piecewise constant interpolation:

$$(P_h u_h)(\mathbf{x}) := \begin{cases} \hat{P}_h u_h(\mathbf{x}) & \text{if } \mathbf{x} \in \Omega, \\ 0 & \text{if } \mathbf{x} \notin \Omega, \end{cases} \tag{9.2.26a}$$

$$\hat{P}_h u_h(x,y) := u_h(x',y'), \ \text{ if } \ (x',y') \in Q_h \ \text{ and} \tag{9.2.26b}$$
$$x' - h/2 < x \le x' + h/2, \ y' - h/2 < y \le y' + h/2,$$

where $u_h(\mathbf{x}) = 0$ for $\mathbf{x} \in Q_h \backslash \Omega_h$ (cf. (2)).

Lemma 9.2.24. *Let $\Omega \in C^{0,1}$. Then we have*

$$|\tilde{R}_h P_h - I|_{-1\leftarrow 0} \le Ch, \tag{9.2.27a}$$

$$|P_h|_{0\leftarrow 0} \le C, \tag{9.2.27b}$$

$$|R_h|_{2\leftarrow 2} \le C. \tag{9.2.27c}$$

PROOF. (a) The estimate (27a) is equivalent to $|P_h^* \tilde{R}_h^* - I|_{0\leftarrow 1} \le Ch$ (cf. (6.3.3)). Thus we must show $|w_h|_0 \le Ch|u_h|_1$ for $w_h := (P_h^* \tilde{R}_h^* - I)u_h$. From the following Exercise 25 we obtain the expression

$$\hat{w}_h(\mathbf{x}) = (\sigma_h^x \sigma_h^y [\sigma_h^x \sigma_h^y - I]\hat{P}_h u_h)(\mathbf{x}) \ \text{ for } \ \hat{w}_h = (\hat{P}_h^* R_h^* - I)u_h, \quad \mathbf{x} \in Q_h.$$

Exercise 12c shows $|\hat{w}_h|_0 \le |w|_{L^2(\mathbb{R}^2)}$ for $w := (\sigma_h^x \sigma_h^y - I)\hat{P}_h u_h$. For every $\xi \in (0,1)\times(0,1)$ define the grid function $w_{h,\xi} \in L_h^2$ with $w_{h,\xi}(\mathbf{x}) := w(\mathbf{x}+\xi h)$ for $\mathbf{x} \in Q_h$. One may check that $w_{h,\xi}(x,y)$ is a weighted sum of first differences $u_h(x,y) - u_h(x',y')$ where $(x',y') \in \{(x \pm h, y), (x, y \pm h), (x \pm h, y \pm h)\}$. Thus we have $|w_{h,\xi}|_0 \le h|u_h|_1$ for all ξ from which one infers

$$|w|_0^2 = \int_{\mathbb{R}^2} |w(\mathbf{x})|^2 \, d\mathbf{x} = h^2 \int_{(0,1)^2} \sum_{\mathbf{x}\in Q_h} |w(\mathbf{x}+\xi h)|^2 \, d\xi$$

$$= \int_{(0,1)^2} |w_{h,\xi}|_0^2 \, d\xi \le h^2 |u_h|_1^2,$$

thus $|\hat{w}_h|_0 \le |w|_0 \le h|u_h|_1$.

From

$$w_h - \hat{w}_h = (P_h^* - \hat{P}_h^*)\tilde{R}_h^* u_h = (P_h^* - \hat{P}_h^*)\sigma_h^x \sigma_h^y \hat{P}_h u_h = (P_h^* - \hat{P}_h^*)\sigma_h^x \sigma_h^y \hat{P}_h (u_h|_{\dot{\gamma}_h})$$

with $\hat{\gamma}_h = \{\mathbf{x} \in \Omega_h \colon K_{3h/2}(\mathbf{x}) \cap \Gamma \neq \emptyset\}$ one infers

$$|w_h - \hat{w}_h|_0 \leq |(P_h^* - \hat{P}_h^*)\sigma_h^x\sigma_h^y|_{0\leftarrow 0}|\hat{P}_h|_{0\leftarrow 0}|u_h|_{\hat{\gamma}_h}|_0 \leq C|u_h|_{\hat{\gamma}_h}|_0 \leq C'h|u_h|_1$$

(cf. part b of the proof and Lemma 15, in which γ_h may be replaced by $\hat{\gamma}_h$). Together with $|\hat{w}_h|_0 \leq h|u_h|_1$ one obtains $|w_h|_0 \leq (C'+1)h|u_h|_1$.
(b) (27b) with $C = 1$ follows from $|P_h u_h|_0 = |u_h|_0$.
(c) Second differences of $\hat{u} = R_h u$ have already been estimated in (19i). (19i) implies $|u_h|_2 \leq C|u|_2$, i.e. (27c). ■

Exercise 9.2.25. Show that the adjoint operators for $\hat{P}_h$ and $\tilde{R}_h$ are $\hat{P}_h^* \colon L^2(\mathbb{R}^2) \to L_h^2$ and $\tilde{R}_h^* \colon L_h^2 \to L^2(\mathbb{R}^2)$ with

$$(\hat{P}_h^* u)(\mathbf{x}) = (\sigma_h^x \sigma_h^y u)(\mathbf{x}) \quad \text{for } \mathbf{x} \in Q_h, \quad \tilde{R}_h^* u_h = \sigma_h^x \sigma_h^y \hat{P}_h u_h.$$

Furthermore, the following holds: $(\sigma_h^x \sigma_h^y \hat{P}_h u_h)(\mathbf{x}) = u_h(\mathbf{x})$ for all $\mathbf{x} \in Q_h$. The identity

$$L_h^{-1} = R_h L^{-1} P_h - L_h^{-1}[(L_h R_h - \tilde{R}_h L)L^{-1}P_h + (\tilde{R}_h P_h - I)]$$

yields the estimate

$$|L_h^{-1}|_{2\leftarrow 0} \leq |R_h|_{2\leftarrow 2}|L^{-1}|_{2\leftarrow 0}|P_h|_{0\leftarrow 0} + |I|_{2\leftarrow 1}|L_h^{-1}|_{1\leftarrow -1} \times$$
$$[|L_h R_h - \tilde{R}_h L|_{-1\leftarrow 2}|L^{-1}|_{2\leftarrow 0}|P_h|_{0\leftarrow 0} + |\tilde{R}_h P_h - I|_{-1\leftarrow 0}].$$

The inverse estimate from Exercise 22 yields $|I|_{2\leftarrow 1} \leq Ch^{-1}$ for the identity $I \colon H_h^1 \to H_h^2$. Together with the inequalities (27a–c) follows

Theorem 9.2.26. *Let L_h be H_h^1-regular and satisfy the consistency condition (9). Let Ω be convex or from C^2 and let L be H^2-regular (i.e., $|L^{-1}|_{2\leftarrow 0} \leq C$). Then L_h is also H_h^2-regular.*

$\Omega \in C^{0,1}$ would be a sufficient condition upon Ω guarantee (27a–c). On the other hand, the condition given agrees with the conditions for Theorem 14 and with the H^2-regularity of L.

Corollary 9.2.27. *In Theorem 26 one can replace (9) and the H_h^1-regularity of L_h by (15) and the H_h^1-regularity of L_h'. The result is the H_h^2-regularity of both L_h and L_h'.*

PROOF. For the proof of the H_h^2-regularity of L_h' define $P_h u_h(\mathbf{x}) := 0$ for $\mathbf{x} \in K_{h/2}(\hat{\mathbf{x}})$, $\hat{\mathbf{x}} \in \gamma_h$, and note $|D_h(\tilde{R}_h P_h - I)D_h^{-1}|_{-1\leftarrow 0} \leq Ch$. ■

Exercise 9.2.28. If $\Omega \in \mathbb{R}^n$ $(n \leq 3)$ is bounded and L_h is H_h^2-regular, then L_h is stable with respect to the row-sum norm: $\|L_h^{-1}\|_\infty \leq C$. Hint: Use $C^{-1}|u_h|_0 \leq \|u_h\|_\infty \leq C|u_h|_2$.

For general Lipschitz domains Ω, such as, for example, the L-shaped domain in Example 2.1.4, L $[L_h]$ cannot be H^2-regular $[H_h^2$-regular$]$. Theorem 9.2.21, however, guarantees $H_0^{1+s}(\Omega)$-regularity for $s \in [0, 1/2)$. For L_h one can define and prove analogously H_h^{1+s}-regularity (cf. Hackbusch [1]).

10 Special Differential Equations

If the boundary value problems have special properties, one often uses special discretisations for them. We give two examples of this in Sections 10.1 and 10.2. For the first example the variational formulation is of decisive importance.

10.1 Differential Equations with Discontinuous Coefficients

10.1.1 Formulation

The self-adjoint differential equation

$$\sum_{i,j=1}^{n} \frac{\partial}{\partial x_i}\left(a_{ij}(\mathbf{x})\frac{\partial}{\partial x_j}u\right) + a(\mathbf{x})u(\mathbf{x}) = f(\mathbf{x}) \quad \text{in } \Omega \tag{10.1.1a}$$

(cf. (5.1.17)) occurs often in physics. It can also be written as

$$\operatorname{div}(\mathbf{A}(\mathbf{x}) \operatorname{grad} u) + au = f$$

with $\mathbf{A}(\mathbf{x}) := (a_{ij}(\mathbf{x}))_{i,j=1,\ldots,n}$, so that for $a = 0$ and $f = 0$ we have the conservation law $\operatorname{div}\phi = 0$ for $\phi := \mathbf{A} \operatorname{grad} u$.

In the applications in physics the coefficients a_{ij} are, in general, constants of the material. The functions a_{ij} can be varying, if the constitution of the material depends on position. As soon as we have several materials in contact with each other, the coefficients $a_{ij}(\mathbf{x})$ may be discontinuous on the boundary of contact γ (see Figure 1).

The equation (1a) can be understood in a classical sense only if $a_{ij} \in C^1(\Omega)$. For discontinuous a_{ij} therefore one uses the variational formulation. If one supplements Equation (1a) with the Dirichlet condition

$$u = 0 \quad \text{on } \Gamma, \tag{10.1.1b}$$

then the weak formulation is written

$$u \in H_0^1(\Omega) \quad \text{with} \quad a(u,v) = (f,v)_{L^2(\Omega)} \quad \text{for all } v \in H_0^1(\Omega), \tag{10.1.2a}$$

where

$$a(u,v) := \int_\Omega [a(\mathbf{x})u(\mathbf{x})v(\mathbf{x}) - \sum_{i,j=1}^n a_{ij}(\mathbf{x})v_{x_i}(\mathbf{x})u_{x_j}(\mathbf{x})]\, d\mathbf{x}. \tag{10.1.2b}$$

Note that Equations (2a,b) are defined for arbitrary $a_{ij} \in L^\infty(\Omega)$.

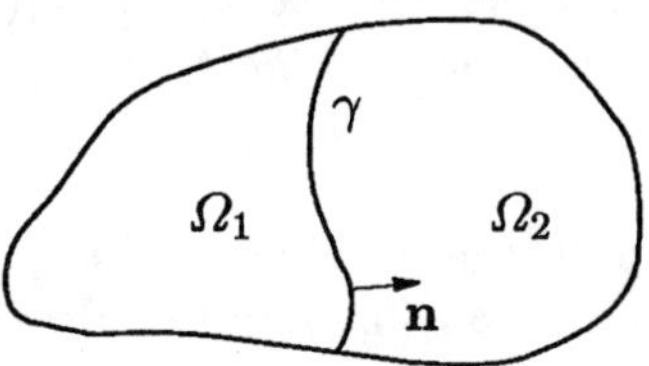

Figure 10.1.1. An interior boundary $\gamma = \partial\Omega_1 \cap \partial\Omega_2$

In the sequel we shall assume the situation to be as in Figure 1: Ω is divided by the boundary line γ into two subregions Ω_1 and Ω_2. Let the coefficients a_{ij} be piecewise smooth: $a_{ij} \in C^1(\Omega_k)$ for $k = 1,2$. Along γ the coefficients may be discontinuous so that the one-sided boundary values

$$a_{ij}^{(1)}(\mathbf{x}) := \lim_{\substack{\mathbf{y}\to\mathbf{x}\\ \mathbf{y}\in\Omega_1}} a_{ij}(\mathbf{y}), \qquad a_{ij}^{(2)}(\mathbf{x}) := \lim_{\substack{\mathbf{y}\to\mathbf{x}\\ \mathbf{y}\in\Omega_2}} a_{ij}(\mathbf{y}) \qquad (\mathbf{x} \in \gamma)$$

may be different. In addition let us assume that the solution u is continuous, $u \in C^0(\overline{\Omega})$, but only that it is piecewise smooth, $u \in C^1(\Omega_1)$ and $u \in C^1(\Omega_2)$. The one-sided boundary values of the derivatives are denoted $u_{x_i}^{(1)}(\mathbf{x})$ and $u_{x_j}^{(2)}(\mathbf{x})$ $(\mathbf{x} \in \gamma)$. With these assumptions integration by parts of $-\int_{\Omega_1} a_{ij}v_{x_i}u_{x_j}\, d\mathbf{x}$ gives the result $\int_{\Omega_1} v(au_{x_j})_{x_i}\, d\mathbf{x} - \int_\gamma va_{ij}^{(1)}u_{x_j}^{(1)}n_i\, d\Gamma$. Since $\mathbf{n} = (n_1, n_2, \ldots)$ in Figure 1 is the outgoing normal to Ω_1, but the ingoing normal with respect to Ω_2, in Ω_2 there results $-\int_{\Omega_2} a_{ij}v_{x_i}u_{x_j}\, d\mathbf{x} = \int_{\Omega_2} v(a_{ij}u_{x_j})_{x_i}\, d\mathbf{x} + \int_\gamma va_{ij}^{(2)}u_{x_j}^{(2)}n_i\, d\Gamma$. Putting this together one has

$$a(u,v) = \int_\Omega \left[au + \sum_{i,j=1}^n \frac{\partial}{\partial x_i}(a_{ij}u_{x_j}) \right] v\, d\mathbf{x} + \int_\gamma \sum_{i,j} (a_{ij}^{(2)}u_{x_j}^{(2)} - a_{ij}^{(1)}u_{x_j}^{(1)})n_i v\, d\Gamma.$$

Thus from the variational equation (2a) there follows in addition to the differential equation (1a) also the transition equation

$$\sum_{i,j=1}^n a_{ij}^{(1)}u_{x_j}^{(1)}n_i = \sum_{i,j=1}^n a_{ij}^{(2)}u_{x_j}^{(2)}n_i \quad \text{on } \gamma. \tag{10.1.3}$$

Thus we have shown the following lemma.

Lemma 10.1.1. *Assume $a_{ij} \in C^1(\Omega_k)$, $k = 1, 2$. If the weak solution u of* (2a) *in Ω is continuous and piecewise differentiable in Ω_1 and Ω_2, then it is the classical solution of the differential equation* (1a) *in $\Omega_1 \cup \Omega_2 = \Omega\backslash\gamma$. In addition to fulfilling the boundary condition* (1b) *the solution also satisfies the transition condition* (3) *on γ.*

Corollary 10.1.2. *If the coefficients are not continuous then the solution of the equation* (2a) *does not, in general, belong to $C^2(\Omega)$, but will have discontinuous derivatives along γ. However, the tangential derivative along γ may be continuous.*

Example 10.1.3. Suppose the coefficient of $a(u,v) := \int_0^1 a(x)u'v'\,dx$ are given by $a(x) = 1$ on the interval $(0,\xi)$ and $a(x) = 2$ on $(\xi, 1)$. For this one-dimensional example the point ξ plays the role of the curve γ. The solution of the equation (2a) with $f(x) = 1$ is

$$u(x) = \frac{1}{2}\left[\frac{1+\xi^2}{1+\xi}x - x^2\right] \quad \text{on } (0,\xi),$$

$$u(x) = \frac{1}{4}\left[\frac{1+2\xi-\xi^2}{1+\xi}(1-x) - (1-x)^2\right] \quad \text{on } (\xi,1).$$

It satisfies $-au'' = 1$ on $(0,\xi)\cup(\xi,1)$, $u(0) = u(1) = 1$ and the transition equation $1 \cdot u'(\xi - 0) = 2 \cdot u'(\xi + 0)$.

As mentioned at the beginning, the differential equation can be written in the form $\operatorname{div}\phi = f$ on $\Omega_1 \cup \Omega_2$, where $\phi := \mathbf{A}(\mathbf{x}) \operatorname{grad} u$ on Ω and $u = 0$ on Γ. The transition condition (3) means that $\langle\phi, \mathbf{n}\rangle$ is continuous on γ. Since $\langle\phi, \mathbf{t}\rangle$ is also continuous for any tangential direction $\mathbf{t}$, ϕ is continuous on γ and thus throughout Ω.

The regularity proofs of Section 9.1 can be carried over to the present situation. The smoothness assumptions upon the coefficients of $a(\cdot,\cdot)$ are in each case to be required piecewise on Ω_1 and Ω_2; in addition the dividing line (or hypersurface) γ must be sufficiently smooth. Then there results the piecewise regularity $u \in H^{m+s}(\Omega_1)$ and $u \in H^{m+s}(\Omega_2)$, instead of $u \in H^{m+s}(\Omega)$. Bringing in the transition condition (3) one obtains $\phi = \mathbf{A}(\mathbf{x}) \operatorname{grad} u \in H^{m+s-1}(\Omega)$ on the whole region.

10.1.2 Discretisation

When one discretises using finite elements there are the following difficulties:

Remark 10.1.4. Linear or bilinear elements give a finite-element solution with the error estimate $|u - u^h|_1 = O(h^{1/2})$. The $L^2(\Omega)$ error bound is in general no better than $|u - u^h|_0 = O(h)$.

PROOF. $\Omega_\gamma := \{t \in \tau : t \cap \gamma \neq \emptyset\}$ consists in all finite elements that are in contact with γ. Since u has discontinuous derivatives on γ, one can have no better estimate on Ω_γ than $\nabla u - \nabla v = O(1)$ $(v \in V_h)$. The surface area of Ω_γ is of the order $O(h)$, so that there results $|u - v|_1 = O(h^{1/2})$. For the bound on $|u - u^h|_0$ see

Example 10.1.5. In Example 3 choose $\xi = (1+h)/2$ and discretise using piecewise linear elements of length h. Then there results an error at $x = 1/2$ of $u(x) - u^h(x) = \alpha h + O(h^2)$ with $\alpha \approx 0.00463$. Since $u - u^h$ behaves linearly in $(0, 1/2)$, one obtains the error $O(h)$ not only for the L^2 norm $|u - u^h|_0$ but also for the maximum norm $|u - u^h|_\infty$ and for the L^1 norm $||u - u^h||_{L^1(0,1)}$.

The usual bounds on the errors $|u - u^h|_1 = O(h)$ and $|u - u^h|_0 = O(h^2)$ can however be attained. To do this one must adjust the geometry of the triangulation to the curve γ. If the dividing line γ is piecewise linear, one must choose the triangulation such that γ concides with sides of (interior) triangles. If γ is curvilinear then it may be approximated by isoparametric elements (cf. §8.6).

Similar assertions hold for the difference scheme (5.1.18).

Example 10.1.6. Let the discontinuity position ξ in Example 3 be a grid point, i.e., $\xi/h \in \mathbb{N}$. Then the difference scheme (5.1.18), which here takes the form

$$h^{-2}\left[a\left(\left(\nu+\tfrac{1}{2}\right)h\right)\left(u_{\nu+1}-u_\nu\right) - a\left(\left(\nu-\tfrac{1}{2}\right)h\right)\left(u_\nu - u_{\nu-1}\right)\right] = 1$$

for $1 \le \nu \le h^{-1} - 1$, is suited to the equation from Example 3. In general the error is of the order $O(h^2)$. Since the solution of the differential equation is piecewise quadratic here, it is in fact exactly given by the difference solution. On the other hand, if ξ is not a grid point then the error is of the order $O(h)$.

In the two-dimensional case one obtains $O(h^2)$ difference solutions if γ coincides with lines of the grid. Otherwise the error worsens to $O(h)$. Another possible form of difference approximation consists in approximating the differential equations separately in the regions Ω_1 and Ω_2, and then, to handle the unknown values on γ to discretise the transition condition (3).

10.2 A Singular Perturbation Problem

10.2.1 The Convection-Diffusion Equation

In the following we shall consider the boundary value problem

$$-\epsilon \Delta u + \sum_{i=1}^{n} c_i u_{x_i} = f \quad \text{in } \Omega, \qquad u = 0 \quad \text{on } \Gamma \tag{10.2.1}$$

for $\epsilon > 0$. One calls $-\epsilon\Delta u$ the diffusion term and $\sum c_i u_{x_i} = \langle \mathbf{c}, \nabla u \rangle$ the convection term. For small ϵ the convection term dominates.

Exercise 10.2.1. Let the coefficients c_i be constants. Equation (1) can be transformed by $v(\mathbf{x}) := u(\mathbf{x}) \exp(-\sum c_i x_i / 2\epsilon)$ into the symmetric $H_0^1(\Omega)$-elliptic equation

$$-\epsilon \Delta v + (\sum c_i^2/(4\epsilon))v = f \exp(-\langle \mathbf{c}, \mathbf{x} \rangle / 2\epsilon) \text{ in } \Omega, \quad v = 0 \text{ on } \Gamma. \quad (10.2.1')$$

Equation (1) is elliptic for all $\epsilon > 0$. That there is a unique solution follows from Exercise 1 (cf. also Theorem 5.1.8). Denote the solution by u_ϵ.

Remark 10.2.2. $u_\epsilon(\mathbf{x})$ and $\partial u_\epsilon(\mathbf{x})/\partial x_i$ cannot converge uniformly on $\overline{\Omega}$ for $\epsilon \to 0$.

PROOF. If u_ϵ and $\partial u_\epsilon / \partial x_i$ were continuous in ϵ, one would be able to take the limit $\epsilon \to 0$ in Equation (1) and one would obtain the first-order differential equation

$$-\sum_{i=1}^{n} c_i \, \partial u_0 / \partial x_i = f \text{ in } \Omega \quad (10.2.2)$$

for $u_0(\mathbf{x}) := \lim_{\epsilon \to 0} u_\epsilon(\mathbf{x})$. This equation is of hyperbolic type, but not compatible with the boundary condition $u = 0$ on Γ (cf. Section 1.4), i.e., Equation (2) has, in general, no solution with $u_0 = 0$ on Γ. ∎

Equation (2) is called the "reduced equation". Equations (1) and (2) differ in the "perturbation term" $-\epsilon \Delta u$. Since the equations, (1) and (2), are of different types one speaks of a "singular" perturbation.

The following example will show that $u_0 = \lim_{\epsilon \to 0} u_\epsilon$ exists and satisfies Equation (2), but that the boundary condition $u_0 = 0$ is only satisfied on a part Γ_0 of the boundary Γ.

Example 10.2.3. (a) The solution of the ordinary boundary value problem

$$-\epsilon u'' + u' = 1 \text{ in } (0,1), \qquad u(0) = u(1) = 0 \quad (10.2.3\text{a})$$

is $u_\epsilon(x) = x - (e^{x/\epsilon} - 1)/(e^{1/\epsilon} - 1)$. On $[0,1)$, $u_\epsilon(x)$ converges to $u_0(x) = x$. This function satisfies the reduced equation (2), $u' = 1$, and the left boundary condition $u_0(0) = 0$, but not $u_0(1) = 0$.
(b) The solution of

$$-\epsilon u'' - u' = 0 \text{ in } (0,1), \qquad u(0) = 0, \; u(1) = 1 \quad (10.2.3\text{b})$$

is $u_\epsilon(x) = (1 - e^{-x/\epsilon})/(1 - e^{-1/\epsilon})$. In this case, $u_0(x) := \lim_{\epsilon \to 0} u_\epsilon(x) = 1$ satisfies Equation (2), $-u' = 0$, and $u_0(1) = 1$, but not the boundary condition at $x = 0$.

Which boundary condition is fulfilled depends for Equations (3a,b) on the sign of the convection term $\pm u'$. In the many-dimensional case the decisive factor is the direction of the vector $\mathbf{c} = (c_1, \ldots, c_n)$.

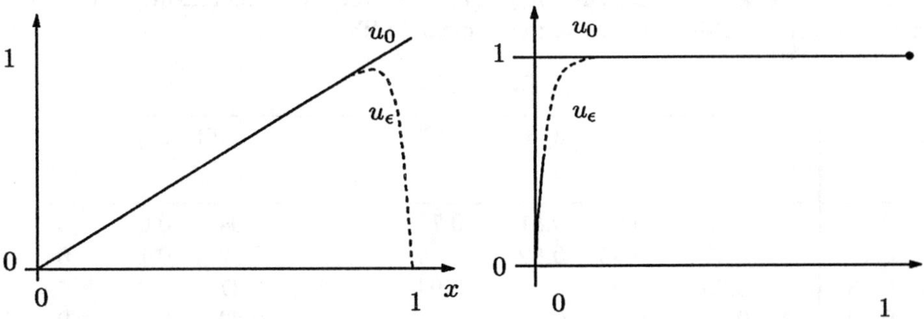

Figure 10.2.1. (a) Solution of Example 3a; (b) Solution of Example 3b

In Figure 1a,b the solutions of Example 3 are sketched. In the interior u_ϵ is close to u_0; only in the neighbourhood of $x = 1$ (Figure 1a) [resp. $x = 0$ (Figure 1b)] does u_ϵ differ from u_0, in order to satisfy the second boundary condition. These neighbourhoods, in which the derivatives of u_ϵ attain the order $O(1/\epsilon)$, are called boundary layers. For Example 3 the thickness of the boundary layer is of order $O(\epsilon)$.

Exercise 10.2.4. The interval $[1-\xi, 1]$, in which the function $(e^{x/\epsilon} - 1)/(e^{1/\epsilon} - 1)$ exceeds the value $0 < \eta < 1$, has the thickness $\xi = O(\epsilon|\log \eta|)$.

An analysis of singularly perturbed problems is to be found, e.g., in Kevorkian-Cole [1] and Goering et al. [1].

10.2.2 Stable Difference Schemes

The special case of Equation (1) for $\mathbf{c} = (1, 0)$ is

$$-\epsilon\Delta u + u_x = f \text{ in } \Omega, \qquad u = \varphi \text{ on } \Gamma. \tag{10.2.4}$$

The symmetric difference formula (5.1.10) for Equation (4) is

$$L_h = h^{-2}\begin{bmatrix} & -\epsilon & \\ -\epsilon - h/2 & 4\epsilon & -\epsilon + h/2 \\ & -\epsilon & \end{bmatrix}. \tag{10.2.5}$$

For a fixed ϵ, L_h is consistent of order 2. From Corollary 5.1.16 there follows

Remark 10.2.5. As soon as $h < 2\epsilon$, the difference scheme (5) leads to an M-matrix.

The property of being an M-matrix was used in Section 5.1.4 to demonstrate the solvability of the discrete equation. It turns out that the difference equations are also solvable for $h \geq 2\epsilon$. However, with increasing h/ϵ there develops an instability which is made clear in Table 1.

Table 10.2.1

x / ϵ	0	1/32	2/32	3/32	4/32	...	30/32	31/32	1
0.5	1	0.93	0.87	0.81	0.75	...	0.04	0.02	0
0.1	1	0.97	0.95	0.92	0.89	...	0.22	0.13	0
0.05	1	0.99	0.97	0.96	0.94	...	0.47	0.31	0
0.02	1	0.99	0.99	0.98	0.98	...	0.82	0.73	0
0.01	1	1.00	0.99	0.99	0.99	...	0.87	1.11	0
10^{-3}	1	1.03	0.99	1.04	0.99	...	0.23	1.89	0
10^{-4}	1	4.95	0.95	5.00	0.90	...	0.08	5.88	0
10^{-5}	1	48.6	0.94	48.7	0.88	...	0.06	49.6	0
10^{-7}	1	4859.5	0.94	4859.6	0.88	...	0.06	4860.4	0

Table 1 shows the values $u_{\epsilon,h}(x, 1/2)$ of the difference solution $u_{\epsilon,h}$ of Equation (4) in the unit square $\Omega = (0,1) \times (0,1)$ with $f = 0$ and $\varphi(x,y) = (1-x)\sin(\pi y)$. The grid step is $h = 1/32$. For $\epsilon > h/2 = 1/64 = 0.015625$, L_h is an (irreducibly diagonally dominant) M-matrix. Because of the maximum principle (cf. Remark 4.4.4) the values lie between $\min \varphi = 0$ and $\max \varphi = 1$. The solution of the reduced equation (2) is $u_0(x,y) = \lim_{\epsilon \to 0} u_\epsilon(x,y) = \sin \pi y$. Accordingly the values $u_{\epsilon,h}(x, 1/2)$ from the first part of Table 1 approach the limiting value $\sin \pi/2 = 1$.

In the second part of the table we have $\epsilon < h/2$. Thus L_h is not an M-matrix. Notice that $u_{\epsilon,h}(31/32, 1/2) > 1$, i.e., even the maximum principle does not hold. In addition one can see that $u_{\epsilon,h}(x, 1/2)$ converges to $1-x$ for the even multiples $x = \nu h$, and thus not to $u_0(x,y) = \sin \pi y$. For the intervening odd multiples $x = \nu h$ an oscillation of amplitude $O(h^2/2\epsilon)$ develops.

The resulting difficulty is similar to that for initial value problems for stiff (ordinary) differential equations: If one keeps ϵ constant and lets h go to zero, the convergence assertions of Section 5.1.4 hold. However, if ϵ is very small, the condition $h < 2\epsilon$, without which one does not obtain a reasonable solution, cannot in practice be satisfied.

One way out of this difficulty has already been described in Section 5.1.4. One must approximate the convection term by suitable one-sided differences (5.1.14)

$$L_h = h^{-2} \begin{bmatrix} & -\epsilon + hc_2^- & \\ -\epsilon - hc_1^+ & 4\epsilon + h|c_1| + h|c_2| & -\epsilon + hc_1^- \\ & -\epsilon - hc_2^+ & \end{bmatrix}, \tag{10.2.6}$$

where $c_i^+ := \max\{0, c_i\}$, $c_i^- := \min\{0, c_i\}$.

Remark 10.2.6. If one discretises Equation (1) using (6) then one obtains, for all $\epsilon > 0$ and $h > 0$, an M-matrix L_h. For fixed ϵ, the scheme has consistency order 1.

Exercise 10.2.7. The discretisation of Equation (3b) corresponding to (6) is $L_h = h^{-2}\,[-\epsilon \quad 2\epsilon + h \quad -\epsilon - h]$ and this gives the discrete solution $u_h(x) = [1 - (1 + h/\epsilon)^{-x/h}]/[1 - (1 + h/\epsilon)^{-1/h}]$.

If one applies the difference operator (6) to a smooth function one obtains the Taylor series

$$L_h u = Lu - \{\,|c_1|u_{xx} + |c_2|u_{yy}\}h + O(h^2). \tag{10.2.7}$$

The $O(h)$ term $h|c_1|u_{xx} + h|c_2|u_{yy}$ is called the numerical viscosity (or the numerical ellipticity) since it amplifies the principal part.

A second remedy consists in replacing the parameter ϵ by a discretisation using $\epsilon_h \geq \epsilon$. If one chooses

$$\epsilon_h := \max\{\epsilon, h|c_1|/2, h|c_2|/2\} \quad \text{or} \quad \epsilon_h = \epsilon + \frac{h}{2}\max|c_i|, \tag{10.2.8}$$

then the symmetric difference method

$$L_h = \epsilon_h h^{-2}\begin{bmatrix} & -1 & \\ -1 & 4 & -1 \\ & -1 & \end{bmatrix} + \frac{1}{2}h^{-1}\begin{bmatrix} & c_2 & \\ -c_1 & 0 & c_1 \\ & -c_2 & \end{bmatrix} \tag{10.2.9}$$

leads to an M-matrix. It is true that the convection term has been discretised to a second order of consistency, but the error of the diffusion term is $O(\epsilon_h - \epsilon)$, which, for the case $h > 2\epsilon/||\mathbf{c}||_\infty$ which is relevant in practice, amounts to $O(h)$. Instead of (7) one has

$$L_h u = Lu - (\epsilon_h - \epsilon)\Delta u + O(h^2). \tag{10.2.10}$$

The difference $-(\epsilon_h - \epsilon)\Delta u$ is called the artificial viscosity.

In the one-dimensional case the methods using numerical and artificial viscosities do not differ:

Remark 10.2.8. If, in the one-dimensional case, on chooses ϵ_h according to the second alternative in (8), then the difference formulae (6) and (9) coincide.

10.2.3 Finite Elements

The difficulties described in the previous section are not restricted to difference methods.

Exercise 10.2.9. Show that

(a) Linear finite elements on a square grid triangulation applied in the case of Equation (4) give a discretisation that is identical with the difference method

$$L_h = \epsilon \begin{bmatrix} & -1 & \\ -1 & 4 & -1 \\ & -1 & \end{bmatrix} + h \begin{bmatrix} 0 & -1/6 & 1/6 \\ -1/3 & 0 & 1/3 \\ -1/6 & 1/6 & 0 \end{bmatrix} \tag{10.2.11a}$$

(cf. Exercise 8.3.13).
(b) For bilinear elements (cf. Exercise 8.3.17) one obtains

$$L_h = \frac{\epsilon}{3} \begin{bmatrix} -1 & -1 & -1 \\ -1 & 8 & -1 \\ -1 & -1 & -1 \end{bmatrix} + \frac{h}{12} \begin{bmatrix} -1 & 0 & +1 \\ -4 & 0 & +4 \\ -1 & 0 & +1 \end{bmatrix}. \tag{10.2.11b}$$

(c) One-dimensional linear elements for $-\epsilon u'' + u' = f$ lead to the central difference formula

$$L_h = \epsilon h^{-1} [-1 \quad 2 \quad -1] + [-1 \quad 0 \quad 1]. \tag{10.2.11c}$$

The Exercises 9b,c show that finite-element methods correspond to central difference formulas, and thus can equally well lead to instability.

The method of artificial viscosity corresponds to the finite-element solution of the equation $-\epsilon_h \Delta u + \langle \mathbf{c}, \operatorname{grad} u\rangle = f$ for appropriate ϵ_h. As in Exercise 9b one may show

Remark 10.2.10. If one sets $\epsilon_h := \max\{\epsilon, |c_1|h, |c_2|h\}$ and uses bilinear elements then the discretisation of Equation (1) leads to an M-matrix.

On the other hand the matrix (11a) has different signs in the sub-diagonal and in the super-diagonal, so that it is not possible to have an M-matrix for any value of ϵ_h.

The analogues of one-sided differences are more difficult to construct. One approach is to combine a finite-element method for the diffusion term with a (one-sided) difference method for the convection term (cf. Thomasset [1, §2.4]).

A second possibility is the generalisation of the Galerkin method to the Petrov-Galerkin method, in which the discrete solution of the general equation (8.1.1) is defined by problem (12):

$$\text{Find } u \in V_h, \quad \text{so that } a(u,v) = f(v) \quad \text{for all } v \in W_h \tag{10.2.12}$$

(cf. Fletcher [1, §7.2], Thomasset[1, §2.2]). Here we have

$$\dim V_h = \dim W_h \quad (\text{but in general } V_h \neq W_h).$$

11 Eigenvalue Problems

11.1 Formulation of Eigenvalue Problems

The classical formulation of an eigenvalue problem reads

$$Le = \lambda e \quad \text{in } \Omega, \quad B_j e = 0 \quad \text{on } \Gamma \quad (j = 1, \cdots, m). \tag{11.1.1}$$

Here L is an elliptic differential operator of order $2m$, and B_j are boundary operators. A solution e of (1) is called an eigenfunction if $e \neq 0$. In this case, λ is the eigenvalue associated with e.

As in Section 7, one can replace the classical representation (1) by a variational formulation, with a suitable bilinear form $a(\cdot,\cdot): V \times V \to \mathbb{R}$ taking the place of $\{L, B_j\}$:

$$\text{Find } e \in V \quad \text{with } a(e,v) = \lambda(e,v)_0 \quad \text{for all } v \in V. \tag{11.1.2a}$$

$(u,v)_0 = \int_\Omega u\bar{v}\,\mathbf{dx}$ is the $L^2(\Omega)$-scalar product. Strictly speaking one ought to replace $(\cdot,\cdot)_0$ by $(\cdot,\cdot)_U$ where U is the Hilbert space of the Gelfand triple $V \subset U \subset V'$ (cf. Section 6.3.3). But here we limit ourselves to the standard case $U = L^2(\Omega)$.

The adjoint eigenvalue problem is formulated as

$$\text{Find } e \in V \quad \text{with } a(v,e) = \overline{\lambda}(v,e)_0 \quad \text{for all } v \in V. \tag{11.1.2b}$$

Definition 11.1.1. Let $\lambda \in \mathbb{C}$. By $E(\lambda)$ one denotes the subspace of all $e \in V$ which satisfy Equation (1)[resp. (2a)]. $E(\lambda)$ is called the eigenspace for λ. With $E^*(\lambda)$ one denotes the corresponding eigenspace of Equation (2b). λ is called an eigenvalue if $\dim E(\lambda) \geq 1$.

Theorems 6.5.15 and 7.2.14 already contain the following statements:

Theorem 11.1.2. *Let $V \subset L^2(\Omega)$ be continuously, densely and compactly embedded (for example, let $V = H_0^m(\Omega)$ with bounded Ω). Let $a(\cdot,\cdot)$ be V-coercive. Then the problems (2a,b) have countably many eigenvalues $\lambda \in \mathbb{C}$ which may only have an accumulation point at ∞. For all $\lambda \in \mathbb{C}$ we have*

$$\dim E(\lambda) = \dim E^*(\lambda) < \infty. \tag{11.1.3}$$

Exercise 11.1.3. In addition to the assumptions of Theorem 2 let $a(\cdot,\cdot)$ be symmetric. Show that all eigenvalues are real and problems (2a) and (2b) are identical so that $E(\lambda) = E^*(\lambda)$.

For ordinary differential equations of second order (i.e., $\Omega \subset \mathbb{R}^1, m = 1$) it is known that all eigenvalues are simple: $\dim E(\lambda) = 1$. This statement is incorrect for partial differential equations as the following example shows.

Example 11.1.4. The Poisson equation $-\Delta e = \lambda e$ in the rectangle $(0,a) \times (0,b)$, with Dirichlet boundary values $e = 0$ on Γ, has the eigenvalues $\lambda = (\nu\pi/a)^2 + (\mu\pi/b)^2$ with $\nu, \mu \in \mathbb{N}$. The associated eigenfunction reads $e(x,y) = e^{\nu,\mu}(x,y) := \sin(\nu\pi x/a)\sin(\mu\pi y/b)$. In the case of the square $a = b$ one obtains for $\nu \neq \mu$ eigenvalues $\lambda = \lambda^{\nu,\mu} = \lambda^{\mu,\nu}$, which have multiplicity at least 2 since $e^{\nu,\mu}$ and $e^{\mu,\nu}$ are linearly independent eigenfunctions for the same eigenvalue. A triple eigenvalue, for example, exists for $a = b$, $\lambda = 50\pi^2/a^2$: $E(\lambda) = \operatorname{span}\{e^{1,7}, e^{7,1}, e^{5,5}\}$.

The eigenfunctions $e \in E(\lambda)$, by definition, belong to V. The regularity investigations of Section 9.1 immediately result in a stronger regularity.

Theorem 11.1.5. *Let $V = H_0^m(\Omega)$ with $m \geq 1$, or $V = H^m(\Omega)$ with $m = 1$. Under the assumptions of Theorems* 9.1.16 [*resp.* 9.1.17] *we have* $E(\lambda) \subset H^{m+s}(\Omega)$.

PROOF. Along with $a(\cdot,\cdot)$, $a_\lambda(u,v) := a(u,v) - \lambda(u,v)_0$ also satisfies the assumptions. Since $a_\lambda(e,v) = 0$ for $e \in E(\lambda)$, $v \in V$, the statement follows from Corollary 9.1.19. ■

Besides the standard form (2a) there are generalised eigenvalue problems. An example is the Steklov problem

$$-\Delta e = 0 \quad \text{in } \Omega, \qquad \partial e/\partial n = \lambda e \quad \text{on } \Gamma,$$

whose variational formulation reads $e \in H^1(\Omega)$, $\int_\Omega \langle \nabla e, \nabla v\rangle\, d\mathbf{x} = \lambda \int_\Gamma ev\, d\Gamma$ $(v \in H^1(\Omega))$. One can show that all eigenvalues are real and that the statements of Theorem 2 hold.

11.2 Finite Element Discretisation

11.2.1 Discretisation

Let $V_h \subset V$ be a (finite-element) subspace. The Ritz-Galerkin (resp. finite-element) discretisations of the eigenvalue problems (1.2a,b) read

$$e^h \in V_h, \quad a(e^h, v) = \lambda_h(e^h, v)_0 \quad \text{for all } v \in V_h, \tag{11.2.1a}$$

$$e^{*h} \in V_h, \quad a(v, e^{*h}) = \overline{\lambda_h}(v, e^{*h})_0 \quad \text{for all } v \in V_h. \tag{11.2.1b}$$

The discrete eigenspaces $E_h(\lambda_h)$, $E_h^*(\lambda_h)$ are spanned by the solutions of the problems (1a) [resp. (1b)]. As in Theorem 11.1.2, $\dim E_h(\lambda_h) = \dim E_h^*(\lambda_h)$ holds. If $a(\cdot,\cdot)$ is symmetric, then $E_h(\lambda_h) = E_h^*(\lambda_h)$.

As in Section 8.1, the formulation (1a,b) can be transcribed into matrix notation.

Remark 11.2.1. Let $\mathbf{e}$ and $\mathbf{e}^*$ be the coefficient vectors for $e^h = \mathbf{Pe}$ and $e^{*h} = \mathbf{Pe}^*$ (cf. (8.1.6)). The eigenvalue problems (1a,b) are equivalent to

$$\mathbf{Le} = \lambda_h \mathbf{Me}, \quad \mathbf{L}^\mathsf{T}\mathbf{e}^* = \overline{\lambda_h}\mathbf{Me}^*, \tag{11.2.1'}$$

where the stiffness matrix $\mathbf{L}$ is defined as in Theorem 8.1.3 and $\mathbf{M}$ by (8.8.7).

Since in general $\mathbf{M} \neq \mathbf{I}$, (1′) involves generalised eigenvalue problems.

Exercise 11.2.2. Show that (a) $\mathbf{M}$ is positive definite and possesses a decomposition $\mathbf{M} = \mathbf{A}^\mathsf{T}\mathbf{A}$ (for example, $\mathbf{A} = \mathbf{M}^{1/2}$ or the Cholesky factor).
(b) The first problem in (1′) is equivalent to the ordinary eigenvalue problem $\tilde{\mathbf{L}}\tilde{\mathbf{e}} = \lambda_h\tilde{\mathbf{e}}$ with $\tilde{\mathbf{L}} := (\mathbf{A}^\mathsf{T})^{-1}\mathbf{L}\mathbf{A}^{-1}$, $\tilde{\mathbf{e}} = \mathbf{Ae}$. The second problem in (1′) corresponds to $\tilde{\mathbf{L}}^\mathsf{T}\tilde{\mathbf{e}}^* = \overline{\lambda_h}\tilde{\mathbf{e}}^*$ with $\tilde{\mathbf{e}}^* = \mathbf{Ae}^*$.

When investigating convergence, one must watch out for the following difficulties:
(i) A uniform approximation of all the eigenvalues and eigenfunctions by discrete eigenvalues and eigenfunctions is impossible since the infinitely many eigenvalues of (1.2a) are set against the only finitely many of (1a). It is only possible to characterise a fixed eigenvalue λ of (1.2a) as an accumulation point of discrete eigenvalues $\{\lambda_h : h < 0\}$, and to set up estimates for $|\lambda - \lambda_h|$.
(ii) If λ and λ_h are the continuous, resp. discrete, eigenvalues then $\dim E(\lambda) = \dim E_h(\lambda_h)$ need not hold. It is preferable to limit oneself to the case of simple eigenvalues where $\dim E(\lambda) = \dim E_h(\lambda_h) = 1$. If $\dim E(\lambda) = k > 1$, it may well be that the multiple eigenvalue λ is approximated by several discrete eigenvalues $\lambda_h^{(i)}$, $i = 1, \cdots, k$: $\dim E(\lambda) = \sum_{i=1}^k \dim E_h(\lambda_h^{(i)})$. The error estimates of $|\lambda - \lambda_h^{(i)}|$ are then generally worse than for simple eigenvalues. Only for the mean value $\hat{\lambda} := \sum_{i=1}^k \lambda_h^{(i)}/k$ does one obtain the usual estimates (cf. Babuška-Aziz [1, p. 338]).

Exercise 11.2.3. Let the eigenvalue problem be as in Example 11.1.4 with $a = b$ and $\lambda = 50\pi^2/a^2$. Let V_h consist of linear elements over a square grid triangulation. Show that to the given triple eigenvalue λ corresponds a double eigenvalue $\lambda_h^{(1)}$ and a single eigenvalue distinct from it, $\lambda_h^{(2)}$, with $\lim_{h\to 0}\lambda_h^{(i)} = \lambda$ $(i = 1, 2)$. Hint: the nodal values of the discrete eigenfunctions agree with the continuous eigenfunctions $e^{1,7}$, $e^{7,1}$, and $e^{5,5}$.

11.2.2 Qualitative Convergence Results

This section concerns the question as to whether $\lambda_h \to \lambda$ and $e^h \to e$ for $h \to 0$. The rate of convergence will be discussed in Section 11.2.3. The basic assumptions are the following:

$$\text{Let } a(\cdot,\cdot)\colon\, V \times V \to \mathbb{R} \text{ be } V\text{-coercive,} \tag{11.2.2a}$$

$$\text{Let } V \subset L^2(\Omega) \text{ be continuously, densely, and compactly embedded,} \tag{11.2.2b}$$

so that the Riesz-Schauder theory is applicable (Theorem 11.1.2). Furthermore, let a sequence of subspaces V_{h_i} ($h_i \to 0$) be given which increasingly approximate V (cf. (8.2.4a)):

$$\lim_{h\to 0} \inf\{\|u - u^h\|_V \colon u^h \in V_h\} = 0 \quad \text{for all } u \in V. \tag{11.2.2c}$$

We define

$$a_\lambda(\cdot,\cdot)\colon V \times V \to \mathbb{C}, \quad a_\lambda(u,v) := a(u,v) - \lambda(u,v)_0, \tag{11.2.3a}$$

$$\omega(\lambda) := \inf_{\substack{u\in V\\ \|u\|_V=1}} \ \sup_{\substack{v\in V\\ \|v\|_V=1}} |a_\lambda(u,v)|, \tag{11.2.3b}$$

$$\omega_h(\lambda) := \inf_{\substack{u\in V_h\\ \|u\|_V=1}} \ \sup_{\substack{v\in V_h\\ \|v\|_V=1}} |a_\lambda(u,v)|. \tag{11.2.3c}$$

Exercise 11.2.4. Let L and L_h be the operators associated with $a(\cdot,\cdot)\colon V \times V \to \mathbb{R}$ and $a(\cdot,\cdot)\colon V_h \times V_h \to \mathbb{R}$ (cf. (8.1.10a)). Show that
(i) If λ is not an eigenvalue we have

$$\omega(\lambda) = \|(L - \lambda I)^{-1}\|^{-1}_{V\leftarrow V'}, \quad \omega_h(\lambda) = \|(L_h - \lambda I)^{-1}\|^{-1}_{V_h\leftarrow V_h'} \tag{11.2.4}$$

(cf. Lemma 6.5.3 and Exercise 8.1.16).
(ii) $\omega(\lambda)$ and $\omega_h(\lambda)$ are continuous in $\lambda \in \mathbb{C}$.
(iii) If in (3b,c) one replaces $a_\lambda(u,v)$ by $a_\lambda(v,u)$ one obtains the variables $\omega^*(\lambda)$ and $\omega_h^*(\lambda)$ which correspond to the adjoint problem. The following holds: $\omega^*(\lambda) = \omega(\lambda)$ and $\omega_h^*(\lambda) = \omega_h(\lambda)$. Hint: Use Lemma 6.5.17.

With the aid of (4) and Theorem 6.5.15 one proves the following connection between $\omega(\lambda)$, $\omega_h(\lambda)$, and the eigenvalue problems.

Remark 11.2.5. Let (2a,b) hold. λ is an eigenvalue of (1.2a) if and only if $\omega(\lambda) = 0$, and λ_h is eigenvalue of (1a) if and only if $\omega_h(\lambda_h) = 0$.

Lemma 11.2.6. *Let $a(\cdot,\cdot)$ be V-coercive (cf. (2a)). Then there exists a $\mu \in \mathbb{R}$ such that $a_\mu(\cdot,\cdot)$ is V-elliptic. In addition, then $\omega(\mu) \ge 1/C_E$ and $\omega_h(\mu) \ge 1/C_E$ hold with $C_E > 0$ from Definition* 6.5.13.

Lemma 11.2.7. *Let $\Lambda \subset \mathbb{C}$ be compact. Let* (2a–c) *hold. Then there exist numbers $C > 0$ and $\eta(h) > 0$, independent of $\lambda \in \Lambda$ with $\lim_{h\to 0}\eta(h) = 0$ such that*

$$\omega_h(\lambda) \geq C\omega(\lambda) - \eta(h), \qquad \omega(\lambda) \geq C\omega_h(\lambda) - \eta(h) \quad \textit{for all } \lambda \in \Lambda. \tag{11.2.5}$$

PROOF. Let the operators $Z = Z(\lambda): V \to V$ and $Z_h = Z_h(\lambda): V \to V_h$ be defined as follows:

$$z := Z(\lambda)u \quad \text{is the solution of} \quad a_\mu(z,v) = (\lambda - \mu)(u,v)_0 \text{ for all } v \in V, \tag{11.2.6a}$$

$$z^h := Z_h(\lambda)u \quad \text{is the solution of} \quad a_\mu(z^h,v) = (\lambda - \mu)(u,v)_0 \text{ for all } v \in V_h, \tag{11.2.6b}$$

where μ is chosen according to Lemma 6. It shows

$$\|Z\|_{V\leftarrow V'} \leq C_Z, \quad \|Z_h\|_{V\leftarrow V'} \leq C_Z \quad \text{for all } \lambda \in \Lambda. \tag{11.2.6c}$$

From $a_\lambda(u,v) = a_\mu(u-z,v)$ and the definition of $\omega(\lambda)$ one infers

$$\omega(\lambda)\|u\|_V \leq \sup_{\substack{v\in V\\ \|v\|_V=1}} |a_\lambda(u,v)| = \sup_{\substack{v\in V\\ \|v\|_V=1}} |a_\mu(u-z,v)| \leq C_S\|u-z\|_V \tag{11.2.6d}$$

with $C_S := \|L - \mu I\|_{V'\leftarrow V}$. For an arbitrary $u \in V_h$ one infers from $a_\lambda(u,v) = a_\mu(u - z^h, v)$, Lemma 6, and (6d) that

$$\sup_{\substack{v\in V_h\\ \|v\|_V=1}} |a_\lambda(u,v)| = \sup_{\substack{v\in V_h\\ \|v\|_V=1}} |a_\mu(u-z^h,v)| \geq C_E^{-1}\|u - z^h\|_V$$
$$\geq C_E^{-1}[\|u-z\|_V - \|z - z^h\|_V] \geq C_E^{-1}[C_S^{-1}\omega(\lambda) - \|Z - Z_h\|_{V\leftarrow V}]\,\|u\|_V;$$

thus $\omega_h(\lambda) \geq (C_E C_S)^{-1}\omega(\lambda) - \|Z - Z_h\|_{V\leftarrow V}/C_E$. From this follows the first part of (5) with $C = (C_E C_S)^{-1} > 0$ and $\eta(h) = \|Z - Z_h\|_{V\leftarrow V}/C_E$, if

$$\lim_{h\to 0}\sup_{\lambda\in\Lambda} \|\, Z(\lambda) - Z_h(\lambda)\|_{V\leftarrow V} = 0. \tag{11.2.6e}$$

The proof of (6e) is carried out indirectly. Its negation reads: there exist $\epsilon > 0$, $\lambda_i \in \Lambda$, $h_i \to 0$ with $\|\, Z(\lambda_i) - Z_{h_i}(\lambda_i)\|_{V\leftarrow V} \geq \epsilon > 0$. Then there exist $u_i \in V$ with

$$\|u_i\|_V = 1, \quad \|[Z(\lambda_i) - Z_{h_i}(\lambda_i)]u_i\|_V \geq \epsilon/2 > 0. \tag{11.2.6f}$$

Due to (2b) and the compactness of Λ there exists a subsequence $\lambda_j \in \Lambda$, $u_j \in V$ with $\lim \lambda_j = \lambda^*$, $\lim u_j = u^* \in V'$ in V'. (2c) and Theorem 8.2.2 show $\|[Z(\lambda^*) - Z_{h_j}(\lambda^*)u^*\|_V \to 0$. Together with (6c) we obtain

$$\begin{aligned}
&||[Z(\lambda_j) - Z_{h_j}(\lambda_j)]u_j||_V \\
&\le ||[Z(\lambda_j) - Z(\lambda^*)]u_j||_V + ||[Z_{h_j}(\lambda_j) - Z_{h_j}(\lambda^*)]u_j||_V \\
&\quad + ||Z(\lambda^*)[u_j - u^*]||_V + ||Z_{h_j}(\lambda^*)[u_j - u^*]||_V + ||[Z(\lambda^*) - Z_{h_j}(\lambda^*)]u^*||_V \\
&\le 2C|\lambda_j - \lambda^*| + 2C_Z||u_j - u^*||_{V'} + ||[Z(\lambda^*) - Z_{h_j}(\lambda^*)]u^*||_V \to 0
\end{aligned}$$

in contradiction to (6f).

For the proof of the second part of (5) one replaces (6d) and the following estimate by

$$\omega_h(\lambda)||u^h||_V \le \sup_{\substack{v\in V_h \\ ||v||_V=1}} |a_\mu(u^h - z^h, v)| \le C_S||u^h - z^h||_V \quad \text{for all } u^h \in V_h,$$

$$\begin{aligned}
\sup_{\substack{v\in V \\ ||v||_V=1}} |a_\lambda(u^h, v)| &= \sup_{\substack{v\in V \\ ||v||_V=1}} |a_\mu(u^h - z, v)| \ge C_E^{-1}||u^h - z||_V \\
&\ge C_E^{-1}[C_S^{-1}\omega_h(\lambda) - ||Z(\lambda) - Z_h(\lambda)||_{V\leftarrow V}]||u^h||_V \quad \text{for all } u^h \in V_h.
\end{aligned}$$

Let $u \in V$ with $||u||_V = 1$ be selected such that

$$\sup_{\substack{v\in V \\ ||v||_V=1}} |a_\lambda(u, v)| = \inf_{\substack{u\in V \\ ||u||_V=1}} \sup_{\substack{v\in V \\ ||v||_V=1}} |a_\lambda(u, v)| = \omega(\lambda).$$

Since $\sup|a_\lambda(u - u^h, v)| \le C_S||u - u^h||_V$, it follows for arbitrary $u^h \in V_h$ that

$$\omega(\lambda) \ge C\omega_h(\lambda) - ||Z(\lambda) - Z_h(\lambda)||_{V\leftarrow V} - C_S||u - u^h||_V.$$

From (2c) and (6e) follows the second part of (5). ■

A corollary to Lemma 7 is Theorem 8.2.8. If Problem (8.1.1) for all $f \in V'$ is solvable, then $\lambda = 0$ cannot be an eigenvalue, i.e., $\omega(0) > 0$. Thus it follows that $\epsilon_{N_i} \ge \omega_{N_i}(0) \ge \epsilon := \frac{1}{2}C\omega(\lambda) > 0$ for sufficiently large i.

A second corollary concerns the convergence of the eigenvalues.

Theorem 11.2.8. *Let* (2a–c) *hold. If* λ_{h_i} $(i \to \infty, h_i \to 0)$ *are discrete eigenvalues of* (1a) *with* $\lambda_{h_i} \to \lambda_0$ *then* λ_0 *is an eigenvalue of* (1.2a).

PROOF. If λ_0 were not an eigenvalue then $\omega(\lambda) \ge \eta_0 > 0$ would be in the ϵ-neighbourhood of $K_\epsilon(\lambda_0)$, since $\omega(\lambda)$ is continuous (cf. Exercise 4(ii)). There would exist $h_0 > 0$ such that $\eta(h) \le C\eta_0/2$ for all $h \le h_0$ (C and $\eta(h)$ from (5)). For all $\lambda_{h_i} \in K_\epsilon(\lambda_0)$ with $h_i \le h_0$ the contradiction follows from (5):

$$0 = \omega_{h_i}(\lambda_{h_i}) \ge C\omega(\lambda_{h_i} - \eta(h_i) \ge C\eta_0 - \frac{1}{2}C\eta_0 = \frac{1}{2}C\eta_0 > 0. \quad ■$$

Lemma 11.2.9. (*Minimum Principle*) *Let* (2a,b) *hold. The functions* $\omega(\lambda)$ *and* $\omega_h(\lambda)$ *in the interior of* $\Lambda \subset \mathbb{C}$ *have no proper positive minimum.*

PROOF. Let L be the operator associated with $a(\cdot,\cdot)$. Let λ^*, with $\omega(\lambda^*) > 0$, be an arbitrary point in the interior of Λ. For sufficiently small $\epsilon > 0$ we have $K_\epsilon(\lambda^*) \subset \Lambda$ and $\omega(\lambda) > 0$ in $K_\epsilon(\lambda^*)$. Thus $(L-\lambda I)^{-1}$ is defined in $K_\epsilon(\lambda^*)$ and holomorphic. Cauchy's integral formula says

$$(L-\lambda I)^{-1} = \frac{1}{2\pi i}\oint_{\partial K_\epsilon(\lambda^*)} (\zeta-\lambda)^{-1}(L-\zeta I)^{-1} d\zeta \quad \text{for all } \lambda \in K_\epsilon(\lambda^*).$$

From this one infers

$$\frac{1}{\omega(\lambda)} = \|(L-\lambda I)^{-1}\|_{V\leftarrow V'} \le \max_{\zeta\in\partial K_\epsilon(\lambda^*)} \|(L-\zeta I)^{-1}\|_{V\leftarrow V'} = \max_{\zeta\in\partial K_\epsilon(\lambda^*)} 1/\omega(\zeta),$$

i.e., $\omega(\lambda) \ge \min\{\omega(\zeta): \zeta \in \partial K_\epsilon(\lambda^*)\}$ (cf. Exercise 4(i)). Thus, $\omega(\lambda)$ cannot assume a proper minimum in $K_\epsilon(\lambda^*)$. For $\omega_h(\lambda)$ the conclusion is the same. ∎

The converse of Theorem 8 is contained in

Theorem 11.2.10. *Let* (2a–c) *hold. Let* λ_0 *be the eigenvalue of* (1.2a). *Then there exist discrete eigenvalues* λ_h *of* (1a) *(for all* h*) such that* $\lim_{h\to 0}\lambda_h = \lambda_0$.

PROOF. Let $\epsilon > 0$ be arbitrary. According to Theorem 11.1.2, λ_0 is an isolated eigenvalue: $\omega(\lambda) > 0$ for $0 < |\lambda - \lambda_0| \le \epsilon$ (ϵ sufficiently small).

Since $\omega(\lambda)$ is continuous and $\partial K_\epsilon(\lambda_0)$ is compact, we have that $\rho_\epsilon := \min\{\omega(\lambda): |\lambda-\lambda_0| = \epsilon\}$ is positive. Because of (5) and $\omega(\lambda_0) = 0$ one obtains for sufficiently small h

$$\omega_h(\lambda) \ge C\omega(\lambda) - \eta(h) \ge C\rho_\epsilon - \eta(h) > \eta(h)/C \ge \omega_h(\lambda_0)$$

for all $\lambda \in \partial K_\epsilon(\lambda_0)$. Thus $\omega_h(\lambda)$ must have a proper minimum in $K_\epsilon(\lambda_0)$.

By Lemma 9 the minimal value is zero. Thus there exists a $\lambda_h \in K_\epsilon(\lambda_0)$ which is a discrete eigenvalue, $\omega_h(\lambda_h) = 0$. ∎

The convergence of the eigenfunctions is obtained from

Theorem 11.2.11. *Let* (2a–c) *hold. Let* $e^h \in E_h(\lambda_h)$ *be discrete eigenfunctions with* $\|e^h\|_V = 1$ *and* $\lim_{h\to 0}\lambda_h = \lambda_0$. *Then there exists a subsequence* e^{h_i} *which converges in* V *to an eigenfunction* $e \in E(\lambda_0)$:

$$e \in E(\lambda_0), \quad \|e^{h_i} - e\|_V \to 0 \quad (i\to\infty), \quad \|e\|_V = 1.$$

PROOF. The functions e^h are uniformly bounded in V. Since $V \subset L^2(\Omega)$ is compactly embedded (cf. (2b)), there exists a subsequence e^{h_i} which converges in $L^2(\Omega)$ to an $e \in L^2(\Omega)$:

$$\|e - e^{h_i}\|_{L^2(\Omega)} \to 0 \quad (i\to\infty). \tag{11.2.7a}$$

We define $z = Z(\lambda_0)e$, $z^{h_i} = Z_{h_i}(\lambda_0)e$ according to (6a,b). According to Theorem 8.2.2 there exists an $h_1(\epsilon) > 0$ such that

$$||z - z^{h_i}||_V \leq \epsilon/2 \tag{11.2.7b}$$

for $h_i \leq h_1(\epsilon)$. The function e^{h_i} is a solution of

$$a_\mu(e^{h_i}, v) = (\lambda_{h_i} - \mu)(e^{h_i}, v)_0 \quad \text{for all } v \in V_{h_i}. \tag{11.2.7c}$$

A combination of $e^{h_i} = Z_{h_i}(\lambda_0)e$ (i.e., (6b) for $\lambda = \lambda_0$) and (7c) yields

$$a_\mu(z^{h_i} - e^{h_i}, v) = F_i(v) := (\lambda_0 - \mu)(e - e^{h_i}, v)_0 - (\lambda_{h_i} - \lambda_0)(e^{h_i}, v)_0 \tag{11.2.7d}$$

for all $v \in V_{h_i}$. Since $||F_i||'_V \to 0$ because $\lambda_{h_i} \to \lambda_0$ and (7a), there exists an $h_2(\epsilon) > 0$ such that $||F_i||_{V'} \leq \epsilon/[2C_E]$ (C_E from Lemma 6) and

$$||z^{h_i} - e^{h_i}||_V \leq \epsilon/2 \tag{11.2.7e}$$

for $h_i \leq h_2(\epsilon)$. (7b) and (7e) show that $||z - e^{h_i}||_V \leq \epsilon$ for $h_i \leq \min\{h_1(\epsilon), h_2(\epsilon)\}$; thus $\lim_{i\to\infty} e^{h_i} = z$ in V. Therefore $\lim e^{h_i} = z$ in $L^2(\Omega) \subset V$ also holds. (7a) proves $z = e \in V$ such that $e = z = Z(\lambda_0)e$ becomes $a(e, v) = \lambda_0(e, v)_0$. Therefore $e = \lim e^{h_i}$ is an eigenfunction of (1.2a). In particular, $||e||_V = \lim ||e^{h_i}||_V = 1$. ■

Exercise 11.2.12. Let (2a–c) hold. Let $\lambda_h, e^h, \lambda_0$, and e be as in Theorem 11. Show that
(a) If $\dim E(\lambda_0) = 1$ then also $\lim_{h\to 0} \dim E_h(\lambda_h) = 1$.
(b) Let $\dim E(\lambda_0) = 1$. Then we have $\lim \hat{e}^h = e$ in V for $\hat{e}^h := e^h/(e^h, e)_V$ if $|(e^h, e)_V| \geq 1/2$ and $\hat{e}^h := e^h$ otherwise.

11.2.3 Quantitative Convergence Results

The geometric and algebraic multiplicities of an eigenvalue λ_0 of (1.2a) agree if

$$\dim \mathrm{Ker}(L - \lambda_0 I) = \dim \mathrm{Ker}(L - \lambda_0 I)^2. \tag{11.2.8}$$

Lemma 11.2.13. *Let* (2a,b) *and* $\dim E(\lambda_0) = 1$ *hold. Then* (8) *is equivalent to* $(e, e^*)_0 \neq 0$ *for* $0 \neq e \in E(\lambda_0)$, $0 \neq e^* \in E^*(\lambda_0)$.

PROOF. We have $\mathrm{Ker}(L - \lambda_0 I) = E(\lambda_0) = \mathrm{span}\,\{e\}$. $\dim \mathrm{Ker}(L - \lambda_0 I)^2 > 1$ holds if and only if there exists a solution $v \in V$ for $(L - \lambda_0 I)v = e$. According to Theorem 6.5.15c this equation has a solution if and only if $(e, e^*)_0 = 0$. ■

Let $E(\lambda_0) = \mathrm{span}\,\{e\}$, $E^*(\lambda_0) = \mathrm{span}\,\{e^*\}$. Under the assumption (8) e and e^* can be normalised so that

$$(e, e^*)_0 = 1. \tag{11.2.8'}$$

We define $\hat{V} := \{v \in V: (v, e^*)_0 = 0\}$, $\hat{V}' = \{v' \in V': (v', e^*)_0 = 0\}$. Let $||\cdot||_{\hat{V}'}$ be the dual norm for $||\cdot||_{\hat{V}} = ||\cdot||_V$. For Problem (9):

$$\text{For } f \in \hat{V}', \text{ find } u \in \hat{V} \text{with } a_\lambda(u,v) = (f,v)_0 \quad \text{for all } v \in \hat{V}, \tag{11.2.9}$$

one defines the variable corresponding to (3b)

$$\hat{\omega}(\lambda) := \inf_{\substack{u\in\hat{V} \\ ||u||_V=1}} \sup_{\substack{v\in\hat{V} \\ ||v||_V=1}} |a_\lambda(u,v)|. \tag{11.2.10}$$

Lemma 11.2.14. *Let* (2a,b), (8), *and* $\dim E(\lambda_0) = 1$ *hold. Then there exists an* $\epsilon > 0$ *such that* $\hat{\omega}(\lambda) \geq C > 0$ *for all* $|\lambda - \lambda_0| \leq \epsilon$. *Problem* (9) *has exactly one solution* $u \in \hat{V}$ *with*

$$||u||_V \leq ||f||_{V'}/\hat{\omega}(\lambda), \text{ if } \hat{\omega}(\lambda) > 0.$$

PROOF. Let $\hat{L}: \hat{V} \to \hat{V}'$ be the operator associated with $a(\cdot,\cdot): \hat{V} \times \hat{V} \to \mathbb{C}$. For $0 < |\lambda - \lambda_0| \leq \epsilon$ (ϵ sufficiently small) $(L - \lambda I)u = f$ has a unique solution $u \in V$. From $f \in \hat{V}'$ follows

$$0 = (f, e^*)_0 = ([L - \lambda I]u, e^*)_0 = (u, [L - \lambda I]^* e^*)_0 = (\overline{\lambda}_0 - \overline{\lambda})(u, e^*)_0,$$

i.e., $u \in \hat{V}$. Thus there exists $(\hat{L} - \lambda I)^{-1}: \hat{V}' \to \hat{V}$ as the restriction of $(L - \lambda I)^{-1}$ to $\hat{V}' \subset V'$. For $\lambda = \lambda_0$ Problem (9) has a unique solution according to Theorem 6.5.15c. Then there exists $(\hat{L} - \lambda I)^{-1}$ for all $\lambda \in K_\epsilon(\lambda_0)$. According to Remark 5 (with $\hat{V}$ instead of V) , $\hat{\omega}(\lambda)$ must be positive in $\overline{K_\epsilon(\lambda_0)}$. The continuity of $\hat{\omega}(\lambda)$ proves $\hat{\omega}(\lambda) \geq C > 0$. In analogy to (4) one has $||u||_V = ||u||_{\hat{V}} \leq ||f||_{\hat{V}'}/\hat{\omega}(\lambda)$. The bound by $\leq ||f||_{V'}/\hat{\omega}(\lambda)$ results from

Exercise 11.2.15. Show that $||f||_{\hat{V}'} \leq ||f||_{V'}$ for all $f \in \hat{V}'$.

Lemma 11.2.16. *Let* (2a–c), $\dim E(\lambda_0) = 1$, *and condition* (8) *hold. Let* λ_h *be discrete eigenvalues with* $\lim_{h\to 0} \lambda_h = \lambda_0$. *According to Exercise* 12b *there exists an* $e^h \in E_h(\lambda_h)$ *and* $e^{*h} \in E_h^*(\lambda_h)$ *with* $e^h \to e \in E(\lambda_0)$, $e^{*h} \to e^* \in E^*(\lambda_0)$, $(e, e^*)_0 = 1$. *This enables one to construct the space* $\hat{V}_h := \{v^h \in V_h: (v^h, e^h)_0 = 0\}$ *and the variable*

$$\hat{\omega}_h(\lambda) := \inf_{\substack{u\in\hat{V}_h \\ ||u||_V=1}} \sup_{\substack{v\in\hat{V}_h \\ ||v||_V=1}} |a_\lambda(u,v)|.$$

Then there exists a $C > 0$ *independent of* h *and* $\lambda \in \mathbb{C}$ *such that* $\hat{\omega}_h(\lambda) \geq C\omega_h(\lambda)$. *For sufficiently small* $\epsilon > 0$ *and* h, $\hat{\omega}_h(\lambda) \geq \eta > 0$ *for all* $|\lambda - \lambda_0| \leq \epsilon$.

PROOF. (a) First statement: there exists $h_0 > 0$ and C such that

$$\min\{||v + \alpha e^{*h}||_V: \alpha \in \mathbb{C}\} \geq ||v||_V/C \quad \text{or all } v \in \hat{V}_h \text{ and all } 0 < h \leq h_0.$$

The proof is carried out indirectly. The negation reads: there exists a sequence $\alpha_i \in \mathbb{C}$, $h_i \to 0$, $v_i \in \hat{V}_{h_i}$ with $||v_i||_V = 1$ and $||v_i + \alpha_i e^{*h_i}||_V \to 0$. Thus there exist subsequences with $\alpha_i \to \alpha^*$, $v_i \to v^*$ in $L^2(\Omega)$. Evidently, $w_i := v_i + \alpha_i e^{*h_i}$ must have the limit $w^* = \lim w_i = v^* + \alpha^* e^*$ in $L^2(\Omega)$. Since $\lim ||w_i||_{L^2(\Omega)} \le C \lim ||w_i||_V = 0$, it follows that $w^* = 0$, and thus $v^* = -\alpha^* e^*$. From $0 = \lim(v_i, e^{*h_i})_0 = (v^*, e^*)_0 = -\alpha^* ||e^*||_0^2$ one infers $\alpha^* = 0$. Thus the contradiction follows from $1 = \lim ||v_i||_V \le \overline{\lim} ||w_i||_V + \overline{\lim} ||\alpha_i e^{*h_i}||_V = \lim ||w_i||_V = 0$.

(b) Second statement: $\hat{\omega}_h(\lambda) \ge C\omega_h(\lambda)$ with C from (a). Because $a_\lambda(u,v) = a_\lambda(u, v + \alpha e^{*h})$ for $u \in \hat{V}_h$ we have

$$\begin{aligned}\hat{\omega}_h(\lambda) &= \inf_{0 \ne u \in \hat{V}_h} \sup_{0 \ne v \in \hat{V}_h} |a_\lambda(u,v)|/(||u||_V ||v||_V) \\ &\ge C \inf_{0 \ne u \in \hat{V}_h} \sup_{0 \ne v \in \hat{V}_h} \max_{\alpha \in \mathbb{C}} |a_\lambda(u, v + \alpha e^{*h})|/[\,||u||_V ||v + \alpha e^{*h}||_V] \\ &= C \inf_{0 \ne u \in \hat{V}_h} \sup_{0 \ne w \in V_h} |a_\lambda(u,w)|/[||u||_V ||w||_V] \\ &\ge C \inf_{0 \ne u \in V_h} \sup_{0 \ne w \in V_h} |a_\lambda(u,w)|/[||u||_V ||w||_V] = C\omega_h(\lambda).\end{aligned}$$

(c) Let $\epsilon > 0$ be chosen such that λ_0 is the only eigenvalue in $\overline{K_\epsilon(\lambda_0)}$. For sufficiently small h, λ_h is the only discrete eigenvalue in $\overline{K_\epsilon(\lambda_0)}$. In the proof of Theorem 10 we have already used $\omega_h(\lambda) \ge \eta' > 0$ for $\lambda \in \partial K(\lambda_0)$, $h \le h_0(\epsilon)$. From Part b follows that $\hat{\omega}_h(\lambda) \ge \eta := \eta' C > 0$ for $\lambda \in \partial K_\epsilon(\lambda_0)$. According to Exercise 12a, $a_\lambda(u,v) = (f,v)_0$ $(v \in \hat{V}_h)$ is solvable for each $f \in \hat{V}_h$ and all $\lambda \in K_\epsilon(\lambda_0)$ such that $\hat{\omega}_h(\lambda) = 0$ is excluded. Lemma 9 shows that $\hat{\omega}_h(\lambda) \ge \eta > 0$ in $K_\epsilon(\lambda_0)$. ■

Exercise 11.2.17. Let (2a,b) hold. The functions $u, v \in V$ satisfy $(u,v)_0 \ne 0$. Let $d(u, V_h)$ be defined as in (8.2.2). Show that if $d(u, V_h)$ is sufficiently small then there exists a $u^h \in V_h$ with

$$||u - u^h||_V \le 2d(u, V_h), \quad (u - u^h, v)_0 = 0.$$

Lemma 11.2.18. *Let (2a–c) hold. Let λ_0 be an eigenvalue with (8) and $\dim E(\lambda_0) = 1$. For sufficiently small h there exists $e^h \in E_h(\lambda_h)$ with*

$$||e - e^h||_V \le C[|\lambda_0 - \lambda_h| + d(e, V_h)].$$

PROOF. Let $z^h := Z_h(\lambda_0)e$ be the solution of (6b). Since $e = Z(\lambda_0)e$, one has

$$||e - z^h||_V \le C_1 d(e, V_h). \tag{11.2.11a}$$

For all $v \in V_h$ we have

$$a_\mu(z^h - e^h, v) = (\lambda_0 - \mu)(e, v)_0 - (\lambda_h - \mu)(e^h, v)_0$$
$$= (\lambda_0 - \lambda_h)(e, v)_0 + (\lambda_h - \mu)(z^h - e^h, v)_0 + (\lambda_h - \mu)(e - z^h, v)_0,$$

so that

$$a_{\lambda_h}(z^h - e^h, v) = (\lambda_0 - \lambda_h)(e, v)_0 + (\lambda_h - \mu)(e - z^h, v)_0 \text{ for all } v \in V_h. \tag{11.2.11b}$$

According to Lemma 16 we have $|(e^*, e^{*h})_0| \geq \eta > 0$ for sufficiently small h. e^h can be scaled so that $(z^h - e^h, e^{*h})_0 = 0$.

(11b) corresponds to Problem (9). Lemma 14 proves

$$||z^h - e^h||_V \leq \hat{\omega}_h(\lambda_h)^{-1} C[|\lambda_0 - \lambda_h| + ||e - z^h||_V] \leq C'[\,|\lambda_0 - \lambda_h| + d(e, V_h)].$$

Together with (11a) one obtains the statement. ∎

Theorem 11.2.19. *Let* (2a–c) *hold. Let* λ_0 *be an eigenvalue with* (8) *and* $\dim E(\lambda_0) = 1$. *Let* $e \in E(\lambda_0)$, $e^* \in E^*(\lambda_0)$, $||e||_V = 1$, $(e, e^*)_0 = 1$. *Then there exist discrete eigenvalues* λ_h *with*

$$|\lambda_0 - \lambda_h| \leq C d(e, V_h) d(e^*, V_h). \tag{11.2.12}$$

PROOF. Choose u^h according to Exercise 17 such that $||e^* - u^h||_V \leq 2d(e^*, V_h)$, $(e^* - u^h, e)_0 = 0$. Discrete eigenvalues $\lambda_h \to \lambda_0$ exist by Theorem 10. From

$$\begin{aligned} 0 &= a_{\lambda_0}(e^h, e^*) = a_{\lambda_h}(e^h, e^*) - (\lambda_0 - \lambda_h)(e^h, e^*)_0 \\ &= a_{\lambda_h}(e^h, e^* - u^h) - (\lambda_0 - \lambda_h)(e^h, e^*)_0 \\ &= a_{\lambda_h}(e^h - e, e^* - u^h) - (\lambda_0 - \lambda_h)(e^h, e^*)_0 \\ &= a_{\lambda_h}(e^h - e, e^* - u^h) - (\lambda_0 - \lambda_h)[(e, e^*)_0 + (e^h - e, e^*)_0] \end{aligned}$$

follows $|\lambda_0 - \lambda_h| \leq C'||e^h - e||_V[||e^* - u^h||_V + |\lambda_0 - \lambda_h|]$. By Lemma 18 there exists an $e^h \in E_h(\lambda_h)$ such that

$$|\lambda_0 - \lambda_h| \leq C'C''[|\lambda_0 - \lambda_h| + d(e, V_h)][|\lambda_0 - \lambda_h| + 2d(e^*, V_h)].$$

From this one obtains (12) with $C = 3C'C''$ for sufficiently small h, since $|\lambda_0 - \lambda_h| \to 0$, $d(e, V_h) \to 0$, $d(e^*, V_h) \to 0$. ∎

Theorem 11.2.20. *Under the assumptions of Theorem* 19 *there exist for* $e \in E(\lambda_0)$, $e^* \in E^*(\lambda_0)$ *discrete eigenfunctions* $e^h \in E_h(\lambda_h)$, $e^{*h} \in E_h^*(\lambda_h)$ *with*

$$||e - e^h||_V \leq C d(e, V_h), \quad ||e^* - e^{*h}||_V \leq C d(e^*, V_h). \tag{11.2.13}$$

PROOF. Insert (12) with $d(e^*, V_h) \leq$ const into Lemma 18. The second estimate in (13) follows analogously. ∎

In the following, let $V \subset H^1(\Omega)$. Theorem 11.1.5 proves

$$E(\lambda_0) \subset H^{1+s}(\Omega), \quad E^*(\lambda_0) \subset H^{1+s}(\Omega). \tag{11.2.14a}$$

Also, let (14b) hold (cf. (8.4.10)):

$$d(u, V_h) \le Ch^s \|u\|_{H^{1+s}(\Omega)} \quad \text{for all } u \in E(\lambda_0) \cup E^*(\lambda_0). \tag{11.2.14b}$$

Corollary 11.2.21. *Let* (2a), (2c), (14a,b) *hold. Let* λ_0 *be the eigenvalue with* (8) *and* $\dim E(\lambda_0) = 1$. *Then there exists* $\lambda_h, e^h \in E_h(\lambda_h)$, $e^{*h} \in E_h^*(\lambda_h)$ *such that*

$$|\lambda_0 - \lambda_h| \le Ch^{2s}, \quad \|e - e^h\|_V \le Ch^s, \quad \|e^* - e^{*h}\|_V \le Ch^s. \tag{11.2.15}$$

Occasionally eigenfunctions may have better regularity than is proven for ordinary boundary value problems. For example, let $-\Delta e = \lambda e$ be in the rectangle $\Omega = (0,1) \times (0,1)$ with $e = 0$ on Γ. First, Theorem 11.1.5 implies $e \in H^2(\Omega) \cap H_0^1(\Omega)$, thus $e \in C^0(\overline{\Omega})$ (cf. Theorem 6.2.30). Thus $e = 0$ holds in the corners of Ω. According to Example 9.1.25 it follows that $e \in H^s(\Omega)$ for $s < 4$.

As in Section 8.4.2 one obtains better error estimates for $e - e^h$ in the L^2-norm. The proof is postponed until after Corollary 29.

Theorem 11.2.22. *Let* (2a–c), (8), $\dim E(\lambda_0) = 1$, *and* (14a,b) *with* $s = 1$ *hold. Let* $a(\cdot,\cdot)$ *and* $a^*(\cdot,\cdot)$ *be* H^2*-regular, i.e., for* $f \in L^2(\Omega)$, $a_\mu(u,v) = (f,v)_0$ *and* $a_\mu(v,u^*) = (v,f)_0$ ($v \in V_h$, μ *from Lemma* 6) *have solutions* $u, u^* \in H^2(\Omega)$. *Let* $e \in E(\lambda_0)$ *and* $e^* \in E^*(\lambda_0)$. *Then there exist* $\lambda_h, e^h \in E_h(\lambda_h)$, *and* $e^{*h} \in E_h^*(\lambda_h)$ *with*

$$\|e - e^h\|_{L^2(\Omega)} \le C'h^2, \quad \|e^* - e^{*h}\|_{L^2(\Omega)} \le C'h^2.$$

If (14a,b) *also hold with some* $s > 1$, *one must replace* $C'h^2$ *by* $C'h^{1+s}$.

11.2.4 Complementary Problems

In Problem (9) we have already encountered a singular equation which nevertheless was solvable. In the following let λ_0 be the only eigenvalue in the disc $\overline{K_r(\lambda_0)}$. The equation

$$a_\lambda(u,v) = (f,v)_0 \quad \text{for all } v \in V \tag{11.2.16a}$$

is singular for $\lambda = \lambda_0$. For $\lambda \approx \lambda_0$ Equation (16a) is ill-conditioned. In the following we are going to show that Equation (16a) is well-defined and well-conditioned if the right-hand side f lies in the orthogonal complement of $E^*(\lambda_0)$:

$$f \perp E^*(\lambda_0) \quad \text{,i.e., } (f, e^*)_0 = 0 \text{ for all } e^* \in E^*(\lambda_0)). \tag{11.2.16b}$$

In the case of $\lambda = \lambda_0$, with u, $u + e$ $(e \in E(\lambda_0))$ is also the solution. The uniqueness of the solution is obtained under the conditions (8) and (16c):

$$u \perp E^*(\lambda_0). \tag{11.2.16c}$$

Remark 11.2.23. Let (2a,b) and (8) hold. Let λ_0 be the only eigenvalue in $K_r(\lambda_0)$. Then (16a,b) has exactly one solution u for all $|\lambda - \lambda_0| \le r$ which satisfies (16c). There exists a C independent of f and λ such that $||u||_V \le C||f||_{V'}$.

PROOF. This follows from Lemma 14 in which the assumption $E(\lambda_0) = 1$ is not necessary. ∎

The finite element discretisation of Equation (16a) reads:

$$\text{Find } u^h \in V_h \quad \text{with } a_\lambda(u^h, v) = (f, v)_0 \quad \text{for all } v \in V_h. \tag{11.2.17}$$

In general, Equation (17) need not be well-defined, even assuming (16b).

For the sake of simplicity we limit ourselves in the following to simple eigenvalues: $\dim E(\lambda_0) = 1$. Equation (17) is replaced by (18a):

$$\text{Find } u^h \in V_h \quad \text{with } a_\lambda(u^h, v) = (f^{(h)}, v)_0 \quad \text{for all } v \in V_h \tag{11.2.18a}$$

$$\text{with } f^{(h)} \perp E_h^*(\lambda_h), \tag{11.2.18b}$$

$$u^h \perp E_h^*(\lambda_h). \tag{11.2.18c}$$

Exercise 11.2.24. Show that (18a–c) is equivalent to: Find $u^h \in \hat{V}_h$ with $a_\lambda(u^h, v) = (f^{(h)}, v)_0$ for all $v \in \hat{V}_h$ with $\hat{V}_h = V_h \cap E_h^*(\lambda_h)^\perp$ as in Lemma 16.

Lemma 16 proves the

Remark 11.2.25. Let (2a–c), (8), $\dim E(\lambda_0) = 1$ hold. Let λ_0 be the only eigenvalue in $K_r(\lambda_0)$. Then there exists an $h_0 > 0$ such that, for all $h \le h_0$ and all $\lambda \in K_r(\lambda_0)$, the Problem (18a,b) has a unique solution $u^h = u^h(\lambda)$ which satisfies the additional conditions (18c). Further there exists a C independent of h, λ, and $f^{(h)}$ such that $||u^h||_V \le C||f^{(h)}||_V$.

If $E_h^*(\lambda_h) \ne E^*(\lambda_0)$, f from (16b) need not satisfy condition (18b). If $e^h \in E_h(\lambda_h)$ and $e^{*h} \in E_h^*(\lambda_h)$ with $(e^h, e^{*h})_0 = 1$ are known, one can define

$$f^{(h)} := Q_h f := f - (f, e^{*h})_0 e^h. \tag{11.2.19}$$

$f^{(h)}$ satisfies (18b) since Q_h represents the projection on $E_h^*(\lambda_h)^\perp$.

Exercise 11.2.26. Let $u \perp E(\lambda_0)$, $\dim E(\lambda_0) = \dim E_h(\lambda_h) = 1$, $\|e_h^*\|_V = 1$, $(e^h, e^{*h})_0 = 1$. Show that

$$d(u, E_h^*(\lambda_h)^\perp \cap V_h) := \inf\{\|u - v^h\|_V,\ v^h \in V_h,\ v^h \perp E_h^*(\lambda_h)\}$$
$$\leq C[d(u, V_h) + \|u\|_V \inf\{\|e^* - e^{*h}\|_{V'} : e^* \in E^*(\lambda_0)\}].$$

Theorem 11.2.27. *Let* (2a–c), (8), $\dim E(\lambda_0) = \dim E_h(\lambda_h) = 1$ *hold. Let* λ_0 *be the only eigenvalue in* $K_r(\lambda_0)$. *Let* h *be sufficiently small such that (following Remark* 25) *the Problem* (18a–c) *is solvable. For the solutions* u *and* u^h *of* (16a–c) *and* (18a–c) *the error estimate*

$$\|u - u^h\|_V \leq C[d(u, V_h) + \|f\|_{V'} \inf\{\|e^* - e^{*h}\|_{V'} : e^* \in E^*(\lambda_0)\} + \|f^{(h)} - f\|_{V'}] \tag{11.2.20}$$

holds, with C *independent of* f, $f^{(h)}$, *and* h.

PROOF. Repeat the proof of Theorem 8.2.1 for $a_\lambda(\cdot,\cdot)$ instead of $a(\cdot,\cdot)$. Here one must choose $w \in V_h$ with $w \perp E_h^*(\lambda_h)$. Furthermore, (8.2.3) becomes

$$a_\lambda(u^h - u, v) = (f^{(h)} - f, v)_0 \quad \text{for all } v \in V_h.$$

ϵ_N agrees with $\hat{\omega}_h(\lambda) \geq \eta > 0$ ($\lambda \in K_r(\lambda_0)$)) (cf. Lemma 16). $\|u - w\|_V$ is estimated with the aid of Exercise 26, with $\|u\|_V \leq \|f\|_{V'}$ being added. ■

Corollary 11.2.28. *If* $f^{(h)}$ *is defined by* (19), *then inequality* (20) *becomes*

$$\|u - u^h\|_V \leq C'[d(u, V_h) + \|f\|_{V'} \inf\{\|e^* - e^{*h}\|_V : e^* \in E^*(\lambda_0)\}]. \tag{12.2.21a}$$

PROOF.
$$|f^{(h)} - f\|_{V'} \leq C|(f, e^{*h})_0| = C|(f, e^* - e^{*h})_0|$$
$$\leq C\|f\|_{V'}\|e^* - e^{*h}\|_V.$$
■

Corollary 11.2.29. *If additionally the assumptions* $u \in H^{1+s}(\Omega)$, (14a), *and* $d(u, V_h) \leq Ch^s|u|_{1+s}$ *hold for* $u \in H^{1+s}(\Omega)$ *then* (21a) *yields the estimate*

$$\|u - u^h\|_V \leq Ch^s\|u\|_{H^{1+s}(\Omega)}. \tag{11.2.21b}$$

It remains to add the

PROOF. (Theorem 22) For $e \in E(\lambda_0)$ there exists $e^h \in E_h(\lambda_h)$ with $f := e - e^h \perp E(\lambda_0)$ and $|f|_1 = \|f\|_V \leq Ch^s = Ch$. According to Remark 25 the problem $a_{\lambda_0}(v, w) = (v, f)_0$ has a solution $w \perp E^*(\lambda_0)$ for all $v \in V$. The assumption of regularity yields $w \in H^2(\Omega)$, $|w|_2 \leq C|f|_0$ such that $w^h \in V_h$ exists with $w^h \perp E_h^*(\lambda_h)$, $|w - w^h|_1 \leq Ch|w|_2 \leq C'h|f|_0$. The value

$$a_{\lambda_0}(f, w^h) = a_{\lambda_0}(e, w^h) - a_{\lambda_0}(e^h, w^h) = 0 - a_{\lambda_0}(e^h, w^h)$$
$$= (\lambda_0 - \lambda_h)(e^h, w^h)_0 - a_{\lambda_h}(e^h, w^h) = (\lambda_0 - \lambda_h)(e^h, w^h)_0$$

can be bounded by $Ch^2|w^h|_0|e^h|_0 \le C'h^2|f|_0$ (cf. (15)). From

$$|f|_0^2 = (f, f)_0 = a_{\lambda_0}(f, w) = a_{\lambda_0}(f, w - w^h) + (\lambda_0 - \lambda_h)(e^h, w^h)_0$$
$$\le C[C'h|f|_1|f|_0 + h^2|f|_0]$$

and $|f|_1 \le Ch$ one infers $|f|_0 \le C'h^2$. The same method is then applied to $|e^* - e^{*h}|_0$. ■

Exercise 11.2.30. Formulate the conditions for and the proof of the error estimate $\|u - u^h\|_{L^2(\Omega)} \le Ch^2|u|_2$ (u, u^h from Corollary 28). Hint: Decompose $u - u^h$ in $f_1 + f_2$ with f_1 with $f_1 \perp E^*(\lambda_0)$ and $f_2 \in E(\lambda_0)$.

11.3 Discretisation by Difference Methods

In the following we limit ourselves to the case of a difference operator of the order $2m = 2$. The differential equation $Lu = f$ with homogeneous Dirichlet boundary condition is replaced, as in Sections 4 and 5, by the difference equation $L_h u_h = f_h$. The eigenvalue equations $Le = \lambda e$, $L^*e^* = \overline{\lambda}e^*$ are discretised by

$$L_h e_h = \lambda_h e_h, \quad L_h^* e_h^* = \overline{\lambda}_h e_h^*. \tag{11.3.1}$$

L_h^* is the transposed and complex-conjugate matrix to L_h. The general assumptions of the following analysis are:

$$V = H_0^1(\Omega), \quad \Omega \in C^{0,1} \text{ bounded,} \tag{11.3.2a}$$

$$a(u, v) = (Lu, v)_0 \quad \text{is } H_0^1(\Omega)\text{-coercive,} \tag{11.3.2b}$$

$$\text{Consistency condition } |L_h R_h - \tilde{R}_h L|_{-1\leftarrow 2} \le Ch. \tag{11.3.2c}$$

Condition (2c) has been discussed in Section 9.2.2. Assume furthermore that L_h is H_h^1-coercive. For suitable $\mu \in \mathbb{R}$

$$L_{\mu,h} := L_h - \mu I \quad (I : \text{identity matrix})$$

is thus H_h^1-regular:

$$(L_{\mu,h} v_h, v_h)_0 \ge C_E |v_h|_1^2. \tag{11.3.2d}$$

Furthermore, let

$$L_\mu := L - \mu I \quad H^2(\Omega)\text{-regular,} \tag{11.3.2e}$$

i.e., $|L_\mu^{-1}|_{2\leftarrow 0} \le C$. The boundedness of L and L_h reads

$$|L|_{-1\leftarrow 1} \le C, \quad |L_h|_{-1\leftarrow 1} \le C. \tag{11.3.2f}$$

In (9.2.26a,b) we have introduced prolongations $\hat{P}_h: L_h^2 \to L^2(\mathbb{R}^2)$ and $P_h: L_h^2 \to L^2(\Omega)$. Now we need a mapping $P_h: H_h^1 \to H_0^1(\Omega)$:

$$\overline{u}(\mathbf{x}) = \begin{cases} \hat{P}_h u_h(\mathbf{x}) & \text{if } K_{h/2}(\mathbf{x}) \subset \Omega, \\ 0 & \text{otherwise,} \end{cases}$$
$$P_h u_h(\mathbf{x}) := (\sigma_h^x \sigma_h^y \overline{u})(\mathbf{x}) \qquad (\mathbf{x} \in \Omega),$$

where $K_{h/2}(\mathbf{x}) = \{\mathbf{y} \in \mathbb{R}^2: \|\mathbf{x}-\mathbf{y}\|_\infty < h/2\}$, $\hat{P}_h$ are defined according to (9.2.26b) and $\sigma_h^x\sigma_h^y$ according to (9.2.11). Check that $P_h u_h \in H_0^1(\Omega)$ and

$$|P_h|_{1\leftarrow 1} \le C. \tag{11.3.2g}$$

Let R_h and $\tilde{R}_h$ be defined as in (9.2.12a,b). They satisfy

$$|R_h|_{0\leftarrow 0} \le C, \quad |R_h|_{1\leftarrow 1} \le C, \quad |\tilde{R}_h|_{0\leftarrow 0} \le C. \tag{11.3.2h}$$

Exercise 11.3.1. Show that :

$$|R_h - \tilde{R}_h|_{0\leftarrow 1} \le Ch, \quad |I - R_h P_h|_{0\leftarrow 1} \le Ch, \quad |P_h^* - R_h|_{0\leftarrow 1} \le Ch, \tag{11.3.2i}$$

$$|I - P_h^* P_h|_{0\leftarrow 1} \le Ch, \quad |I - \tilde{R}_h P_h|_{0\leftarrow 1} \le Ch, \quad |I - \tilde{R}_h^* R_h|_{0\leftarrow 1} \le Ch. \tag{11.3.2j}$$

The first inequality in (2j) is equivalent to

$$|(P_h u_h, P_h v_h)_0 - (u_h, v_h)_0| \le Ch|u_h|_0 |v_h|_1. \tag{11.3.2j*}$$

Hint: Exercise 9.2.12 and Lemma 9.2.15.

Lemma 11.3.2. *Let* (2a,c,g,h,i) *hold.* (a) *It is true that*

$$\lim_{h\to 0} |u - P_h R_h u|_1 = \lim_{h\to 0} |u - \tilde{R}_h^* R_h u|_1 = 0 \quad \textit{for all } u \in H_0^1(\Omega), \tag{11.3.2k}$$

$$\lim_{h\to 0} |(\tilde{R}_h - R_h)u|_0 = 0 \quad \textit{for all } u \in L^2(\Omega), \tag{11.3.2l}$$

$$\lim_{h\to 0} |\,[L_{h,\lambda} R_h - \tilde{R}_h L_\lambda]u|_{-1} = 0 \quad \textit{for all } u \in H_0^1(\Omega), \lambda \in \mathbb{C}. \tag{11.3.2m}$$

(b) *If* $u \in H_0^1(\Omega)$ *and* $\lim_{h\to 0} |R_h u|_0 = 0$ *then* $u = 0$.

PROOF. (a) The proof of (2k,l,m) follows the same pattern, which will be demonstrated for the case of (2l). For $\epsilon > 0$ one has to show $|(\tilde{R}_h - R_h)u| \le \epsilon$ for $h \le h(\epsilon)$. Since $H_0^1(\Omega)$ is dense in $L^2(\Omega)$ there exists a $\tilde{u} \in H_0^1(\Omega)$ with $|u - \tilde{u}|_0 \le \epsilon/[2|\tilde{R}_h - R_h|_{0\leftarrow 0}]$ such that $|(\tilde{R}_h - R_h)(u-\tilde{u})|_0 \le \epsilon/2$. By (2i) there follows $|(\tilde{R}_h - R_h)\tilde{u}|_0 \le Ch|\tilde{u}|_1 \le \epsilon/2$ for $h \le h(\epsilon) := \epsilon/[2C|\tilde{u}|_1]$. Altogether this gives $|(\tilde{R}_h - R_h)u|_0 \le \epsilon$.
(b) From (2k) one infers $0 = \lim_{h\to 0}(R_h u, P_h^* u)_0 = \lim_{h\to 0}(P_h R_h u, u)_0 = (u,u)_0$ thus $u = 0$. ∎

Exercise 11.3.3. Let $\Lambda_h := I - \partial_x^+\partial_x^- - \partial_y^+\partial_y^-$ and $\Lambda := I - \Delta$. Show that

$$
\begin{aligned}
&|u|_1^2 = (\Lambda u, u)_0, \qquad |u_h|_1^2 = (\Lambda_h u_h, u_h)_0,\\
&|L_\lambda u|_{-1}^2 = (v, L_\lambda u)_0 \quad \text{for} \quad v = \Lambda^{-1} L_\lambda u \in H_0^1(\Omega),\\
&|L_{\lambda,h} u_h|_{-1}^2 = (v_h, L_{\lambda,h} u_h)_0 \quad \text{for} \quad v_h = \Lambda_h^{-1} L_{\lambda,h} u_h,\\
&\lim_{h\to 0} |(\Lambda_h R_h - \tilde{R}_h \Lambda) u|_{-1} = 0 \quad \text{for all} \quad u \in H_0^1(\Omega).
\end{aligned}
$$

The following analysis is tailored to the properties of the difference methods under discussion. We have tried to avoid a more abstract theory that is applicable to finite elements as well as difference methods. This kind of approach can be found in Stummel [1] and Chatelin [1].

The variable $\omega_h(\lambda)$ is now defined by

$$
\omega_h(\lambda) := \inf_{|u_h|_1=1} \sup_{|v_h|_1=1} |(L_{\lambda,h} u_h, v_h)_0| = \inf_{|u_h|_1=1} |L_{\lambda,h} u_h|_{-1}. \tag{11.3.3}
$$

As in Exercise 11.2.4 we have

$$
\begin{aligned}
\omega_h(\lambda) &= 0, \quad \text{if } \lambda \text{ is an eigenvalue of } L_h;\\
\omega_h(\lambda) &= 1/|L_{\lambda,h}^{-1}|_{1\leftarrow -1} \quad \text{otherwise}.
\end{aligned} \tag{11.3.3'}
$$

The analogue of Lemma 11.2.7 reads

Lemma 11.3.4. *Let* $\lambda \subset \mathbb{C}$ *be compact. Let* (2a-m) *hold. Then there exist variables* $C > 0$ *and* $\eta(h) \to 0$ $(h \to 0)$, *independent of* $\lambda \in \Lambda$, *such that*

$$
\omega_h(\lambda) \ge C\omega(\lambda) - \eta(h), \quad \omega(\lambda) \ge C\omega_h(\lambda) - \eta(h) \quad \textit{for all } \lambda \in \Lambda,\, h > 0. \tag{11.3.4}
$$

PROOF. (a) Since Λ is compact, $\omega(\lambda)$ is continuous, and $\omega_h(\lambda)$ is equicontinuous in λ, it is sufficient to show that $\underline{\lim}_{h\to 0}\omega_h(\lambda) \ge C\omega(\lambda)$, and $\omega(\lambda) \ge C\,\overline{\lim}_{h\to 0}\omega_h(\lambda)$ for all $\lambda \in \Lambda$ with $C > 0$.
(b) Define for $\lambda \in \Lambda$ and u_h with $|u_h|_1 = 1$ and $|L_{\lambda,h} u_h|_{-1} = \omega_h((\lambda)$

$$
u := P_h u_h, \quad z_h := (\lambda - \mu) L_{\mu,h}^{-1} u_h, \quad z = (\lambda - \mu) L_\mu^{-1} u
$$

with μ from (2e). Without loss of generality it can be assumed that $\mu \notin \Lambda$. We have

$$
\begin{aligned}
|u - z|_1 &= |P_h(u_h - z_h) + P_h z_h - z|_1 \le |P_h|_{1\leftarrow 1} |u_h - z_h|_1 + |P_h z_h - z|_1,\\
|P_h z_h - z|_1 &= |P_h[z_h - R_h z] - [I - P_h R_h] z|_1\\
&\le |P_h|_{1\leftarrow 1} |\lambda - \mu|\, |L_{\mu,h}^{-1}[\tilde{R}_h L_\mu - L_{\mu,h} R_h] L_\mu^{-1} u + L_{\mu,h}^{-1}[I - \tilde{R}_h P_h] u_h|_1\\
&\quad + |(I - P_h R_h) z|_1 \to 0 \quad (h \to 0)
\end{aligned}
$$

(cf. (2g,j,m)) so that

$$
|u_h - z_h|_1 \ge C_1 |u - z|_1 - o(1), \quad C_1 > 0. \tag{11.3.5a}
$$

As in (2.6d) one obtains

$$|u - z|_1 \geq C_2 \omega(\lambda)|u|_1, \quad C_2 > 0. \tag{11.3.5b}$$

From

$$\begin{aligned}\omega_h(\lambda) &= |L_{\lambda,h}u_h|_{-1} = |L_{\mu,h}u_h + (\mu - \lambda)u_h|_{-1} \\ &\geq (L_{\mu,h}u_h + (\mu - \lambda)u_h, u_h)_0 \geq -|\mu - \lambda|\,|u_h|_0^2 + (L_{\mu,h}u_h, u_h)_0 \\ &\geq -C_\Lambda |u_h|_0^2 + C_E|u_h|_1^2 = -C_\Lambda|u_h|_0^2 + C_E\end{aligned}$$

with $C_\Lambda := \max\{|\mu - \lambda| : \lambda \in \Lambda\} > 0$ follows

$$|u_h|_0^2 \geq C_\Lambda^{-1}[C_E - \omega_h(\lambda)].$$

Either $\omega_h(\lambda) \geq C_E/2$ from which the statement results directly, or $\omega_h(\lambda) \leq C_E/2$ which yields

$$|u_h|_0 \geq C_0 = C_0|u_h|_1 \quad \text{with } C_0 := \sqrt{C_\Lambda^{-1}C_E/2}. \tag{11.3.5c}$$

We want to show that there exists $h_0 > 0$ and $C_p = C_p(C_0)$ such that

$$|u_h|_1 \leq C_p|P_h u_h|_1 \quad \text{for all } u_h \text{ with } |u_h|_0 \geq C_0|u_h|_1 \text{ and } h \leq h_0. \tag{11.3.5d}$$

The negation of (5d) reads: there exists u_h with $|u_h|_1 = 1$, $|u_h|_0 \geq C_0$ and $|P_h u_h|_1 \to 0$ $(h \to 0)$. From $|R_h P_h u_h|_0 \leq |R_h P_h u_h|_1 \leq C|P_h u_h|_1 \to 0$ and $|u_h - R_h P_h u_h|_0 \leq |I - R_h P_h|_{0\leftarrow 1}|u_h|_1 \leq Ch \to 0$ follows $|u_h|_0 \to 0$ in contrast to $|u_h|_0 \geq C_0$. Thus we have (5d).

Together with $L_{\mu,h}(u_h - z_h) = L_{\lambda,h}u_h$, (2d) and (5a,b,d) yield the first of the inequalities (4):

$$\begin{aligned}\omega_h(\lambda) &= |L_{\lambda,h}u_h|_{-1} = |L_{\mu,h}(u_h - z_h)|_{-1} \geq C_E|u_h - z_h|_1 \\ &\geq C_E C_1|u - z|_1 - o(1) \geq C_E C_1 C_2 \omega(\lambda)|u|_1 - o(1) \\ &\geq C\omega(\lambda) - o(1) \quad \text{with } C := C_E C_1 C_2/C_p > 0.\end{aligned}$$

(c) Let $\epsilon > 0$ be arbitrary; $u \in H_0^1(\Omega)$ with $|u|_1 = 1$ can be selected such that $\omega(\lambda) \geq |L_\lambda u|_{-1} - \epsilon$. Set $u_h := R_h u$. According to Exercise 3 it holds that

$$\begin{aligned}&|L_\lambda u|_{-1}^2 = (v, L_\lambda u)_0 \quad \text{for } v = \Lambda^{-1}L_\lambda u \in H_0^1(\Omega), \quad \Lambda = I - \Delta, \\ &|L_{\lambda,h}u_h|_{-1}^2 = (v_h, L_{\lambda,h}u_h)_0 \quad \text{for } v_h := \Lambda_h^{-1}L_{\lambda,h}u_h.\end{aligned}$$

From

$$\begin{aligned}&R_h v - v_h = \Lambda_h^{-1}[\Lambda_h R_h - \tilde{R}_h\Lambda]\Lambda^{-1}L_\lambda u + \Lambda_h^{-1}[\tilde{R}_h L_\lambda - L_{\lambda,h}R_h]u \to 0 \\ &\tilde{R}_h L_\lambda u - L_{\lambda,h}u_h = [\tilde{R}_h L_\lambda - L_{\lambda,h}R_h]u \to 0, \\ &(v, L_\lambda u)_0 - (R_h v, \tilde{R}_h L_\lambda u)_0 = ([I - \tilde{R}_h^* R_h]v, L_\lambda u)_0 \to 0,\end{aligned}$$

for $h \to 0$ (cf. (2m), (2k)), one infers $|L_{\lambda,h}u_h|_{-1} \to |L_\lambda u|_{-1}$ and

$$\omega(\lambda) \geq |L_\lambda u|_{-1} - \epsilon \geq |L_{\lambda,h}u_h|_{-1} - \epsilon - o(1) \geq \omega_h(\lambda) - \epsilon - o(1) \quad (h \to 0)$$

for each $\epsilon > 0$. Thus $\omega(\lambda) \geq \overline{\lim}_{h\to 0}\omega_h(\lambda)$ has been proved. ∎

Corollary 11.3.5. *Under the assumptions of Lemma 4 the following holds: If there exists L_λ^{-1} for all $\lambda \in \Lambda$ then there exists an $h_0 > 0$ such that $L_{\lambda,h}$ for all $\lambda \in \Lambda$ and $h \leq h_0$ is H_h^1-regular:* $\sup\{|L_{\lambda,h}^{-1}|_{1\leftarrow -1} : \lambda \in \Lambda, h \leq h_0\} \leq C$.

PROOF. We have assumed $\omega(\lambda) > 0$ in Λ so $\max\{\omega(\lambda) : \lambda \in \Lambda\} =: \eta > 0$. Choose h_0 according to Lemma 4 such that $\omega_h(\lambda) \geq C\omega(\lambda) - \frac{1}{2}C\eta \geq \frac{1}{2}C\eta$ for $h \leq h_0$. Then $|L_{\lambda,h}^{-1}|_{1\to -1} \leq 2/(C\eta)$ for all $\lambda \in \Lambda, h \leq h_0$. ∎

The proof of Theorem 11.2.8 can be carried over without change and results in

Theorem 11.3.6. *Assume* (2a–m). *If λ_{h_i} $(h_i \to 0)$ are discrete eigenvalues of Problem* (1) *with $\lambda_{h_i} \to \lambda_0$ then λ_0 is an eigenvalue of* (1.2a).

Lemma 11.2.9 and Theorem 11.2.10 can also be carried over without changes.

Theorem 11.3.7. *Let* (2a–m) *hold. Let λ_0 be an eigenvalue of* (1.2a). *Then there exist discrete eigenvalues λ_h of* (1) *[for all h] such that* $\lim_{h\to 0}\lambda_h = \lambda_0$.

Theorem 11.3.8. *Let* (2a–m) *hold. Let e_h be discrete eigenfunctions with $|e_h|_1 = 1$ for λ_h, where $\lambda_h \to \lambda_0$ $(h \to 0)$. Then there exists a subsequence e_{h_i} such that $P_{h_i}e_{h_i}$ converges in $H_0^1(\Omega)$ to an eigenfunction $0 \notin e \in E(\lambda_0)$. Further, we have $|e_{h_i} - R_{h_i}e|_1 \to 0$.*

PROOF. Because $|P_he_h|_1 \leq C$ (cf. (2g) the functions $e^h := P_he_h$ are uniformly bounded. $H_0^1(\Omega)$ is compactly embedded in $L^2(\Omega)$ (cf. (2a) and Theorem 6.4.8a) such that a subsequence e^{h_i} converges in $L^2(\Omega)$ to an $e \in L^2(\Omega)$: $|e^{h_i} - e|_0 \to 0$. We have in particular

$$|R_he - e_h|_0 \leq |R_h(e - e^h) - (R_hP_h - I)e_h|_0 \to 0 \quad (h = h_i \to 0).$$

Estimate (2c) yields

$$\begin{aligned}&|R_hz - z_h|_0 \leq |R_hz - z_h|_1 \leq Ch|e|_0 \to 0\\ &\text{for } z = (\lambda_0 - \mu)L_\mu^{-1}e, \quad z_h := (\lambda_0 - \mu)L_{\mu,h}^{-1}\tilde{R}_he.\end{aligned}$$

From $L_{\mu,h}(z_h - e_h) = (\lambda_0 - \mu)(\tilde{R}_he - e_h) + (\lambda_h - \lambda_0)e_h \to 0$ in H_h^{-1} follows $|z_h - e_h|_1 \to 0$ such that $|R_h(e - z)|_0 \to 0$ $(h = h_i \to 0)$ results. Lemma 2b shows that $e = z \in H_0^1(\Omega)$, and $0 = e \in E(\lambda_0)$ is excluded because of $|\lim R_{h_i}e|_1 = |\lim e_h|_1 = 1$. ∎

Theorem 11.3.9. *Let* (2a–m) *hold. Let e_h^* be the solution of the discrete eigenvalue problem $L_h^*e_h^* = \overline{\lambda}_he_h^*$ with $|e_h^*|_1 = 1$,* $\lim_{h\to 0}\lambda_h = \lambda_0$. *Then there* *exists a subsequence $e^*_{h_i}$ such that $P_{h_i}e^*_{h_i}$ in $H^1_0(\Omega)$ converges to an eigenfunction $0 \neq e^* \in E^*(\lambda_0)$. Furthermore, $|e^*_{h_i} - R_{h_i}e^*|_1 \to 0$.*

Proof Sketch. The proof is not analogous to that of Theorem 8 since the consistency condition (2a,m) does not necessarily imply the corresponding statements for the adjoint operators. One has to carry out the following steps:
(a) $e^{*h} := P_h e^*_h \to e^*$ converges in $L^2(\Omega)$ for a subsequence $h = h_i \to 0$.
(b) For $z = (\overline{\lambda_0 - \mu})L^{*-1}_\mu e^*$ and $z_h := (\overline{\lambda_0 - \mu})L^{*-1}_{\mu,h} e^*_h$ the following holds:

$$z - \tilde{R}^*_h z_h = (\overline{\lambda_0 - \mu})L^{*-1}_\mu[(e^* - R^*_h e^*_h) + (R^*_h L^*_{\mu,h} - L_\mu \tilde{R}^*_h)]L^{*-1}_{\mu,h} e^*_h.$$

For each $v \in L^2(\Omega)$ we obtain

$$(v, z - \tilde{R}^*_h z_h)_0 = (\lambda_0 - \mu)\{(L^{-1}_\mu v, e^* - e^{*h})_0 + ([P^*_h - R_h]L^{-1}_\mu v, e^*_h)_0$$
$$+(L^{-1}_{\mu,h}[L_{\mu,h}R_h - \tilde{R}_h L_\mu]L^{-1}_\mu v, e^*_h)_0\} \to 0$$

for $h = h_i \to 0$ (cf. (2i), (2c)).
(c) $|z_h - e^*_h|_1 \to 0$ may be inferred from $L^*_{\mu,h}(z_h - e^*_h) = (\overline{\lambda_0 - \lambda_h})e^*_h \to 0$. In particular, $(v, \tilde{R}^*_h z_h - R^*_h e^*_h)_0 \to 0$ for each $v \in L^2(\Omega)$.
(d) $(v, \tilde{R}^*_h e^*_h - R^*_h e^*_h)_0 = ([\tilde{R}_h - R_h|v, e^*_h)_0 \to 0$ for each $v \in L^2(\Omega)$ (cf. (2l)).
(e) $(v, R^*_h e^*_h - e^*)_0 \to 0$ for each $h = h_i \to 0$.
(f) From (b) and (e) follows $(v, z - e^*)_0 = 0$, thus $z = e^* \in E^*(\lambda_0)$, where $e^* \neq 0$.

Exercise 11.3.10. Carry over Exercise 11.2.12 to the present situation.

In Lemma 11.2.16 we defined $\hat{\omega}_h(\lambda)$. Now set

$$\hat{V}_h := \{u_h : (u_h, e^*_h)_h = 0\} = \{e^*_h\}^\perp, \quad L^*_h e^*_h = \overline{\lambda}_h e^*_h, \quad \lambda_h \to \lambda_0,$$
$$\hat{\omega}(\lambda) := \inf_{0 \neq u_h \in \hat{V}_h} \sup_{0 \neq v_h \in \hat{V}_h} |(L_{\lambda,h}u_h, v_h)_0|/(|u_h|_1\,|v_h|_1). \tag{11.3.6}$$

A basic condition for the following is

$$\dim E_h(\lambda_h) = 1, \quad E_h(\lambda_h) = \text{span}\ \{e_h\}, \quad E^*_h(\lambda_h) = \text{span}\ \{e^*_h\}. \tag{11.3.7}$$

Here

$$E(\lambda_h) := \{u_h \in H^1_h : L_h u_h = \lambda_h u_h\}, \quad E^*_h(\lambda_h) := \{u_h \in H^1_h : L^*_h u_h = \overline{\lambda}_h u_h\}.$$

Exercise 10 shows that (7) is valid for $h \le h_0$ if $\dim E(\lambda_0) = 1$.

Lemma 11.3.11. *Let (2a–m), $\dim E(\lambda_0) = 1$, and (2.8) hold. Then there exist $h_0 > 0$ and a $C > 0$ that is independent of $h \le h_0$ and $\lambda \in \mathbb{C}$ such that $\hat{\omega}_h(\lambda) \ge C\omega_h(\lambda)$ for $h \le h_0$. For sufficiently small $\epsilon > 0$ and h, $\hat{\omega}_h(\lambda) \ge \eta > 0$ for all $|\lambda - \lambda_0| \le \epsilon$.*

PROOF. (a) There exists a $C > 0$ such that

$$|v_h + \alpha e_h^*|_1 \geq |v_h|_1/C \quad \text{for all } v_h \in \hat{V}_h,\ \alpha \in \mathbb{C}, h > 0.$$

Indirect Proof: Assume that there exists a sequence v_{h_i} with $h_i \to 0$, $|v_{h_i}|_1 = 1$, $\alpha_i \in \mathbb{C}$, $w_{h_i} := v_{h_i} + \alpha_i e_{h_i}^*$, $|w_{h_i}|_1 \to 0$. For a subsequence of $\{h_i\}$

$$\alpha_i \to \alpha^*, \quad P_{h_i} v_{h_i} \to v^* \quad \text{and } P_{h_i} e_{h_i}^* \to e^* \neq 0 \quad \text{in } L^2(\Omega),$$

$$P_{h_i} w_{h_i} \to w^* := v^* + \alpha^* e^* = 0$$

converge. From $0 = (v_{h_i}, e_{h_i}^*)_0 = (v_{h_i}[I - P_{h_i}^* P_{h_i}]e_{h_i}^*)_0 + (P_{h_i} v_{h_i}, P_{h_i} e_{h_i}^*)_0 \to (v^*, e^*)_0$ one infers $(v^*, e^*)_0 = 0$, $\alpha^* = (w^*, e^*)_0/(e^*, e^*)_0 = 0$, $v^* = 0$. The contradiction results from $1 = \lim |v_{h_i}|_1 = \lim |w_{h_i}|_1 = 0$.
(b) The rest of the proof runs as in Lemma 11.2.16. ∎

Lemma 11.3.12. *Let* (2a–m), $\dim E(\lambda_0) = 1$, *and* (2.8) *hold. One may choose* $0 \neq e \in E(\lambda_0)$ *and* $e_h \in E_h(\lambda_h)$ *so that* $|R_h e - e_h|_1 \leq C[h + |\lambda_0 - \lambda_h|]$.

PROOF. We have that $e = (\lambda_0 - \mu)L_\mu^{-1} e \in H^2(\Omega) \cap H_0^1(\Omega)$. Assume also that $z_h := (\lambda_0 - \mu)L_{\mu,h}^{-1} \tilde{R}_h e$. The inequality (9.2.22) implies $|R_h e - z_h|_1 \leq Ch|e|_2$. For sufficiently small h we have $(e_h, e_h^*) \neq 0$ so that one can scale e_h such that $(e_h - z_h, e_h^*)_0 = 0$. From

$$\begin{aligned} L_{\lambda_h,h}(e_h - z_h) &= (\lambda_h - \lambda_0)\tilde{R}_h e + (\lambda_h - \mu)(z_h - \tilde{R}_h e) \\ &= (\lambda_h - \lambda_0)\tilde{R}_h e + (\lambda_h - \mu)[(z_h - R_h e) + (R_h - \tilde{R}_h)e] \end{aligned}$$

one infers that $|e_h - z_h|_1 \leq C[|\lambda_h - \lambda_0| + h]$ (cf. Lemma 11) since $|(R_h - \tilde{R}_h)e|_{-1} = O(h)$. The statement follows from this. ∎

Lemma 11.3.13. *Under the assumptions of Lemma* 12 *we have the inequality* $|\lambda_h - \lambda_0| \leq Ch$.

PROOF. Choose e_h, e_h^* such that $(R_h e - e_h, e_h^*)_0 = (e_h, R_h e^* - e_h^*)_0 = 0$. For the Rayleigh quotient $\tilde{\lambda}_h := (L_h R_h e, R_h e^*)_0/(R_h e, R_h e^*)_0$ we then have

$$|\tilde{\lambda}_h - \lambda_h| \leq C|R_h e - e_h|_1 |R_h e^* - e_h^*|_1 \leq \epsilon_h[h + |\lambda_0 - \lambda_h|]$$

$$\text{with } \epsilon_h := C|R_h e^* - e_h^*|_1.$$

From (2c, j) one infers that

$$\begin{aligned} &(L_h R_h e, R_h e^*)_0 - (Le, e^*)_0 \\ &= ([L_h R_h - \tilde{R}_h L]e, R_h e^*)_0 + (Le, [\tilde{R}_h^* R_h - I]e^*)_0 = O(h) \end{aligned}$$

and $|\tilde{\lambda}_h - \lambda_0| = O(h)$ such that $|\lambda_h - \lambda_0| \leq Ch + \epsilon_h|\lambda_h - \lambda_0|$. For sufficiently small h we have $\epsilon_h \leq \frac{1}{2}$ (cf. Theorem 9, Exercise 10) thus $|\lambda_h - \lambda_0| \leq 2Ch$. ∎

Lemmata 12 and 13 give

Theorem 11.3.14. *Let* (2a–m), $E(\lambda_0) = \operatorname{span}\{e\}$, *and* (2.8) *hold. There exists* $e_h \in E_h(\lambda_h)$ *with* $|R_h e - e_h|_1 \le Ch$.

Since $|R_h e^* - e_h^*|_1 = o(1)$ or even $|R_h e^* - e_h^*|_1 = O(h)$, according to Theorem 11.2.19 one expects that $|\lambda_0 - \lambda_h| = o(h)$ [resp. $O(h^2)$]. In general, this estimate is false as the following counterexample shows.

Example 11.3.15. The eigenvalue problem $-u'' + u' = \lambda u$ in $(0,1)$ with $u(0) = u(1) = 0$ has the solution $u(x) = \exp(\Lambda x)\sin(\pi x)$ with $\Lambda = 1/2$. The eigenvalue is $\lambda_0 = \pi^2 + 1/4$. One calculates that the discretisation $-\partial^-\partial^+ u + \partial^+ u = \lambda u$ has the eigenvalue

$$\begin{aligned}\lambda_h &= h^{-2}[2 - \cos(\pi h)(e^{\Lambda' h} + e^{-\Lambda' h}) - i\sin(\pi h)(e^{\Lambda' h} - e^{-\Lambda' h})] \\ &\quad + h^{-1}[\cos(\pi h)e^{\Lambda' h} - 1 + i\sin(\pi h)e^{\Lambda' h}] \\ &= \pi^2 + \frac{1}{4} + \left(\frac{1}{8} - \frac{3\pi^2}{8}\right)h + O(h^2) \quad \text{with } \Lambda' := \log(1-h)/(2h)\end{aligned}$$

such that $|\lambda_0 - \lambda_h|$ turns out no better than $O(h)$.

12 Stokes Equations

12.1 Systems of Elliptic Differential Equations

In Example 1.1.11 we have already stated the Stokes equations for $\Omega \subset \mathbb{R}^2$:

$$-\Delta u_1 + \partial p/\partial x_1 = f_1, \tag{12.1.1a$_1$}$$
$$-\Delta u_2 + \partial p/\partial x_2 = f_2, \tag{12.1.1a$_2$}$$
$$-\partial u_1/\partial x_1 - \partial u_2/\partial x_2 = 0. \tag{12.1.1b}$$

In the case of $\Omega \subset \mathbb{R}^3$ another equation $-\Delta u_3 + \partial p/\partial x_3 = f_3$ needs to be added, and the left-hand side of (1b) must be supplemented by $-\partial u_3/\partial x_3$. A representation independent of the dimension can be obtained if one takes together (u_1, u_2) [resp. (u_1, u_2, u_3)] as a vector u satisfying, in Ω, the equations

$$-\Delta u + \nabla p = f \tag{12.1.2a}$$
$$-\operatorname{div} u = 0. \tag{12.1.2b}$$

Here, div is the divergence operator

$$\operatorname{div} u = \sum_{i=1}^{n} \partial u_i/\partial x_i,$$

and n is both the dimension of $\Omega \subset \mathbb{R}^n$ and the number of components of $u(\mathbf{x}) \in \mathbb{R}^n$. Only $n \le 3$ is of physical interest here. In fluid mechanics the Stokes equation describes the flow of an incompressible medium (neglecting the inertial terms) and describes the velocity field. With $\mathbf{x} \in \mathbb{R}^n$, $u_i(\mathbf{x})$ is the velocity of the medium in the x_i direction; the function p denotes the pressure.

Up to now we have not formulated any boundary conditions. In the following we limit ourselves to Dirichlet boundary values:

$$u = 0 \quad \text{on } \Gamma. \tag{12.1.3}$$

This implies that the flow vanishes at the boundary. No boundary condition is given for p. Since both the pairs (u, p) and $(u, p + \text{const})$ satisfy the Stokes equations (2a,b), (3), one obtains:

Remark 12.1.1. *p is determined only up to a constant by the Stokes equation (2a,b) and the boundary conditions (3).*

The Stokes equations have been chosen as an example of a system of differential equations. It remains to investigate whether Equation (2a,b) is elliptic in a sense yet to be defined. Even though the functions u_i, for given p, are solutions of the elliptic Poisson equations $-\Delta u_i = f_i - \partial p/\partial x_i$, in determining p, (2a,b) do not provide an elliptic equation, in any current sense of elliptic.

A general system of q differential equations for q functions $u_1, \cdots, u_q$ can be written in the form

$$\sum_{j=1}^{q} L_{ij} u_j = f_i \quad \text{in } \Omega \subset \mathbb{R}^n \quad (1 \le i \le q) \tag{12.1.4a}$$

with the differential operators

$$L_{ij} = \sum_{|\alpha| \le k_{ij}} c_\alpha D^\alpha \quad (1 \le i, j \le q). \tag{12.1.4b}$$

The equations (4a) are summarised as $Lu = f$ where L is the matrix (L_{ij}) of differential operators and $u = (u_1, \cdots, u_q)^\mathsf{T}$, $f = (f_1, \cdots, f_q)^\mathsf{T}$. The order of the operator L_{ij} is at most k_{ij}. Let the numbers $m_1, \cdots, m_q, m'_1, \cdots, m'_q$ be chosen so that

$$k_{ij} \le m_i + m'_j \quad (1 \le i, j \le q). \tag{12.1.5}$$

As the *principal part* of L_{ij} one defines

$$L^P_{ij} := \sum_{|\alpha| = m_i + m'_j} c_\alpha D^\alpha.$$

If $k_{ij} < m_i + m'_j$, $c_\alpha = 0$ holds for $|\alpha| = m_i + m'_j$, and thus $L^P_{ij} = 0$. The characteristic polynomial associated with L^P_{ij} reads:

$$L^P_{ij}(\xi; \mathbf{x}) := \sum_{|\alpha| = m_i + m'_j} c_\alpha(\mathbf{x}) \xi^\alpha \quad (\xi \in \mathbb{R}^q, \mathbf{x} \in \Omega)$$

and forms the matrix function

$$L^P(\xi; \mathbf{x}) := (L^P_{ij}(\xi; \mathbf{x}))_{i,j=1,\cdots,q}.$$

Definition 12.1.2. (Agmon-Douglis-Nirenberg [1]) Let (5) hold for m_i, m'_j. The differential operator $L = (L_{ij})$ is said to be *elliptic* in $\mathbf{x} \in \Omega$ if

$$\det L^P(\xi; \mathbf{x}) \ne 0 \quad \text{for all } \mathbf{0} \ne \xi \in \mathbb{R}^q. \tag{12.1.6a}$$

L is said to be *uniformly elliptic* in Ω if there exists an $\epsilon > 0$ such that

$$|\det L^P(\xi; \mathbf{x})| \ge \epsilon |\xi|^{2m} \quad \text{for all } \mathbf{x} \in \Omega, \xi \in \mathbb{R}^q \tag{12.1.6b}$$

with $2m := \sum_{i=1}^{q} (m_i + m'_i)$. To be more precise one should call L elliptic with indices m_i, m'_j, since the definition does depend on m_i, m'_j. In connection with this problem, as well as for an additional condition for $\mathbf{x} \in \Gamma, q = 2$, see the original paper of Agmon–Douglas–Nirenberg [1]. Further information on this subject can be found in Cosner [1].

Exercise 12.1.3. (a) Show that the numbers m_i, m'_j are not unique. If m_i and m'_j satisfy the inequality (5), then so do $m_i - k$ and $m'_j + k$. The definition of L^P_{ij} is independent of k.
(b) Show that for $q = 1$, i.e., for the case of a single equation one recovers from (6a) the Definitions 1.2.3a [resp. (5.1.3a)]; (6b) corresponds to (5.1.3a′).
(c) For a first-order system (i.e., $k_{ij} = 1$, $m_i = 1$, $m'_j = 0$), (6a) coincides with Definition 1.3.2.

In order to describe the Stokes equations in the form (4a,b) we set

$$q := n+1, \quad u = (u_1, \cdots, u_n, p)^\mathsf{T}, \quad f = (f_1, \cdots, f_n, 0)^\mathsf{T},$$

$$L_{ii} = -\Delta, \quad L_{iq} = -L_{qi} = \partial/\partial x_i \quad \text{for } 1 \le i \le n, \quad L_{ij} = 0 \text{ otherwise.}$$

The orders are $k_{ii} = 2$, $k_{iq} = k_{qi} = 1$ $(i \le n)$, and $k_{ij} = 0$ otherwise. The numbers

$$m_i = m'_i = 1 \quad \text{for } 1 \le i \le n = q-1, \quad m_q = m'_q = 0$$

satisfy (5). L^P_{ij} coincides with L_{ij} and is independent of $\mathbf{x}$:

$$L^P_{ii}(\xi) = -|\xi|^2, \quad L^P_{iq}(\xi) = -L^P_{qi}(\xi) = \xi_i \quad \text{for } i \le n, \quad L^P_{ij}(\xi) = 0 \quad \text{otherwise.}$$

From this we see

$$|\det L^P(\xi)| = \left| \det \begin{bmatrix} -|\xi|^2 & & 0 & \xi_1 \\ 0 & \ddots & & \vdots \\ & & -|\xi|^2 & \xi_n \\ -\xi_1 & \dots & -\xi_n & 0 \end{bmatrix} \right| = |\xi|^{2m} \text{ with } 2m = 2n,$$

so that (6b) is satisfied, with $\epsilon = 1$.

Exercise 12.1.4. In elasticity theory the so-called displacement function $u\colon \Omega \subset \mathbb{R}^3 \to \mathbb{R}^3$ is described by the Lamé system:

$$\mu\Delta u + (\lambda + \mu)\nabla \operatorname{div} u = f \quad \text{in } \Omega, \quad u = \varphi \quad \text{on } \Gamma$$

$(\mu, \lambda > 0)$. Show that this system of three equations is uniformly elliptic: $|\det L^P(\xi)| = \mu^2(2\mu + \lambda)|\xi|^6$.

For the treatment of Stokes equations we will use a variational formulation in the next section. For reasons of completeness we point out the following transformation.

Remark 12.1.5. Let $n = 2$ and thus $u = (u_1, u_2)$. Because $\operatorname{div} u = 0$ there exists a so-called stream function Φ with $u_1 = \partial\Phi/\partial x_2$, $u_2 = -\partial\Phi/\partial x_1$.

Insertion in Equation ($1a_{1,2}$) results in the biharmonic equation $\Delta^2\Phi = \partial f_2/\partial x_1 - \partial f_1/\partial x_2$. The boundary condition (3) means $\nabla\Phi = 0$ on Γ. This is equivalent to $\partial\Phi/\partial n = 0$ and $\partial\Phi/\partial t = 0$ on Γ where $\partial/\partial t$ is the tangential derivative. $\partial\Phi/\partial t = 0$ implies $\Phi =$ const on Γ. Since the constant may be chosen arbitrarily, one sets $\Phi = \partial\Phi/\partial n = 0$ on Γ.

12.2 Variational Formulation

12.2.1 Weak Formulation of the Stokes Equations

Since $u = (u_1, \cdots, u_n)$ is a vector-valued function, we introduce

$$\mathbf{H}_0^1(\Omega) := H_0^1(\Omega) \times H_0^1(\Omega) \times \cdots \times H_0^1(\Omega) \quad (n\text{-fold product}).$$

A corresponding definition holds for $\mathbf{H}^{-1}(\Omega)$, $\mathbf{H}^2(\Omega)$, etc. The norm associated with $\mathbf{H}_0^1(\Omega)$ will again be denoted by $|\cdot|_1$ in the following.

According to Remark 12.1.1 the pressure component p of the Stokes problem is not uniquely determined. In order to determine uniquely the constant in $p = \tilde{p}+\text{const}$, we standardise p by the requirement $\int_\Omega p\, d\mathbf{x} = 0$. That is the reason why in the following p will always belong to the subspace $L_*^2(\Omega) \subset L^2(\Omega)$:

$$L_*^2(\Omega) := \{p \in L^2(\Omega) \colon \int_\Omega p(\mathbf{x})\, d\mathbf{x} = 0\}.$$

To derive the weak formulation we proceed as in Section 7.1 and assume that u and p are classical solutions of the Stokes problem (1.2a,b). Multiplication of the ith equation in (1.2a) with $v_i \in C_0^\infty(\Omega)$ and subsequent integration implies that

$$\int_\Omega f_i(\mathbf{x}) v_i(\mathbf{x})\, d(\mathbf{x}) = \int_\Omega [-\Delta u_i(\mathbf{x}) + \partial p/\partial x_i] v_i(\mathbf{x})\, d\mathbf{x} \tag{12.2.1a}$$

$$= \int_\Omega \{\langle \nabla u_i(\mathbf{x}), \nabla v_i(\mathbf{x})\rangle - p(\mathbf{x})\partial v_i(\mathbf{x})/\partial x_i\}\, d\mathbf{x} \text{ for } v_i \in C_0^\infty(\Omega),\ 1 \le i \le n.$$

Summation over i now gives

$$\int_\Omega \{\langle \nabla u(\mathbf{x}), \nabla v(\mathbf{x})\rangle - p(\mathbf{x})\,\mathrm{div}, v(\mathbf{x})\}\, d\mathbf{x} = \int_\Omega \langle f(\mathbf{x}), v(\mathbf{x})\rangle\, d\mathbf{x}, \tag{12.2.1a'}$$

where the abbreviation

$$\langle \nabla u, \nabla v\rangle := \sum_{i=1}^n \langle \nabla u_i, \nabla v_i\rangle = \sum_{i,j=1}^n \frac{\partial u_i}{\partial x_j}\frac{\partial v_i}{\partial x_j}$$

is used. Equation (1.2b) is then multiplied with some $q \in L_*^2(\Omega)$ and integrated, giving

$$-\int_\Omega q(\mathbf{x}) \operatorname{div} u(\mathbf{x})\, d\mathbf{x} = 0 \quad \text{for all } q \in L^2_*(\Omega). \tag{12.2.1b}$$

With the bilinear forms

$$a(u,v) := \int_\Omega \langle \nabla u(\mathbf{x}), \nabla v(\mathbf{x})\rangle\, d\mathbf{x} \quad \text{for } u,v \in \mathbf{H}^1_0(\Omega), \tag{12.2.2a}$$

$$b(p,v) := -\int_\Omega p(\mathbf{x}) \operatorname{div} v(\mathbf{x})\, d\mathbf{x} \quad \text{for } p \in L^2_*(\Omega), v \in \mathbf{H}^1_0(\Omega), \tag{12.2.2b}$$

we obtain the 1Vweak formulation of the Stokes problem as (3a–c):

$$\text{Find } u \in \mathbf{H}^1_0(\Omega) \text{ and } p \in L^2_*(\Omega) \text{ such that} \tag{12.2.3a}$$

$$a(u,v) + b(p,v) = f(v) := \int_\Omega \langle f, v\rangle d\mathbf{x} \quad \text{for all } v \in \mathbf{H}^1_0(\Omega), \tag{12.2.3b}$$

$$b(q,u) = 0 \quad \text{for all } q \in L^2_*(\Omega). \tag{12.2.3c}$$

In (3b) we first replace "$v \in \mathbf{H}^1_0(\Omega)$" by "$v \in C_0^\infty(\Omega)$". Since both sides of (3b) depend continuously on $v \in \mathbf{H}^1_0(\Omega)$ and since $C_0^\infty(\Omega)$ is dense in $\mathbf{H}^1_0(\Omega)$, (3b) follows for all $v \in H^1_0(\Omega)$.

Remark 12.2.1. A classical solution $u \in C^2(\overline{\Omega}) \cap \mathbf{H}^1_0(\Omega)$, $p \in C^1(\Omega) \cap L^2_*(\Omega)$ of the Stokes problem (1.2a,b), (1.3) is also a weak solution, i.e., a solution of (3a–c). If conversely (3a–c) has a solution with $u \in C^2(\overline{\Omega})$, $p \in C^1(\overline{\Omega})$, then it is also the classical solution of the boundary value problem (1.2a,b), (1.3).

PROOF. (a) The above considerations prove the first part.
(b) Equation (3c) implies $\operatorname{div} u = 0$. Let $i \in \{1, \cdots, n\}$. In Equation (3b) one can choose v with $v_i \in C_0^\infty(\Omega)$, $v_j = 0$ for $j \neq i$. Integration by parts recovers (1a) and hence the ith equation in (1.2a). ■

12.2.2 Saddlepoint Problems

The situation in (3a–c) is a special case of the following problem. We replace the spaces $\mathbf{H}^1_0(\Omega)$ and $L^2_*(\Omega)$ in (3a–c) by two Hilbert spaces V and W. Let

$$a(\cdot,\cdot)\colon V \times V \to \mathbb{R} \quad \text{a continuous bilinear form on } V \times V, \tag{12.2.4a}$$

$$b(\cdot,\cdot)\colon W \times V \to \mathbb{R} \quad \text{a continuous bilinear form on } W \times V, \tag{12.2.4b}$$

$$f_1 \in V', \quad f_2 \in W'. \tag{12.2.4c}$$

In generalisation of (6.5.1) we call $b(\cdot,\cdot)\colon W \times V \to \mathbb{R}$ continuous (or bounded), if there exists a $C_b \in \mathbb{R}$ such that

$$|b(w,v)| \le C_b \|w\|_W \|v\|_V \quad \text{for all } w \in W, v \in V.$$

The objective of this chapter is to solve the problem (5a–c):

$$\text{Find } v \in V \quad \text{and} \quad w \in W \quad \text{with} \tag{12.2.5a}$$
$$a(v,x) + b(w,x) = f_1(x) \quad \text{for all } x \in V, \tag{12.2.5b}$$
$$b(y,v) \quad = f_2(y) \quad \text{for all } y \in W. \tag{12.2.5c}$$

Formally, (5a–c) can be transformed to the form

$$\text{Find } u \in X \quad \text{with } c(u,z) = f(z) \quad \text{for all } z \in X \tag{12.2.6a}$$

if one sets:

$$X = V \times W, \quad c(u,z) := a(v,x) + b(w,x) + b(y,v) \quad \text{and}$$

$$f(z) = f_1(x) + f_2(y) \quad \text{for} \quad u = \begin{pmatrix} v \\ w \end{pmatrix}, z = \begin{pmatrix} x \\ y \end{pmatrix}. \tag{12.2.6b}$$

Exercise 12.2.2. Show that (a) $c(\cdot,\cdot)\colon X \times X \to \mathbb{R}$ is a continuous bilinear form.
(b) Problems (5a–c) and (6a,b) are equivalent.

That the variational problems (5) and (6) must be handled differently than in Section 7 is made clear by the following remark.

Remark 12.2.3. The bilinear form $c(\cdot,\cdot)$ in (6b) cannot be X-elliptic.

PROOF. We have $c(u,u) = 0$ for all $u = \begin{pmatrix} 0 \\ x \end{pmatrix}$. ∎

In analogy to (6.5.9) we set

$$J(v,w) := a(v,v) + 2b(w,v) - 2f_1(v) - 2f_2(w),$$

and therefore $J(v,w) = c(u,u) - 2f(u)$ for $u = \begin{pmatrix} v \\ w \end{pmatrix}$. For $v \in V$, $w \in W$ we know $J(v,w)$ is neither bounded below nor above. Therefore the solution v^*, w^* of (5a–c) does not give a minimum of J; however, under suitable conditions, v^*, w^* may be a saddle point.

Theorem 12.2.4. *Let* (4a–c) *hold. Let* $a(\cdot,\cdot)$ *be symmetric and* V*-elliptic. The pair* $v^* \in V$, $w^* \in W$ *is a solution of the problem* (5a–c) *if and only if*

$$J(v^*,w) \le J(v^*,w^*) \le J(v,w^*) \quad \text{for all } v \in V, w \in W. \tag{12.2.7}$$

Another equivalent characterisation is

$$J(v^*,w^*) = \min_{v\in V} J(v,w^*) = \max_{w\in W} \min_{v\in V} J(v,w). \tag{12.2.8}$$

PROOF. (aa) Let v^*, w^* solve (5). The expression in brackets in

$$J(v,w^*) - J(v^*,w^*) = a(v^*-v, v^*-v)$$

$$-2[a(v^*,v^*-v) + b(w^*,v^*-v) - f_1(v^*-v)]$$

vanishes because of (5b). Since $a(v^*-v,v^*-v) > 0$ for all $v^* \neq v \in V$, the second inequality in (7) follows. One also proves the converse as for Theorem 6.5.12: If $J(v,w^*)$ is minimal for $v = v^*$ then (5b) holds.

(ab) If v^* is a solution of (5c), then

$$J(v^*,w^*) - J(v^*,w) = 2[b(w^*-w,v^*) - f_2(w^*-w)]$$

vanishes for all w, which proves the first part of (7) in the stronger form $J(v^*,w) = J(v^*,w^*)$. If, however, v^* is not a solution of (5c), there exists a $w \in W$ such that $\delta := J(v^*,w^*) - J(v^*,w) \neq 0$. Since $J(v^*,w^*) - J(v^*,\hat{w}) = -\delta$ for $\hat{w} := 2w^* - w$, the first inequality cannot be valid. For the converse define $w_\pm = w^* \mp w$. The first part of (7) implies

$$0 \le J(v^*,w^*) - J(v^*,w_\pm) = \pm 2[b(w,v^*) - f_2(w)]$$

for both signs, and so $b(w,v^*) = f_2(w)$. Since $w \in W$ is arbitrary, one obtains (5c).

(ba) We set $j(w) := \min_{v\in V} J(v,w)$. According to Theorem 6.5.12, $j(w) = J(v_w,w)$, where $v_w \in V$ is the solution of (5b). If v_w and $v_{w'}$ are the solutions for w and w', it follows that

$$a(v_w - v_{w'}, x) = F(x) := b(w-w',x) \quad \text{for all } x \in V.$$

Since $\|F\|_{V'} \le C_b\|w-w'\|_V$ and $\|v_w - v_{w'}\|_V \le C'\|F\|_{V'}$ one obtains

$$\|v_w - v_{w'}\|_V \le C\|w-w'\|_V \quad \text{for all } w,w' \in W. \tag{12.2.9a}$$

(bb) By using the definition of v_w in (5b) we can write:

$$\begin{aligned} J(v^*,w^*) - J(v_w,w) &= a(v^*,v^*) + 2b(w^*,v^*) - 2f_1(v^*) - 2f_2(w^*) \\ &\quad - [a(v_w,v_w) + 2b(w,v_w) - 2f_1(v_w) - 2f_2(w)] \\ &= 2[f_1(v_w - v^*) - b(w,v_w - v^*) - a(v_w,v_w - v^*)] \\ &\quad + a(v_w - v^*, v_w - v^*) + 2[b(w^*-w,v^*) - f_2(w^*-w)] \\ &= a(v_w - v^*, v_w - v^*) + 2[b(w^*-w,v^*) - f_2(w^*-w)]. \end{aligned} \tag{12.2.9b}$$

(bc) Let v^*, w^* be a solution of (5a–c). Because of (5c) the expression in brackets in (9b) vanishes and we have

$$J(v^*,w^*) = J(v_w,w) + a(v_w - v^*, v_w - v^*) \ge J(v_w,w) = j(w).$$

(5b) gives $v_{w^*} = v^*$, and so

$$J(v^*,w^*) = j(w^*) = \max_{w\in W} j(w); \tag{12.2.9c}$$

i.e., (8) holds.

(bd) Now let v^*, w^* be a solution of (8). If in (9b) one sets $w = w^*, v_w = v_{w^*}$ one obtains from $J(v^*, w^*) = j(w^*)$ that $v^* = v_{w^*}$. Hence $a(v_w - v^*, v_w - v^*) = a(v_w - v_{w^*}, v_w - v_{w^*})$ depends quadratically on $\|w - w^*\|_V$ (cf. (9a)). The variation over $w := w^* - \lambda y$ ($\lambda \in \mathbb{R}, y \in W$ arbitrary) gives

$$0 = \frac{d}{d\lambda} j(w^* - \lambda y)|_{\lambda=0} = 2[b(y, v^*) - f(y)],$$

and so (5c) is proved. (5b) has already been established with $v^* = v_{w^*}$. ∎

12.2.3 Existence and Uniqueness of the Solution of a Saddlepoint Problem

To make the saddlepoint problem (5a–c) somewhat more transparent we introduce the operators associated with the bilinear forms:

$$\begin{aligned} &A \in L(V, V') \quad \text{with} \quad a(v, x) = \langle Av, x \rangle_{V' \times V} \text{ for } v, x \in V, && (12.2.10a) \\ &B \in L(W, V'), \qquad B^* \in L(V, W') \\ &\qquad\qquad \text{with } b(w, x) = \langle Bw, x \rangle_{V' \times V} = \langle w, B^* x \rangle_{W \times W'}, && (12.2.10b) \\ &C \in L(X, X') \text{ with } c(u, z) = \langle Cu, z \rangle_{X' \times X} \text{ for } u, z \in X. && (12.2.10c) \end{aligned}$$

Thus problem (6a,b) now has the form $Cu = f$, while (5a–c) can be written as

$$Av + Bw = f_1 \tag{12.2.11a}$$

$$B^* v = f_2. \tag{12.2.11b}$$

If one assumes the existence of $A^{-1} \in L(V', V)$, one can solve (11a) for v:

$$v = A^{-1}(f_1 - Bw) \tag{12.2.12a}$$

and substitute in (11b):

$$B^* A^{-1} B w = B^* A^{-1} f_1 - f_2. \tag{12.2.12b}$$

The invertibility of A is in no way necessary for the solvability of the saddlepoint problem. However, it does simplify the analysis, and does hold true in the case of the Stokes problem.

Remark 12.2.5. (a) Under the assumptions

$$A^{-1} \in L(V', V), \quad (B^* A^{-1} B)^{-1} \in L(W', W) \tag{12.2.13}$$

the saddlepoint problem (5a–c) [resp. Equations (11a,b)] are uniquely solvable.
(b) A necessary condition for the existence of $(B^* A^{-1} B)^{-1}$ is

$$B \in L(W, V') \text{ is injective.} \tag{12.2.14}$$

PROOF. (a) Under the assumption of $(B^*A^{-1}B)^{-1} \in L(W', W)$, (12b) is uniquely solvable for w, and then (12a) yields v. (b) The injectivity of $B^*A^{-1}B$ implies (14). ■

Attention. In general, $B: W \to V'$ is not bijective so that the representation of $(B^*A^{-1}B)^{-1}$ as $B^{-1}AB^{*-1}$ is not possible. The example of the 3×3 matrix $\mathbf{C} = \begin{pmatrix} \mathbf{A} & \mathbf{B} \\ \mathbf{B}^\mathsf{T} & \mathbf{0} \end{pmatrix}$ with $\mathbf{A} = \begin{pmatrix} 1 & -1 \\ -1 & 1 \end{pmatrix}$ and $\mathbf{B} = \binom{1}{1}$ shows that a system of the form (11a,b) can be solvable even with a singular matrix $\mathbf{A}$. Therefore the assumption $\mathbf{A}^{-1} \in L(V', V)$ is not necessary. A closer look reveals the subspace

$$V_0 := \ker B^* = \{v \in V: B^*v = 0\} = \{v \in V: b(y, v) = 0 \text{ for all } y \in W\} \subset V, \tag{12.2.15}$$

which, as we noted before, in general is not trivial. The kernel of a continuous mapping is closed so that V, according to Lemma 6.1.17, can be represented as the sum of orthogonal spaces:

$$V = V_0 \oplus V_\perp \quad \text{with} \quad V_\perp := (V_0)^\perp. \tag{12.2.16a}$$

Exercise 12.2.6. Let (16a) hold. Show that (a) V' can be represented as

$$V' = V_0' \oplus V_\perp', \text{ where} \tag{12.2.16b}$$

$$\begin{aligned} V_0' &:= \{v' \in V': v'(v) = 0 \text{ for all } v \in V_\perp\}, \\ V_\perp' &:= \{v' \in V': v'(v) = 0 \text{ for all } v \in V_0\}. \end{aligned} \tag{12.2.16c}$$

As a norm on V_0' and $V_\perp'$ one uses $\|\cdot\|_{V'}$.
(b) The Riesz isomorphism $J_V: V \to V'$ maps V_0 onto V_0' and $V_\perp$ onto $V_\perp'$.
(c) V_0' and $V_\perp'$ are orthogonal spaces with respect to $\|\cdot\|_{V'}$.
(d) The following holds:

$$\|v'\|_{V'}^2 = \|v_0'\|_{V_0'}^2 + \|v_\perp'\|_{V_\perp'}^2 \quad \text{for } v' = v_0' + v_\perp', v_0' \in V_0', v_\perp' \in V_\perp'. \tag{12.2.16d}$$

The decompositions (16a,b) of V and V' define a block decomposition of the operator A:

$$A = \begin{bmatrix} A_{00} & A_{0\perp} \\ A_{\perp 0} & A_{\perp\perp} \end{bmatrix},$$

$$A_{00} \in L(V_0, V_0'), \quad A_{0\perp} \in L(V_\perp, V_0'), \quad A_{\perp 0} \in L(V_0, V_\perp'), \quad A_{\perp\perp} \in L(V_\perp, V_\perp').$$

Here, for example, A_{00} is defined as follows:

$$A_{00}v_0 = v_0' \quad \text{for } v_0 \in V_0 \quad \text{if } Av_0 = v_0' + v_\perp' \text{ with } v_0' \in V_0', v_\perp' \in V_\perp'.$$

The corresponding decomposition of B^* into $(B_0^*, B_\perp^*)$ is written as $(0, B^*)$, since $B_0^* = 0$ according to the definition of V_0. Conversely, we have range $(B) \subset V_\perp'$, so that $B = \binom{0}{B}$. System (11a,b) thus becomes

$$A_{00}v_0 + A_{0\perp}v_\perp = f_{10} \tag{12.2.17a}$$
$$A_{\perp 0}v_0 + A_{\perp\perp}v_\perp + Bw = f_{1\perp} \tag{12.2.17b}$$
$$B^*v_\perp = f_2 \tag{12.2.17c}$$

where $v = v_0 + v_\perp$, $v_0 \in V_0$, $v_\perp \in V_\perp$, $f_1 = f_{10} + f_{1\perp}$, $f_{10} \in V_0'$, $f_{1\perp} \in V_\perp'$.

Theorem 12.2.7. *Let* (4a–c) *hold. Let* V_0 *be defined by* (15). *A necessary and sufficient condition for the unique solvability of the saddlepoint problem* (5a–c) *for all* $f_1 \in V'$ *is the existence of the inverses*

$$A_{00}^{-1} \in L(V_0', V_0), \quad B^{-1} \in L(V_\perp', W). \tag{12.2.18}$$

PROOF. (a) (17a–c) represents a staggered system of equations. (18) implies $B^{*-1} = (B^{-1})^* \in L(W', V_\perp)$ so that one can solve (17c) for $v_\perp = B^{*-1}f_2$. $v_0 \in V_0$ is obtained from (17a): $v_0 = A_{00}^{-1}(f_{10} - A_{0\perp}v_\perp)$. Finally, w results from (17b).

(b) In order to show that (18) is necessary, we take $f_{10} \in V_0'$ arbitrary, $f_{1\perp} = 0$ and $f_2 = 0$. By hypothesis here we have a solution $(v_0, v_\perp, w) \in V_0 \times V_\perp \times W$. $B^*v_\perp = 0$ implies $v_\perp \in V_0$, so that $v_\perp = 0$ because $V_0 \cap V_\perp = \{0\}$. Thus $A_{00}v_0 = f_{10}$ has a unique solution $v_0 \in V_0$ for each $f_{10} \in V_0'$. Since $A_{00}: V_0 \to V_0'$ is bijective and bounded, Theorem 6.1.13 shows that $A_{00}^{-1} \in L(V_0', V_0)$. If one takes $f_{1\perp} \in V_\perp'$ arbitrary and $f_{10} = 0$, $f_2 = 0$, one infers $v_\perp = 0$ and $v_0 = 0$, so that $Bw = f_{1\perp}$ has a unique solution $w \in W$. As we did for A_{00}, one also infers that $B^{-1} \in L(V_\perp', W)$. ∎

The formulation of conditions (18) in terms of the bilinear forms results in the Babuška-Brezzi conditions:

$$\inf\{\sup\{|a(v_0, x_0)|: x_0 \in V_0, \|x_0\|_V = 1\}: v_0 \in V_0, \|v_0\|_V = 1\} \geq \alpha > 0, \tag{12.2.19a}$$
$$\sup\{|a(x_0, v_0)|: x_0 \in V_0, \|x_0\|_V = 1\} > 0 \quad \text{for all } 0 \neq v_0 \in V_0, \tag{12.2.19b}$$
$$\inf\{\sup\{|b(w, x)|: x \in V, \|x\|_V = 1\}: w \in W, \|w\|_W = 1\} \geq \beta > 0. \tag{12.2.19c}$$

Exercise 12.2.8. Show that (19a) [resp. (19c)] are equivalent to (19a') [resp. (19c')]:

$$\sup\{|a(v_0, x_0)|: x_0 \in V_0, \|x_0\|_V = 1\} \geq \alpha\|v_0\|_V \quad \text{for all } v_0 \in V_0, \tag{12.2.19a$'$}$$
$$\sup\{|b(w, x)|: x \in V, \|x\|_V = 1\} \geq \beta\|w\|_W \quad \text{for all } w \in W. \tag{12.2.19c$'$}$$

Lemma 12.2.9. *Let* (4a,b) *hold. Let* V_0 *be defined by* (15). *Then conditions* (18) *and* (19a–c) *are equivalent. Here we have* $\|A_{00}^{-1}\|_{V_0 \leftarrow V_0'} \leq 1/\alpha$, $\|B^{-1}\|_{W \leftarrow V_\perp'} \leq 1/\beta$.

PROOF. Because of (16d) and $b(w,x)=0$ for $x\in V_0$ one can write (19c) in the form

$$\inf\{\sup\{|b(w,x)|: x\in V_\perp, \|x\|_V=1\}: w\in W, \|w\|_W=1\}\geq\beta>0. \tag{12.2.19d}$$

For $0\neq x\in V_\perp$ one has that $x\notin V_0$, therefore according to (15):

$$\sup\{|b(w,x)|: w\in W, \|w\|_W=1\}>0 \quad \text{for all } 0\neq x\in V_\perp. \tag{12.2.19e}$$

As in the proof of Lemma 6.5.3, we obtain the equivalence of (19a,b) with $A_{00}^{-1}\in L(V_0',V_0)$ and of (19d,e) with $B^{-1}\in L(V_\perp',W)$. ■

Corollary 12.2.10. (a) *Condition* (19b) *becomes superfluous if* $a(\cdot,\cdot)$ *is symmetric on* $V_0\times V_0$ *or if Lemma* 6.5.17 *applies.*
(b) *Each of the following conditions is sufficient for* (19a,b) *and hence also for* $A_{00}^{-1}\in L(V_0',V_0)$*:*

$$a(\cdot,\cdot): V_0\times V_0\longrightarrow \mathbb{R} \;\; \textit{is } V_0\textit{-elliptic:}$$
$$a(v_0,v_0)\geq\alpha\|v_0\|_V^2 \quad \textit{for all } v_0\in V_0, \tag{12.2.20a}$$
$$a(\cdot,\cdot): V\times V\longrightarrow \mathbb{R} \;\; \textit{is } V\textit{-elliptic.} \tag{12.2.20b}$$

PROOF. (a) As in Lemma 6.5.17. (b) (20b) implies (20a); (20a) yields (19a,b). ■

Exercise 12.2.11. Show that under the assumptions (4a–c), (18) is also equivalent to the existence of $C^{-1}\in L(X',X)$ (cf. (10c)). Find a bound for $\|C^{-1}\|_{X\leftarrow X'}$ in terms of $\|A_{00}^{-1}\|_{V_0\leftarrow V_0'}$, $\|A\|_{V'\leftarrow V'}$, and $\|B^{-1}\|_{W\leftarrow V_\perp'}$.

12.2.4 Solvability and Regularity of the Stokes Problem

Conditions (19a,b) (i.e., $A_{00}^{-1}\in L(V_0',V_0)$) are easy to satisfy for the Stokes problem:

Lemma 12.2.12. *Let Ω be bounded. Then the forms* (2a,b) *describing the Stokes problem satisfy the conditions* (4a,b) *and* (19a,b).

PROOF. (4a,b) is self-evident. According to Example 7.2.10, $\int_\Omega\langle\nabla u,\nabla v\rangle\,dx$ is $H_0^1(\Omega)$-elliptic. From this follows the $\mathbf{H}_0^1(\Omega)$-ellipticity of $a(\cdot,\cdot)$. Corollary 10b proves (19a,b). ■

It remains to prove condition (19c), which for the Stokes problem assumes the form

$$\sup\{|\int_\Omega w(\mathbf{x})\operatorname{div}u(\mathbf{x})\,d\mathbf{x}|: u\in\mathbf{H}_0^1(\Omega), |u|_1=1\}\geq\beta|w|_0 \quad \text{for all } w\in L_*^2(\Omega), \tag{12.2.21}$$

$$\text{or} \quad \|\nabla w\|_{\mathbf{H}^{-1}(\Omega)} \geq \beta \|w\|_{L^2(\Omega)} \quad \text{for all } w \in L^2_*(\Omega). \tag{12.2.21$'$}$$

Lemma 12.2.13. *Sufficient and necessary for* (21) *is that for each* $w \in L^2_*(\Omega)$ *there exists* $u \in \mathbf{H}^1_0(\Omega)$ *such that*

$$w = \operatorname{div} u, \quad |u|_1 \leq \beta^{-1} |w|_0. \tag{12.2.21$''$}$$

PROOF. (a) For $w \in L^2_*(\Omega)$ select u such that (21″) holds and set $\tilde{u} := u/|u|_1$. The left-hand side in (21) is $\geq \int_\Omega w(\mathbf{x}) \operatorname{div} \tilde{u}(\mathbf{x})\, d\mathbf{x} = |w|_0^2/|u|_1 \geq \beta |w|_0$.
(b) If (21) holds one infers as in Section 12.2.3 the bijectivity of $B^*: V_\perp \to W$ with $\|B^{*-1}\|_{V_\perp \leftarrow W} \leq 1/\beta$. Therefore $u := B^{*-1} w$ satisfies condition (21″). ∎

Necǎs [2] proves

Theorem 12.2.14. *Condition* (21) *is satisfied if* $\Omega \in C^{0,1}$ *is bounded. Under this assumption, the Stokes problem*

$$-\Delta u + \nabla p = f, \quad -\operatorname{div} u = g \text{ in } \Omega, \quad u = 0 \quad \text{on } \Gamma \tag{12.2.22}$$

has a unique solution $(u, p) \in \mathbf{H}^1_0(\Omega) \times L^2_*(\Omega)$ *for* $f \in \mathbf{H}^{-1}(\Omega)$ *and* $g \in L^2_*(\Omega)$, *with*

$$|u|_1 + |p|_0 \leq C_\Omega [\, |f|_{-1} + |g|_0]. \tag{12.2.23}$$

Remark 12.2.15. Under the conditions that $n = 2$ and that $\Omega \in C^2$ is a bounded domain, the existence proof can be carried out as follows.

PROOF. We need to prove (21″). For $w \in L^2_*(\Omega)$ solve $-\Delta\varphi = w$ in Ω, $\varphi = 0$ on Γ. Theorem 9.1.16 shows that $\varphi \in H^2(\Omega)$. Since $\nabla\varphi \in \mathbf{H}^1(\Omega)$ and $\mathbf{n}(\mathbf{x}) \in C^2(\Gamma)$ it follows that $g := \partial u/\partial n \in H^{1/2}(\Gamma)$ (cf. Theorem 6.2.40a). From (3.4.2) one infers that $\int_\Gamma g\, d\Gamma = \int_\Omega w\, d\mathbf{x} = 0$ since $w \in L^2_0(\Omega)$. Integration of g over Γ yields $G \in H^{3/2}(\Gamma)$ with $\partial G/\partial t = g$, where $\partial/\partial t$ is the tangential derivative. There exists a function $\psi \in H^2_0(\Omega)$ with $\psi = G$ and $\partial\psi/\partial n = 0$ on Γ and $|\psi|_2 \leq C|G|_{3/2} \leq C'|g|_{1/2} \leq C''|u|_2 \leq C'''|w|_0$. We set

$$u_1 := -\varphi_x - \psi_y, \quad u_2 := -\varphi_y + \psi_x.$$

Clearly, $u_1, u_2 \in H^1(\Omega)$. Let the normal direction at $\mathbf{x} \in \Gamma$ be $\mathbf{n}(\mathbf{x}) = (n_1(\mathbf{x}), n_2(\mathbf{x}))^\mathsf{T}$. The tangent direction is thus $\mathbf{t}(\mathbf{x}) := (n_2(\mathbf{x}), -n_1(\mathbf{x}))^\mathsf{T}$. For $u = (u_1, u_2)^\mathsf{T}$ one obtains

$$\langle u, \mathbf{n}\rangle = -\varphi_x n_1 - \psi_y n_1 - \varphi_y n_2 + \psi_x n_2 = -\partial\varphi/\partial n + \partial\psi/\partial t = -g + \partial G/\partial t = 0,$$

$$\langle u, \mathbf{t}\rangle = -\varphi_x n_2 - \psi_y n_2 + \varphi_y n_1 - \psi_x n_1 = -\partial\varphi/\partial t - \partial\psi/\partial n = 0,$$

since $\varphi = 0$ on Γ implies $\partial\varphi/\partial t = 0$. $\langle u, \mathbf{n}\rangle = \langle u, \mathbf{t}\rangle = 0$ yields $u = 0$ on Γ such that $u = (u_1, u_2) \in \mathbf{H}^1_0(\Omega)$ is proved. One verifies that

$$\operatorname{div} u = \partial u_1/\partial x + \partial u_2/\partial y = -\varphi_{xx} - \psi_{yx} - \varphi_{yy} + \psi_{xy} = -\Delta\varphi = w$$

with $|u|_1 \le |\varphi|_2 + |\psi|_2 \le C|w|_0$. ■

The above proof uses the H^2-regularity of the Poisson problem and requires corresponding assumptions on Ω. Theorem 14 also assumes $\Omega \in C^{0,1}$. Since the Poisson equation $-\Delta u = f$ is solvable for any domain Ω which is contained in the disk $K_R(0)$, or at least in a strip $\{\mathbf{x} \in \mathbb{R}^n : |x_1| < R\}$, resulting in the inequality $|u|_1 \le C_R|f|_{-1}$, one might conjecture that a similar result holds for the Stokes problem.

Counterexample 12.2.16. For $\epsilon \in (0,1)$ let $\Omega_\epsilon = \{(x,y): -1 < x < 1, 0 < y < \epsilon + (1-\epsilon)|x|\}$ (cf. Figure 1). All $\Omega_\epsilon \in C^{0,1}$ are located in the rectangle $(-1,1) \times (0,1)$. Nevertheless, there exists no $\beta > 0$ such that (21″) holds for all $\epsilon > 0$, $w \in L^2(\Omega_\epsilon)$.

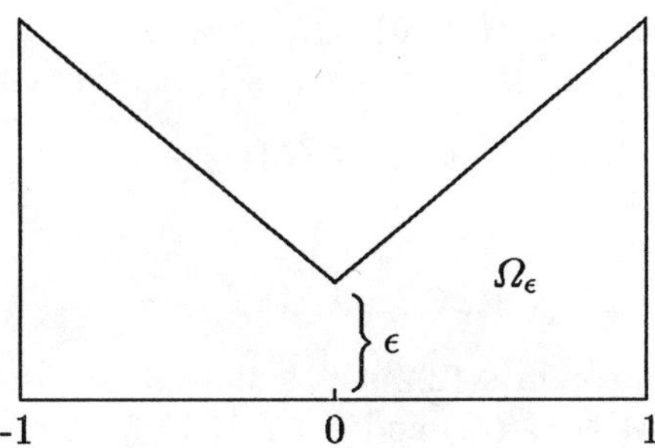

Figure 12.2.1. Ω_ϵ

PROOF. We select $w \in L^2_*(\Omega_\epsilon)$ such that $w(x,y) = 1$ for $x > 0$, $w(x,y) = -1$ for $x \le 0$. Let the inequality (21″) hold for Ω_ϵ with $\beta_\epsilon > 0$. Let $u \in \mathbf{H}^1_0(\Omega_\epsilon)$ with $|u|_1 \le |w|_0/\beta_\epsilon$ be chosen according to Lemma 13. We continue u by $u = 0$ onto $\mathbb{R}^2$. For the restriction on $x = 0$ we have according to Theorem 6.2.28

$$\|u_1(0,\cdot)\|_{L^2(\mathbb{R})} \le \|u_1(0,\cdot)\|_{H^{1/2}(\mathbb{R})} \le C|u_1|_1 \le C|w|_0/\beta_\epsilon \le 2C/\beta_\epsilon.$$

Let $\chi(y) = 1$ for $0 < y < \epsilon$ and $\chi(y) = 0$ otherwise. Since $u_1(0,y) = u_1(0,y)\chi(y)$ and $|\chi|_0 = \sqrt{\epsilon}$, we have

$$\left|\int_0^\epsilon u_1(0,y)\,dy\right| = \left|\int_{\mathbb{R}} u_1(0,y)\chi(y)\,dy\right| \le |u_1(0,\cdot)|_0|\chi|_0 \le 2C\sqrt{\epsilon}/\beta_\epsilon.$$

Let $\Omega_\epsilon^+ = \{(x,y) \in \Omega_\epsilon : x > 0\}$ and $\gamma := \{(x,y) : x = 0,\ 0 < y < \epsilon\} = \partial\Omega_\epsilon^+ \backslash \partial\Omega_\epsilon$. Because $w = 1$ in Ω_ϵ^+, and because $\operatorname{div} u = w$ we have

$$1/2 \le \int_{\Omega_\epsilon^+} |w(\mathbf{x})|^2\,d\mathbf{x} = \int_{\Omega_\epsilon^+} w \operatorname{div} u\,d\mathbf{x} = \int_{\Omega_\epsilon^+} \operatorname{div} u\,d\mathbf{x} = \int_\gamma \langle u, \mathbf{n}\rangle\,d\Gamma$$
$$= -\int_\gamma u_1\,d\Gamma = -\int_0^\epsilon u_1(0,y)\,dy.$$

The last two inequalities result in $1/2 \le 2C\sqrt{\epsilon}/\beta_\epsilon$, from which we infer that $\beta_\epsilon > 0$ cannot be independent of ϵ. ∎

Exercise 12.2.17. Construct a domain Ω located on the strip $\mathbb{R} \times (0,1)$ in which the Stokes equations are not solvable. Hint: Join the domains $\Omega_{1/\nu}$ ($\nu \in \mathbb{N}$) from Figure 1.

As for the case of scalar differential equations one obtains stronger regularity of the Stokes solution u, p if one assumes more than $f \in H^{-1}(\Omega)$.

Theorem 12.2.18. *Let Ω be bounded and sufficiently smooth. Let u and p be the (weak) solution of the Stokes problem (22) with $f \in \mathbf{H}^k(\Omega)$, $g \in H^{k+1}(\Omega) \cap L^2_*(\Omega)$ for $k \in \mathbb{N} \cup \{0\}$. Then we have $u \in \mathbf{H}^{k+2}(\Omega) \cap \mathbf{H}^1_0(\Omega)$, $p \in H^{k+1}(\Omega) \cap L^2_*(\Omega)$ and there exists a C depending only on Ω such that*

$$|u|_{k+2} + |p|_{k+1} \le C[|f|_k + |g|_{k+1}]. \tag{12.2.23b}$$

PROOF. Cf. Ladyženskaja [1, Chap.III, §5] ∎

In analogy to Theorem 9.1.22 it is sufficient to assume the convexity of Ω in order to obtain $u \in \mathbf{H}^2(\Omega)$ and $p \in H^1(\Omega)$ from $f \in \mathbf{L}^2(\Omega)$.

Theorem 12.2.19. (*Cf. Kellogg-Osborn* [1]) *Let $\Omega \subset \mathbb{R}^2$ be bounded and convex. If $f \in \mathbf{L}^2(\Omega)$, then the Stokes equation* (1.2a,b) *has a unique solution $u \in \mathbf{H}^2(\Omega) \cap \mathbf{H}^1_0(\Omega)$, $p \in H^1(\Omega) \cap L^2_*(\Omega)$, which satisfy the estimate*

$$|u|_2 + |p|_1 \le C|f|_0. \tag{12.2.23c}$$

For the more general problem (22) *with $g \ne 0$ in a convex polygonal domain the solution satisfies*

$$|u|_2 + |p|_1 \le C[\,|f|_{\mathbf{L}^2(\Omega)} + |g|_{\mathbf{H}^1_\delta(\Omega)}], \tag{12.2.23d}$$

if $f \in \mathbf{L}^2(\Omega)$ and $g \in L^2_*(\Omega) \cap H^1_\delta(\Omega)$. Here, $H^1_\delta(\Omega)$ is the subspace of $H^1(\Omega)$ with the following (stronger) norm:

$$\|g\|_{H^1_\delta(\Omega)} := \Big|\sum_{|\alpha|=1} \|D^\alpha g\|^2_{L^2(\Omega)} + \|\delta^{-1} g\|^2_{L^2(\Omega)}\Big|^{1/2} \quad \textit{with}$$

$$\delta(\mathbf{x}) := \min\,\{|\mathbf{x} - \mathbf{e}| : \mathbf{e} \in \Gamma \ \text{ corners of the polygon } \Omega\}.$$

12.2.5 A V_0-elliptic Variational Formulation of the Stokes Problem

$V_0 \subset \mathbf{H}_0^1(\Omega)$ has been defined in (15) by $V_0 := \{u \in \mathbf{H}_0^1(\Omega) : \operatorname{div} u = 0\}$. As kernel for the mapping $B^* = -\operatorname{div} \in L(\mathbf{H}_0^1(\Omega), L_*^2(\Omega))$, V_0 is a closed subspace of $\mathbf{H}_0^1(\Omega)$, i.e., again a Hilbert space for the same norm $|\cdot|_1$. In the following we investigate the problem

$$\text{Find } u \in V_0 \quad \text{with } a(u,v) = f(v) \quad \text{for all } v \in V_0, \tag{12.2.24}$$

where $a(u,v) := \int_\Omega \langle \nabla u(\mathbf{x}), \nabla v(\mathbf{x}) \rangle \, d\mathbf{x}$. Problem (24) has the same form as the weak formulation of the Poisson equations $-\Delta u_i = f_i$ $(i = 1, \cdots, n)$, only here $\mathbf{H}_0^1(\Omega)$ has been replaced by V_0.

Lemma 12.2.20. *Let Ω be a bounded domain (or bounded in one direction; cf. Exercise* 6.2.12b). *The form $a(\cdot,\cdot)$ is V_0-elliptic. The constant $C_E > 0$ in $a(u,u) \ge C_E |u|_1^2$ depends only on the diameter of Ω. In particular, problem* (24) *has a unique solution $u \in V_0$ with $|u|_1 \le C_E^{-1} |f|_{V_0'}$ for $f \in V_0'$.*

PROOF. The $\mathbf{H}_0^1(\Omega)$-ellipticity of $a(\cdot,\cdot)$ (cf. Lemma 12) carries over to $V_0 \subset \mathbf{H}_0^1(\Omega)$ (cf. Exercise 6.5.6a). This implies the other statements (cf. Theorem 6.5.9) ■

Theorem 12.2.21. *Let $\Omega \in C^{0,1}$ be a bounded domain. Assume $f \in \mathbf{H}^{-1}(\Omega)$. Then the solution $u \in V_0$ of problem* (24) *coincides with the solution component u of the mixed formulation* (3a–c).

PROOF. According to Exercise 6 one can split $f \in \mathbf{H}_0^1(\Omega)$ in such a way that

$$f = f_0 + f_\perp,\ f_0 \in V_0',\ f_\perp \in V',\ f_0(v) = 0 \text{ for } v \in V_\perp,\ f_\perp(v) = 0 \text{ for } v \in V_0.$$

In (24) one can replace $f(v)$ by $f_0(v)$. $u \in V_0 \subset \mathbf{H}_0^1(\Omega)$ results in $-\Delta u \in \mathbf{H}^{-1}(\Omega)$. The part of $-\Delta u$ that belongs to $V_\perp'$ is $g_\perp \in V_\perp'$ with

$$g_\perp(v_\perp) := a(u, v_\perp) \quad \text{for all } v_\perp \in V_\perp = (V_0)^\perp.$$

Theorem 14 proves the condition (21′), which results in the bijectivity of $B : L_*^2(\Omega) \to V_\perp'$ for $B = \nabla$ (cf. Lemma 13). $p := B^{-1}(f_\perp - g_\perp)$, by definition, satisfies

$$b(p, v_\perp) = \langle Bp, v_\perp \rangle_{\mathbf{H}^{-1}(\Omega) \times \mathbf{H}_0^1(\Omega)} = f_\perp(v_\perp) - g_\perp(v_\perp) = f_\perp(v_\perp) - a(u, v_\perp)$$

for all $v_\perp \in V_\perp$. Furthermore, since $\operatorname{div} v_0 = 0$ for $v_0 \in V_0$, it follows that

$$b(p, v_0) = 0 \text{ for all } v_0 \in V_0.$$

For arbitrary $v \in \mathbf{H}_0^1(\Omega)$, to be split into $v = v_0 + v_\perp$ with $v_0 \in V_0$ and $v_\perp \in V_\perp$, one obtains

$$b(p,v) = b(p,v_0) + b(p,v_\perp) = f_\perp(v_\perp) - a(u,v_\perp) = f(v_\perp) - a(u,v) + a(u,v_0)$$

$$= f(v_\perp) - a(u,v) + f(v_0) = f(v) - a(u,v),$$

by (24). Since also $b(w,u) = 0$ for all $w \in L^2_*(\Omega)$ because $u \in V_0$, u and p satisfy the variational formulation (3a–c) of the Stokes problem. ■

Note that Problem (24) is solvable for all bounded domains although Problem (3a–c) depends more sensitively on Ω (cf. Counterexample 16). Theorem 21 shows that only the component p has a domain-dependent bound $|p|_0 \le C_\Omega |f|_{-1}$ while $|u|_1 \le C|f_0|_{-1} \le C|f|_{-1}$ holds for all $\Omega \subset K_R(0)$.

Strictly speaking, the variational problem (24) is not equivalent to the Stokes problem since, for example, for Ω from Exercise 17 the Stokes equations are not solvable, whereas Problem (24) definitely has a solution.

The original formulation (3a–c) may be interpreted as Equation (24), into which one has introduced the "side condition" $\operatorname{div} u = 0$ using the Lagrange function p (cf. Section 8.3.6).

12.3 Mixed Finite-Element Method for the Stokes Problem

12.3.1 Finite-Element Discretisation of a Saddlepoint Problem

One would have an ordinary Ritz-Galerkin discretisation if in the variational formulation (2.24) one were to replace the space V_0 by a finite-dimensional subspace $V_h \subset V_0$. But this is not as easy as it sounds.

Exercise 12.3.1. Let the square $\Omega = (0,1) \times (0,1)$ be triangulated regularly as in Figure 8.3.2, 5a or decomposed into grid squares. Let $V_h^{(1)} \subset H_0^1(\Omega)$ be the space of the finite triangular elements (cf. (8.3.8)) [resp. of bilinear elements (cf. (8.3.12b))]. Define the corresponding subspace for the Stokes problem as $V_h := \{u = (u_1, u_2)\colon u_1, u_2 \in V_h^{(1)} \text{ and } \operatorname{div} u = 0\} \subset V_0$. Show that V_h contains only the null function.

The remaining procedure is thus oriented toward the weak formulation (2.3a–c). The space $X = V \times W$ is replaced with $X_h = V_h \times W_h$. The discrete problem

$$\text{Find } v^h \in V_h \text{ and } w^h \in W_h \text{ with} \tag{12.3.1a}$$

$$a(v^h, x) + b(w^h, x) = f_1(x) \quad \text{for all } x \in V_h, \tag{12.3.1b}$$

$$b(y, v^h) = f_2(y) \quad \text{for all } y \in W_h, \tag{12.3.1c}$$

is called a mixed Ritz-Galerkin problem [resp. a mixed finite-element problem if V_h and W_h are formed from finite elements]. To the formulation (2.6a,b) corresponds the equivalent way of writing (1a–c)

$$\text{Find } x^h \in X_h \text{ with } c(x^h, z) = f(z) \text{ for all } z \in X_h. \tag{12.3.1'}$$

In the case of the Stokes equations, (2.3a–c) provides the desired solution which satisfies the side condition $\operatorname{div} u = 0$. However, the finite-element solution of (1a–c) generally does not satisfy the condition $\operatorname{div} u^h = 0$. One can view v^h from (1a–c) as the solution of a nonconformal finite-element discretisation of (2.24), as is shown in the following exercise.

Exercise 12.3.2. Let $f_2 = 0$ in (1c); set $V_{0,h} := \{x \in V_h\colon b(y,x) = 0$ for all $y \in W_h\}$. Show that each solution v^h in (1a–c) is also a solution of

$$\text{Find}\quad v^h \in V_{0,h} \text{ with } a(v^h, x) = f_1(x) \text{ for all } x \in V_{0,h}. \tag{12.3.2}$$

Since in general $V_{0,h} \not\subset V_0$ (cf. (2.15)), (2) is a nonconformal discretisation of (2.24).

Let $\{b_1^V, \cdots, b_{N_{V,h}}^V\}$ and $\{b_1^W, \cdots, b_{N_{W,h}}^W\}$ be suitable bases of V_h and W_h, where

$$N_{V,h} = \dim V_h, \quad N_{W,h} := \dim W_h. \tag{12.3.3a}$$

Let the coefficients of $v \in V_h$ and $w \in W_h$ be $\mathbf{v}$ and $\mathbf{w}$:

$$v = \mathbf{P}_V \mathbf{v} := \sum_{i=1}^{N_{V,h}} v_i b_i^V, \quad w = \mathbf{P}_W \mathbf{w} := \sum_{i=1}^{N_{W,h}} w_i b_i^W. \tag{12.3.3b}$$

As in Section 8.1, we prove

Theorem 12.3.3. *The variational problem* (1a–c) *is equivalent to the system of equations*

$$\begin{bmatrix} \mathbf{A}_h & \mathbf{B}_h \\ \mathbf{B}_h^\mathsf{T} & \mathbf{0} \end{bmatrix} \begin{bmatrix} \mathbf{v} \\ \mathbf{w} \end{bmatrix} = \begin{bmatrix} \mathbf{f}_1 \\ \mathbf{f}_2 \end{bmatrix} \tag{12.3.4a}$$

where the matrices and vectors are given by

$$\mathbf{A}_{h,ij} := a(b_j^V, b_i^V), \quad \mathbf{B}_{h,ik} := b(b_k^W, b_i^V) \quad (1 \le i,j \le N_{V,h}; 1 \le k \le N_{W,h}) \tag{12.3.4b}$$

$$f_{1,i} := f_1(b_i^V), \quad f_{2,k} := f_2(b_k^W) \quad (1 \le i \le N_{V,h}; 1 \le k \le N_{W,h}). \tag{12.3.4c}$$

The connection between (1a–c) *and* (4a–c) *is given by* $v^h = \mathbf{P}_V \mathbf{v}$, $w^h = \mathbf{P}_W \mathbf{w}$. *For* $\mathbf{u} := \binom{\mathbf{v}}{\mathbf{w}}, \mathbf{f} := \binom{\mathbf{f}_1}{\mathbf{f}_2}$ *one obtains the system of equations*

$$\mathbf{C}_h \mathbf{u} = \mathbf{f}; \quad \mathbf{C}_h := \begin{bmatrix} \mathbf{A}_h & \mathbf{B}_h \\ \mathbf{B}_h^\mathsf{T} & \mathbf{0} \end{bmatrix}, \tag{12.3.4a'}$$

which corresponds to the formulation (1′).

12.3.2 Stability Conditions

When selecting the subspaces V_h and W_h one has to be careful, for even seemingly reasonable spaces lead to singular or unstable systems of equations (1a–c). The former occurs in the following example.

Example 12.3.4. Let the Stokes problem be posed for the L-shaped domain in Figure 8.3.2. For all three components u_1, u_2, and p we use piecewise linear triangular elements over the triangulation τ which is given by the second or third picture of Figure 8.3.2. Here, let $u_1 = u_2 = 0$ on Γ and let $p \in W_h$ satisfy the side condition $p \in L^2_*(\Omega)$, i.e., $\int_\Omega p\,d\mathbf{x} = 0$. Then the discrete variational problem (1a–c) is not solvable.

PROOF. The second [third] triangulation in Figure 8.3.2 has 5 [10] inner nodes $\mathbf{x}^i$, with each of them carrying values $u_1(\mathbf{x}^i)$ and $u_2(\mathbf{x}^i)$. Thus $\dim V_h = 2 \cdot 5 = 10$ [$\dim V_h = 20$]. Because of the side condition $\int_\Omega p\,d\mathbf{x} = 0$ the dimension of W_h is smaller by one than the number of inner and boundary nodes: $\dim W_h = 21 - 1 = 20$ [$\dim W_h = 21$]. In both cases the statement results from

Lemma 12.3.5. *It is necessary for the solvability of* (1a–c) *that* $N_{V,h} \geq N_{W,h}$, *i.e.,* $\dim V_h \geq \dim W_h$.

PROOF. According to Theorem 3 the solvability of (1a–c) is equivalent to the nonsingularity of the matrix $\mathbf{C}_h$ in $(4a')$. Elementary considerations show that $\operatorname{rank}(\mathbf{C}_h) \leq N_{V,h} + \operatorname{rank}(\mathbf{B}_h)$. Since $\mathbf{B}_h$ is an $N_{V,h} \times N_{W,h}$-matrix it follows from $N_{V,h} < N_{W,h}$ that $\operatorname{rank}(\mathbf{C}_h) < N_{V,h} + N_{W,h}$ and hence $\mathbf{C}_h$ is singular. ■

Thus an increase in the dimension of W_h does not always lead to an better approximation of $w \in W$. The choices of V_h and W_h must be mutually adjusted. The inequality $\dim V_h \geq \dim W_h$ corresponds to the requirement that B in (2.10b) needs to be injective, but not necessarily to be surjective.

In order to formulate the necessary stability conditions, we define

$$V_{0,h} := \{v \in V_h : b(y, v) = 0 \text{ for all } y \in W_h\} \tag{12.3.5}$$

(cf. Exercise 2). $V_{0,h}$ is the discrete analogue of the space V_0 in (2.15). The conditions, which can be traced back to Brezzi [1], read:

$$\inf\{\sup\{|a(v_0, x_0)| : x_0 \in V_{0,h}, \|x_0\|_V = 1\} : v_0 \in V_{0,h}, \|v_0\|_V = 1\} \geq \alpha_h > 0, \tag{12.3.6a}$$

$$\inf\{\sup\{|b(w, x)| : x \in V_h, \|x\|_v = 1\} : w \in W_h, \|w\|_w = 1\} \geq \beta_h > 0. \tag{12.3.6b}$$

Theorem 12.3.6. *Let* (2.4a–c) *hold and* $\dim V_h < \infty$. *The Brezzi conditions* (6a–b) *are necessary and sufficient for the solvability of the discrete problem* (1a–c). *The solution* $u^h = (v^h, w^h) \in V_h \times W_h$ *satisfies*

$$\|u^h\|_X := [\,\|v^h\|_V^2 + \|w^h\|_W^2\,]^{1/2} \leq C_h \|f\|_{X'} := C_h[\,\|f_1\|_{V'} + \|f_2\|_{W'}^2\,]^{1/2}, \tag{12.3.7}$$

where C_h *depends on* α_h, β_h *and on the bounds* C_a *and* C_b *in* $|a(v, x)| \leq C_a \|v\|_V \|x\|_V$ *and* $|b(w, x)| \leq C_b \|w\|_W \|x\|_V$. *If*

$$\alpha_h \geq \underline{\alpha} > 0, \quad \beta_h \geq \underline{\beta} > 0 \tag{12.3.6c}$$

holds for all parameters h of a sequence of discretisations, then the discretisation is said to be stable *and C_h remains bounded: $C_h \leq \overline{C}$ for all h.*

PROOF. Theorem 12.2.7 and Lemma 12.2.9 are applicable with $V_{0,h}, V_h, W_h$ instead of V_0, V, W. (2.19a,c) then are the same as (6a,b), while (2.19b) follows from (2.19a) because $\dim V_{0,h} < \infty$ (cf. Exercise 6.5.4). ■

Condition (6a) is trivial for the Stokes problem:

Exercise 12.3.7. Show that Condition (6a) is always satisfied with a constant α_h independent of h, if $a(\cdot,\cdot): V \times V \to \mathbb{R}$ is V-elliptic.

12.3.3 Stable Finite-Element Spaces for the Stokes Problem

12.3.3.1 Stability Criterion

For the Stokes problem $V_h \subset \mathbf{H}_0^1(\Omega)$, $W_h \subset L_*^2(\Omega)$ must hold. In a bounded domain $a(\cdot,\cdot)$ is $\mathbf{H}_0^1(\Omega)$-elliptic so that (6a) is satisfied with $\alpha_h \geq C_E > 0$. It is somewhat more difficult to prove the conditions (6b,c), which for the Stokes problem assume the form

$$\sup\{|b(p,u)|: u \in V_h, |u|_1 = 1\} \geq \underline{\beta}|p|_0 \quad \text{for all } p \in W_h \tag{12.3.6d}$$

wherein $\underline{\beta} > 0$ must be independent of p and h. It is simpler to prove the modified condition

$$\sup\{|b(p,u)|: u \in V_h, |u|_0 = 1\} \geq \tilde{\beta}|p|_1 \quad \text{for all } p \in W_h \tag{12.3.8}$$

($\tilde{\beta} > 0$ independent of p and h). Since in (8) the norm $|p|_1$ occurs, the latter requires $W_h \subset H^1(\Omega)$. This excludes, for example, piecewise constant finite elements.

Together with the usual assumptions, Condition (8) is sufficient for stability.

Theorem 12.3.8. *Let $\Omega \in C^{0,1}$ be bounded. Let the Poisson problem be H^2-regular (it is sufficient that Ω be convex; cf. Theorem 9.1.22). Let V_h satisfy the approximation property*

$$\inf\{|u - u^h|_1 : u^h \in V_h\} \leq C_A h |u|_2 \quad \textit{for all } u \in H^2(\Omega) \cap H_0^1(\Omega) \tag{12.3.9a}$$

and the inverse estimate

$$|u^h|_1 \leq C_1 h^{-1} |u^h|_0 \quad \textit{for all } u^h \in V_h. \tag{12.3.9b}$$

Then Condition (8) is sufficient for the Brezzi condition (6b,c).

PROOF. (a) For given $p \in W_h$ there exists $u \in \mathbf{H}_0^1(\Omega)$ with $|u|_1 = 1$ and $b(p,u) \ge \beta |p|_0$ (cf. Theorem 12.2.14 and (2.21)). According to Exercise 9.1.24, the orthogonal $\mathbf{H}_0^1(\Omega)$-projection u^h of u on V_h satisfies the conditions

$$u = u^h + e, \quad |u^h|_1 \le |u|_1 = 1, \quad |e|_0 \le Ch|u|_1 = Ch.$$

From $b(p,u^h) = b(p,u) - b(p,e) \ge \beta|p|_0 - b(p,e) \ge \beta|p|_0 - |p|_1|e|_0 \ge \beta|p|_0 - Ch|p|_1$ and $|u^h|_1 \le 1$ one infers

$$\sup\{|b(p,v^h)|: v^h \in V_h, |v^h|_1 = 1\} \ge \beta|p|_0 - Ch|p|_1. \tag{12.3.10}$$

(b) Because of (8) and (9b) there exists $u^* \in V_h$ with $|u^*|_0 = 1$ and

$$|b(p,u^*)| \ge \tilde\beta|p|_1 = \tilde\beta|p|_1|u^*|_0 \ge (\tilde\beta/C_I)h|p|_1|u^*|_1.$$

From this follows

$$\sup\{|b(p,v^h)|: v^h \in V_h, |v^h|_1 = 1\} \ge \hat\beta h|p|_1 \quad \text{with } \hat\beta := \tilde\beta/C_I. \tag{12.3.11}$$

If one multiplies (10) with $\hat\beta/(C+\hat\beta)$ and (11) with $C/(C+\hat\beta)$ the sum reads:

$$\sup\{|b(p,v^h)|: v^h \in V_h, |v^h|_1 = 1\} \ge \underline{\beta} := \beta\hat\beta/(C+\hat\beta). \tag{12.3.12}$$

Since $\underline{\beta}$ is independent of p and h, (6d) [i.e., (6b,c)] has been proved. ■

12.3.3.2 Finite-Element Discretisations with the Bubble Function

In the following let Ω be a polygonal domain and τ_h an admissible triangulation. Example 4 shows that linear triangular elements make no sense for u and p. We increase the dimension of V_h by adding to it the so-called "bubble functions" or "bulb functions".

On the reference triangle $T = \{(\xi,\eta): \xi,\eta > 0, \xi+\eta < 1\}$ the bubble function is defined by

$$u(\xi,\eta) := \xi\eta(1-\xi-\eta) \quad \text{in } T, \quad u = 0 \text{ otherwise.} \tag{12.3.13a}$$

The name derives from the fact that u is positive only in T and vanishes on ∂T and outside. The map $\phi: T \to \tilde T$ to a general $\tilde T \in \tau_h$ (cf. Exercise 8.3.14) results in the expression

$$\tilde u_{\tilde T}(x,y) := u(\phi^{-1}(x,y)) \tag{12.3.13b}$$

for the bubble function on $\tilde T$.

Exercise 12.3.10. Let τ_h be a quasiuniform triangulation. Show that there exists a $C > 0$ independent of h such that $\int_{\tilde T} \tilde u_{\tilde T}(x,y)\,dx\,dy \ge C\int_{\tilde T} dx\,dy$.

We set

$$\begin{aligned} &V_h^1\text{: linear combinations of the linear elements } \in H_0^1(\Omega) \\ &\qquad \text{and the bubble functions for } \tilde{T} \in \tau_h, \\ &V_h := V_h^1 \times V_h^1, \quad W_h\text{: linear elements } \in L_*^2(\Omega). \end{aligned} \tag{12.3.14}$$

For the side condition $W_h \subset L_*^2(\Omega)$ see Section 8.3.6. Since $\tilde{u}_{\tilde{T}} \in H_0^1(\Omega)$, we have $V_h \subset \mathbf{H}_0^1(\Omega)$.

Theorem 12.3.11. *Let τ_h be the quasiuniform triangulation on a bounded polygonal domain Ω. Let V_h and W_h be given by* (14). *Then the stability condition* (8) *is satisfied. Under the further conditions that $\Omega \in C^{0,1}$ and that the Poisson problem be $H^2(\Omega)$-regular the Brezzi condition* (6b,c) *also holds.*

PROOF. (a) Choose an arbitrary $p \in W_h$. On every $\tilde{T} \in \tau_h$, ∇p is constant: $\nabla p = (p_x|_{\tilde{T}}, p_y|_{\tilde{T}})$. We set

$$v := \sum_{\tilde{T}\in\tau_h} \begin{pmatrix} p_x|_{\tilde{T}}\tilde{u}_{\tilde{T}} \\ p_y|_{\tilde{T}}\tilde{u}_{\tilde{T}} \end{pmatrix} \in V_h, \quad \tilde{v} := v/|v|_0 \quad (\tilde{u}_{\tilde{T}} \text{ bubble function (13b) on } \tilde{T}),$$

so that $|\tilde{v}|_0 = 1$. Exercise 10 yields

$$\begin{aligned} b(p,v) = \int_\Omega \langle \nabla p, v\rangle \, d\mathbf{x} &= \sum_{\tilde{T}\in\tau_h} \{|p_x|_{\tilde{T}}|^2 + |p_y|_{\tilde{T}}|^2\} \int_{\tilde{T}} \tilde{u}_{\tilde{T}} \, dx\, dy \\ &\geq C \sum_{\tilde{T}\in\tau_h} \int_{\tilde{T}} (p_x^2 + p_y^2)\, dx\, dy = C|\nabla p|_0^2. \end{aligned}$$

Since $|\nabla p|_0$ and $|p|_1$ are equivalent norms on the subspace $H^1(\Omega) \cap L_*^2(\Omega)$ it follows that $|b(p,v)| \geq C'|p|_1|\nabla p|_0$ and $|b(p,\tilde{v})| \geq C'|p|_1|\nabla p|_0/|v|_0$. In a similar way one shows that $|v|_0 \leq C''|\nabla p|_0$ and obtains $|b(p,\tilde{v})| \geq \tilde{\beta}|p|_1$ with $\tilde{\beta} := C'/C''$ independent of h. The left-hand side in (8) is $\geq |b(p,\tilde{v})|$, so that (8) follows.
(b) (9a) is satisfied for V_h (cf. Theorem 8.4.5). The same holds for the inverse estimate (9b) (cf. Theorem 8.8.5). Therefore (6b,c) follows from Theorem 8. ■

A general result on the stabilisation by bubble functions can be found in Brezzi-Pitkäranta [1].

12.3.3.3 Stable Discretisations with Linear Elements in V_h

If one wants to avoid bubble functions, one must increase the dimension of V_h in some other way. In this section we shall consider for V_h and W_h two different triangulations $\tau_{h/2}$ and τ_h. By decomposing each $\tilde{T} \in \tau_h$ as in Figure 1, through halving the sides, into four similar triangles, one obtains $\tau_{h/2}$. We define:

$$\begin{aligned} &V_h \subset \mathbf{H}_0^1(\Omega) : \text{linear elements for triangulation } \tau_{h/2}, \\ &W_h \subset H^1(\Omega) \cap L_*^2(\Omega) : \text{linear elements for triangulation } \tau_h, \end{aligned} \tag{12.3.15}$$

or

$$\begin{aligned} &V_h \subset \mathbf{H}_0^1(\Omega) : \text{quadratic elements for triangulation } \tau_h, \\ &W_h \subset H^1(\Omega) \cap L_*^2(\Omega) : \text{linear elements for triangulation } \tau_h. \end{aligned} \tag{12.3.16}$$

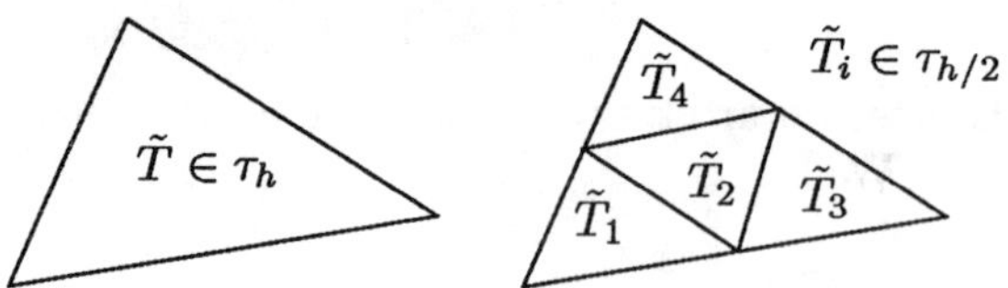

Figure 12.3.1. Triangulations τ_h and $\tau_{h/2}$

Theorem 12.3.12. *Theorem* 11 *holds analogously for* $V_h \times W_h$ *in* (15) *or* (16).

PROOF. (a) Let $V_h \times W_h$ be given by (15). For each inner triangular side γ of the triangulation τ_h there exist two triangles $T_{1\gamma}, T_{2\gamma} \in \tau_h$ with $\gamma = \overline{T_{1\gamma}} \cap \overline{T_{2\gamma}}$ (cf. Figure 2). Let $\partial/\partial t$ be the derivative in the direction of γ, let $\partial/\partial n$ be the directional derivative perpendicular to it. There exists a_γ and b_γ with

$$a_\gamma^2 + b_\gamma^2 = 1, \quad \partial/\partial x = a_\gamma \partial/\partial n + b_\gamma \partial/\partial t, \quad \partial/\partial y = b_\gamma \partial/\partial n - a_\gamma \partial/\partial t.$$

In contrast to $\partial p/\partial n$, $\partial p/\partial t$ is constant on $T_{1\gamma} \cup T_{2\gamma} \cup \gamma$. We denote its value by $p_{t|\gamma}$. The mid-point $\mathbf{x}^\gamma$ of γ is a node of $\tau_{h/2}$. We define the piecewise linear function u_γ over $\tau_{h/2}$ by its values at the nodes

$$u_\gamma(\mathbf{x}^\gamma) = p_{t|\gamma}, \quad u_\gamma(\mathbf{x}^j) = 0 \quad \text{at the remaining nodes} \tag{12.3.17a}$$

and set

$$v := \sum_\gamma \begin{pmatrix} b_\gamma \\ -a_\gamma \end{pmatrix} p_{t|\gamma} u_\gamma \in V_h, \quad \tilde{v} := v/|v|_0. \tag{12.3.17b}$$

The sum $\sum_\gamma$ extends over all interior sides of τ_h. In $T_{1\gamma} \cup T_{2\gamma}$ we have

$$\langle \nabla p, \begin{pmatrix} b_\gamma \\ -a_\gamma \end{pmatrix} \rangle p_{t|\gamma} = |p_{t|\gamma}|^2,$$

so that

$$b(p,v) = \sum_\gamma \int_{T_{1\gamma}\cup T_{2\gamma}} \langle \nabla p, \begin{pmatrix} b_\gamma \\ -a_\gamma \end{pmatrix} p_{t|\gamma} u_\gamma \rangle \, d\mathbf{x} = \sum_\gamma |p_{t|\gamma}|^2 \int_{T_{1\gamma}\cup T_{2\gamma}} u_\gamma \, d\mathbf{x}$$
$$\geq Ch^2 \sum_\gamma |p_{t|\gamma}|^2.$$

If $\gamma_1, \gamma_2, \gamma_3$ are the sides of $\tilde{T} \in \tau_h$ (τ_h is quasiuniform!), then $\int_{\tilde{T}} |\nabla p|^2 \, d\mathbf{x} \leq C' h^2 \sum_{i=1}^3 |p_{t|\gamma_i}|^2$.

From this one infers that $b(p,v) \geq C''|p|_1 |\nabla p|_0$, as in the proof of Theorem 11, and finishes the proof analogously.

(b) In the case of quadratic elements given by (16) one has the same nodes as in (a) (cf. Figure 8.3.8a and Figure 1). Use (17a,b) to define the quadratic function $v \in V_h$ and carry out the proof as in (a). ∎

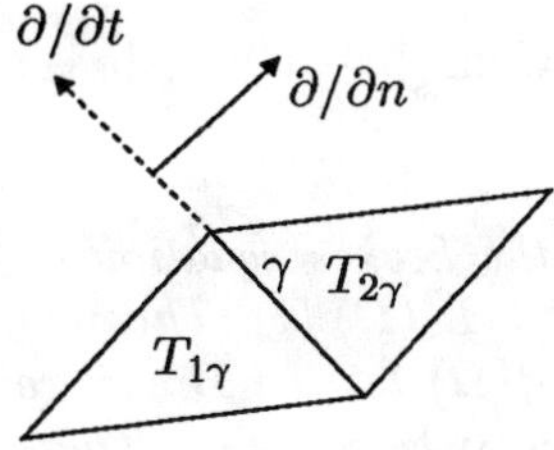

Figure 12.3.2. γ, $T_{1\gamma}$, $T_{2\gamma}$

12.3.3.4 Error Estimates

In the following, the condition (8.1.12) should be replaced by the stability condition (6a–c). In the place of the approximation property (8.4.6′) we now have the inequalities

$$\inf\{|v - v^h|_1 : v^h \in V_h\} \leq Ch|v|_2 \text{ for all } v \in V = \mathbf{H}_0^1(\Omega), \quad (12.3.18a)$$

$$\inf\{|p - p^h|_0 : p^h \in W_h\} \leq Ch|p|_1 \text{ for all } p \in W = L_*^2(\Omega). \quad (12.3.18b)$$

Condition (18a) is equivalent to condition (9a) in Theorem 8.

The following theorem applies to general saddlepoint problems, and can be reduced to Theorem 8.2.1.

Theorem 12.3.13. *Let* $u = \binom{v}{w} \in X = V \times W$ *be the solution of* (2.5a–c) [*resp.* (6a)]. *Let the discrete problem* (1a–c) *with* $X_h = V_h \times W_h \subset X$ *satisfy* *the Brezzi condition* (6a–c) *and have the solution* $u^h = (v^h, w^h)$. *Then there exists a variable* C *independent of* h *such that*

$$||u - u^h||_X \le C \inf\{||u - x^h||_X : x^h \in X_h\}. \tag{12.3.19}$$

PROOF. The Brezzi condition (cf. Theorem 6) yields $||u^h||_X \le \overline{C}||f||_{X'}$ for all right-hand sides $f \in X'$ in (2.5a–c), in particular for all $f \in X_h = X_h'$. The above inequality means $||L_h^{-1}||_{X_h \leftarrow X_h'} \le \overline{C}$ for the operator $L_h: X_h \to X_h'$ which belongs to $c(\cdot,\cdot): X_h \times X_h \to \mathbb{R}$. According to Exercise 8.1.16 $||L_h^{-1}||_{X_h \leftarrow X_h'} \le \overline{C}$ is equivalent to Condition (8.1.12) for $c(\cdot,\cdot): X \times X \to \mathbb{R}$ [instead of $a(\cdot,\cdot): V \times V \to \mathbb{R}$] with $\epsilon_N = 1/\overline{C}$. Theorem 8.2.1 yields the statement (19). ■

For the Stokes problem with $\binom{u}{p}, \binom{u^h}{p^h}$ instead of $u = \binom{v}{w}, u^h = \binom{v^h}{w^h}$, inequality (19) is now rewritten as follows:

$$|u - u^h|_1^2 + |p - p^h|_0^2 \le C^2 \inf\{|u - u^h|_1^2 + |p - p^h|_0^2 : u^h \in V_h, p^h \in W_h\}$$

or

$$|u - u^h|_1 + |p - p^h|_0 \le \sqrt{2} C \inf\{|u - u^h|_1 + |p - p^h|_0 : u^h \in V_h, p^h \in W_h\}. \tag{12.3.20}$$

Theorem 12.3.14. *Let the Stokes equation* (1.2a,b) *have a solution* $u \in \mathbf{H}^2(\Omega) \cap \mathbf{H}_0^1(\Omega)$, $p \in H^1(\Omega) \cap L_*^2(\Omega)$ *(cf. Theorem* 12.2.19*). For the subspaces* $V_h \subset \mathbf{H}_0^1(\Omega)$ *and* $W_h \subset L_*^2(\Omega)$ *let the Brezzi condition* (6a–c) *and the approximation conditions* (18a,b) *be satisfied. Then the discrete solution* u^h, p^h *satisfies the estimate*

$$|u - u^h|_1 + |p - p^h|_0 \le C' h [|u|_2 + |p|_1]. \tag{12.3.21}$$

PROOF. Combine inequalities (20) and (18a,b). ■

Using the same reasoning as in the second proof for Theorem 8.4.11 with $c(\cdot,\cdot)$ instead of $a(\cdot,\cdot)$ one proves

Theorem 12.3.15. *For each* $f \in \mathbf{L}^2(\Omega)$, $g \in L_*^2(\Omega) \cap H^1(\Omega)$ *let the Stokes problem* $-\Delta u + \nabla p = f$, $-\operatorname{div} u = g$ *have a solution* $u \in \mathbf{H}^2(\Omega) \cap \mathbf{H}_0^1(\Omega)$, $p \in H^1(\Omega) \cap L_*^2(\Omega)$ *with* $|u|_2 + |p|_1 \le C[|f|_o + |g|_1]$. *Under the assumptions of Theorem* 14 *we then have the estimates*

$$|u - u^h|_0 + |p - p^h|_{-1} \le C' h [|u|_1 + |p|_0], \tag{12.3.22a}$$
$$|u - u^h|_0 + |p - p^h|_{-1} \le C'' h^2 [|u|_2 + |p|_1] \tag{12.3.22b}$$

for the finite element solutions. Here $|p|_{-1}$ *is the dual norm for* $H^1(\Omega) \cap L_*^2(\Omega)$.

Corollary 12.3.16. *Combining* (22b) *and* (21) *one obtains*

$$|u - u^h|_0 + h|p - p^h|_0 \leq Ch^2[|u|_2 + |p|_1]. \tag{12.3.22c}$$

A new exposition of the mixed finite-element discretisation can be found in Brezzi–Fortin [1].

Bibliography

Adams, R. A.

[1] *Sobolev spaces,* Academic Press, New York (1975).

Agmon, S., A. Douglis, and L. Nirenberg

[1] Estimates near the boundary for solutions of elliptic partial differential equations satisfying general boundary conditions II, *Comm. Pure Appl. Math.*, **17** (1964) 35–92.

Aziz, A. K. (ed.)

[1] *The mathematical foundation of the finite element method with applications to partial differential equations,* Academic Press, New York (1972).

Babuška, I. and A. K. Aziz

[1] Survey lectures on the mathematical foundations of the finite element method, In: Aziz [1] 3–363.

Bank, R. E., T. F. Dupont, and H. Yserentant

[1] The hierarchical basis multigrid method, *Numer. Math.*, **52** (1988) 427–458.

Blum, H. and M. Dobrowolski

[1] On finite element methods for elliptic equations on domains with corners, *Computing*, **28** (1982) 53–63.

Blum, H. and R. Rannacher

[1] On the boundary value problem of the biharmonic operator on domains with angular corners, *Math. Meth. Appl. Sci.*, **2** (1980) 556–581.

[2] Extrapolation techniques for reducing the pollution effect of reentrant corners in the finite element method, *Numer. Math.*, **52** (1988) 539–564.

Bramble, J. H.

[1] A second order finite difference analog of the first biharmonic boundary value problem, *Numer. Math.*, **9** (1966) 236–249.

Bramble, J. H. and B. E. Hubbard

[1] A finite difference analog of the Neumann problem for Poisson's equation, *SIAM J. Numer. Anal.*, **2** (1964) 1–14.

[2] Approximation of solutions of mixed boundary value problems for Poisson's equation by finite differences, *J. Assoc. Comput. Mach.*, **12** (1965) 114–123.

Bramble, J. H. and A. H. Schatz

[1] Higher order local accuracy by averaging in the finite element method, *Math. Comp.*, **31** (1977) 94–111.

Brezzi, F.

[1] On the existence, uniqueness and approximation of saddle-point problems arising from Lagrangian multipliers, *RAIRO Anal. Numer.*, **8** (1974) 129–151.

Brezzi, F. and M. Fortin

[1] *Mixed and hybrid finite element methods,* Springer, Berlin (1991).

Brezzi, F. and J. Pitkäranta

[1] On the stabilization of finite element approximations of the Stokes equations, In: Hackbusch [6] 11–19.

Chatelin, F.

[1] *Spectral approximation of linear operators,* Academic Press, New York (1983).

Ciarlet, Ph. G.

[1] *The finite element method for elliptic problems,* North-Holland, Amsterdam (1978) (paperback edition 1980).

Collatz, L.

[1] *The numerical treatment of differential equations,* Springer, Berlin (1966).

Cosner, C.

[1] On the definition of ellipticity for systems of partial differential equations, *J. Math. Anal. Appl.*, **158** (1991) 80–93.

Dieudonné, J.

[1] *Foundations of Modern Analysis,* Academic Press, New York (1960).

Eriksson, K. and C. Johnson

[1] An adaptive finite element method for linear elliptic problems, *Math. Comp.*, **50** (1988) 361–383.

Fletcher, C. A. J.

[1] *Computational Galerkin methods,* Springer, New York (1984).

Gantmacher, F. R.

[1] *The theory of matrices. Vol. 1,* Chelsea Publishing, New York (1977) German edition: Matrizenrechnung. Band I. Deutscher Verlag der Wissenschaften, Berlin, 1958

Gladwell, I. and R. Wait (eds.)

[1] *A survey of numerical methods for partial differential equations,* Clarendon Press, Oxford (1979).

Goering, H., A. Felgenhauer, G. Lube, H.-G. Roos, and L. Tobiska
[1] *Singularly perturbed differential equations,* Mathematische Forschung, Vol. 13, Akademie-Verlag, Berlin (1983).

Green, G.
[1] *An essay on the application of mathematical analysis to the theories of electricity and magnetism,* Nottingham (1828).

Grisvard, P.
[1] *Elliptic problems in nonsmooth domains,* Pitman, Boston (1985).

Hackbusch, W.
[1] On the regularity of difference schemes, *Ark. Mat.*, **19** (1981) 71–95.
[2] On the regularity of difference schemes. Part II: Regularity estimates for linear and nonlinear problems, *Ark. Mat.*, **21** (1983) 3–28.
[3] A note on the penalty correction method, *Zeitschrift für Analysis und ihre Anwendungen,* **2** (1982) 59–69.
[4] Local defect correction method and domain decomposition techniques, *Computing,* **5** Supplement(1984) 89–113.
[5] *Multi-grid methods and applications,* Springer, Berlin (1985).
[6] *Efficient solvers for elliptic systems,* Notes on Numerical Fluid Mechanics, 10, Vieweg, Braunschweig (1984).
[7] *Integralgleichungen. Theorie und Numerik,* Teubner, Stuttgart (1989).
[8] On first and second order box schemes, *Computing,* **41** (1989) 277–296.
[9] *Iterative Lösung großer schwachbesetzter Gleichungssysteme,* Teubner, Stuttgart (1991) English translation: Iterative solution of large sparse systems. Springer, Berlin (to appear in 1992)
[10] *Parallel algorithms for PDEs,* Proceedings. Kiel, January 1990. Notes on Numerical Fluid Mechanics, 31, Vieweg, Braunschweig (1991).

Hackbusch, W. and R. Rannacher (eds.)
[1] *Numerical treatment of the Navier-Stokes equations,* Proceedings. Kiel, January 1989. Notes on Numerical Fluid Mechanics, 30, Vieweg, Braunschweig (1990).

Heinrich, B.
[1] *Finite difference methods on irregular networks. A generalized approach to second order problems,* ISNM, Vol. 82, Birkhäuser, Basel (1987).

Hellwig, G.
[1] *Partial differential equations,* 2nd ed., Teubner, Stuttgart (1977) German edition: Partielle Differentialgleichungen. Teubner, Stuttgart, 1960

Heuser, H.
[1] *Funktionalanalysis,* 2nd ed., Teubner, Stuttgart (1986).

Kadlec, J.
[1] O reguljarnosti rešenija zadači Puassona na oblasti s granicej, lokol'no podobno granice vypukloj oblasti (On the regularity of the solution

of the Poisson equation on a domain with boundary locally similar to the boundary of a convex domain), *Czechoslovak Math. J.*, **14** (1964) 386–393.

Kato, T.
[1] *Perturbation theory for linear operators,* Springer, Berlin (1966).

Kellogg, O. D.
[1] *Foundations of potential theory,* Springer, Berlin (1929).

Kellogg, R. B. and J. E. Osborn
[1] A regularity result for the Stokes problem in a convex polygon, *J. Funct. Anal.*, **21** (1976) 397–431.

Kevorkian, J. and J. D. Cole
[1] *Perturbation methods in applied mathematics,* Springer, New York (1981).

Kondrat′ev, V. A.
[1] Boundary value problems for elliptic equations in domains with conical or angular points, *Trans. Moscow Math. Society,* **16** (1967) 227–313.

Ladyženskaja, O. A.
[1] *Funktionalanalytische Untersuchungen der Navier-Stokesschen Gleichungen,* Akademie-Verlag, Berlin (1965).

Ladyženskaja, O. A. and N. N. Ural′ceva
[1] *Linear and quasilinear elliptic equations,* Academic Press, New York (1968).

Layton, W. and T. D. Morley
[1] On central difference approximations to general second order elliptic equations, *LAA*, **97** (1987) 65–75.

Leis, R.
[1] *Vorlesungen über partielle Differentialgleichungen zweiter Ordnung,* Bibliographisches Institut, Mannheim (1967).
[2] *Initial boundary value problems in mathematical physics,* Teubner, Stuttgart and Wiley, Chichester (1986).

Lesaint, P. and M. Zlámal
[1] Superconvergence of the gradient of finite element solutions, *RAIRO Analyse Numer.*, **13** (1979) 139–166.

Lions, J. L. and E. Magenes
[1] *Non-homogeneous boundary value problems and applications. Vol. I,* Springer, Berlin (1972).

Louis, A.
[1] Acceleration of converenge for finite element solutions of the Poisson equation, *Numer. Math.*, **33** (1979) 43–53.

Marchuk, G. I. and V. V. Shaidurov
[1] *Difference methods and their extrapolation,* Springer, New York (1983).

Meis, Th. and U. Marcowitz

[1] *Numerical solution of partial differebtial equations,* Springer, New York (1981) German edition: Numerische Behandlung partieller Differentialgleichungen. Springer, Berlin, 1978

Miranda, C.

[1] *Partial differential equations of elliptic type,* Springer, Berlin (1970).

Mitchell, A. R.

[1] *Computational methods in partial differential equations,* Wiley, London (1969).

Mitchell, A. R. and R. Wait

[1] *The finite element method in partial differential equations,* Wiley, London (1977).

Mizohata, S.

[1] *The theory of partial differential equations,* University Press, Cambridge (1973).

Nečas, J.

[1] Sur la coercivité des formes sesqui-linéaires elliptiques, *Rev. Roumaine Math. Pures Appl.*, **9** (1964) 47–69.

[2] *Equations aux dérivée partielles,* Presses de l'Université de Montréal .

[3] *Les Méthodes directes en théorie des équations elliptiques,* Masson, Paris (1967).

Nitsche, J.

[1] Ein Kriterium für die Quasi-Optimalität des Ritzschen Verfahrens, *Numer. Math.*, **11** (1968) 346–348.

Oden, J. R, and J. N. Reddy

[1] *An introduction to the mathematical theory of finite elements,* Wiley, New York (1976).

Pereyra, V.

[1] Highly accurate numerical solution of quasi-linear elliptic systems in n dimensions, *Math. Comp.*, **24** (1970) 771–783.

Pereyra, V., W. Proskurowski, and O. Widlund

[1] High order fast Laplace solvers for the Dirichlet problem on general regions, *Math. Comp.*, **31** (1977) 1–16.

Rheinboldt, W. C.

[1] Adaptive mesh refinement processes for finite element solutions, *Int. J. Numer. Meth. Engrg.*, **17** (1981) 649–662.

Schatz, A. H.

[1] A weak discrete maximum principle and stability in the finite element method in L^∞ on plane polygonal domains, *Math. Comp.*, **34** (1980) 77–91.

Schatz, A. H. and L. B. Wahlbin

[1] Maximum norm estimates in the finite element method on plane polygonal domains. Part 2: Refinements, *Math. Comp.*, **33** (1979) 465–492.

Schwarz, H. R.

[1] *Finite element methods,* Academic Press, London (1988) German edition: Methode der finiten Elemente. 3rd edition. Teubner, Stuttgart 1991.

Shortley, G. H. and R. Weller

[1] Numerical solution of Laplace's equation, *J. Appl. Phys.*, **9** (1938) 334–348.

Stoer, J.

[1] *Einführung in die Numerische Mathematik I.,* 5 Aufl., Springer, Berlin (1989).

Stoer, J. and R. Bulirsch

[1] *Introduction to numerical analysis,* Springer, Berlin (1980).

Strang, G. and G. J. Fix

[1] *An analysis of the finite element method,* Prentice-Hall, Englewood Cliffs, N.J. (1973).

Stummel, F.

[1] Diskrete Konvergenz linearer Operatoren. I, *Math. Ann.*, **190** (1970) 45–92.

[2] The generalized patch test, *SIAM J. Numer. Anal.*, **16** (1979) 449–471.

Thomasset, F.

[1] *Implementation of finite element methods for Navier-Stokes equations,* Springer, New York (1981).

Thomée, V.

[1] Discrete interior Schauder estimates for elliptic difference operators, *SIAM J. Numer. Anal.*, **5** (1968) 626–645.

[2] High order local approximations to derivatives in the finite element method, *Math. Comp.*, **31** (1977) 652–660.

Varga, R. S.

[1] *Matrix iterative analysis,* Prentice Hall, Englewood Clifs, N. J. (1962).

Velte, W.

[1] *Direkte Methoden der Variationsrechnung,* Teubner, Stuttgart (1976).

Walter, W.

[1] *Einführung in die Potentialtheorie,* Bibliographisches Institut, Mannheim (1971).

Wathen A. J.

[1] Realistic eigenvalue bounds for the Galerkin mass matrix, *J. of Numer. Analysis*, **7** (1987) 449–457.

Whiteman, J. R. (ed.)

[1] *The mathematics of finite elements and applications,* VII – MAFELAP 1990. Proceedings. Uxbridge, April 1990, Academic Press, London (1991).

Wloka, J.

[1] *Partial differential equations,* Cambridge University Press, Cambridge (1987) German edition: Partielle Differentialgleichungen. Teubner, Stuttgart, 1982

Yosida, K.

[1] *Functional analysis,* Springer, Berlin (1974).

Yserentant, H.

[1] On the multi-level splitting of finite element spaces, *Numer. Math.*, **49** (1986) 379–412.

Zienkiewicz, O. C.

[1] *The finite element method in engineering science,* McGraw-Hill, New York (1977) German translation: Methode der finiten Elemente. Carl Hanser Verlag, München, 1984

Zlámal, M.

[1] Discretization and error estimates for elliptic boundary value problems of the fourth order, *SIAM J. Numer. Anal.*, **4** (1967) 626–639.

Index

Note: The page numbers given in bold-face refer to occurrences of the term printed emphatically on the page indicated.

Springer-Verlag
Berlin
Heidelberg
New York
London
Paris
Tokyo
Hong Kong
Barcelona
Budapest